Sources and Studies
in the History of Mathematics and
Physical Sciences

Editorial Board

J.Z. Buchwald   J. Lützen   J. Hogendijk

Advisory Board

P.J. Davis   T. Hawkins
A.E. Shapiro   D. Whiteside

Fibonacci's *De Practica Geometrie*

Edited by

Barnabas Hughes

Barnabas Hughes
4966 Alonzo Avenue
Encino, California
U.S.A. 91316-3608
Email, personal: barnabashughes@hotmail.com

ISBN-13: 978-0-387-72930-5   e-ISBN-13: 978-0-387-72931-2

Library of Congress Control Number: 2007934985

Mathematics Subject Classification (2000): 01-xx

© 2008 Springer Science+Business Media, LLC
All rights reserved. This work may not be translated or copied in whole or in part without the written permission of the publisher (Springer Science+Business Media, LLC, 233 Spring Street, New York, NY 10013, USA), except for brief excerpts in connection with reviews or scholarly analysis. Use in connection with any form of information storage and retrieval, electronic adaptation, computer software, or by similar or dissimilar methodology now known or hereafter developed is forbidden.
The use in this publication of trade names, trademarks, service marks, and similar terms, even if they are not identified as such, is not to be taken as an expression of opinion as to whether or not they are subject to proprietary rights.

Printed on acid-free paper

9 8 7 6 5 4 3 2 1

springer.com

In homage to the medieval Franciscan masters
who used mathematics to strengthen their scholastic writings:
Alexander of Hales, Saint Bonaventure, Peter John Olivi,
Roger Marston, William of Alnwick, Gonsalvus Hispanus,
Blessed John Duns Scotus

Foreword

Leonardo of Pisa (Fibonacci) is the preeminent European mathematician of the Middle Ages. Besides being, in his own right, a creative mathematician, he was also a compiler of and an evangelist for the mathematics he believed was needed for his times. Building upon Greek, Byzantine, and Arabic sources, Leonardo published works that combined theory with practice and established a new scientific genre of "practicae:" brief works, written in a popular style, emphasizing applications, and intended to reach a large audience. But despite his importance as a mathematical innovator, English language access to much of Leonardo of Pisa's work has long been lacking. Finally, in 1983, Laurence Sigler, a Latinist, published a translation of Leonardo's *Liber quadratorum* (1225) and his English language translation of *Liber abaci* (1202) was released posthumously in 2002. Now Barnabas Hughes, O.F.M., has provided readers with a critical, English language translation of Leonardo's *De practica geometrie* (1223), completing the trilogy of the major works of this important medieval mathematician. I am pleased and honored at being asked to comment on this new valuable resource for the history of mathematics.

Whereas Europe's great Gothic cathedrals stand in moot testimony to the geometric ingenuity and ability of their builders, little documented evidence is available as to the specific geometric principles and techniques included in their designs. True, much of the theory was "trade secrets" of the architects and artisans that constructed the great buildings. However, for this knowledge to be exploited for broader applications it had to be exposed and communicated to a larger audience. Thus, Leonardo's *De practica geometrie* was conceived and written. It was not merely a passive manual recording what had been done, but rather a "call to action" suggesting how geometric applications could be expanded and improved. In this book, Euclidian rigor in the form of demonstrations and proofs, content from Euclid's *Elements* and *On the Division of Figures* are obvious, but Leonardo goes beyond planimetry to include the use of trigonometry and algebra to solve geometrical problems. He provides directions on the use of measuring instruments and gives advice to surveyors and architects.

Barnabas Hughes, a respected historian of mathematics, an accomplished researcher, Latinist, and teacher, has brought all of his skills and talents to the undertaking of this translation task. Working mainly from Boncompagni's nineteenth century transcription of *De practica geometrie*, he has also tracked down and compared other relevant descendant transcriptions of the original text. Where possible, he has affirmed Leonardo's references. What Hughes has tried to produce is a translation true to Leonardo of Pisa's original intent and style. What he has accomplished is a highly readable, scholarly insight into the genius of Fibonacci and the status of medieval geometry. This book will serve as a valuable reference in the history of mathematics for many years to come.

Frank Swetz,
Harrisburg, Pennsylvania
April 2007

Preface

Had the family and friends of L.E. Sigler not seen to the publication of his translation of Fibonacci's *Liber abaci*, I would not have thought to translate *De practica geometrie*. Earlier in 1987 Professor Sigler had translated Fibonacci's *Book of Squares* into English. Perhaps this prompted him to begin the more ambitious project, one that he was not destined to see through to publication. Regardless, Professor Sigler accomplished a remarkable feat for Fibonacciana that had received new momentum with the incisive analysis of *Liber abaci* by Heinz Lüneburg in 1993. The international conference held in Pisa in 2002 honoring the eighth centenary of the first edition of *Liber abaci* brought together outstanding scholars to laud, explain, and expostulate upon the work of the greatest continental mathematician of the Middle Ages. This critical translation of *De practica geometrie* would advance the appreciation of Fibonacci's pre-eminent talents.

Most readers are familiar with what the word *translation* implies, although I have much to say about this farther on, yet their curiosity about the modifier *critical* deserves satisfaction here. I have compared the printed transcription of Baldassarre Boncompagni with many of the extant manuscripts to establish what I offer as the closest approximation in American English to what Fibonacci presented in Latin, in the year 1220. On the other hand, and with the help of others, I have located most of the resources from which he drew, arranged this gargantuan work, and determined what he contributed. The results of these activities define the word *critical*. To these I have added my observations as commentary.

Two able historians have monitored my work at various times. Jeffrey Oaks of he University of Indianapolis read and critiqued the entire translation. Mahdi Abdeljaouad of the University of Tunis taught me a great deal about medieval Arabic mathematicians and their works. Regardless of the few times I did not follow their advice, their irreplaceable assistance improved this research. I am immensely grateful. Many thanks also to Frank Swetz for the foreword. The Ruth and Karl Bjorkman Fund financed the project. A Vatican Film Library Mellon Fellowship at Saint Louis University, Missouri, provided me with the time and place to study microfilm of several manuscripts used in

x   Preface

this research as well as the invaluable assistance of the staff there. I am grateful to both. I thank too the staff at Springer/New York who saw this work through its production, including the unknown referee who kept me from "flogging a dead horse." Finally, as Fibonacci remarked in the introduction to *Liber quadratorum*, "Cum omnibus habere memoriam et in nullo peccare sit divinitatis potius quam humanitatis, nemo sit vitio carens et undique circumspectus." I hope my readers enjoy Fibonacci's *Practical Geometry* as much as I do.

Barnabas B. Hughes, O.F.M.
California State University, Northridge
Spring 2004 to Spring 2007

Table of Contents

Foreword vii
Preface ix
Notation xv

Background xvii
   Fibonacci's Knowledge of Arabic xviii
   Fibonacci's Schooling xxi
   Fibonacci's Basic Resources xxii
   Sources for the Translation xxvi
   The Translation xxviii
   Italian Translations xxx
   Conclusion xxxiv

Prologue and Introduction 1
   Commentary and Sources 1
   Text 4
      Definitions [1] 5
      Properties of Figures [2] 5
      Geometric Constructions [3] 6
      Axioms [4] 6
      Pisan Measures [5] 7
      Computing with Measures [6–8] 7

1 Measuring Areas of Rectangular Fields 11
   Commentary and Sources 11
   Text 14
      1.1 Area of Squares [1] 14
      1.2 Areas of Rectangles 14
         Method 1 [2–30] 14
         Method 2 [31–45] 26
      1.2 Keeping Count with Feet [13] 17

2 Finding Roots of Numbers 35
   Commentary and Sources 35
   Text 38
      2.1 Finding Square Roots 38

Integral Roots [1–22].	38
Irrational Roots [23–24].	48
Fractional Roots [40–42].	55
2.2 Operating with Roots.	49
Multiplication [25–27].	49
Addition [28–32].	50
Subtraction [33–37].	53
Division [38–39].	54

3 Measuring All Kinds of Fields 57
Commentary and Sources.	57
Text.	65
3.1 Measuring Triangles.	65
General [1–6].	65
Pythagorean Theorem [7–8].	68
Right Triangles [9–13].	69
Acute Triangles [14–25].	71
Oblique Triangles [26–41].	77
Hero's Theorem [31].	80
Surveyors' Method [42–43].	87
Ratios/Properties of Triangles [44].	88
Lines Falling Within a Single Triangle [44–49].	88
Lines Falling Outside a Single Triangle [50–67].	90
Composition of Ratios [68].	99
Excision of Ratios [69].	100
Conjunction of Ratios [70–78].	100
Combination of Ratios [79–82].	104
3.2 Measuring Quadrilaterals.	106
General [83].	106
Algebraic/Geometric Model [84–94].	106
Squares [95–96].	112
Algebraic Method [97–106].	113
Rectangles [107–138].	116
Multiple Solutions [139–146].	128
Other Quadrilaterals [147].	131
Rhombus [148–164].	131
Rhomboids [165–168].	137
Trapezoids.	139
Concave Quadrilaterals [182].	147
Convex Quadrilaterals [182].	147
3.3 Measuring Multisided Fields [183–187].	147
3.4 Measuring the Circle and Its Parts.	151
Areas [188–193].	151
π [194–200].	154
Arc Lengths and Chords [201–207, 210].	158
Ptolemy's Theorem [208–209, 232].	162

	Sectors and Segments [220–226]	163
	Inscribed Figures [227–231, 233–239]	166
3.5	Measuring Fields on Mountain Sides [240–247]	174
	Archipendium [242]	174

4 Dividing Fields Among Partners 181
Commentary and Sources........................... 181
Text... 185
    4.1 Multisided Figures 186
        Triangles [1–26] 186
        Parallelograms [27–31] 205
        Trapezoids [32–56] 211
        Quadrilaterals With Unequal Sides [57–64, 66–69] 230
        Squares [65] 237
        Pentagons [70–75] 242
    4.2 Circles 246
        General [76–81] 246
        Semicircles [82–83, 85] 250
        Segments [84, 86] 251

5 Finding Cube Roots 255
Commentary and Sources........................... 255
Text... 259
    5.1 Finding Cube Roots [1–11] 259
    5.2 Finding Numbers in Continued Proportions 265
        Archytas' Method [12] 265
        Philo's Method [13] 267
        Plato's Method [14–15] 268
    5.3 Computing with Cube Roots 270
        Multiplication [16] 270
        Division [17] 271
        Addition and Subtraction [18–23] 271

6 Finding Dimensions of Bodies 275
Commentary and Sources........................... 275
Text... 277
    6.1 Definitions [1–3] 277
        Euclidean Resources [4–10] 278
        Various Areas and Volumes 282
        Parallelepipeds [11–18] 282
        Wedge [19–20] 287
        Column [21–25] 289
    6.2 Pyramids [26–41, 44] 292
        Cones [42–43] 305
    6.3 Spheres [45–53] 308

xiv   Table of Contents

	Surface Area and Volume [54–60]............................	319
	Inscribed Cube [61–67]	324
	Ratios of Volumes [68–73].................................	330
	Other Solids [74. 76–84]	333
6.4	Divide a Line in Mean and Extreme Ratio [75]	335

7 Measuring Heights, Depths, and Longitude of Planets......... 343
Commentary and Sources................................... 343
Text.. 346
   7.1 Different Heights [1–3] 346
   7.2 Tools: Triangle [4] 348
        Quadrant [5–9]..................................... 349
   7.3 Table of Arcs and Chords [211–219]..................... 354

8 Geometric Subtleties 361
Commentary and Sources................................... 361
Text.. 365
   8.1 Pentagons [1–2], [6–7], [10–12], [16–18], [21–22], [25–26] 365
   8.2 Decagons [3–5], [8–9], [13–15], [19], [23–24[, [27] 367
   8.3 Triangles [20–33*]..................................... 377

Appendix   Problem with Many Solutions..................... 395
Commentary and Sources................................... 395
Text.. 396

Bibliography... 399

Index... 407

Notation

With some 800 footnotes notes many of which are citations from or references to the same three or four works, abbreviations became imperative to avoid cumbersome repetitions. I employ the following.

- Sources are identified only by author and year of publication followed by page numbers, for example, Folkerts (2004), 98–102, except for Euclid's principal work which is cited only by *Elements* with Book and Proposition numbers.
- *Elements citations are from Folkerts (2006) IX, 8 *ff.*
- *Liber abaci* always refers to the transcription published in Boncompagni (1857).
- [45] identifies paragraph 45 within a chapter, all paragraphs being so numbered.
- (124.16) identifies page 125 line 16 in Boncompagni (1862); Arrighi (1966) is cited with the same notation.
- (124.16–32) refers to page 124 lines 16 to 32.
- (124.16–125.6) cites pages 124 lines 16 into page 125 line 6.
- (124.16, *f.* 102r.23) states that what is found on page 124 line 16 of Boncompagni (1862) is matched with Vatican Library Urbino 292 folio 102 recto line 23.
- **{p. 32}** in bold print identifies a page number in Boncompagni's transcription.
- (triangles *abg* and *def*) identify material that is in the text.
- [as follows] identifies an addition I made to clarify the text.
- *Italics* are used to identify rules, theorems, and geometric objects such as line-segment *ag* or point *b*.

Background

Leonardo da Pisa, perhaps better known as Fibonacci (ca. 1170–ca. 1240),[1] selected the most useful parts of Greco–Arabic geometry for the book known as *De practica geometrie*. Practical Geometry, a term created by Hugh of Saint Victor, is the name of the craft for medieval land-measurers, known as *agrimensores* in Roman times, and as surveyors in modern times. Fibonacci wrote *De practica geometrie* for these artisans, a fitting complement to *Liber abaci*. Beginning with the definitions and constructions found early on in Euclid's *Elements*, Fibonacci instructed his reader how to compute with Pisan units of measure, find square and cube roots, determine dimensions of both rectilinear and curved surfaces and solids, work with tables for indirect measurement, and perhaps finally fire the imagination of builders with analyses of pentagons and decagons. His work exceeded what readers would expect for the topic, particularly with the lists of units of measurement and the many practical problems that exemplify their use.

Fibonacci had been at work on the geometry project for some time when a friend encouraged him to complete the task. This he did, going beyond the merely practical, as he remarked, "Some parts are presented according to geometric demonstrations, other parts in dimensions after a lay fashion, with which they wish to engage according to the more common practice." The translation offers a reconstruction of *De practica geometrie* as I judge that Fibonacci wrote it.[2] In order to appreciate what Fibonacci created, we consider the ambiance: his command of Arabic, his schooling, and the resources available to him. To these are added my view on translation and remarks about the Italian translations. A bibliography of primary and secondary resources follows the translation, completed by an index of names and special words.

---

[1] The most current and preferable biography is by Franci (2002). Vogel's article (1971) is still worthwhile for its descriptions of Fibonacci's works.

[2] See the introductions to Chapters 3 and 7 and to Chapter 8 and the Appendix for replacements of lengthy texts that are briefly described here.

FIBONACCI'S KNOWLEDGE OF ARABIC

An issue to discuss is how much Arabic Fibonacci knew. This is important for gathering a list of books that he did or might have consulted. Inasmuch as the issue appears to be moot, we must consider it were we to credit Fibonacci with a knowledge of Arabic whereby he might have consulted mathematics books written in Arabic. There are several possible answers: (1) for all practical purposes he knew none at all; (2) he knew enough to find his way about an Arabic community anywhere on the Mediterranean rim; (3) he was completely fluent in reading, writing, and speaking Arabic. The second is included in the third, and as with the first is of no use for reading, much less studying an Arabic text. Hence, I ignore the second possibility and attend to the first and third possibilities.

Contemporary research has shown the first option untenable, regardless of its base. Allard set a position that appeared in the 1980s, "... Fibonacci's work does not show his knowledge of Arabic."[3] Was he saying that Fibonacci knew Arabic but that there are no traces in his work, or that Fibonacci knew no Arabic at all? If the former, the statement is ambiguous and must be clarified. If the latter, then I have not been able to discover the basis of his thinking and why he placed it in an endnote.[4] Could he have sought Arabisms in Fibonacci's work and found none? (I want to think that there are no Latinisms in my translation of his autobiography, in contrast to Sigler's corresponding translation in *Liber abaci*.) Furthermore, his lack of familiarity with *De practica geometrie* completely confounds his statement to which the endnote is attached, "Whatever the state of ignorance in which we find ourselves with regards to the true sources of Leonardo Fibonacci's work" (he had just referred to *De practica geometrie*). If by "true" he meant "exact", then he was clearly asking too much. If he meant "identifiable," then he was mistaken. Two works of Euclid are present: *Elements* (especially the two "Arabic" books at the end) are cited by name or proposition and *On Division* (which he mentioned. Hence, he could not have meant "true" in the strict sense) upon which Fibonacci expands. Furthermore, Eutocius' *Commentary on the Sphere and Cylinder of Archimedes* might have been used as the source for examples requiring cube roots, Abū Kāmil's *Algebra* is the matrix for Chapter 8 on subtle problems, Ptolemy's *Almagest* for at least one theorem, and Banū Mūsā ibn Shākir's *On Geometry* for more theorems. All these resources are clearly identifiable, as I discuss below. Although it is important to spell out all the factors disproving Allard's initial statement, lest readers of that text alone form an erroneous position, it is to his credit that he has clarified his position. He apparently modified his stance when he wrote, "Fibonacci used his knowledge of the Arabic Euclid to solve the problem posed above (*Elements*, II,I)."[5]

---

[3] Allard (1996), 576 n. 114. In this the English translation of the original French edition (1984) his opinion was unchanged.
[4] Rashed (1996) II, 578 n. 114.
[5] Allard (2001) $12_{1,2}$:88.

Finally, I looked at Fibonacci's quotations from Euclid's *Elements*. Menso Folkerts had already done much of the work here. After a careful inspection of excerpts and citations from Books I, II, XI, XII, and XIII, he concluded that there is "no significant agreement with the known Latin texts of Euclid."[6] Where he did find a nearly word-for-word agreement was with quotations from Books XIV and XV that Busard suggested were translated by Fibonacci.[7] Folkerts further states as "certain" that Fibonacci at least paraphrased passages from a direct Latin translation of Greek *Elements*, citing seven parallel passages that are clearly similar.[8] Extrapolating from the thinking that he and Busard are comfortable with, namely that Fibonacci is the translator of the proposition in the spurious Books XIV and XV of the *Elements*,[9] I would offer an alternative position. Mindful of the fact that Fibonacci cited many of the defintions and axioms from Book I of the *Elements* in his introductory chapter, that he listed many of the propositions from Books XI, XII, and XIII in his Chapter 6, and cited or referred to numerous propositions from throughout the *Elements* in his arguments, I suggest this: in the course of his studies somewhere sometime Fibonacci had access to an Arabic *Elements*. From this he translated into Latin at least the propositions of all fifteen Books, to provide himself with a *vademecum* of Euclid's *Elements*. Later he had access to a Latin translation taken directly from the Greek, from which he might have adapted some Greek words.[10] Regardless, the issue is no longer moot: Fibonacci was proficient in Arabic.

The third possibility, complete fluency in Arabic, now accepted as correct, leads us to inspect closely certain remarks in Fibonacci's autobiography that might offer us an idea of the schooling he received in Bougie, Algeria:

My father was a notary for Pisan merchants doing business with the customs office in Bougie. He sent for me when I was a youth, with an eye on me being useful to him now and in the future. He enrolled me in an abacus school for some time where I was taught to compute; a wonderful teacher instructed me in the art of the nine Indian digits. I was so delighted with this knowledge that I preferred it to all other subjects.[11]

---

[6] Folkerts (2004), 109–110.

[7] Busard (1987), 20.

[8] Ibid., 110–112.

[9] Ibid., 110.

[10] Ibid., 112, but not all the words cited here. Both *dodrans* and *dextans* had been in Latin use for some 1000 years.

[11] "Cvm genitor meus a patria publicus scriba in duana bugee pro pisanis mercatoribus ad eam confluentibus constitutus preesset, me in pueritia mea ad se uenire faciens, inspecta utilitate et commoditate future, ibi me studio abbaci per aliquot dies stare uoluit et doceri. Vbi ex mirabili magisterio in arte per nouem figuras indorum introductus, scientia artis in tantum mihi pre certeris placuit" in *Liber abaci*, 1.24–31. Both political and commercial conditions in the Western Mediterranean were severally disrupted during the early 1180s. Peace and harmony were restored by treaty between the Comune of Pisa and the Caliph, Almohade, on November 15, 1186 (Banti [1988], 51–55). This might be the time when Bonacci Senior sent for his son, although in consideration of the amount of time it would have taken to assemble and write *Liber abaci*, a time six or eight years earlier may be a better guess.

First, regarding Pisa and Bougie. Bougie became the capital of an autonomous province of the Almohad Empire.[12] The city of Pisa had been commercially involved with the Moslems in the Maghrib before 1100 A.D.,[13] and business was carried out in a Pisan *fondaco*. Originally, a *fondaco* was a hostelry, the name being an Italian adaptation from the Arabic *funduq* that described a building for travelers and transient merchants. "[I]t particularly catered to merchants, not only lodging traveling traders but providing storage for their goods, places for sales and negotiation, and a locus for governmental taxation." All of this was regulated between Pisa and the rulers of each city or area by a treaty that included details about safe conduct, access to and from the *fondaco*, legal jurisdiction, and such matters.

The first such treaty was signed on July 2, 1133, between the Pisans and Ali ibn Yusof, King of Morocco and Tlemcen, and renewable after periods of ten years.[14] The treaties were crucial for survival of trade because they made commerce possible even during periods of political unrest. Among the duties of the notary were to maintain contact with Moslem authorities, to safeguard the rights of the *fondaco*, to maintain records of the goods imported and stored in the *fondaco*, and to ensure correct taxation, obligations that required a knowledge of Arabic. For instance, in Tunis sometime before 15 March 1289, the notary Enrico Merio translated a document from Arabic into Latin.[15] Granted that practices in the Eastern Mediterranean did not dictate procedures in the Maghreb, it is worth observing an account from 1183 when a dignitary arrived at a royal *fonde* in Damascus, he found "Christian clerks of the customs with their ebony ink-stands ornamented with gold. They write in Arabic, which they also speak."[16]

Because Fibonacci was brought to Bougie to be of assistance to his father and later to make his way about the Mediterranean rim as a young adult, it stands to reason that as a youth he took advantage of every opportunity to learn about Bougie, its people, and its language. No one questions the ease with which the young, especially the talented, learn foreign languages. Before Fibonacci went to Bougie, he already knew his native dialect, presumably also Latin, and perhaps another Italian dialect. We may assume that he learned Arabic, particularly to be of assistance to his father who had to negotiate daily with Moslems and Pisan merchants and to indulge developing his knowledge of mathematics. With the latter purpose in mind, let us consider the mathematical environment in which he might have immersed himself.

---

[12] Aïssani (1994), 77–78.

[13] Abulafia (1994), 5.

[14] Tangheroni (1994), 17 and 20.

[15] "... secundum quod continetur in quodam instrumento, scripto in arabico et exemplicato de arabico in latino per Enricum Merium ..." Pistarino (1986), 42; see also pp. 58 and 187.

[16] Constable (2003), 41 and 221. This is my principal source (an in-depth historical and commercial study) for my remarks about the *fondaco*.

FIBONACCI'S SCHOOLING

The kind of education that Fibonacci might have received would be based on the intellectual milieu in Western Islam. The following offers an overview.

Under the rule of the Almohad Dynasty (1147–1269), Marrakech became the capital of an empire ruling part of Spain and almost all Maghreb. Six cities became important: Seville in Spain, Ceuta and Marrakech (in actual Morocco), Tlemcen and Bougie (in actual Algeria), and Tunis. In that period, men of science, teachers and students (religion law, mathematics, medicine, ...) are known to have moved from Spain to North Africa and back. As an example, al-Hassār moved from Ceuta to Marrakech and back to Spain (fl. 1175), Ibn 'Aqnūn worked and died in Ceuta in 1225, Ibn 'Aqnūn from Marrakech to Seville and back to Marrakech where he died in 1204, ibn al-Mun'im who was born in al-Andalus, worked and died in Marrakech in 1228, the Andalusian al-Qurashi worked in Seville and in Bougie where he died in 1184. All of these mathematicians had strong links with al-Andalus, but worked in North Africa. All these dates make one conjecture that the works of these eminent authors of mathematics were circulating among the scholars of these six cities.[17]

With this preparation, let us focus on his education.

Fluent in Arabic, Fibonacci and his schooling are not impossible to imagine. In his brief autobiography, he remarked that besides learning how to compute and use the Indian numerals, he became enthralled with mathematics, "I was so delighted with this knowledge that I preferred it to all other subjects." We do not know which textbooks he actually used, however, it is possible to develop a scenario that might well fit his situation, keeping in mind that his father expected him to be of assistance in the notary's office. Inasmuch as religion was the prevailing factor in Arab teaching,[18] so much so that the initial education of boys that began about age 5 was devoted entirely to memorizing the Koran, Fibonacci's schooling in Bougie more probably began at the second level with a regular teacher. Any one of these books might possibly have been the one he studied:

- *Kitāb al-jam 'wal tafrīq bi hisāb al-Hind* (Book of Hindu Reckoning) by al-Khwārizmī (R-I, n. 41-M1), which, although it no longer exists in an Arabic copy, is available in a medieval Latin translation
- *Kitāb al-Bayān wa at-tudhkār* (Book of Demonstration and Recollection) by al-Hassās (ca. 1175)
- *Kitāb uul-kāfi fi 'Ilm al- hisāb* (Sufficient Book on the Science of Arithmetic) by al-Karāji (R-I, n. 309-M1)
- *al-Urjuza fīl-jabr wa-l muqābala* (Poem on Algebra) by Ibn al-Yāsamin (d. 1204, R-I n. 521), a concise memorabilia for recalling propositions and algorithms

---

[17] Abdeljaouad (private communication).
[18] Abdeljaouad (2004_2), "Issues", 8.

- *Kitāb fīl-jabr wa'l muqābala* of Abū Kāmil (850–930, R-I, n. 124), particularly useful because al-Khwārizmī's *al-jabr* was not available in the Maghreb despite wide circulation in al-Andulus
- Euclid's *Elements*: translations and revisions by al-Haǧǧāj ibn Matar (ca. 786–833, R-I, n. 34) and by Ishaq al-'Ibādī (830–910, R-I n. 114) revised by Thabit ibn Qurra (836–901, R-I n. 103)
- Not to overlook the no longer available *Kitāb al-missāha wa'l-handasa* (Book of Measurement and Geometry) by Abū Kāmil (R-I, n. 124-M9)

Whichever the texts for whatever course, Fibonacci memorized what he was taught.[19] A strong memory was a necessary tool for every medieval student, and Fibonacci became a student of the mathematics Islam offered. When we consider the plethora in the contents of *De practica geometrie*, we are not surprised, neither with its breadth nor its depth. He laid a solid foundation for composing his own treatises. We now turn to his resources.

FIBONACCI'S BASIC RESOURCES

Any discussion of books that Fibonacci had at hand as he composed *De practica geometrie* does well to consider again this part of the autobiographical account of his early education.

... a wonderful teacher instructed me in the art of the nine Indian digits. I was so delighted with this knowledge that I preferred it to all other subjects. In fact, as I journeyed about Egypt, Syria, Greece, Sicily, and the Provence on business, I studied whatever I could about this multifaceted subject, even discussing it with others.[20]

I hazard the thought that my reader has experienced something of the thrill of mathematics, sometime in the reader's own education. Math *is* fascinating! Although Fibonacci referred only to "the art of the nine Indian digits," it is clear from his remarks that he got caught up in all the mathematics available in his day. Equally worthy of our attention is the remark "with an eye to be useful to him now and in the future." Having learned Arabic he could meander about the Mediterranean rim, even though coastal Levant was under the control of the legatees of the Second Crusade. The success of Fibonacci's transit about the largely Moslem Mediterranean rim was due most probably to his experiences in Bougie. Where the teenager was educated, Moslems and Christians worked in harmony. Although each considered the other as heretics,

---

[19] Ibid., 33.
[20] "... ex mirabili magisterio in arte per nouem figuras indorum introductus, scientia artis in tantum mihi pre certeris placuit, et intellexi ad illam, quod quicquid studebatur ex ea apud egyptum, syriam, greciam, siciliam et prouinciam cum suis uariis modis, ad que loca negotiationis tam postea peragraui per multum studium et disputationis didici conflicturm." *Liber abaci*, 1.26–31.

they worshipped the same one God. Fibonacci must have been aware of the personal characteristics of the class of people with whom he was enriched during his travels: prudent, God fearing, religious, and just.

Having found his way up into Constantinople where he visited with others interested in mathematics, he might have had copies or made copies of the texts used by his instructors. Furthermore, as he traveled the Mediterranean rim, I contend that he gathered those Arabic texts in mathematics that suited his interests, nor did he neglect treatises in Latin as the evidence shows. It is quite possible that when he was in Provence, he happened upon a copy of the *Istikmāl* (*Comprehensive Treatise*) of Yūsuf al-Mu'aman ibn Hūd (d. 1085). The treatise is a compilation of at least these works: Euclid, *Elements*, Books I–XIV, Ptolemy, *Almagest* Book I, Eutocius, *Commentary on Archimedes' Sphere and Cylinder* II, Menelaus, *On Spheres*, Theodosius, *On Spheres*, and Banū Mūsā ibn Shākir, *Measurement of Plane and Spherical Figures*.[21] To have this single volume in one's library would be a blessing without compare.

The most influential book Fibonacci chanced on, probably in Provence, was *Liber embadorum* (*Book of Areas*), Plato of Tivoli's translation of *Hibbūr ha-Meshūhah ve-ha-Tishboret* by his contemporary Abraham bar Hiyya (d. 1136). There is little doubt that he was impressed with its structure because there are some 80 points of similarity both large and small between *De practica geometrie* and *Liber embadorum* as detailed by Curtze,[22] but little congruence save for common procedures and figures that might be found in any treatise on geometry. For a significant example of similarity, both treatises have tables of arcs and chords but they differ significantly. Abraham's table is based on a circle with diameter divided in 28 equal parts that given the length of a chord the measure of its arc is easily found in partes, minutes, and seconds and vice versa.[23] Fibonacci constructed a table from a circle with diameter measuring 42 rods and that lists 65 minor arcs with their complementary major arcs, and the 65 corresponding chords measured in rods, feet, inches, and points; given the measure of either arc or chord, the measure of the other must be computed. Although Abraham's table is less precise than Fibonacci's because of fewer divisions in the diameter, his may be more practical because of the generality of measurements, undefined parts versus Pisan units. Unlike Plato who for an unknown reason used both *alinuarum* and *rhumbos* in the definition of rhombus,[24] *almuncharif* in the definition after next, and *almugesem* for solid numbers,[25] Fibonacci worked with the Latin vocabulary he had developed. The strange thing is that, although Fibonacci named many of his resources, he did not mention Plato's source. Both incipit

---

[21] Hogendijk (1991), *passim*.
[22] Curtze (1902), *passim*.
[23] Ibid., 108–118.
[24] Ibid., 14. n.25.
[25] Ibid., 18. n.20.

and explicit of Plato's translation contain the same phrase, "liber embadorum a Savasorda." Fibonacci made no reference to this. The simplest explanation is that he recognized the name Savasorda as titular rather than patronymic, derived from the Arabic "Sahib al-Shorta" (Chief of the Bodyguard) Abraham's title in Arabic and hardly helpful for identifying an author.

The table of contents of *De practica geometrie* may serve to list his resources, however, three parameters should be kept in mind: some passages mention books by name or author, others contain identical or very comparable passages without naming the source, and still others contain passages that are merely similar and are part of an oral tradition. Above all, let me remark that the list is probably incomplete. As anyone who has rummaged in the mathematics section of a used book store and has picked up a few gems knows so well, it is entirely conceivable that Fibonacci had a larger library or access to more codices than are listed here. Nor should one neglect to recognize the prodigious memories of medieval scholars who had read widely. With these *cautelae* let us look at the sources the text suggests.

LIST OF PROBABLE SOURCES

Introduction
  Euclid, *Elements*, I.
  Abraham bar Hiyya, *On Mensuration and Calculation*, Ch. 1.

Ch. 1: Measuring Areas of Rectangular Fields
  Euclid, *Elements*, II, III, V.
  Abraham bar Hiyya, *op.cit.*, Ch. 1.

Ch. 2: Finding Roots of Numbers
  Fibonacci, *Liber abbaci*, Ch. 14.

Ch. 3: Measuring of Fields of All Kinds
  Abū Bekr, *On Mensuration*.
  Banū Mūsā, *On Geometry*, IV, VI. (R 74, M-3).
  Ptolemy, *Almagest*, Ch. 1.
  Liber abaci, Ch. 9.
  Ahmed ibn Yusuf, *On Ratio and Proportion* (R 119, M-1).

Ch. 4: Dividing Fields Among Partners
  Euclid, *On Division of Figures*.
  Euclid, *Elements*, I, II, III, IV, V, VI, VI, XII, XIII, XIV.
  Abū Bekr, *On Mensuration*.

Ch. 5: Finding Cube Roots
  Fibonacci, *Liber abbaci*, Ch. 14.
  Eutocius, *Commentary on Archimedes' Sphere and Cylinder*, II.
  Banū Mūsā, *On Geometry*, XVI–XVII.

Ch. 6: Finding Dimensions of Bodies
  Euclid, *Elements*, XI, XII, XIII, XIV, XV.
  Theodosius, *On Spheres*.
  Menelaus, *On Spheres*.
  Banū Mūsā, *On Geometry*, VIII–XV.

Ch. 7: Measuring Heights, Depths, and Widths
  (Gerbert ?) *Geometria incerti auctoris*.

Ch. 8: Geometric Specialties
  Abū Kāmil, *Algebra*, Ch. 2.

Appendix: Indeterminate Problems
  Abū Kāmil, *Algebra*, Appendix.

My introductory remarks to each chapter discuss in detail Fibonacci's use of and contribution to each resource.

What he intended to compile, however, is another issue. Any comparison of his text with most other practical geometries would distill striking commonalities as well as differences. Among the commonalities are the three chapters on measuring heights, areas, and volumes, appropriately titled as *altimetria, planimetria,* and *steriometria*.[26] A well-known exception to this tripartition is the practical geometry of Hugh St. Victor, in which the measurement of the heavens (*cosmimetria*) replaces volume measurement. All of them, in one way or another, invoke and employ the theory of proportion to solve most of their problems. Reading the preface to Fibonacci's *De practica geometrie*, however, sees a promise of more than what other texts offer. His chapter titles could be rearranged and titled as *altimetria, planimetria, steriometria,* and *cosmimetria*, together with examples of multiplication (in local units, as the reader will learn), finding of square and cube roots, and geometric challenges. He wanted to offer more than was currently available in a single text. On the other hand, he assembled the best resources, and after making those modifications he thought necessary translated the whole into Latin. His objective was made from a practical viewpoint: to improve the reader's expertise as a practioner and to give the reader a solid theoretical background for the work. In a sense, Fibonacci did what Euclid accomplished, although at a lower level. His practical geometry superceded all others, by placing in one text all the basic concepts, theory, and skills necessary for the problem solver, the measurer of fields and hills and containers. Mindful of the contents and from the advantage of the twenty-first century, I would add one more thought: *De practica geometrie* is a useful theoretical anthology of Grecian and Arabic mathematics.

Despite the foregoing laudatory description I must offer a caution. How sure can we be that we have Fibonacci's tract before us? The major difficulty in evaluating, even ascribing, much of the contents of Fibonacci's treatise is

---

[26] I rely on Victor (1979), 14–23, for this information.

the lack of at least a near contemporary manuscript of *De practica geometrie*. The earliest that I have been able to find was written more than a hundred years after Fibonacci had completed his work. Moreover, as I discuss below, it is not a copy of the text that he gave his friend, Dominic, who in the first place had requested Fibonacci to write the treatise. Furthermore, consider the chapter *On Divisions*: are the proofs there Fibonacci's creations or did he have a complete Arabic manuscript, propositions with proofs, that has not been located? Nor should be ignored the required repositioning of a section in the *Appendix*, as I note in my next paragraph. In addition, the Italian translations gave me pause. Those that I have read[27] contain a longer section in Chapter 7 than is found in the Latin manuscripts, specifically, a section on finding depths that is completing lacking in the received edition, and they omit Chapter 8, *Geometric Subtleties*. And so on. Although I would advocate the originality of Fibonacci's work in many respects and the ingenuity with which he collected material for the several chapters, I must also keep an open mind about describing original contents and crediting originality, awaiting the discovery of a (near) contemporary manuscript.

SOURCES FOR THE ENGLISH TRANSLATION

The primary source for the translation is Boncompagni's transcription of the Vatican manuscript, Urbino Latin 292, 14th c., ff. 133r–145v. Compared to the other Latin copies in the Vatican collection of *De practica geometrie*, Urbino 292 was cleaner, easier to read, and apparently complete. Beginning on folio 118*r* (page 184 in Boncompagni) and apparently in a contemporary hand is a series of twelve marginal notes prefaced with the phrase "in alio. ..." They offer alternate readings from another manuscript not among those which I have seen. The addenda are clearly distinct from any correction/addition to the manuscript made by the scribe of the exemplary copy. In addition, in Chapter 4, the solution of problem [69] is incorrect in development and false in conclusion. I think that it was added by someone else; Fibonacci could not have made these errors. Further yet, there is a final appendix to the codex that requires its reader to substitute a lengthy passage back to a previous quire. This does not reflect a presentation copy of the autograph. I used the other Latin manuscripts marked with an asterisk in the bibliography as auxiliary guides in my work. Despite the good intentions of Federico Commandino,[28] the transcription by Boncompagni is the first printed transcription of *De practica geometrie*.

---

[27] See bibliography.
[28] "Commandin, il es vrai, avait voulu publier la *Pratique de la Géometrié;* mais la mort l'empêcha d'effectuer ce projet," in Libri, *Histoire*, 26–27, his resource: Baldi, *Cronica de Matematici*, Urbino 1707, in –4, p. 89.

A few comments about other manuscripts that I studied together with their media.

Florence BNC II.III. 22 (16 c.) begins with "Incipit practica geometrie composita de Leonardo bigollosie filio Bonacci Pisano In Anno m.cc.xxi" (*sic*). The copy belongs to the same family as Urbino Latin 292. The figures in the early part of the text have numerical values assigned to the line-segments. Toward the end of Chapter 6 there are no more figures. There are many lengthy, empty spaces in the text. The text is incomplete, ceasing at the end of Chapter 7, "... uenient cubitas $\frac{1}{4}$ 68 pro altitudine *oiq*"[29] (*Microfilm*).

Florence BNC II.III. 24 (14 c.) begins with four folios that are impossible to read; in fact, they are not counted in the foliation that begins "1" with the fifth folio in the set. Water stains appear on many of the folios. The readable text (*f.* 1r) begins "... duarum linearum equalium et equidistantium ..."[30] in a hand suggestive of the early fourteenth century. At the bottom of *f.* 1r appears, "Abbatie florentini S Lxxiii. A. C." in a later hand. Even though the text belongs to the Urbino 292 family because of the "in alio" marginalia, it lacks many figures. The last appears in Chapter 8 at [24]. The text continues through the second appendix (the part that I moved forward) to end at *f.* 147r, "... et hoc est quod uolui demonstrare. Explicit etc," the last two words in another hand (*Microfilm*).

Paris BN 7223 (15th c.) is noteworthy for lacking all of the "in alio" additions in the margins, although it has the same reading at those locations as the Urbino 292 family. Also, in Book IV paragraphs [17], [18], and [19] were not included. In addition, the copy concludes at the end of Chapter 7. Chapter 8 and the Appendices are missing (*Photocopy*).

Paris BN 10258 (17th c.) written in three clearly definable hands is almost devoid of abbreviations and contractions; it is replete with errors, particularly spelling mistakes that are corrected in the margins. It has some of the "in alio" additions both within the text and in the margins. It is a descendant of Urbino 292 (*Photocopy*).

Princeton Scheide Coll. 32 (16th c.) concludes at the end of paragraph [3] in Chapter VII. A series of hash marks (/ / /) suggests that the second copyist (second because the first folio is in an entirely different contemporary hand) had come to the end of the exemplary text that belonged to the Urbino 292 family. The copy is interesting for two reasons. First, a note on the manuscript, "Tuscani xvi s.," would locate the place and time of copying. Second, fractions are written in the modern form: $3\frac{1}{7}$ instead of $\frac{1}{7}3$, with a consequent confusing of $1152\frac{21}{37}$ for $\frac{2}{3}\frac{1}{7}1152$[31] (*Microfilm*).

Vatican Latin 4962 (15th c.) is almost impossible to read because the ink had bled through the paper to mix with the lines on the recto page. The text

---

[29] Op.cit., 206.31.
[30] Op.cit., 2.25.
[31] Boncompagni (1862), 187.41; Princeton *f.* 188v.10.

is incomplete. It lacks almost all the figures even though the scribe provided space for them. It is a descendant of Vatican Urbino Latin 292 (*Microfilm*).

Vatican Latin 11589 (16th c.) clearly shows four hands, the script of the marginalia often differing from that of the text. Although it appears to be a descendant of Vatican Urbino Latin 292, it does not contain all the marginalia of the latter and in some places enlarges on the text (*Microfilm*).

Vatican Ottobonian Latin 1545 and 1546 (17th c.) divide the treatise between themselves, about half and half. They show figures not found in the earlier manuscripts. Furthermore, the scribe incorporated many of the marginalia directly into the treatise. The text is a descendant of Vatican Urbino Latin 292 (*Microfilm*).

Vatican Urbino Latin 259 (17th c.) is a handsomely formatted copy, possibly of Paris BN 10258 because, although lacking the *in alio* marginalia, some are found within the text, for example, at *f.* 147r.7 but not at line 10 where another *in alio* would be expected. There are numerous corrections to the text, evidence of close reading by someone, for example, in *f.* 148r where *similia* is written over a crossed-out *equalia*. The text is a descendant of Vatican Urbino Latin 292 (*Video disk*).

THE TRANSLATION

A common remark is, "This does not translate well." Translating is a tricky task. Any translation except in the simplest case is an interpretation that may even lead to a meaningless phrase. In short, *traduttore tradittore*.[32] Consider this case: in the discussion in Chapter 5 of finding the cube root of 987654, Fibonacci led the reader through the positioning of the digits of partial products according to the digital positions the multiplier and multiplicand have. The problem 234 times 56 requires the reader to consider the positions of the 5 in the multiplier and the 2 in the multiplicand in order to determine where to place the 1 of the 10 in the product of 5 and 2. Fibonacci expressed this placement succinctly: "quia cum secundus gradus multiplicat tertium, quartum gradum facit."[33] The literal translation, of course, is "because since the second place multiplies the third, it makes the fourth place." The modern translation, adopted for this work, is "because the product of the digit in the second place by the digit in the third place goes into the fourth place." By a careful melding of *a verbo ad verbum* and *a sensu ad sensum*, I translated *De practica geometrie* into 21st century American English within the boundaries of two criteria: fidelity to the meaning of the text and fidelity to the format of the text.

Fidelity to the meaning of the text depends upon the text chosen for translation and an experience of the thinking of the author. (The autograph

---

[32] The phrase appears in Hays (1988), 183.
[33] Boncompagni (1862), 151.40–41.

manuscript is not known to exist.) My chosen text, which I call *the received text*, was Boncompagni's printed transcription (1862) of a single manuscript copy, Vatican Codex Urbino Latin 292, which I call *the exemplary text*. Unclear passages in the transcription were clarified by consulting either the exemplary text or other manuscript copies of the treatise.[34] The thinking of the author can be experienced by the prevalence of certain words and phrases typical to his era and uncommon today. Among these are five common words that may require a modern reader to pause momentarily: equidistant, equal as an adjective, cathete, *census*, and thing. Equidistant always means parallel. In adjectival use equal means congruent, except that equiangular means that corresponding angles of a figure are equal. Cathete is taken directly from the Greek and means a perpendicular line, most often an altitude. *Census* always means x^2, the unknown in an equation. In the sense of x, I kept Fibonacci's *res* as *thing*. Hence, a *thing* multiplied by itself is a *census*. An English word that does not appear in the text is *square* as a verb where one reads *multiplicatur in se* (multiplied by itself). Simply for economy I chose to ignore the literal translation and use the verb *square*. Furthermore, the reader will see mixed numbers expressed as Fibonacci and many of his contemporaries wrote them. Seven and a half was written and here printed as $\frac{1}{2}$ 7. This notation is Arabic from the Maghreb, and it reflects the Arabic method of writing from right to left. Less comfortable are larger fractions; six and five eighteenths is written $\frac{1}{2}\frac{2}{9}$ 6. This becomes $\frac{1}{(2)(9)} + \frac{2}{9}$ 6 in expanded format.

Another Arabism that is included is the repetitive use of the conjunction *and*, as in 3 *census* and 4 roots and 6 drachmas equal to 12 dirhams. The modern mind would tend to read and translate *and* as *plus*, as in $3x^2 + 4x + 6 = 12$. Not so the early medieval Arab! He saw 13 objects in a group balanced against 12 other objects. The thinking of al-Khwārizmī and Abū Kāmil was to reduce these groups until there was only one object (x^2) in the lead group. Such was the thinking of Fibonacci, too. It is met in Chapter 6 [14] and [15] and elsewhere. This is not to say that, inasmuch as Fibonacci thought of groups on each side of the verb *equals*, he also thought of an equation as a balancing. First, I must observe that he did not use the verb *balances* as is found in earlier tracts, such as *Liber augmenti et diminutionis*. He used either the verb or adjective for *equals*. Secondly, he used the word *equation*, one time each in *Liber abbaci* and *De practica geometrie*. By adopting the noun, I think that he made a step forward from *balances* to *equals*.

Two oddities resulted during the course of the translation. Although described at length in the introductions to the chapters in which they occur, a few remarks will alert the reader. I shifted two sections of considerable size from their positions in the received text. One transfer of ten paragraphs including a table of arcs and chords is from Chapter 3 to Chapter 7. In the appropriate place I offer reasons for this move done on my own initiative. The

---

[34] The manuscripts that I actually consulted are marked with an asterisk in the bibliography.

other transfer is from the Appendix to Chapter 8, as directed by a rubric in the text. In both instances, I did not change the page numbers that located the texts in Boncompagni's edition.

Being faithful to the format of the text assumes that the format reflects the thinking of the author, specifically the manner in which he wished to convey his ideas to his readers. With respect to *De practica geometrie* the format reflects a lecture, including an interchange between teacher and students, signaled by the pronouns *we, I,* and *you.* Fibonacci seemed to have been speaking to his readers. There is more for the modern reader than just feeling like a member of Fibonacci's audience, one of the "you" to whom he is speaking. The modern reader must be prepared for an apparent ambiguity in some terminology. For instance and as discussed at length in the commentary on Chapter 3, the geometric meaning of the word "root" is "rectangular area." Furthermore, in the same context, the word "side" may mean "geometric root." "Caveat lector!"

Without damaging the format, I made minor changes in wording and structure. For example: the text shows several ways of saying that a proposition has to be proven, using such words as "probatio" and "demonstratio." The variation was usually standardized in the statement, "The proof follows," and similarly with the conclusion of propositions where one sees "ut oportet" or "ut oportebat ostendere." I frequently used the phrase "as required." Structurally, there is great variation among the manuscripts regarding punctuation and separation of recognizable sections. I punctuated statements and separated paragraphs as context seemed to suggest. Furthermore, I numbered nearly every paragraph, enclosing the numbers in brackets. Finally, regarding the figures: first, I numbered them, a better way to align them with the context to which they pertain. Secondly, I redrew many in order to relate them better to the text. The given data and constructions required for some propositions were squeezed into one figure, as seen in the received text. I separated this material into two figures. Finally, I added a few figures. Because I did so much editing on the figures, I saw no purpose in recording the changes; Boncompagni's edition can always be consulted for comparisons. Toward this end I incorporated the corresponding page numbers in Boncompagni's text into the translation; they are in bold type within brackets.

ITALIAN TRANSLATIONS

Late thirteenth and early fourteenth century Italy witnessed the propagation of scholarly treatises in the vernacular. Two developments of *De practica geometrie* arose almost contemporaneously, one a set of apparent translations, the other a group of compilations. The former are considerably shorter than the Latin text, the latter recognize *De practica geometrie* as the major component with material from "molti altri" geometers. The first group has at least three members:

Florence, BNC Fondo Principale II. III. 198 (ca. 1400)
Florence, Biblioteca Riccardiana 2186 [*olim* R. III. 25] (1443)
Vatican Chigi E.VII. 234 (16th c.)
The second set recognizes two members:
Florence, BNC Fondo Palatino 577 (ca. 1400)
Vatican Ottoboniani 3307 (ca. 1465)

Beyond remarking that the title of the Ottoboniani was penned by the scribe of the Palatino manuscript (exactly the same words) and the Vatican text differs appreciably from the Florentine, I leave further meritorious comparison and contrast of the two treatises to the interested. The first group merits our attention.

The Riccardiana text is clearly the work of one Crsitofano di Gherardo di Dino. He identified himself by name, where he lived, and May 1, 1442, the date of writing "lo presente Libbro d'anbaco," eventually to commence writing "la Pratica della Geometria di M^0 Lunardo Pisano."[35] The codex consists of 132 folios of which folios 92r to 131r contain his edition of Fibonacci's *De practica geometrie*. That the latter is an translation and not a slavish copy of *De practica geometrie* is inferred from his reference to Fibonacci's work, "Et noi sappiamo, per la quarte parte della tersa distintione della *Practica di Lunardo*...."[36] The citation of the sectional numbers and titles in *De practica geometrie*, which Crsitofano did not use, informs the reader that at least he had a copy of the Latin text at hand. Crsitofano grouped the contents according to his own schedule, namely

INTRODUCTORY MATERIAL

Measurement of Triangles
Measurement of Quadrilaterals
Measurement of Other Polygons
Measurement of Circles Including the *Table*
Measurement of Hills and Valley and Uneven Plains
Division of Land Held in Partnership
Measurement of Heights of Towers and Lengths at a Distance.

A few specifics: nearly the same material that introduces *De practica geometrie* appears in the Riccardiana, definitions and figures but not the constructions. Fibonacci then discusses units of measurement found in various parts of Italy and then settles on those common to Pisa. Crsitofano named some different units of measurement: *braccia, passi, ccorda, carubbe, rinpennj,* and *pumera*.[37] Both proceed with exemplary problems but with

---

[35] Arrighi (1966), 8–9.
[36] Ibid., 71.
[37] Only the first three are found in Zupko (1981).

different numbers utilizing methods for performing all the measurements mentioned above.

A collation of the contents of Riccardiana with that of *De practica geometrie* displays almost complete overlapping. That is, for the most part what you read in Riccardiana reflects similar content in *De practica geometrie*. I found sixty-five passages that correspond to sections in *De practica geometrie*. Some corresponding passages are either abbreviated or expanded. On the other hand, there are several passages from Crsitofano's own resources, such as the use of a mirror to find the height of a tower. The archipendium is explained, however, the quadrant is ignored. The noticeable difference between the two from the Riccardiana position lies in its final chapter on indirect measurement by similar triangles. There are solved problems that measure heights of towers and hills, depths of valleys and wells, and inaccessible widths of rivers. None of these are in *De practica geometrie*. On the other hand, the two chapters on finding roots, the chapter on solid geometry, and the final chapter on geometric subtleties in *De practica geometrie* are not in Riccardiana. Nor is algebra used to solve any problems in Riccardiana. Finally, the contents of three chapters, on triangle, polygons, and circles, are combined with other material in Chapter 3 of *De practica geometrie*.

Crsitofano's work was useful for my translation. Under the genus quadrilateral Fibonacci had identified four kinds of trapezoids, "de diuisione quatuor figurarum caput abscisarum que duo tantum latera habent equidistantia." The origin of the figure with its head cut off is a triangle decapitated by a section through two sides and parallel to the third. Then he enumerated these four: *semicaput abscisa*, altera *eque caput abscisa*, tertia *diuerse caput abscisa*, quarta *caput abscisa declinans*.[38] My initial attempt at translation ended in simply leaving the Latin words mixed with the English text. In his transcription of Crsitofano's text, Arrighi included labeled figures; and there are the four trapezoids identified by name: *mezzo capo tagliato, eghualmente capo tagliato, diversamente capo tagliato, capo chinante*.[39] Pairing the names of the figures with the Latin words (they are in the same order), the following translations obtain:

semicaput abscisa = mezzo capo tagliato = right (angled) trapezoid
eghualmente capo tagliato = eque caput abscisa = isosceles trapezoid
diversamente capo tagliato = diuerse caput abscisa = scalene trapezoid
capo chinante = caput abscisa declinans = slanting trapezoid

In the first three species, altitudes dropped from the endpoints of the upper base either fall within the figure or one falls along a side. One altitude of the slanting trapezoid falls outside on an extension of the lower base. I would never have understood this had I not seen the figures in Arrighi's

---

[38] (125.27–29). Note: here and henceforward this form of citation refers exclusively to Boncompagni (1862), *De practica geometrie*.
[39] Arrighi (1966), 84.

transcription. In a sense of completeness to the quattro-cento translator: he named a fifth trapezoid, *pescie* or *fish*. It resembles the slanting trapezoid except that it is taller and narrower. Fibonacci did not write about this one. Furthermore, I suspect that the name *mezzo capo tagliato* suggests an isosceles trapezoid that had been bisected by a line perpendicular to the bases. In addition, this translation does not support my hypothesis about the misplacement of Fibonacci's table of chords; the table appears in Chapter 3 as it does in the exemplary edition. Finally, because the Riccardiana treatise is more accessible than the others, it is the norm against which the other two Italian translations are compared.

The information about Crsitofano's work may suggest that he excerpted from *De practica geometrie* those parts he judged most useful for the geometric practitioner. This indeed is the opinion of Arrighi who characterized the treatise as a "riduzine" that fits practical needs, "la misura delle terre, siano esse in piano o in declivione, e i vari modi per la loro divisone in parti fra 'consorti'."[40] This viewpoint is reasonable, but it is not the only viewpoint. Elsewhere I develop the stance that Crsitofano translated the original practical geometry to which Fibonacci referred in his dedicatory preface to *De practica geometrie*. Without displaying my chain of reasoning here,[41] I would observe that as a Pisan Crsitofano was the right person in the right place to find an original manuscript by Fibonacci. The fact that he had a copy of *De practica geometrie* at hand bolsters my thesis.

An unknown scribe penned the Chigi manuscript (16th c.).[42] Contentwise it seems to be a copy of the Riccardiana, for instance,
Riccardiana: "Et la prima si chiama mezo capo tagliato" (p. 50).
Chigi: "Et la prima se chiama mezo capo talhato" (f. 14r).
Riccardiana: "Et se vuoi trovare per arismetrica lo punto" (p. 78).
Chigi: "Et se vuoi trouare per abbaco lo punto" (f. 23v).
Riccardiana: "Lo partimento de quadranghuli si accade in 3 modi" (p. 81).
Chigi: "Lo partimento de quadrangli scade in 3 guise" (*f*. 25r).

There is a major difference: the *Table* and surrounding propositions in Riccardiana are not present in Chigi, from "Et se, per altro più soctil modo ..." (p. 68) in Riccardiana to "... li loro archi non saputj trovare" (p. 73). This section includes instructions for constructing, using, and practicing with the *Table*. The missing section belongs between lines 12 and 13 on *f*. 21r. Hence, it is not a matter of a folio having been removed from the codex; the section was not there to be translated and copied. Consequently, one may wonder if the Chigi translation not only antedates the Riccardiana but is a more faithful reflection of what Fibonacci initially wrote.

---

[40] Arrighi (1966), 11.
[41] Hughes (2007).
[42] Catalogue *Biblioteca Chigiana* (Sala Cons. Mss. 190 xxix), #161.

xxxiv   Background

The Principale tract is a short excerpt from *De practica geometrie* that concludes with "Deo gratias." Between the title and *explicit* lie defintions of point, line, two angles, four triangles, two quadrilaterals, circle with parts, and π equal to $3\frac{1}{7}$. Depending on how you separate the texts, there are eight exercises about circles, an inscribed square, and triangles. Unlike the writers of the two previous editions, this editor used both geometric and algebraic tools. The latter are exemplified in the final problem, to find the altitude of a 13-14-15 triangle (see Figure 0.1; the figure is from the text.) The reader is led through all the *cosa/censo* steps of $13^2 - x^2 = 15^2 - (14 - x)^2$, to find the segment of the base, $x = 5 = cosa$, whereby the Pythagorean theorem determines that measure of the altitude. A geometric solution for the same problem 3 [18] in *De practica geometrie* closes the tract.

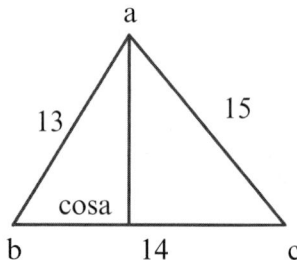

CONCLUSION

The foregoing has been expository, to acquaint the reader with Fibonacci, the contents of *De practica geometrie* and its sources, and some particulars. Finally, I would consider the veracity of the text. By veracity I mean, is the text in Codex Urbino 292 as transcribed in Boncompagni's edition and corrected by me the treatise produced by Fibonacci in 1220? The following facts help shape an answer.

First, the Latin manuscripts that are nearly complete and known today belong to the same family, even if some are cousins rather than direct descendents of the original exemplar. The reason for grouping them together is that they refer to one other manuscript outside the family, identified by the phrase "in alio ..." where different readings are found. Manuscript "in alio" has not been located. Furthermore, it may be closer to what Fibonacci wrote. For my purposes, the original exemplar is probably reflected in Codex Urbino 292.

Second, problem [68] in Chapter 4, for which a proof [69] is provided, presents a difficulty. The statement of the problem is apparently incomplete and the proof is faulty. What seems incomplete is an unannounced line-segment required for an operation. The fault lies in concluding that because corresponding sides of two triangles are proportional, the triangles are congruent, as the Latin text states. In fact they are only similar, as I corrected the text. The error is so obvious that it suggests to me that someone else added the two paragraphs. Inasmuch as both are at the end of a major section, I wonder if they were not slipped in by an over-eager instructor or copyist who thought he had successfully captured Fibonacci's method.

Third, the second appendix, paragraphs [21*] through [33*], carries the prefatorial remark that it belongs earlier in the manuscript, with directions on where to place it: "The following problems on triangles belong with the antecedent quire after the problem on an equilateral triangle whose measure with perpendicular is 10." Fibonacci compiled *De practica geometrie* as a gift to a good friend. I cannot imagine him sending on a treatise with obvious errors and misplaced sections.

The foregoing suggests that all the located manuscripts are flawed. Neither the autograph copy nor copies taken directly from it have been located. What we have, despite its value as a select compilation of medieval Arabic and ancient Greek mathematics, is an edition of an earlier work yet to be located.

With this background, I offer my evaluation: despite textual weaknesses at hand, we can conclude that Fibonacci provided the surveyors of his day with many useful propositions and tools. The outstanding value of this work for both practioner and theoretician arises from his compiling those treatises he considered as among the best mathematics in the Greco–Arabic world.

Prologue and Introduction

COMMENTARY

The treatise begins in epistolary fashion addressed to Leonardo's friend, Dominic,[1] apparently a Spanish cleric who had introduced him to Frederick II. Because the emperor was well known for advancing scholarship,[2] the meeting was most advantageous for Fibonacci. After acknowledging his friendship, he described the eight chapters (*distinctiones* in Latin). Thereafter Fibonacci proceeded to the *Introduction* that starts immediately with twenty-eight definitions. "Point is that which lacks dimension; that is, it cannot be divided" is in the first place. After the definitions follow statements about five constructions, ten theorems, and seven axioms, all Euclidean in nature. The next section lists and defines the various units for linear and areal measurements. Although recognizing the diversity of units of measurement throughout his world, Fibonacci opted for the Pisan system that clearly distinguished between liner and areal units of measurement. The unit that bridges both dimensions is the rod (*pertica*), at once a measuring instrument of about 2 m 96 cm in length and the name of the unit.[3]

Table 1 shows the linear units by name together with their equivalents (as found in the text) in italics, the other numbers having been extrapolated. The smallest unit, as we learn from a fourteenth century translation of *De practica geometrie*, is the grain of wheat, two of which determine the size of the point.[4] Crucial definitions in the text are linear foot consisting of 18 linear [inches] and linear rod being six linear feet long. Finally, all linear measurements of fields and buildings were made in rods, feet, and inches.[5]

---

[1] Franci (2002), 298; *see also* Arrighi (1966), 10.

[2] Toubert–Bagliani (1994), *passim*.

[3] Dilke (1987), 58.

[4] "Est enim pertica sex pedum; et pes est decem et octo unciarum; et uncia uiginti punctorum" (3.37–38). Units of measurement were not constant throughout Italy, as Zupko (1981) lists so clearly. One of the Italian translations of *De practica geometrie* has, "'l piè è uncie 18 e l'uncia è puntj 18 et li 2 puntj sono di grossesa di uno granello di grano" Arrighi (1966), 26.

[5] The table is based on what Fibonacci wrote in Chapter 7 [211] (95.38). Arrighi's source has 18 points to the inch (see footnote 4 above). The metric equivalents were adjusted from the value of the Pisan rod, 2.918 m, in Zupko (1981), 190.

Table of Linear Measurements

Rod 3 m	Foot 50 cm	Inch 2.78 cm	Point 1.4 mm	Grain .7 mm
				1
			1	1/2
		1	*20*	10
	1	*18*	360	180
1	*6*	108	2160	1080

A similar set of terms is used for areal measurement. Here we need be cautious, because areas are computed by rectangular strips, a bilateral requirement made clear only in an Italian translation.[6] First, each strip is made up of geometric squares, of which there are two sizes, smaller and larger. The smaller is one foot on each side, is called a denier of measure (*denario di misura*), and is used to measure strips smaller than an areal foot. The larger is six linear feet on each side that I name an areal rod (*pertica superficiale*). Neither of these is the basic unit of areal measurement. The basic unit is the areal foot equal to an area 1 linear rod by 1 linear foot, because six areal feet are used to describe an areal rod. The areal foot is used to measure areas as large as or larger than an areal foot. Second, all rectangular strips have one dimension in common: one side measured by 1 rod. Hence, an areal inch is 1 rod by 1 inch. Since there are 6 feet in a rod and an inch is $\frac{1}{18}$ of a rod, the measure of the areal inch is $\frac{1}{3}$ a denier of measure. An areal foot is 1 rod by 1 foot. Because there are 6 feet in 1 rod, then the measure of an areal foot is 6 deniers. Obviously, the measure of an areal rod is 36 deniers of measure. In measuring larger strips, one side is always 1 rod and the other side is so many rods. Hence, the measure of 1 scala is 1 rod by 4 rods equal to 4 areal rods or 144 deniers of measure. The Table of Areal Measurements attempts to organize these ideas. Again, numbers in italics are from the text; the others were created by extrapolation.

Table of Areal Measurements

Modium	Starium	Panis	Scala	Rod	Soldus	Foot	Denier	Inch
								1
							1	*3*
						1	6	*18*
					1	*2*	12	36
				1	3	6	*36*	*108*
			1	*4*	*12*	24	144	432
		1	1.375	5.5	16.5	33	198	594
	1	*12*	16.5	66	198	396	2 376	7 128
1	*24*	288	396	1 584	4 752	9 504	57 024	171 072

The area, often called *embadum* in the text, had its own set of measurements. For area of homes, the chosen units were scala, soldus, and denier of measure.

---

[6] Arrighi (1966), 26.8–37.

Areas of fields were measured in staria, panes, soldi, and deniers, or in parts of one panis, all defined as follows with metric equivalents based on the Pisan areal rod.[7]

An areal inch is one rod long and an eighteenth part of a linear foot wide. It is the third part of a denier, $\frac{1}{18}$ of an areal foot, or $\frac{1}{108}$ of an areal rod (7.87 sq cm).

A denier of measure is one foot wide and one foot long. It is the thirty-sixth part of a whole areal rod. There are six deniers for each areal foot (23.61 sq cm).

An areal foot is one rod long and a sixth part of a rod wide. As a whole it is a sixth part of an areal rod (1.467 sq m).

A soldus of measure is two areal feet (2.934 sq m).

A square or areal rod consists of six areal feet or thirty-six deniers, and is an eleventh part of two thirds of a starium (8.5 sq m).

A scala contains four areal rods or 12 soldi of measure (39 sq m).

A panis contains five and a half areal rods or 16 soldi and is the twelfth part of a starium (46.75 sq m).

A starium contains sixty-six square rods or 12 panes or 198 soldi (561 sq m).

A modium contains 24 staria (1.346 ha).

Throughout all multiplication there is a running substitution of a higher unit of measure for the appropriate number of lower units, even if the higher unit is not a recognized lower unit. For instance, 8 rods times 11 rods becomes 8 rods times 2 panes which equals 16 panes. The product becomes 1 starium and 4 panes. Note that the word *rod* is linear as a factor and *areal* as a product. The common products are these:

1 inch by 1 inch = $\frac{1}{324}$ denier of measure

1 inch by 1 foot = $\frac{1}{18}$ soldus of measure

1 inch by 1 rod = $\frac{1}{3}$ denier of measure

1 foot by 1 foot = 1 denier of measure

1 foot by 1 rod = $\frac{1}{6}$ (areal) rod

n rods by n rods => may become so many rods, panes, and/or staria with the requisite number of lower units.

After guiding the reader through a multitude of examples, switching from rods to staria with occasional retroglances at soldi and deniers, all of which he asserted would be quite useful as the reader would come to realize, Fibonacci began his introduction to *De practica geometrie*.

SOURCES

The prevailing assumption throughout the discussion of sources is that for the most part Fibonacci had his own Arabic copies of the various works, whether they are identified by English or Latin titles, not to overlook Latin translations

---

[7] Zupko (1981), 190.

of some texts. For example, a side-by-side comparison of *De practica geometrie* and *Liber embadorum* strongly suggests that Fibonacci used Abraham bar Hiyya's *On Mensuration and Calculation* as a guide during the preparation of the *Introduction* of his treatise. There is no cogent reason to believe that he cut-and-pasted for his own work after considerably rewriting Plato of Tivoli's translation. The material itself, numerous definitions, axioms, but not the postulates, together with crucial constructions, was taken from Euclid's *Elements*, Book I. Of the fourteen statements regarding constructions in [3] below and allowing for a measure of variation in translation, twelve statements represent word-for-word statements from Euclid's *Elements*, Book I, as translated by al-Hajjaj. As for the many problems, Fibonacci crafted them, either from personal resources or a common treasury of problems.

PROLOGUE AND INTRODUCTION[8]

{p. 1} Dominic my friend and reverend Master! You asked me to write you a book on practical geometry. So prompted by your friendship and yielding to your petition, I have reworked for your sake a tract already begun so that others might find a finished treatise. Some parts are presented according to geometric demonstrations, other parts in dimensions with which they wish to engage according to a more common lay practice. The topics of the eight chapters follow.

1. How the sides of fields having four equal angles can be multiplied by their lengths in three ways
2. Certain rules of geometry, and how to find rational square roots solely by geometric methods
3. How to find the square measures of all fields of whichever form
4. How to divide all areas among partners
5. How to find cube roots
6. How to find the square measures of all bodies of whichever figure that are contained by three dimensions, namely, length, width, and depth
7. How to find the longitude of planets and the altitude of high things
8. Certain geometric subtleties

Before I proceed to develop these chapters, some introductory remarks are in order. To do this as well as I can and confident of your capability to make suitable modifications, I prepared this work for your consideration, so that your expertise can adjust whatever needs correction.

---

[8] "Incipit practica geometrie composita a Fibonacci pisano de filiis bonaccii anno Mccxx" (1.1–2, *f*. 1r.1–2). Note: all folio references *f* are to Urbino 292 unless otherwise noted.

INTRODUCTORY MATERIAL[9]

0.1 Definitions

[1] Point is that which lacks dimension; that is, it cannot be divided. Line is length lacking width, with points as its ends. Straight line is that which is drawn from point to point. Surface has both width and length, with lines for its boundaries. It is a plane if it can be covered with straight lines everywhere within its boundaries. A plane angle is an inclination of two lines touching one another in a plane and not crossing one another. A rectilinear angle is an angle contained by straight lines. When a straight line stands atop a straight line and makes of itself two angles equal to each other, both are called right angles. The line that is standing on the other is called cathete or perpendicular. An ample or obtuse angle is greater than a right angle. An acute angle is less than a right angle. A boundary is the end of whatever. Any figure lies under or within one or more boundaries. A rectilinear figure is bounded by straight lines. A trilateral figure is bounded by three straight lines. A quadrilateral is bounded by four straight lines. A multisided figure is bounded by more than four straight lines. A circle is a plane figure lying within a single line called the circumference or periphery within which is a point from which all straight lines drawn to the circumference are equal to each other. The point is called the center of the circle. Now when some straight line is drawn through the center and ends on some part of the circumference {p. 2}, that straight line is called the diameter of the circle and divides it into two equal parts of which either is called a semicircle. The portion or segment of the circle is a figure contained between the circumference of the circle and a straight line; it may be larger or less than a semicircle. A sector of a circle is a plane figure contained between two straight lines drawn from the center to the circumference and the enclosing arc of the circumference.

0.2 Properties of Figures

[2] The straight lines which are in the same surface and extended infinitely from both ends and never intersect are called equidistant. If a straight line intersects them, it makes on the one part two right interior angles or angles equal to two right angles. The exterior angle is equal to its opposite interior angle. And the angles that are permuted[10] are equal to one another. For instance, if a straight lines *ez* falls on two equidistant lines *ab* and *dg*, angles *biz* and *izd* are either right angles or are equal to two right angles. The same holds for angles *aiz* and *izg*. And exterior angle *eib* equals its opposite interior angle *izd*. And angle *aiz* equals angle *izd* that lies permuted. It follows further

---

[9] "Incipiunt introductoria" (1.22; *f.* 1r. 24).
[10] That is, opposite interior and exterior angles.

6   Fibonacci's *De Practica Geometrie*

that angle *biz* equals angle *izg*, as is shown in geometry.[11] There is a great deal to know for those who wish to advance about measure and division of bodies according to geometric expertise, as is shown clearly in Euclid.

0.3 Construction of Figures

[3] Among the constructions are the following: to construct an equilateral triangle upon a given straight line; to bisect a given angle or line; to erect a perpendicular on a given point of a given line; to drop a perpendicular from a given point to a given line. Two straight lines make two right angles. When two right lines intersect each other, the two opposite angles with a common vertex are equal to each other [see Figure a.1]. If a straight line intersects two other straight lines and makes the two interior angles less than two right angles then the two straight lines on the side of the two aforementioned interior angles will intersect if they are extended. A pair of equal and equidistant lines are always equal and equidistant from one another. To construct a rectilinear angle at a given point on a given line equal to another rectilinear angle. Triangles and parallelograms constructed on equal bases and between equidistant lines are respectively equal, triangle to triangle and parallelogram to parallelogram, and that the parallelogram is twice the triangle. The opposite angles of a parallelogram are equal. The diameter of a parallelogram cuts it into two equal triangles. If any side of a triangle is extended, the angle it makes outside the triangle with the side equals the sum of the other two angles. The three angles of a triangle equal two right angles.

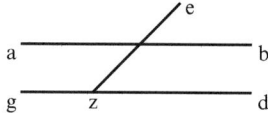

0.4 Axioms

[4] Things equal to the same thing are equal to one another. If equals are removed from equals, what remain are equal. If equals are added to unequals, the sums are unequal. If equals are subtracted from unequals, the remainders are unequal. If the same things are doubled, they are still equal to one another; similarly if they are halved, and likewise if they are increased by the same amount. The whole is greater than its part. And two straight lines cannot contain space. With the foregoing and knowing how to find square roots,

---
[11] *Elements*, I, 29.

those who would proceed along the common way {**p. 3**} with these things, which I shall show in their proper place below, can easily move on. With this settled, I propose to explain briefly the measures for determining the dimensions of fields.

0.5 Pisan Measures

[5] Some measure fields by cubits or ells or paces; others use rods or some instrument for measuring length or area. Linear is strictly by length. Area is measure by length and width in square units which require four right angles. Some gather by multiplying from these areal measures a certain quantity which they call iugerum or aripennium or carrucam or tornaturam or culturam or other quantities which require other words. I, however, follow the custom at Pisa beginning with the rod. The Pisan linear rod is six linear feet long. A linear foot consists of 18 linear points. The square or areal rod consists of six areal feet. An areal foot is one rod long and six parts of a rod wide. An areal inch is one rod long and an eighteenth part of a long foot wide. Hence an areal foot is a sixth part of an areal rod. An areal inch is the eighteenth part of an areal foot or one hundred eighth part of an areal rod. Likewise an areal rod consists of thirty-six deniers by measure, and so there are six deniers for each foot and an areal inch is the third part of a denier. A denier is one foot wide and one foot long. And so a square denier has four right angles. Thus a denier is the thirty-sixth part of a whole areal rod. Four areal rods make a certain measure that is called scala. Five and a half areal rods make one panis. Sixty-six square rods make a certain measure called starium according to which fields are bought and sold in the Pisan bishopric, and for which I will show how to measure an embada for areas of fields. From the product of the above-mentioned starium another sum or quantity can be had which is called a modium. One modium contains 24 staria. Likewise, $\frac{1}{2}$ 16 scale equal one starium. One scala is an eleventh part of two thirds of a starium or a scala is 12 soldi by measure. Again, 12 panes are one starium; hence a panis is the twelfth part of a starium. And a panis is 16 soldi and a starium is 198 soldi. Fields and the space in homes are measured in rods, feet, and linear inches. But the embada of fields are measured in staria, panes, soldi, and deniers or in parts of one panes. The area of homes, however, are measured in scale and parts of a scala and in soldi and deniers.

0.6 Computing with Measures

[6] Now that this has been explained, what arises from the multiplication of rods by rods according to the above-mentioned measures must be shown. For whatever is multiplied by one or more rods, the same proceeds from the multiplication, as when one or several rods are multiplied by several rods, whatever comes from the multiplication will be rods. For example, three

rods multiplied by nine rods makes 27 rods. And if one rod or several rods are multiplied by panes, or panes in several rods, whatever arises from the multiplication are panes. For example, three rods multiplied by nine panes or nine panes by three {p. 4} rods make 27 panes. And thus it must be understood about the multiplication of rods by scale, staria, and modiora, or of the multiplication of scalas, staria, and modiora by rods. Again, feet multiplied by rods or rods by feet are feet or halves of solda. For example: three feet multiplied by nine rods or nine rods by three feet are 27 feet, that is, 13 solda and $\frac{1}{4}$ of a measure. Likewise inches multiplied by rods or rods multiplied by inches make inches or thirds of one denier. For example: from four inches multiplied by nine rods or conversely there arise 37 inches, that is 12 measures of deniers. Further, feet multiplied by feet produce deniers. For example: three feet multiplied by five feet produce 15 deniers. Again feet multiplied by inches produce eighteenths of one denier. And inches multiplied by inches make eighteenth parts of one eighteenth part of one denier, namely three hundred twenty-fourths of one denier. Note that all the measures that are multiplied are linear, and what result from the multiplication are planar. It must also be known that 100 planar rods are one and half starium. Moreover one rod and a hundredth of a scala are six staria. Further, one scala and a hundred panes is $\frac{1}{3}$ 8 staria. Likewise $\frac{1}{2}$ 16 soldi are one panis. Therefore 33 soldi are 2 panes, and $\frac{1}{2}$ 49 soldi are 3 panes. 66 solid are 4 panes, and $\frac{1}{2}$ 82 soldi are 5 panes. 99 soldi are 6 panes. Likewise $\frac{1}{2}$ 5 scale are 4 panes, and 11 scale are 8 panes. Note that we mention scale because some people use them to compute the measures of land in scale from which they determine the panes.

[7] Whence we leave work of this sort about scale to teach how to assign the measure of earth by staria, panes, soldi, and deniers. Whence it must be said that $\frac{1}{2}$ 5 rods multiplied by however many rods, so many panes arise from the multiplication, as there are rods by which the $\frac{1}{2}$ 5 rods were multiplied. Because 11 rods multiplied by however many rods returns so many panes twice as many as there are rods. Whence $\frac{1}{2}$ 16 rods multiplied by however many rods returns three times as many panes, 22 rods four times as many, $\frac{1}{2}$ 37 rods five times as many, 33 rods six times as many, $\frac{1}{2}$ 38 rods seven times as many, 44 rods eight times as many, $\frac{1}{2}$ 49 nine times as many, 55 tens times as many, and $\frac{1}{2}$ 60 eleven times as many. And so on: if 66 rods are multiplied by however so many rods, 12 times as many panes arise from the multiplication of those rods by which the 66 rods had been multiplied. That is, there are as many staria as there were rods. Because of this 132 rods multiplied by however so many rods make twice as many staria. The same is true in the similar case. This is why we said above about panes, staria, and modia multiplied by rods, because it is the same as multiplying 22 rods by 3 rods as 4 panes by 3 rods. For it is the same as multiplying 132 rods by however so many rods as to multiply 2 staria by the same number of rods.

[8] And it must be noted that a rod whose width is one has a length of $\frac{1}{2}$ 5 panes. And a rod of width one and twice of $\frac{1}{2}$ 5 rods in length is twice one

panis. So you are to understand about thrice and fourfold and other multiples of $\frac{1}{2}$ rod in length and one rod in width. Likewise one rod in width and 66 rods in length is one starium. Wherefore {p. 5} two rods in width[12] and 66 rods in length are 2 staria. And 3 rods in width and 66 rods in length are 3 staria. And so on in similar cases. Again double of one rod in width and half of 66 rods in length is one starium. Triple of one rod in width and a third of 66 rods in length is similarly one starium. Understand this from some other multiple of one rod if it is the width and from the same fraction of 66 rods in length. Whence if it is asked with 6 rods in width of some square field, how many rods are in its length in order to have [so many] staria? Then because 6 is six times one rods, then take a sixth of 66 rods, namely 11, and you have the length of the field. Likewise if there are 7 rods in width, find the starium. If you have a seventh of 63 rods for the length, then 9 rods are found. If the difference between 63 and 66 rods is three rods, then you have 18 feet. Divide these by 7 and you have 2 feet with a remainder of 4 to be divided. Change these to inches and there are 72 inches. Divide these by 7 to obtain $\frac{2}{7}$ 10 inches. And thus we have 9 rods, 2 feet, and $\frac{2}{7}$ 10 inches for $\frac{1}{7}$ of 66 rods. All of this is very useful as will be shown in its own place.

---

[12] "in capite" (5.1).

1
Measuring Areas of Rectangular Fields

COMMENTARY

A set of twenty-five solved problems and twelve theorems comprise the two parts or Methods of Chapter 1. In Part I all of the problems focus on finding areas of fields given dimensions in one, two, and/or three different units of measurement, which make the multiplication complex. Fibonacci's method for multiplication most probably reflects the method common to Pisa, if not much of the Mediterranean world. A crucial factor is one's ability to move rapidly among the various units, just as a modern person would be expected to move easily among the various metric or English units.

The best way to be introduced to Fibonacci's method of multiplication is to walk through one of the problems, as follows.

[15] Likewise, if you wish to multiply 26 rods 4 feet by 43 rods 5 feet, first multiply the 26 rods by the 43 rods to get 16 staria, 11 panes, and 4 soldi. Likewise multiply 4 feet, namely 2 soldi, by 43 rods to get 86 soldi, namely 5 panes and 3 soldi. Likewise multiply 5 feet, namely $\frac{1}{2}$ 2 soldi, by 26 rods to get 65 soldi. Likewise multiply 4 feet by 5 feet to get 20 deniers. By combining these four products into one you have in sum 17 staria, 8 panes, 8 soldi, and 8 deniers.

The problem is shown in its medieval format in Figure 1.1. The vertical and diagonal lines connect the pairs of numbers that will be multiplied together; imagine the "FOIL method" for multiplying a pair of binomials. The first step recorded by Fibonacci, "multiply the 26 rods by the 43 rods to get 16 staria, 11 panes and $\frac{1}{2}$ 4 soldi," may be expanded as follows: 26 rods by 43 rods is clearly 1118 rods. According to the table above, there are 66 rods in 1 starium. Hence,

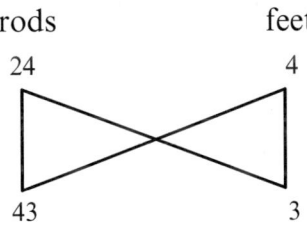

Figure 1.1

there are 16 staria in 1118 rods, with 62 rods left over. With 5.5 rods in each panis, there are 11 panes in the 62 rods, with 1.5 rods remaining. There are 3 soldi to the rods. Hence, we have 4.5 soldi. The final answer to the first multiplication is as shown above: 16 staria, 11 panes, and $\frac{1}{2}$ 4 soldi, a computation performed mostly in the head. With this preparation, the remaining three products and their sum may be found. The combined answer raises the question, why do not rods and feet appear in the answer? It would seem to be a matter of taste, often keeping rods, feet, and inches for linear measurement.

At this point it would be well to make a few remarks about Fibonacci's fractions. The first thing to note is the format, $\frac{1}{2}$ 4, which means four and a half. The format is unique to Andalusia and the Maghrib and reflects the Arabic method of writing from right to left,[1] something Fibonacci most probably learned as a student in a Moslem school in Bougie. Secondly, the initial meeting is with fractions to which we are accustomed. Later, however, we meet a strange format, such as $\frac{1\ 14}{3\ 18}$ 4.[2] This is a mixed number that needs to be read from right to left. Upon expansion it becomes

$$4 + \frac{14}{18} + \frac{1}{3}\left(\frac{1}{18}\right) \quad \text{or} \quad 4\,\frac{43}{54}.$$

Much has been written about this format for fractions in which Fibonacci showed himself well versed. The format first arose where he discussed division of whole numbers, in *Liber abaci*.[3]

The second representation of fractions requires more attention. Aptly described as "parts of parts"[4] this format is uncommon today. Perhaps wishing not to overburden readers of *Liber abaci* with too much on fractions, he waited until a more difficult division problem required further development.[5] Here he offered a table for division by composite numbers from 12 to 100, coupled with the strong recommendation "to learn this by heart." For our purposes Fibonacci closed the exposition by demonstrating how to create unit fractions for any rational numbers.[6] The last presentation can suggest the second way of denoting fractional numbers, multiplicative-additive, or ascending continued fractions.[7] An example[8] illustrates both formats:

$$\frac{1\ 5\ 3}{2\ 6\ 10} = \frac{1}{2(6)(10)} + \frac{1}{6(10)} + \frac{3}{10} = \frac{\frac{\frac{1}{2}+5}{6}+3}{10}$$

---

[1] See, for instance, *Talkih al-Afkar* of ibn al-Yasamin (d. 1204) that was written in Andalusia or in Morocco around the middle of the twelfth century.
[2] Boncompagni (1862), (10 [21]).
[3] *Liber abaci*, 21ff; Sigler (2002), 49ff.
[4] Hoyrup (1990), 293–297.
[5] *Liber abaci*, 36.38–37; Sigler (2002), 64–65.
[6] *Liber abaci*, 77.39–83.28; Sigler (2002), 119–126.
[7] Hoyrup, ibid; Dutton and Grim (1966).
[8] Tropfke (1980), 113–114.

While accepting the representation of rational numbers in the ascending continued format, one does well to be cautious about crediting Leonardo with knowledge of this process. As a matter of fact his fractions are either as regular as are ours or are clearly parts of parts. Fibonacci learned his fractions under Maghribean influence. Although today the Arabic format can be adjusted to that of ascending continued fractions, Leonardo thought of his fractions and wrote of them as parts of parts.

The calculation may appear fortuitous. However, Fibonacci would use the format to join fractional parts of a denier to larger units. Thus, a long verbal answer can be read from a short presentation in a kind of positional notation. For example, in problem [2], the abbreviated answer is

$$\frac{0\ 4\ 0\ 2}{6\ 6\ 11\ 6}4.$$

The lone whole number indicates staria. Now reading the denominator from right-to-left (the direction in which the Arabic language is written), the first 6 names panes, the 11 identifies rods, the 6 to the immediate left is for soldi, and the last 6 is for deniers. This arrangement is entirely in keeping with the customary naming of square units or embada. The actual number of units is found in this way. We begin with the 4 staria. Next, the number of panes is four for twice the number in the numerator above the rightmost 6. Then to the left: there are zero rods above the 11. The number of soldi is always half the upper number appearing in the third left place, or 2. And the number of deniers is found by multiply the number the 4 above the 6 in the third place by the lower 6 in the fourth place and adding the number above it (here zero) to get 24. The verbal answer is 4 staria, 2 soldi, and 24 deniers. He closed Part 1 of Chapter 1 with the advice that if you could not reduced inches to sixths of sixths, then forget them and compute with ordinary fractions.

Part II is quite theoretical, offering a Euclidean framework onto which one may base any practical work in measurement. Leonardo remarked initially that these propositions would be used for finding square roots in the first part of Chapter 2.

SOURCES

Part I was drawn from common knowledge about rectangular areas, Pisan units of measurement, and typical method of multiplication. Part II lists and/or proves twelve propositions from Euclid's *Elements*, the first nine being from Book II but not in the same order. In contrast with Plato's translation, they affirm that Fibonacci had at hand an Arabic translation of the *Elements*.

TEXT[9]

1.1 Areas of Squares

[1] If you wish to measure *abcd* as a quadrilateral, equilateral, and equiangular field having 2 rods on each side, I say that its area is to be found by multiplying side *ac* by its adjacent side *ab*, namely 2 rods by 2 rods. Thus there is a square of 4 plane rods. Let lines *ab* and *cd* be divided into two equal parts at points *e* and *f*, and draw line *ef*. Likewise divide *ac* and *bd* in two equal parts at points *g* and *h*, and draw line *gh*. Thus quadrilateral *abcd* has been divided into four perpendicular squares, each of which is measured by one rods on a side. Thus there are 4 plane rods in the entire square quadrilateral *abcd* [see Figure 1.2]. For the straight line *ae* is equal to and equidistant from line *c* because line *ef* is equal to and equidistant from line *ac*. And *ef* is equal to and equidistant from line *bd*, since it is equal to and equidistant from line *ac*. Similarly, line *gh* can be found equidistant and equal to both lines, *ab* and *cd*. And because line *ef* is equidistant from line *a* so is line *ae* from line *cd*. Therefore angle *aef* is equal to its opposite angle *acf*, the right angle under angle *acf*, and the right angle under angle *aef* for the same reason it is shown that right angle *efc* is equal to right angle *cae*. Therefore equilateral *eacd* is orthogonal and contains in itself two square rods; one is the quadrilateral *eag*, the other is *igcf*. The aforementioned shows that all the angles are right angles, such as those about *i* and the angles that are at *g* and *h*. For the plane *ec* contains the half plane *bc* with plane *ec* equal to plane *ed*. Therefore the square measurement of the whole plane *ad* is 4 rods, as we said.

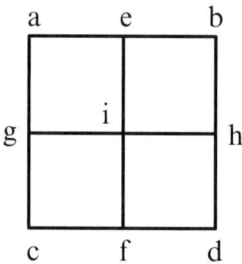

Figure 1.2

1.2 Areas of Rectangles: Method 1

[2] Likewise, let the long quadrilateral be *abgd* in whose opposite boundary, namely in .*ag*. and .*bd*. are two rods. In each of the remaining sides are three

---

[9] "Incipit distinctio prima de multiplicatione latitudinum camporum quadratorum rectos angulos habentium in eorum longitudine, in quibus multiplicationibus eorum embada continentur" (5.16–18, *f*. 3r.11–12).

rods. I say that its square measure is 6 rods because it is created by multiplying one boundary into one side, namely 2 by 3. In order to prove this, divide straight line ag and bd into two equal parts {**p. 6**} at points e and z and join e and z. Divide both straight lines ab and gd into three equal parts at points i and t and points k and l. Straight lines are formed by joining points i and t with points k and l [see Figure 1.3]. Then show by what was said above about the square that the entire quadrilateral $.abgd$ has been divided into six equal and orthogonal quadrilaterals of which any one contains on its single side one rods. Then, with all this demonstrated, we show how the widths of the given fields must be multiplied by their lengths.

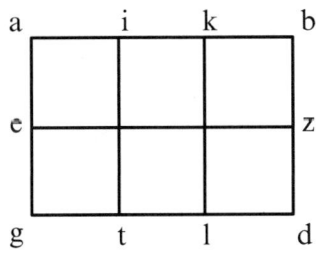

Figure 1.3

[3] If we wish to multiply 7 rods in width by 23 rods in length, first we multiply 7 rods by 22 rods, namely by 4 panes to obtain 28 panes in accord with the above stated reasoning. Then multiply the 1 rods remaining from the 23 by 7 to obtain the square rods of the 7. From all of this take $\frac{1}{2}$ 5 rods or one panis, and the $\frac{1}{2}$ 1 rods that remains is $\frac{1}{2}$ 4 soldi. Thus we have in sum 29 panes, which is $\frac{1}{2}$ 4 soldi of measure; that is, 2 staria, 5 panes, and $\frac{1}{2}$ 4 soldi.

[4] Likewise, if you wish to multiply 13 rods by 31 rods, first multiply 11 rods [namely, 2 panes] by 31 rods to obtain 62 panes, namely twice the 31. Then multiply the 2 rods remaining from the 13 rods by 31 rods to obtain the square 62 rods that is 11 panes and 18 deniers. Add these to the 62 panes already found, and you have 73 panes and 18 deniers. Divide these by 12 because 12 panes equals 1 starium. The answer is 6 staria, 1 panis, and 18 deniers of measure.

[5] Likewise, if you wish to multiply 19 rods by 41 rods, first multiply the 19 by 33 rods to obtain half of 19 staria, namely $\frac{1}{2}$ 9 staria. Then multiply the 8 that remain from the 41 by $\frac{1}{2}$ 16 rods to obtain 24 panes. Add to this $\frac{1}{2}$ 9 staria to get $\frac{1}{2}$ 11 staria. Having subtracted $\frac{1}{2}$ 16 rods from $\frac{1}{2}$ 19 rods, what remains is $\frac{1}{2}$ 2 rods. Multiply again these $\frac{1}{2}$ 2 rods by the above noted 8 rods to obtain 20 planar rods, of which $\frac{1}{2}$ 16 rods make 3 panes. The remaining $\frac{1}{2}$ 3 rods are $\frac{1}{2}$ 10 soldi. Add to these 3 panes, $\frac{1}{2}$ 10 soldi, and $\frac{1}{2}$ 1 staria to make in sum 11 staria, 9 panes,[10] and $\frac{1}{2}$ 10 soldi.

---

[10] Text and manuscript have 9; context requires 3 (6.26, *f.* 4r.26).

16 Fibonacci's *De Practica Geometrie*

[6] I want to explain the reason for the multiplication in a demonstration. Let there be a long quadrilateral *abgd* having boundaries *ab* and *gd* of 19 rods. Let the long sides *ag* and *bd* measure 41 rods. Take from side *ag* line *ae* measuring 33 rods; the remaining line *eg* has 8 rods. Similarly, take from line *bd* line *bz* equal to line *ae*. Draw right line *ez* that has 19 rods since it is equal to and equidistant from both lines *ad* and *gd* [Figure 1.4]. Now *zd* is equal to line *eg*, namely, to 8 rods. Again, from lines *ez* and *gd* take straight lines *ei* and *gt*, each of which has $\frac{1}{2}$ 16 rods Now join line *it* and it equals both lines *eg* and *zd*. Now when lines *ei* and *gt* are removed from lines *ez* and *gd*, each of which measures $\frac{1}{2}$ 16 rods, there remains for each of the straight lines *iz* and *td* $\frac{1}{2}$ 2 rods. And so when we shall have multiplied the 19 rods from above into the 33, then we have the square measure of quadrilateral *az* and there remains for us the quadrilateral *ezdg* out of the whole quadrilateral *abgd* and the area of quadrilateral *eabz* was $\frac{1}{2}$ 9 staria. Then when we have multiplied 8 by $\frac{1}{2}$ 16 rods, namely into 3 panes, we have 2 staria for the square measure of quadrilateral *eitg*. Again, when we multiplied $\frac{1}{2}$ 2 rods into 8 rods, then we have 20 square rods, namely 3 panes and $\frac{1}{2}$ 10 soldi as the measure of the square quadrilateral *izdt*. For the three quadrilaterals *az*, *et*, and *tz* are equal to the whole quadrilateral *abgd*. And this is what we wanted to demonstrate.

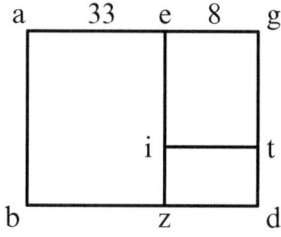

Figure 1.4

[7] Likewise if {p. 7} you wish to multiply 25 rods by 52 rods, multiply first 22 of the 25 rods, namely 4 panes by 52 rods to obtain a third of one staria on account of the 4 panes that are a third part of one staria. For a third of 52 staria are $\frac{1}{3}$ 17 staria. Then multiply the 3 rods [that remain from taking 22 from 25] by $\frac{1}{2}$ 49 of the 52 rods, namely by 9 panes and there are 27 panes. Add to this the $\frac{1}{3}$ 17 staria, to reach 19 staria and 7 panes. After this multiply the 3 rods by the $\frac{1}{2}$ 2 rods [that is the difference between 52 and $\frac{1}{2}$ 49 rods] to obtain $\frac{1}{2}$ 7 rods or one panis and 6 soldi. And thus you have in sum 19 staria, 8 panes, and 6 soldi. Another way: multiply the 25 by the 52 to obtain 1300 rods that equals 13 and so many halves of staria and moreover 13 rods or 19 staria, 8 panes, and 6 soldi, as I said before.

[8] Likewise, if you wish to multiply 31 rods by 69 rods, multiply the 31 rods first by 66 rods or 1 starium to obtain 31 staria. Then multiply the 3 rods [that remained as the difference of the 66 and 69 rods] by 31 rods to obtain 93

rods. These are 17 panes less 18 deniers. And thus we have in sum 32 staria, 5 panes, less 18 deniers.

[9] Likewise, if you wish to multiply 43 rods by 85 rods, first multiply the 43 rods by $\frac{1}{2}$ 82 rods or 15 panes. This is 15 panes times 43 staria with so many fourths, that is, $\frac{3}{4}$ 53 staria. Then multiply the $\frac{1}{2}$ 2 rods by 43 [that remains as the difference of $\frac{1}{2}$ 82 and 83]. Nevertheless first multiply those $\frac{1}{2}$ 2 rods by 33 rods or 6 panes that yields 15 panes. Add this to $\frac{3}{4}$ 53 staria to get 55 staria. Multiply again those $\frac{1}{2}$ 2 rods by 10 rods [the difference between 43 and 33], and you have 25 rods that amounts to 4 panes and 3 soldi. And thus you will have for the desired product 55 staria, 4 panes, and 9 soldi.

[10] Likewise, if you wish to multiply 54 rods by 113 rods, first multiply the 54 by 110 rods or 20 panes to obtain 1080 panes. Divide this by 12 to obtain 90 staria. Then multiply the 3 rods [that remain from subtracting 110 from 113] by 54 rods, that is by 9 panes and $\frac{1}{2}$ 4 rods, to obtain 27 panes and $\frac{1}{2}$ 13 rods that are 29 panes and $\frac{1}{2}$ 7 soldi. Add this to 90 staria to obtain 92 staria, 5 panes, and $\frac{1}{2}$ 7 soldi.

[11] Likewise if you wish to multiply 72 rods by 149 rods, first multiply 66 rods by 149 rods to obtain 149 staria. Then multiply 149 rods by 6 rods [the difference between 66 and 72]. Then first multiply by 132 rods or 2 staria and afterwards by 17 rods [the difference between 132 and 149] to obtain 12 staria and 102 rods; that is 13 staria, 6 panes, and 9 soldi. Combine this with 149 staria to obtain 162 staria, 6 panes, and 9 soldi. Enough has been said about the multiplication of whole rods into whole rods.

[12] Now we wish to speak of the multiplication of the same with feet. For if a foot is multiplied into however many rods, as has been said, you have so many halves of a soldus as there are rods. If two feet are multiplied by rods, you have as many soldi as there are rods. And if 3 feet are multiplied by rods, there come as many soldi as there are rods together with so many halves of soldi. And if 4 feet are multiplied by rods, you have twice as many soldi. And from 5 feet multiplied by rods there result likewise twice the number of soldi of the same rods and moreover so many halves. And from the multiplication of feet into feet come deniers, as we said above.

[13] If you wish to multiply 12 rods and 1 foot by 25 rods and 2 feet, multiply {p. 8} the 12 rods by the 25 rods to get 54 panes and 9 soldi. Keep the panes in your right hand, the soldi in your left. Then multiply 1 foot by 25 rods and you have $\frac{1}{2}$ 12 rods. Add these to the 9 soldi that are in your left hand to get 21 and with your feet keep the count of 6 deniers.

Now keeping count with your feet is done this way. Placing the tip of the left foot above the tip of the right signifies 1. Placing the tip of the left above the right ankle bone means 2. Touching the heel of the right with the front of the left means 3. To lead the left foot beyond the right and touch the tip of the right foot on the exterior part with the tip of the left is 4. To touch the right ankle bone from the same side with the tip of the left foot means 5. The

18 Fibonacci's *De Practica Geometrie*

other five signs are made in the same order with the right foot on the left. The sign for 11 is made by placing the heel of the right foot over the left ankle bone. We do not need additional signs because 12 deniers make one soldus that can be kept in the left hand. Nothing need be said about manual signs because everyone who knows how to calculate knows them.

We return to what was proposed: you will multiply again 2 feet by 12 rods to get 12 solmdi. To these add the $\frac{1}{2}$ 21 soldi that you had kept to get $\frac{1}{2}$ 33 soldi. Likewise multiply 1 foot into 2 feet to get 2 deniers. Add these to the 6 that you kept on your feet to get 8 deniers. Save these with your feet. And thus you will have 54 panes, 33 soldi, and 8 deniers; namely, 4 staria, 8 panes, and 8 deniers for the desired product.

[14] Likewise, if you wish to multiply 17 rods and 2 feet by 36 rods and 3 feet, first multiply the 17 rods by the 36 rods and you have 9 staria, 3 panes, and $\frac{1}{2}$ 4 soldi. Keep the staria in your heart, the panes in your right hand, the soldi in your left, and the deniers on your feet. Over all this add the product of 2 feet by 36 rods, namely 36 soldi, and the product of 3 feet by 17 rods, or $\frac{1}{2}$ 25 soldi, and the product of 3 feet by 2 feet, namely 6 deniers. And however much the soldi in your left hand had increased for you, change them to as many panes as possible. And thus you will have in sum 9 staria, 7 panes, and 6 deniers.

[15] Likewise, if you wish to multiply 26 rods 4 feet by 43 rods 5 feet, first multiply the 26 rods by the 43 rods to get 16 staria, 11 panes, and $\frac{1}{2}$ 4 soldi. Likewise multiply 4 feet, namely 2 soldi, by 43 rods to get 86 soldi, namely 5 panes and $\frac{1}{2}$ 3 soldi. Likewise multiply 5 feet, namely $\frac{1}{2}$ 2 soldi, by 26 rods to get 65 soldi. Likewise multiply 4 feet by 5 feet to get 20 deniers. By combining these four products into one you have in sum 17 staria, 8 panes, 8 soldi, and 8 deniers.

[16] If you wish to multiply 28 rods, 1 foot, and 7 inches by 53 rods, 5 feet, and 12 inches, first multiply the 28 rods and 1 foot by 53 rods and 5 feet to get 22 staria, 11 panes, 11 soldi, and 5 deniers. Then multiply 7 inches, namely $\frac{1}{2}$ 2 deniers, by 53 rods. Nevertheless, first multiply by 48, namely 4 soldi, and afterwards by 5 deniers, to get 10 soldi and $\frac{2}{3}$ 3 deniers. Likewise multiply 12 inches, namely 4 deniers, by 28 rods. First multiply by 24, namely 2 soldi, and afterwards by 4 to get 9 soldi and 4 deniers. After this multiply 7 inches by 5 feet and 12 inches by 1 foot to obtain a sixth of 47 inches, namely $\frac{11}{18}$ 2. And multiply 7 inches by 12 inches and they are $\frac{84}{324}$ of one denier. There is no need to make any mention of this small amount since it is not one inch. Add the remaining products for the total sum of 23 staria, 14 soldi, and 9 deniers.

[17] Likewise, if you wish to multiply 37 rods, 2 feet, and $\frac{1}{3}$ 5 inches by 67 rods, 4 feet, and $\frac{1}{4}$ 10 inches, multiply first 3 rods {p. 9}, 2 feet, and 5 inches by 67 rods, 5 feet, and 10 inches to produce 38 staria, 5 panes, 8 soldi, 5 deniers, and a bit more. To this add the product of a third of one inch by 67 rods, 4 feet, and $\frac{1}{4}$ 10 inches, namely a little more than $\frac{1}{2}$ 22 inches. For you will add the product of a fourth of one inch by 37 rods, 2 feet, and $\frac{1}{3}$ 5 inches, namely a little less than $\frac{1}{3}$ 9 inches, that are in sum a little less than 38 staria,

4 panes, 9 soldi, and 5 deniers. Thus by diligently studying all of the above, you will be able to do similar cases.

[18] If you wish to multiply 17 rods and 3 feet into 28 rods and 4 feet, put the feet of the longer side under the feet of the shorter side, and similarly the rods under the rods, namely 4 feet under 3 feet. Afterward put the 28 rods under the 17 rods back toward the left, as is shown [see Figure 1.5]. Multiply 3 feet by 4 feet to get 12 deniers. For when feet are multiplied by feet, deniers arise out of the product, as was said. Put 1 soldus in your left hand to represent the 12 deniers. Now multiply 3 feet by 28 rods and 4 feet by 17 rods crosswise, as I showed you how in the multiplication of two digits by two digits in my larger book, *On Calculations*.[11] You will halve the product after multiplying. That is, you will multiply half of 3 feet by 28 or half of 28 by 3. Similarly you will multiply half of 4 feet by 17 or half of 17 by 4. And you will do this because in the multiplication of a foot by rods, the outcome is a foot or half of one soldus. Whence the multiplication of 3 feet in half of 28 rods, namely 14, makes 42 soldi. These you add to the soldus you kept in your hand to make 43 soldi in the same hand. And the multiplication of half of 4 feet by 17 rods makes 34 soldi. To these add the 43 soldi to make 77 soldi. Change these to panes, namely 4 panes and 11 soldi. Keep two of the panes in your right hand and the soldi in your left. After this multiply 17 rods by 28 rods according to the usual method because the multiplication of rods by rods has already been described above. Thus you will have at the end 7 staria, 7 panes, 3 soldi, and 6 deniers.

rods	feet
17	3
28	4

Figure 1.5

[19] And if you wish to multiply only 19 rods by 41 rods and 4 feet, put the 41 rods under the 19 rods. And in place of the feet next to the 19 put a zero, and next to the 41 rods put the 4 feet, as shown here [see Figure 1.6]. Now multiply zero by 4 feet to get zero, and zero by half of 41 is again zero. Half of 4 feet by 19 makes 38 soldi that are 2 panes in the right hand and 5 soldi in the left. 19 rods by 33 rods makes $\frac{1}{2}$ 9 staria. And you keep the 9 staria in your heart [*sic*], 8 panes in your right hand, and 5 soldi in your left. Multiply 8 rods [the difference of 33 and 41] by 19 rods, namely 11 in 8 that make 16 panes and 64 rods. The 64 rods are 11 panes and 9 soldi. These sum to 12 staria less 12 deniers.

---

[11] This is the translation of *Liber abaci*.

rods	feet
19	0
41	4

Figure 1.6

[20] Likewise if you wish to multiply 20 rods, 4 feet, and 11 inches by 46 rods, 5 feet, and 12 inches, place the rods under the rods, the feet under the feet, and the inches under the inches, as shown in the margin [see Figure 1.7]. Then multiply after the manner of multiplying three figures by three figures, as though inches were in place of the first figure for units, and feet in the second place for tens, and rods in the third place for hundreds. You will multiply 11 inches by 12 inches to make $\frac{132}{324}$ of one denier, because the product of an inch by an inch is $\frac{1}{324}$ of one denier, the rule of 324 being $\frac{1\ 0}{18\ 18}$. Whence if we divide 132 by the first 18, which is under {p. 10} the bar, the answer is $\frac{1}{3}$ 7. Or multiply a sixth of 12 inches by 11 and divide what results by a sixth of 18 to get similarly $\frac{1}{3}$ 7 that is $\frac{\text{so many}}{18}$ of one denier. Save the 7 in your hand and keep the fractions on the tablet or in your heart. And you will cross multiply[12] 11 inches by 5 feet and 12 inches by 4 feet. Add the two products to the 7 you saved to become 110 that are $\frac{some}{18}$ of one denier, because $\frac{some}{18}$ of one denier comes from the product of inches by feet. Whence $\frac{1}{3}$ 110 divided by 18 makes $\frac{1\ 2}{3\ 18}$ 6 deniers [6 $\frac{7}{54}$]. Keep the 6 deniers in your hand and the fractions on the tablet or in your heart. Add to all of these the product of the third part of 11 inches by 46 rods or the third part of 46 rods by 11. Since the product of an inch by a rods makes $\frac{1}{3}$ of one denier, you have $\frac{2}{3}$ 174 deniers that equals 14 soldi and $\frac{2}{3}$ 6 deniers. Add to this again the product of the third part of 12 inches by 20 rods and the product of 4 feet by 5 feet, and the sum is 22 soldi and $\frac{2}{3}$ 10 deniers. Now add $\frac{2}{3}$ of one denier that you had saved to obtain $\frac{1\ 14}{3\ 18}$ of one denier. And to the 22 soldi and 10 deniers that you saved in your right hand, the 6 soldi in the left hand, and the 4 deniers on your feet, add half the cross product of feet by rods and rods by rods as we did before. And you will have in sum 14 staria, 9 panes, 4 soldi, and $\frac{1\ 14}{3\ 18}$ 4 deniers.

rods	feet	inches
20	4	11
46	5	12

Figure 1.7

---

[12] Cross multiply is translated from *multiplicare in cruce* (10.4).

[21] Again, if you wish to multiply 21 rods by 47 rods and 2 feet by 10 inches, place the 10 inches under zero and 2 feet under another zero, and the 47 rods under 21 rods as shown [see Figure 1.8]. Multiply inches by inches, namely zero by 10 to get zero that you set aside. Cross multiply 0 by 2 and 10 by 0 making zero that you again set aside. Multiply the zero in the place of the inches by $\frac{1}{3}$, by 47 rods, and 10 by $\frac{1}{3}$ of the 21, and the zero in the place of feet by 2 feet to make 5 soldi and 10 deniers. Then multiply the zero in the place of the feet by $\frac{1}{2}$ of the 47 rods, and $\frac{1}{2}$ of the two feet by 21 rods to make 21 soldi. And thus you have 26 soldi and 10 deniers. This is one panis, 10 soldi, and 4 deniers. Add to this the product of 21 rods by 47 rods and you have 15 staria, 1 panis, 1 soldus, and 4 deniers of measure, the sum of the desired product.

rods	feet	inches
21	0	0
47	2	10

Figure 1.8

[22] Thus be careful to put zeros in the proper places for multiplication, so that there are as many places on one level, as there are on the other. For we say that in the first place are inches, in the second feet, and in the third place are rods. For example, we wish to multiply 13 rods and 11 inches by 28 rods and 14 inches [see Figure 1.9]. We place the rods below the rods in the third place and the inches below the inches in the first place. And in the second place we put zeros between the inches and the rods[13] as shown in the marginal diagram. Multiply the numbers the way we taught and you will have in sum 5 staria, 7 panes, and $\frac{5}{9}\frac{14}{18}$ 1 deniers.

rods	feet	inches
13	0	11
29	0	14

Figure 1.9

[23] Likewise, we wish to multiply 14 rods, 2 feet by 31 rods, 15 inches. Write them as usual [see Figure 1.10]. Multiply zero inches by 15 inches

---

[13] Text and manuscript have *pedes*; context requires *perca* (10.33, f. 6v.23).

to make 0. Divide the 0 by 18 to make 0 again for what is left is nothing. Multiply 0 inches by 0 feet and 15 inches by 2 feet to make 30 that divided by 18 yield $\frac{2}{3}$ 1 deniers. Multiply $\frac{1}{3}$ of zero inches by 31 peach and $\frac{2}{3}$ of 15 inches by 14 rods, and 2 feet by 0 feet. Add $\frac{2}{3}$ 1 deniers and you have 6 soldi less $\frac{2}{3}$ of one denier. Now multiply half the 2 feet by 31 rods and half the zero by 14 rods, an add the 6 soldi that you had saved to make 37 soldi less $\frac{2}{3}$ of one denier. These make 2 panes, (4) soldi, less $\frac{2}{3}$ of one denier. And from the 14 rods take 11 rods, and multiply {p. 11} them by 31 rods to get 62 panes. Multiply the 3 remaining rods by 41 rods, by 31 rods, and first by 22 and afterwards by 9. Or in one multiplication by 31 you get 16 panes and 15 soldi. Thus you will have in sum 6 staria, 9 panes, 2 soldi, and $\frac{2}{3}$ 5 deniers of measure.

rods	feet	inches
14	2	0
31	0	15

Figure 1.10

[24] Again, if you wish to multiply 17 rods, 4 feet, and $\frac{1}{2}$ 9 inches by 32 rods, 5 feet, and $\frac{3}{4}$ 14 inches, then place the rods under the rods, feet under feet, and inches under inches [see Figure 1.11]. Multiply $\frac{1}{2}$ 9 inches by $\frac{3}{4}$ 14 inches thus: first 9 by 14 makes 126 to which you add half of the 14 and $\frac{3}{4}$ of the 9 by the cross to make more than 139. Divide this by 18 to get more than 7 and keep the 7 in your hand. Cross multiply $\frac{1}{2}$ 9 inches by 5 feet, $\frac{3}{4}$ 14 inches by 4 feet, and add the 7 that you kept to make more than 113. Divide this by 18 to get $\frac{1}{3}$ 6 deniers. Now multiply $\frac{1}{3}$ of $\frac{1}{2}$ 9 or $\frac{1}{6}$ 3 by 32 rods to get $\frac{1}{3}$ 101 deniers. After adding these to the $\frac{1}{3}$ 6 deniers that you kept, you have 9 soldi less $\frac{1}{3}$ of one denier. To this add the product of the third part of $\frac{3}{4}$ 14 inches by 17. Or do the multiplication this way: take 12 from the $\frac{3}{4}$ 14, $\frac{1}{3}$ of which is 4. Multiply 4 by 17 rods to get 68 deniers. After this take 2 from $\frac{3}{4}$ 2 that you multiply by 17 to get 34. Make deniers out of this by dividing by 3 and you have $\frac{1}{3}$ 11 deniers. And thus you have $\frac{1}{3}$ 79 deniers. Afterwards take a $\frac{1}{3}$ of the remaining $\frac{3}{4}$ to get $\frac{1}{4}$. Multiply by 17 to get $\frac{1}{4}$ 4 deniers. And thus you have $\frac{1}{4}$ of $\frac{1}{3}$ 83 for multiplying by $\frac{1}{3}$ of $\frac{3}{4}$ 14 inches by 17. Or, take $\frac{1}{3}$ of 17 that is $\frac{2}{3}$ 5. Then multiply this by $\frac{3}{4}$ 14 thus: 5 times 14 makes 70 deniers, and $\frac{2}{3}$ of 14 make $\frac{1}{3}$ 9 deniers, and $\frac{3}{4}$ of 5 are $\frac{3}{4}$ 3 deniers, and $\frac{2}{3}$ of $\frac{3}{4}$ are $\frac{1}{2}$ of one denier. And thus you have $\frac{7}{12}$ 83 deniers, as above. Another way: add $\frac{1}{4}$ to $\frac{3}{4}$ 14 to make 15. Take $\frac{1}{3}$ of this or 5 and multiply by 17 to get 85 deniers. Take $\frac{1}{3}$ from the $\frac{1}{4}$ that you added to get $\frac{1}{12}$. Remove this part of 17 or $\frac{5}{12}$ 1 deniers that you subtracted from 85. As we predicted: $\frac{7}{12}$ 83 deniers remain.

rods	feet	inches
17	4	$\frac{1}{2}$ 9
32	5	$\frac{3}{4}$ 75

Figure 1.11

[25] Thus strive to proceed along a similar way that seems better to you according to the number of inches. Add therefore $\frac{7}{12}$ 83 deniers to the 9 soldi less $\frac{1}{3}$ of one denier to make 16 soldi less $\frac{3}{4}$ of one denier. Add to this the product of feet by feet, namely 4 by 5, and the 20 denier become 17 soldi and $\frac{1}{4}$ 7 deniers. To this add as above the cross product of half the feet by the rods, and of rods by rods. You will have in sum 8 staria, 10 panes, 7 soldi, and $\frac{1}{4}$ 1 deniers

[26] You can do it differently with fractions of the inches, namely at the beginning before you start the multiplication. Take the fractions of the inches above from the rods below, and the fractions of the inches below from the rods above. Then multiply in this way: for the $\frac{1}{2}$ inch which is in the upper inches take half of the 32; for the $\frac{3}{4}$ which is in the lower inches, take $\frac{3}{4}$ of the 17 to make *in cruce* 16 and $\frac{3}{4}$ 12. And this is $\frac{3}{4}$ 28 inches that are almost 10 denier that you saved. Delete the fractions of the inches from the multiplication. Multiply only the 17 rods by the 4 feet and the 9 inches by 32 rods, and the 5 feet by 14 inches. Add to their sum the 10 deniers that you saved.

[27] If you wish to multiply 13 rods and 2 feet by 21 rods and 3 feet, since a foot is $\frac{1}{6}$ of a rods, substitute so many sixths for the rods as there are feet placed with the rods, and you will have $\frac{2}{6}$ 13 to multiply by $\frac{3}{6}$ 21 rods [see Figure 1.12]. Put the sixths under the sixths, the rods under the rods as shown here. And put two sixes under the bar with nothing above, thus: $\frac{}{6\,6}$. Then multiply 2 feet by 3 which are above both sixes and you have 6 [deniers]. Divide this {p. 12} first 6 that is on the left side under the bar as we placed it, and 1 appears with a remainder of 0. Keep the 1 in your hand and place the 0 above as in $\frac{0}{6\,6}$. Now cross multiply the 2 above the 6 by 21 and the 3 above the other 6 by 13. Add their product with the one which you saved and you have 82. Divide this by 6 to obtain 13 rods with 4 remaining. Place the 4 above the 6 thus $\frac{0\,4}{6\,6}$, and keep the 13 in your hand.[14] Add to the above the product of 13 rods by 21 rods that is 286 rods. Divide this by $\frac{1\,0}{11\,6}$, the rule of the relation of rods to staria,[15] and you obtain $\frac{0\,2}{11\,6}$ 4 staria. Join this to $\frac{0\,4}{6\,6}$, and you have $\frac{0\,4\,0\,2}{6\,6\,11\,6}$ 4. You know that

---

[14] Keep in mind that the remainder is in rods. Hence the $\frac{0\,4}{6\,6}$ are so many denier. For this reason the fraction can be simply appended to the next larger fraction, as Leonardo would instruct.

[15] The relation is 1 starium : 66 rods. Hence, the fraction should be $\frac{1\,0}{11\,6}$ a.

when you have done all this, what remains outside the bar is staria. After the integer[16] and above the 6 and atop the bar are two panes. Rods are above the 11. Understand that each of these is 3 soldi. Deniers are above the other two sixes. This is how you do it: multiply the number above the 6 by the other 6 that follows. Then add the number that is above the 6. For example: for the 4 that is outside the bar we have 4 staria, and for the 2 that is above the 6 we have 4 panes, and for the 4 that is above the other 6 we have 2 soldi. By multiplying 4 by 6 and adding the zero above the 6 we obtain 24 deniers.

$$\boxed{\begin{array}{l} \text{rods} \\ \frac{2}{6}\ 13 \\ \frac{3}{6}\ 21 \end{array}}$$

Figure 1.12

[28] Likewise, if we wish to multiply 14 rods and 5 feet by 41 rods, put the 41 under the 14 and $\frac{0}{6}$ under the $\frac{5}{6}$. You know therefore that we place the $\frac{0}{6}$ after the 41 so that the lower fraction can be paired with the upper fraction. Put $\frac{10\ 0\ 0}{66\ 11\ 6}$ to one side. Now multiply the 5 above the 6 by 0 above the other 6 to make 0, because you put 0 above the first 6. Now cross multiply 5 by 41 and 0 by 14 to make 205. Divide this by the following 6 to obtain 34 with a remainder of 1. Put the 1 above the 6 and keep the 34 in your hand to which you add the product of 14 by 41 to make 608. Divide this by $\frac{1\ 0}{11\ 6}$, which remained in the bar of division, to yield 3 above 11 and 1 above 6 and 9 outside the bar, as appears in the problem. Note that 2 staria, 9 soldi, and 6 deniers of measure appear in the sum.[17]

[29] Likewise, if you wish to multiply 15 rods, 3 feet, and 12 inches by 42 rods, 2 feet, and 15 inches, it is necessary to change the inches to sixths of one foot, namely to sixths of a sixth of one rods.[18] Because 18 inches make 1 foot, it follows that 3 inches are $\frac{1}{6}$ of a foot and 12 inches are $\frac{4}{6}$ of one foot. Put the sixths under one bar after the $\frac{3}{6}$. Consequently, put $\frac{5}{6}$ for the 15 inches after the $\frac{2}{6}$, and you will have $\frac{4\ 3}{6\ 6}$ 15 to multiply by $\frac{5\ 2}{6\ 6}$ 42 [see Figure 1.13]. After this extend the bar of $\frac{1\ 0}{11\ 6}$ and place four sixths under it to create $\overline{6666116}$. Now multiply

---

[16] *Post integrum* refers placing the fraction *after the integer*. The preposition indicates that Leonardo wrote his numerals from right to left; otherwise he would have had to write *ante integrum* for numerals written from left to right.

[17] The final sum is 9 staria, 2 panes, 9 soldi, and 6 deniers of measure.

[18] Although Fibonacci instructed the reader to change inches to feet, the given change is to so many thirty-sixths of a rod (12.27).

4 by 5 that are above the first sixes, and divide by the first 6 under the bar to get 3, and 2 remains above that 6. Cross multiply 4 by 2 and 5 by 3 that are above the bar and add the 3 that was saved to get 26. Divide this by the second 6 to get 4 so that 2 remains above that 6. Now multiply 4 by 42, 5 by 15, and 3 by 2. Multiply as before where we multiplied three figures by three figures, and add the 4 that we saved to get 253. Divide this by the third 6 to obtain 42, and 1 remains above that 6. Add the cross product of 3 by 42 and of 2 by 15 to make 198. Divide by the fourth 6 to yield 33 and 0 remains above that 6. Add the product of 15 by 42 to 323 to obtain 663 rods. Divide this by 11 to get 60 with a remainder of 3 above the 11. Then divide the 60 by 6 that remains above the bar and 10 results before the bar and 0 above that 6. And thus you have in {**p. 13**} sum 10 staria, 9 soldi, and $\frac{1\,2\,3}{6\,6\,6}$ 1 deniers of measure. Because we have part of one denier above the first two sixes, and we have deniers above the other two sixes, and above 11 we have rods, namely three soldi. And we have two panes above 6, that is, above the bar.

$$\begin{array}{|c|}\hline \text{rods} \\ \frac{3\ 4}{6\ 6}\ 15 \\ \frac{2\ 5}{6\ 6}\ 42 \\ \hline\end{array}$$

Figure 1.13

[30] Again, if we wish to multiply 16 rods, 1 foot, and 10 inches by 43 rods, $\frac{1}{2}$ 14 inches, reduce the inches to sixths of one foot to become sixths of $\frac{1}{3}$ 3. Make two sixths of the third. And thus you have in the upper part $\frac{2\,3\,1}{6\,6\,6}$ 16 rods [see Figure 1.14]. We can make sixths of one foot from 10 inches in another way. Since 10 inches are $\frac{10}{18}$ of one foot, therefore 10 inches are $\frac{20}{36}$ of a foot. For which reason if we divide 20 by the rule of 36, we obtain $\frac{2\,3}{6\,6}$ of one foot, as we said. Similarly you double $\frac{1}{2}$ 14 inches to become $\frac{29}{36}$ of one foot. And thus $\frac{5\,4\,0}{6\,6\,6}$ 43 is in the lower side as you see in the margin. Put the 6 sixes under one bar before the $\frac{1\,0}{11\,6}$ thus: $\overline{666666116}$. You will multiply the numbers above the bar after the manner of the four figures as before. For example, multiply 2 by 3 that are above the first 6 and divide by the first 6. Then multiply 2 by 4 and 5 by 3, and divide by the second 6. After this multiply 2 by 0, and 5 by 1, and 3 by 4, and divide by the third 6. And 2 by 43, and 5 by 16, and 3 by 0, and 4 by 1, and divided by the fourth 6. And you will have above the four given sixth parts only of one denier. Then multiply 4 by 43 and 4 by 16, and 1 by 0, and divide by the fifth 6. And you will have above it the deniers. Then multiply 1 by 43 and 0 by 16, and divided by the sixth 6. And you will have above it 6 half soldi. Finally multiply the 16 rods by 43 rods and divided by

$\frac{1}{11}\frac{0}{6}$ and you will have 11 triple soldi, and above the 6 twice the panes, and before the bar the staria. And thus with sixths of sixths of a foot we can proceed indefinitely by reducing the fractions of inches that were placed in the multiplication by sixths of sixths, if the fractions can be so changed. And if the number of sixths of one side is less than the number of sixths on the other, you can add to them with zeros above those sixths; that is, if two sixths are in one place, and two are in another, and if three are in a third, and so on. If it is impossible to reduce the fractions of the inches to sixths of sixths, then by no means can you work with these fractions. So abandon them, and multiply the rest by the fractions themselves that cannot be made into sixths of sixths. And proceed as we said above.

$$\begin{array}{|c|}\hline \text{rods} \\ \frac{2\ 3\ 1}{6\ 6\ 6}\ 16 \\ \frac{5\ 4\ 0}{6\ 6\ 6}\ 43 \\ \hline \end{array}$$

Figure 1.14

METHOD 2[19]

The method of multiplying discussed above will help you to learn how to multiply with different measures from other regions. And because we have promised an expansion of this as the second method,[20] we explain here what is necessary to do the work well.

[31] *If any number is divided into however so many parts, and each part is multiplied by the whole number, the sum of these products is equal to the square of the whole number, namely the multiplication of the number by itself.*[21] If the number *ab* is divided into parts such as *ag*, *gd*, and *db*, I say that sum of the products of *ag* by *ab*, *gd* by *ab*, and *db* by *ab* equals the product of *ab* by itself. For the number of units in part *ag* with those in *ab* will produce the product of *ag* by *ad* [see Figure 1.15]. Similarly, the number of units in part *gd* with those in *ab* will produce the product of *gd* by *ad*. For the same reason therefore, the units {**p. 14**} in the number *db* with those in *ab* create the product of *db* by *ab*. Because there are as many units in the number *ab*,

---

[19] "In modo secundo" (13.31, *f.* 8v.4).
[20] *in hac secunda distinctione* is ambiguous (13.34, *f.* 8v.7). The sentence makes sense only if "second distinction" refers to "second method," the title of this section.
[21] *Elements*, II.1.

namely in the parts *ag*, *gd*, and *db*, so many are united in the number *ab* from the multiplication of *ag*, *gd*, and *db* by *ab*. Truly, as many units as there are in *ab*, so many arise from the multiplication of *ab* by itself. Hence, the sum of the product of *ag*, *gd*, and *db* by *ab* equals the product of *ab* by itself, as required. So that this may become clearer, let right line *ab* be 10 ells long. For its parts let *ag* be 2, *gd* be 3, and *db* be 5. Now multiply 10 by 2, 10 by 3, and 10 by 5. That is, *ag*, *gd*, and *db* by the whole *ab*. Their products equal 100 that is the product of 10 by 10.

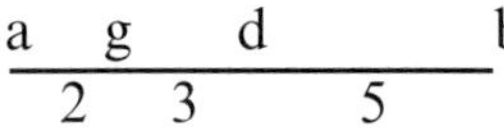

$$\begin{array}{cccc} a & g & d & b \\ \hline & 2 & 3 & 5 \end{array}$$

Figure 1.15

[32] *If a straight line is divided into however so many parts, and each is multiplied by another line, the sum of all the products equals the product of the whole divided line by the other line.*[22] We show this by numbers. Let a line 10 ells long by divided into lengths of 2, 3, and 5. Let some other line be 12 ells long. If the lengths 2, 3, and 5 are multiplied by 12 and their products added together, the sum is the same as the product of 10 by 12, namely 120.

[33] *Again: if a straight line is divided wherever in two parts, the product of one part by the other with the product of the same part by itself equals the product of the same part by the whole line*[23] [see Figure 1.16]. Let *ab* be divided at point *g*. The product of *ag* by itself with the product of *ag* by *gb* equals the product of *ag* by all of *ab*. Let some line *d* equal line *ag*. The product of line *d* by *ag* and by *gb* is equal to the product of *d* by all of *ab*. Truly, the product of *d* by *ag* is as the product of *ag* by itself because line *d* is equal to line *ag*. And the product of line *d* by *gd* is as the product of line *ag* by line *gb*. Because the product of *ag* by itself together with the product of *ag* by *gb* equals the product of *ag* by ab, as required. If we want to show this with numbers, then let *ag* be 3 and *bg* be 7 and the whole line *ab* 10. The product of *ag* by itself or 3 by 3 with the product of *ag* by *gb* or 3 by 7 gives 30 that equals the product of *ag* by *ab* or 3 by 10, as I said.

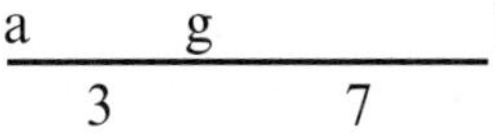

$$\begin{array}{ccc} a & g & b \\ \hline & 3 & 7 \end{array}$$

Figure 1.16

[34] Moreover: *if a straight line were divided into two parts, the squares of the two parts together with the twice the product of one part by the other equal the*

---

[22] *Elements*, II.2.
[23] *Elements*, II.3.

square, namely the product of the whole line by itself.[24] Let line *ag* be divided in two parts, which are *ab* and *bg*. I say that the product of *ab* by itself with the product of *bg* by itself together with twice the product of *ab* by *bg* equals the product of the whole line *ag* by itself. Because the line *ag* by divided in two parts at point *b*, the products of *ab* by itself and *ab* by *bg* is the same as the product of *ab* by the whole line *ag*. Similarly, the product of *bg* in itself with the product of *gb* by *ba* is as the product of *gb* by the whole line *ga*. Therefore the product of *ab* by itself with *bg* by itself together with twice the product of *ab* by *gb* equals the two products *ab* by *ag* and *bg* by *ag*. Truly, the products of *ab* by *ag* and *gb* by *ag* equal the product of *ag* by itself. Wherefore the square on *ab* with the square on *bg* together with twice the product of *ab* by *bg* equals the square on the line ag, as required [see Figure 1.17]. This can be shown with numbers. Let lines *ab* be 3 and *bg* be 7, so that the whole line *ag* is 10. Whence if we take the square on line *ab* (9) and the square **{p. 15}** of *bg* (49) and twice *ab* by *bg* (42), they equal 100, the product of *ag* by itself.

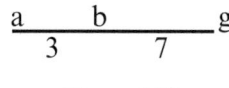

Figure 1.17

[35] Again: *if a straight line is divided anywhere in two parts, twice the product of one part by the whole line with the square on the other part equals two squares, namely the square on the whole line and the square on the part that was multiplied twice by the whole line.*[25] Let line *bg* be divided wherever you choose at point *a*. I say that twice the product of *ab* by *bg* with the square on line *ag* equals the two squares on lines *bg* and *ba*. Because line *bg* was divided in two parts at point *a*, the product of *ba* by itself with the product of *ba* by *ag* is as the product of *ba* by *bg*. Whence twice the product of *ba* by *bg* is twice the square on line *ba* and twice the product of *ba* by *ag*. Truly, the one square *ba* with the square *ag* and twice the product of *ba* by *ag* equals the square on the whole line *bg*. Whence two squares of line *ba* with the square on line *ag* and with twice the product of line *ba* by *ag* equals the two squares of lines *bg* and *ba*. But two squares of line *ba* with twice the product of *ba* by *ag* have been shown equal to twice the product of line *ba* by the whole line *bg*. Consequently twice the product of line *ba* by the whole line *bg* with the square on line *ag* equals the two squares of lines *bg* and *ba*, as required. We can show the same thing with numbers [see Figure 1.18]. Let the whole line *bg* be 10 and its parts *ba* be 3 and *ag* 7. Twice the product of *ba* by *bg* or 3 by 10 with the product of line *ag* by itself produces 49. This rises to 109 that is equal to the product of line *bg* by itself, namely 100 and line *ab* by itself or 9.

---

[24] *Elements*, II.4.
[25] *Elements*, II.7.

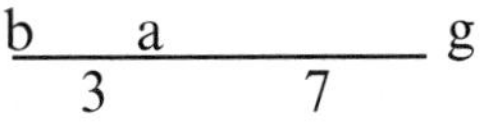

1 Measuring Areas of Rectangular Fields   29

```
    b    a              g
    ─────┼──────────────
      3         7
```

Figure 1.18

[36] *If a straight line is divided into two equal parts and divided again into two unequal parts, then the product of the unequal parts, namely of one by the other, with the square on the line that lies between the two sections is equal to the square on half of the divided line.*[26] So, if line *ab* is divided in two equal parts at point *g* and unequal parts at point *d*, I say that the product of *ad* by *db* with the square on line *dg* is equal to the square on line *ag*. Because line *db* was divided wherever at point *g* to which a certain straight line *ad* is adjacent, the product of *ad* by *db* is equal to twice the product of line *ad* by *bg* and by *gd*. Now, line *bg* equals line *ga*. Whence the product of line *ad* by line *ag* and by line *gd* equals the product of line *ad* by line *db*. Again, because line *ag* was divided wherever in two parts by point *d*, the product of the parts *ad* and *gd* by *ag* equals the product of *ad* by itself. Now *gd* by *ag* equals the square on *dg* and the product of *ad* by *dg*. Therefore the products of *ad* by *ag* and *ad* by *dg* with the square on line *dg* equal the product of *ag* by itself. We found above that the products of *ad* by *ag* and *dg* by *da* are equal to the product of *ad* by *db*. Whence *ad* by *db* with the square on *dg* equals the square on line *ag*, that was required. We can show this with numbers [see Figure 1.19]. Let line *ab* be 10 ells and divided into 5 and 5 at point *g* and into 3 and 7 at point *d*. Then the product of *ad* by *db* or 3 by 7 with the square on *dg* is equal to the product of *ag* by itself or 5 by 5.

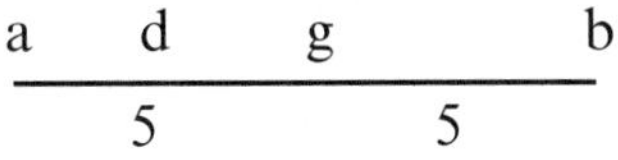

```
    a    d    g         b
    ─────┼────┼──────────
       5           5
```

Figure 1.19

[37] *If a straight line is divided into two equal parts and another line of whatever length is joined to it to form a longer line, the product of the whole line formed from the original line and its added length together with the square on half the divided line equals the square on the line made from half the first line and the added line.*[27] So, if line *ab* is divided into two equal parts at point *g* and is extended {p. 16} by a certain other line *bd* of any length, then I say that the product of lines *ad* by *bd*, that is *bd* by *ad* with the square on line *gb* or *ga* equals the square on line *gd*. Because line *ad* was divided wherever at points *g* and *b*, the product of *bd* by the whole line *ad* is equal to the sum of the three products: *bd* by *ag*, by *gb*, and by *bd*. Now because line *ag* is equal

---

[26] *Elements*, II.5.
[27] *Elements*, II.6.

to line *bg*, the product of line *bd* by *ag* equals the product of *bd* by *gb*. Whence the two products *bd* by *ag* and *bd* by *bg* make a sum that is twice the product of *bd* by *gb*. Hence, the square on line *bd* with twice the product of *bd* by *gb* is equal to the product of line *bd* by the whole line *ad*. Let the square on line *gb* be adjoined to both groups.[28] Then the two squares on lines *bd* and *bg* with twice the product of *bd* by *gb* equal the sum of the product of line *bd* by *ad* and the square on line *bg*. But the two squares on lines *bd* and *bg* with twice *bd* by *bg* equal the square on line *gd*. Whence the product of line *bd* by *ad* with the square on line *bg* equals the square on line *gd*, as required. Now this may be shown with numbers [see Figure 1.20]. Let line *ab* of length 10 be divided in two equals parts, 5 and 5, at point *g*. Extend *ab* by 3 ells so that the whole line *ad* is 13 long and *gd* 8. The product of *bd* by *ad* or 3 by 13 with the square on line *gb* or 25 equals 64, the square on line *gd*.

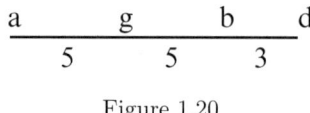

Figure 1.20

[38] *If a straight line is divided into equal and unequal parts, the squares on the unequal parts are twice the square on the half line and the square on the difference of the unequal lines.*[29] Let line *gd* be divided in equal parts at point *a* and unequal parts at point *b*. I say the square on the parts *gb* and *bd* are twice the two squares on lines *da* and *ab*. Since line *ag* was bisected at point *b*, the two squares on lines *bg* and *ba* with twice the product of *ab* by *bg* equal the square on line *ga*, that is, the square on line *ad*. Now the square on line *ba* with just one product of *ba* by *bg* equals *ba* by *ag*. Whence the square *bg* with the product of *ba* by *ag* and with that of *ba* by *bg* equals the square on line *ad*. Let the square on line *ba* be commonly adjoined [see Figure 1.21]. Then the two squares *gb* and *ba* with the products of *ba* by *ga* and *ba* by *bg* are equal to the squares on lines *ad* and *ab*. But the square on line *ba* with the product of *ab* by *bg* is as the product of *ba* by *ag*. Therefore the square *gb* with twice the product of *ba* by *ga* is equal to the two squared lines *da* and *ab*. When the square on line *da* with the square on line *ab* exceeds the square on line *gb* by twice the product of line *ba* by *ga*; that is, *ba* by *ad*. Likewise, because line *bd* was bisected at point *a*, the squares on lines *da* and *ab* with twice the product of *ba* by *ad* equal the tetragon or square on line *bd*. Therefore tetragon *bd* exceeds tetragons *da* and *ab* by twice the product of *ba* by *ad*. Therefore, by as much as tetragons *da* and *ab* exceed the tetragon *bg*, by so much are they exceeded by tetragon *bd*. Whence tetragons *gb* and *bd* are twice the squares *da* and *ab*. And this is what we had to prove. This can be shown in numbers. Let line *gd* be 10 and divided in two equal parts or 5 and 5 at point *a* and at

---

[28] In effect, Fibonacci is adding the same quantity to both sides of an equation.
[29] *Elements*, II.9.

point b into two unequal parts or 3 and 7. Let the tetragons gb be 9 and bd be 49. Joined together they sum to 58. This is twice the squares da and ab. Because the square on da is 25 and that of ab is 4.

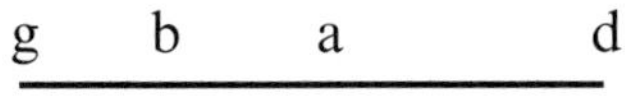

g   b   a   d
―――――――――――――――――――――

Figure 1.21

[39] *Likewise, if a straight line is divided into two equal parts and one part is extended, the tetragon on the increased line is twice the square on half the divided line and the square on the half line and the added part.*[30] So {p. 17} if line ab is divided in two equal parts at point g and extend by section bd, I say that the tetragon on line ad together with the tetragon on line bd is twice the tetragon on line ag and line gd. Since line ad has been divided at point b, therefore the square on line ad with the square on line bd is equal to the square ab and twice the product of bd by ad [see Figure 1.22]. But twice bd by ad is as two squares on line bd with twice the product of bd by ba. Therefore the tetragon ab with twice the square on line bd and with twice bd by ab is equal to the squares on lines ad and bd. In fact, the product of bd by ab is twice that of bd by gb, since it is half of bg by ab. Whence twice bd by ab is four times bd by gb. Therefore the tetragon ab with the two squares on line bd and the fourfold product of bd by gb are equal to the two squares on lines ad and bd. But the tetragon on line ab is four times the tetragon that is described on half of ab (line gb). Therefore four times the tetragon gb with the two tetragons on line bd and the fourfold product bd by gb are equal to the two squares on lines ad and bd.

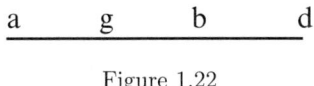

Figure 1.22

[40] Again, because line gd was bisected at b, the two squares on the parts gb and bd with twice the product of bd by bg equal the square on line gd. Whence twice the square on line gb with twice the square on line bd together with the fourfold bd by gb are equal to twice the square on line gd. The remaining twice the square on line gb is twice the square on line ga. Therefore, four times the square on line gb and twice the square on line bd with the fourfold product of bd by gb are twice the squares on the parts ab and bd. But four times the square on line gb and twice the square on line bd together with four times bd by gb has been shown equal to the two squares on lines ad and db. Whence the squares on lines ad and db are twice the squares on the parts ag and gb. And is what we had to demonstrate. We can show the same thing with numbers. Let line ab of 10 units be divided in two parts (5 by 5) at point g. Extend the

---

[30] *Elements*, II.10.

line by *bd* equal to 3 ells for a total length of 13. Further, *gd* equals 8 ells. The square on line *ad* is 169 and that on *bd* is 9. Adding the two together is 178 that is twice 89 the sum of the squares *ag* and *gd* or 25 and 64.

[41] Since it seem superfluous for us to demonstrate what Euclid has proved, in my book I shall use whatever is necessary from his book, without further proof.

[42] *If three numbers or quantities are proportional, then the first is to the second as the second is to the third.*[31] Then the product of the first number by the third equals the product of the second by itself. For example, given the numbers 4, 6, and 9, then 4 is to 6 as 6 is to 9; or, 9 is to 6 as 6 is to 4. Hence, the product of 4 by 9 equals the product of 6 by 6. Hence, when of three given numbers the product of the first by the third equals the square of the second number, then the three numbers are in continued proportion.

[43] Likewise: *when four numbers are proportional, the first to the second as the third is to the fourth, then the product of the first by the fourth equals the product of the second by the third.*[32] For example, let the first number be 6, the second 8, the third 9, and the fourth 12. Hence, the product of 6 by 12 equals the product of 8 by 9 or 72. From this it is clear that when the product of the first number by the fourth equals the product of the second by the third, then the four numbers are proportional by the very proportion in which {p. 18} the first is to the second as the third is to the fourth. Note that if any number is divided by another and the quotient is multiplied by the divisor, the original number will return. For instance, if 20 is divided by 4, the quotient is 5 that if multiplied by 4 will return to 20. If however, from three numbers in proportion if two are known, namely the first and second, with the third unknown, then the square of the second divided by the first will yield the third number. For instance, let the first number be 4, the second 6, and the third unknown. The square of 6 or 36 divided by 4 produces 9, the third number. Now if we multiply the quotient 9 by the divisor 4, 36 returns as the divided number. Consequently when the product of the first number by the third produces the square of the second, then those three numbers are proportional. It is as though 4 is to 6 as 6 is to 9. If we do not know the first number, then by dividing the same 36 by the third number or 9, we have the first number or 4. Similarly, if we do not know the second number, then we multiply the first by the third (4 by 9) to find 36 whose root, 6, is the second number.

[44] Likewise: *as the first number a is to the second b, so the third g is to the fourth d.* Hence, the product of the first by the fourth (*a* by *d*) equals the product of the second by the third (*b* by *g*) [see Figure 1.23]. If *a*, *b*, and *g* are known with *d* unknown, by dividing the product of *b* and *g* by *a*, we have *d*. Because of that same procedure of dividing the product of *b* and *g* by *a* to obtain *d*, we can

---

[31] *Elements*, VI.13.
[32] **Elements*, VII.19.

divide the product of *a* and *d* by *b* to obtain *g*. And if the same product were divided by *g*, we would have *b*. For example, let *a* be 4, *b* 6, *g* 8, and *d* 12. If the product of *b* and *g* (48) were divided by *a* (4), *d* or 12 is found. Further, if 48 were divided by 12 (the fourth number), then 4 the first number arises. Similarly, if the product of the first by the fourth (4 by 12) is divided by the second (6), then the third (8) emerges. Again, if 48 is divided by the third number (8), then the second number appears. All of this must be committed to memory.

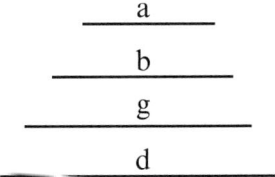

Figure 1.23

[45] Further: *if within a circle there are two intersecting lines, then the product of the first part of the first line by its other part equals the product of the first part of the other line by its other part*[33] [see Figure 1.24]. As in circle *abgd* two straight lines *ag* and *bd* intersect at some point *e*. The product of *ae* by *eg* equals the product of *be* by *ed*, as shown in Euclid's elements. Now we move to finding roots of numbers.

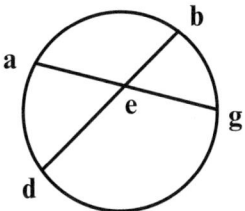

Figure 1.24

---

[33] *Elements*, III.35.

2
Finding Roots of Numbers

COMMENTARY

The scope of Chapter 2 is square roots of numbers: how to find them and how to compute with them. (Cube roots are discussed in Chapter 5.) Not only are the readers instructed on how to find square roots of up to six digit numbers, each clearly exemplified, but they are returned to the world of Pisan units of measurement and shown how to find the roots of square units of rod, feet, and inches, not to overlook a digression into the astronomical world of square degrees, minutes, seconds, and thirds (!). Fibonacci offered two ways of proving that a root is what it claims to be, the obvious method of multiplying a number by itself and one that suggest modular arithmetic [11]. Noteworthy is the fact that he made no reference to any of the geometric theorems discussed at length in Chapter 1, even if the reader can see where some are applied. In grappling with large numbers such as 9876543 he advised the reader to find first the root of the last five digits (98765). Next and after joining any remainder to the remaining digits (43), find its root. And then simply put the two roots together as one root. Finally, and after a practical definition of *radix*, the readers are instructed on how to add, subtract, multiply, and divide radicals.

Three remarks may make the reading of the root procedures easier. First, the method is essentially the tremendously tedious technique of the Hindus.[1] It is apparently based on Euclid's *Elements*, II.4, say applied to a three-digit whole number $n = 100a^2 + 20ab + b^2$. Each term on the right is subtracted successively from n beginning with $100a^2$, any remainder being equal to or greater than zero but less than $(10a + b)$. Secondly, subtraction is always performed from the left, another Arabic characteristic.[2] Because the technique depends on division, the medieval format is followed that requires placing digits of numbers above and below the number being rooted.[3] The following example is an enlargement of [4] in which the square root of 864 is sought, the numbers in parentheses indicating the sequence of steps in finding the square root.

---

[1] Datta and Singh (1935–1938) I, 169–175.
[2] Levey and Petruck (1982), 50.
[3] D.E. Smith (1953) II, 136–139.

(1)	(2)	(3)	(4)	(5)	(6)
				1	1
	4	4	4	4 0	4 0
8 6 4	8 6 4	8 6 4	8 6 4	8 6 4	8 6 4
2	2	2	2 9	2 9	2 9
		4	4	4	4

[23]

(1) The root will be a two-digit number whose square is immediately less than the third digit 8. So place 2 under the 6.
(2) Subtract 2^2 from 8 and put the remainder 4 over the 8.
(3) Double 2 and put it under 4.
(4) Readers forced to learn the Hindu Method in elementary or primary school may recall that the critical step was finding the unit's digit for a two-digit root. Fibonacci's method for "guessing" this number follows. On a diagonal you can see 46. Divide it by the 4 from (3) to get 11. The largest single digit less than 11 is 9, probably the unit's digit of the root. Put 9 in the first place under 4.
(5) Multiply 9 by 4 (under 2) and subtract 36 from 46 to get 10. Place 10 on the diagonal.
(6) Square 9, subtract 81 from 104 (on the diagonal again), to get the remainder 23.
(7) The square root of 864 is 29 with the remainder marked as [23].

Fibonacci's curious method for proving that a derived root is correct is well met in a nearly literal translation; see [11].[4] A modern representation of the method showing that the root of the 12345 is 111 with remainder 24 suggests an operation in mod 7:

[1] 111 ÷ 7 leaves remainder 6.   [2] $6^2 = 36$.
[3] 36 ÷ 7 leaves remainder 1.   [4] 24 ÷ 7 leaves remainder 3.
[5] (From [3] and [4]) 1 + 3 = 4. [6] 12345 ÷ 7 leaves remainder 4.
[7] Because the results of steps [5] and [6] are identical, the root is correct.

Fibonacci closed the section on finding square roots with a discussion on finding the square roots of numbers that cannot be expressed as rational numbers but can be found as lines. The method is based on Euclid's *Elements*, II.13, "To find a mean proportional between two straight lines." In Figure 2.a let AB = 67, BD = 1, then BD = $\sqrt{67}$. We turn now to operations on radical numbers.

If nothing else, the sections on operating with roots make us eternally grateful for modern notation. We have symbols; they had only words. Addition of some roots (and their subtraction, *mutatis mutandis*) is based on the ratio of squares, the squares hidden within the radicands. For instance, to add the root of 27 to

---

[4] The method also appears in *Liber abaci*, 39.35–41; Sigler (2002), 67.

the root of 48, notice first that they have the ratio of 9 to 16 whose roots are 3 and 4. The sum of these last numbers is 7 whose square is 49. Because 27 is thrice 9 and 48 is three times 16, multiply 49 by three to get 147 whose root is the required sum. Multiplication and division of roots are performed "within the radical sign;" for example, $(4\sqrt{5})(3\sqrt{2}) = (\sqrt{50})(\sqrt{18}) = \sqrt{(50)(18)} = \sqrt{900}$. Fibonacci accompanied all operations in the margins of the manuscript with geometric procedures, if not proofs. Fibonacci announced two ways to find the roots of fractions, "as accurately as possible." Inasmuch as the second way is marred by manuscript errors (*q.v.*), his example of the first method follows, to find the square root of two-thirds in modern dress but retaining Fibonacci's auxiliary number 60:

$$\sqrt{\frac{2}{3}} \to \frac{2}{3}(60) = 40 \to 40(60) = 2400 \to \sqrt{2400} \approx 49 \to 49 \div 60 = 0.816667,$$

a very reasonable approximation.

The method of finding square roots is discussed further under Sources. Here I want to consider why Fibonacci included the method at all. Not that it was not needed, but rather he might have discovered a simpler, actually a very old, technique, then verbal but now in symbols:
To find

$$\sqrt{N} \text{ let } N = a^2 + r.$$

Then

$$\sqrt{N} = a + \frac{r}{2a+1}.$$

My thinking was prompted by his original method for finding cube roots as expressed in a modern formula:
To find

$$\sqrt[3]{N} \text{ let } N = a^3 + r.$$

Then

$$\sqrt[3]{N} = a + \frac{r}{3(a[a+1])+1}.$$

The analogy is so close: 3 represents three dimensions for the cube; so 2 would be introduced for the square. The additional term $[a+1]$ contributes nothing to the square root, so it would be dropped. Thus Fibonacci might have made a universally accepted boon to all who extracted square roots down to modern times.

Unlike Chapter 5 on finding cube roots, which for the most part is an excerpt from Chapter 14 from *Liber abaci*, the present chapter reflects only the method and two of the several examples of the first part of Chapter 14. For this chapter, Fibonacci apparently rewrote nearly the entire section on square roots with a tighter explanation and more examples. Compare, for example, how he found the square root of 12345 here and in *Liber abaci* where the technique is not so clear.[5] Because *De practica geometrie* was written before the revision of *Liber abaci* in 1228, one may wonder why Fibonacci did not incorporate the clearer, more succinct version into the revision. The material on binomina and computing with them as studied in *Liber abaci* is not discussed here.

TEXT[6]

2.1 Finding Roots

Because we want to measure fields by geometric rules, we will have to find the roots of five numbers in certain sizes. Consequently I now describe how to find these roots efficiently enough for the purposes of this work.

Integral Roots

[1] The root of a number is the side of its square, because when you multiply a side by itself, you obtain the area of the square. Whence the sum[7] of the multiplication of the side is properly called a square because it encloses a squared surface. Note further that the root of a single or double digit number is a single digit number. There are two {p. 19} digits for the root of a three or four digit number, three digits for the root of a five or six digit number, four digits for the root of a seven or eight digit number, and so on. It is quite necessary to know by heart the roots of numbers of one and two digits, for only thereby can we proceed efficiently to find the roots of numbers with more digits. Hence, 1 is the root of 1 because one times one makes one: 2 is the root of 4, 3 the root of 9, 4 the root of 16, 5 the root of 25, 6 the root of 36, 7 the root of 49, 8 the root of 64, 9 the root of 81, and 10 the root of 100. The other numbers below these square numbers do not have roots. If necessary, however, we can approximate them closely by many methods that I shall show in their own place.

[2] If you want to find the integral root of a three digit number, you begin with the root of the last digit.[8] Place this under the digit in the second place.[9] You do this because you know that the root of a three digit number has two digits. Thus one digit is under the second digit and the other under the first.

---

[5] *Liber abaci* 354; Sigler (2002), 492.
[6] "Distinctio secunda, incipit capitulum de inuencione radicum" (18.35–36, f. 11r.35–11v.3).
[7] This is probably an Arabicism. The word comes from *jama`a* which means "what is gathered," and is often used to speak of the result of a multiplication (Oaks, 2 February 2006).
[8] Often a modern reader needs to accommodate one's thinking to Fibonacci's right-to-left description of the digits of any number. We may consider the 3 in 369 as the first digit because we write it first; 3 is the last digit for Fibonacci because he wrote it last, after the Arabic fashion.
[9] This is a in the expansion of the binomial $(10a + b)^2$ which governs the search for the root.

Whatever remains [from subtracting the square] from the third digit, you put above the third digit. Now, join the remainder with the second digit [to form a new number]. Put the other digit [of the root] under the first digit, that is, before the digit that you put under the second digit. Now the product of this[10] by twice the digit[11] you found as a root is so close to the new number that from the difference of the new number and the first, you can subtract the square of the first digit.[12] And nothing remains beyond twice the root you found. If nothing remains from [the subtraction of] the last digit, understand that it was due to the union of the remainder from the third digit with the second digit.

[3] For example, we wish to find the root of 153. You will find the root of 1 in the third place to be 1. Place it under the 5 and put 2 before it under the 3. Multiply 2 by 2, which is twice the root that you found, to get 4 [see Figure 2.1]. Subtract this from 5 and put the remainder 1 above the 5. Join the 1 to the 3 in the first place to form 13. Subtract from this the square of 2 to get 9 that is less than 24, twice the root which has been found. Consequently, the whole root of 153 is 12 with remainder 9.

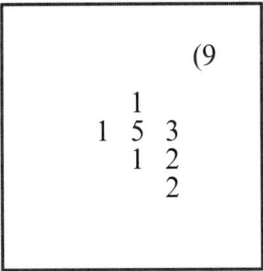

Figure 2.1

[4] If you wish to find the root of 864, put 2 under the 6 because 2 is the whole root of 8. Put the remainder 4 above the 8. Then double 2 to get 4 placing it under the 2. Form 46 from the 4 above the 8 and the 6 in the second place. Now divide the new number 46 by 4 to get 11 [see Figure 2.2]. From this division we get an idea[13] of the following first digit[14] which must be multiplied by twice the digit[15] you already found. Afterwards, square it.[16] The digit is a little less or exactly as much as what comes from the division. Practice with

---

[10] b.
[11] $20a$.
[12] Subtract b^2.
[13] "possumus habere arbitrium" (19.28).
[14] b. The phrase Fibonacci uses here is *figura ponenda* (19.29). He means the single digit that will be put in the unit's place to complete the finding of the square root. I uniformly translate this as *first digit*.
[15] $20a$.
[16] b^2.

this procedure will perfect you. So we choose 9 since it is less than 11, and put it under the first digit. Multiply 9 by 4 (twice the second term) and subtract the product from 46. The remainder is 10. Put 0 over the 6 and 1 above the 4. Join 10 with 4 in the first place to make 104. Subtract the square of 9 from it to get 23. This is less than 29 the root that has been found.

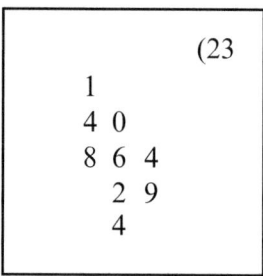

Figure 2.2

[5] Again, if you wish to find the root of 960, put 3 the root of 9 under the 6. Double 3 to get 6 and place it under the 3. Now multiply 6 by the first digit. Subtract the product from the upper 6 from which a digit remains. Join it with 0 [see Figure 2.3]. You can subtract the square of the digit. There will be no remainder beyond twice the root that had been found. And that digit is 0. Now the product of 0 with 6 (twice 3) subtracted from 6 leaves the same {p. 20} 6, Join this with 0 in the first place to make 60. Subtracting the square of 0 from this leaves 60, twice 30 the root that was found.

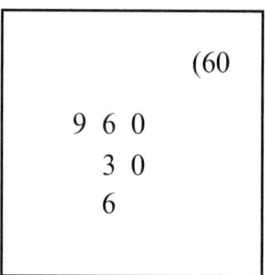

Figure 2.3

[6] If you wish to know the root of a four digit number, find first the root of the last two digits. Join the remainder with the remaining two digits and proceed according to what we said above for three digit numbers. For instance, we want to find the root of 1234 [see Figure 2.4]. Put 3 the root of the square less than 12 under the 3 and the remainder 3 over the 2 of 12. Join 3 with the following digits to make 334. Then double the root which you found namely 6 and place it under the root 3. Now think: how many multiplications are to be made according to the instructions given above,

and how many digits will there be in the number from which we must subtract those multiplications? There will be two multiplications. The first is the product of the first digit by twice the [partial] root or 6. The second is the square of the first digit. Both of these products are to be subtracted from 33 in order to finish the last multiplication under the digit in the first place. Hence, the first product will be subtracted from 33 the union of the last two digits, and the other product from the union of the remainder and the 4 under the first digit. Consequently, put 5 under the first digit because 33 divided by 6 leaves 5. Multiply 5 by 6 and subtract the product from 33 for a remainder of 3 in the second place. Join 3 with 4 in the first place to make 34. From this subtract the square of 5 to leave 9. Thus, you have 35 as the root of 1234 with remainder 9.

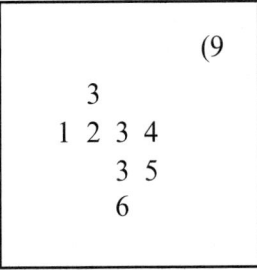

Figure 2.4

[7] Again, if you wish to find the root of 6142, find the root of 61. It is 7 with remainder 12. Put the 7 under the 4 and the 12 over 61. Double 7 to get 14. Put 4 under the 7 and 1 after the 7. You know about the union of the remainder 12 with 42 to make 1242. As a four digit number three subtractions are required [see Figure 2.5]. The first of these is the first digit by the [last][17] digit. The second is [the first digit] by 4 that is before the 1. The third is the square [of the first digit] itself. We must gradually subtract the three products from the four given digits so that the final product falls in the first place. Because the number of digits exceeds the number of products in the first figure, one digit must be joined to the last digits of 1242 with what follows, namely 12. From this 12 subtract the first product. Then the second product falls under the second place, and the last under the first. Whence put 8 before 7. Multiply 8 by the given 1 and subtract [the product] from 12 for a remainder of 4. Join this with the 4 that follows in the second place to get 44. Subtract from this the product of 8 and 4 (the 4 being under the 7) to yield a remainder of 12. Place this over the 44 and join it with 2 in the first place to make 122. Subtract from this 64 the

---

[17] The text and manuscript have *per positum vnum* (20.26, f. 12v.9). However, the procedure requires one times eight, as seen six lines below. Hence, *vnum* should be *ultimum*.

42   Fibonacci's *De Practica Geometrie*

square of 8 for a remainder of 58. And thus you have 78 as the root of 6142 with remainder 58.

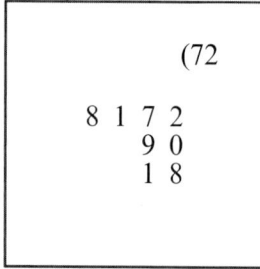

Figure 2.5

[8] If you wish to find the root of 8172,[18] put the root of 81 or 9 under the 7. Double the 9 and put the 8 under the 9 and the 1 after it to the left. Now the 1 and 8 must be multiplied by the first digit, one at a time. Then square the first digit [see Figure 2.6]. And thus there are three products to be subtracted gradually from 72, the remainder from the 81 after finding of the root of 81. Whence, as we obviously know, nothing comes after it except 0. Since a step is lacking, it is the first product that can be subtracted. Because if the first product is subtracted from 7, the second needs be {p. 21} subtracted from 2. But then there is no place from where to subtract the third product. Or in another way: because the first place is a factor with any step, that step arises from the multiplication. Since the product of the digit in the first place and the digit in the third place, namely by 1, fills the third place, there is no place for 72. Therefore the root of 8172 is 90 and the remainder is 72.

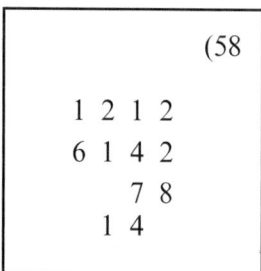

Figure 2.6

[9] If you wish to find the root of any number of 5 digits, find the root by the foregoing instructions for the last three digits. If anything is left over, put the excess over that place or the place where the excess occurred. Then join the excess with

---

[18] The text and manuscript show 8172 (20.37, *f.* 12v.20) but the accompanying illustration has 81.71 which is obviously incorrect.

the two remaining digits. Put twice the root you found under that root, like places under like places. By studying the preceding instructions, put some other digit in the first place. Because the first figure in the root of a five digit number belongs in the third place, the root of a five digit number has three digits.

[10] For example: we wish to find the root of 12345 [see Figure 2.7]. First, find the root of 123 which is 11 with remainder 2. Put the first 1 of 11 under the 3 and the other 1 under the 4. Double 11 to get 22 and place it under the 11. Place that remaining 2 above the 3. Join it to the following digits to make 245, as a three-digit number. There will be the usual three multiplications. There will be a single multiplication from each single digit. Hence, there needs be placed such a digit before the 11 already in position. This will be multiplied by the first binomial, then by the second, and finally by itself. The first multiplication can be taken from the 2. Put what remains over the 3. And then another from the 4 and finally another from the 5 which is in the first position. And that digit is 1. Having subtracted the product of 1 by the first binomium and the product by the 2 over the 3, nothing remains. Likewise, having multiplied 1 by the following binomium and subtracted it from 4, what remains is 2 over the 4. Having joined the 2 with the 5 in the first place, 25 is made. Having squared 1 and subtracted it from 25, 24 remains. And thus you have 111 as the root of 12345 with remainder 24.

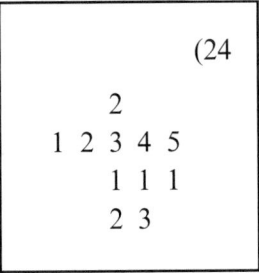

Figure 2.7

[11] There is a way to prove what you did is correct. Simply multiply 111 by itself. Accept the product as the proof. Add 24 to the product. Thus you have the proof that what you did for 12345 was correct. Another example: the remainder[19] from 111 divided by 7 is 6. The quotient of 111 by 7 has a remainder of 6 that multiplied by itself gives 36. This divided by 7 produces a remainder of 1. Add this to 3, the remainder of 24 [divided by 7], and you have 4. This is the proof [that $\frac{24}{111}$ 111 is the root of] 12345, because 12345 divided by 7 has a remainder of 4. This is what we want. By this technique you can always prove your answers in finding roots.

---

[19] Fibonacci used the word *proba*, which means "a sample taken for an examination" (Latham [1965], 373), whereas in *Liber abaci* he used *pensa*, a unit of weight (*Liber abaci*, 356), which Sigler translated "residue" adding "modulo seven" for *per septemnarium* ([2002], 494).

[12] Again, if you wish to find the root of 98765, find first the root of 987. It is 31 with a remainder of 26. Put the 3 under the 7, the 1 under the 6, and the remainder 26 over the 87, the 2 above the 8 and the 6 above the 7. Join 26 with the other two digits to make 2665. Double the 31 and put the 6 under the 3 and the 2 under the 1 [see Figure 2.8]. And because four digits remain in the number, we must gradually delete them by three multiplications that have to be done with the first digit before 31. The multiplications are first by 6, second by 2, and the third is the square of the first digit. We must therefore take the first product from 26. Whence we divide 26 by 6 (twice the 3 found in the partial root and placed under the 3) to get 4. Therefore the first digit is 4 to be placed before the 31. So, put 4 before 31. Since it can be done, multiply 4 by 6. Subtract the product from 26 for a remainder of 2 that you put over the 6 above the 7. Join it with the 6 in the second place to make 26. Then subtract from this number the product of 4 by 2 {p. 22} which placed under the 1 leaves 18, 1 over the 2 and 8 over the 6. Join the 18 with the 5 in the first place to form 185. From this subtract the square of 4 to leave 169. Thus you have 314 for the root of 98765 with a remainder of 169.

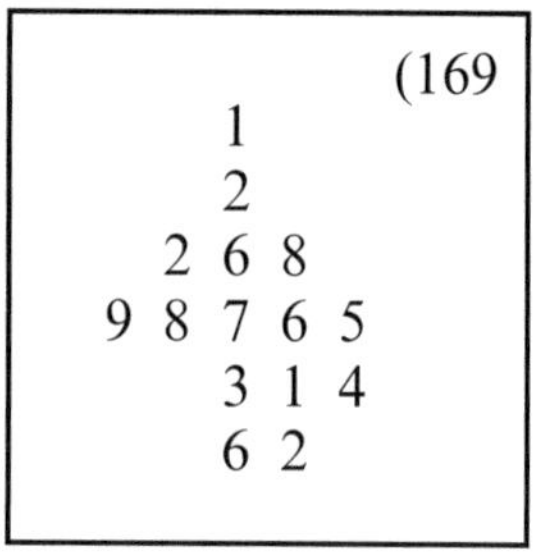

```
                        (169
          1
          2
       2 6 8
     9 8 7 6 5
         3 1 4
         6 2
```

Figure 2.8

[13] If you wish to find the root of a number with six digits, first find the root of the last four digits and join the remainder with the following two digits, and continue as before. For example, if we want to find the root of 123456, first find the root of 1234, which is 35 with remainder 9. Put 35 under 45 and the remainder 9 above[20] the 4 [see Figure 2.9]. Join 9 to the 56 to make 956. Because there are three digits here and there are three multiplications. The first must be by 7, the second by zero, and the third the square of the number itself. Then we know that the first multiplication must come out of the 9. So divide 9 by 7 to produce 1, the first digit before 35. Multiply 1 by 7 and subtract the product from 9 to leave the remainder 2 above the 9. Join the 2 to the 5 to make 25. Multiply 1 by 0, subtract the product from 25, and 25 remains. Join this to the following digit to make 256. From this subtract the square of 1 to leave 255. Thus we have 351 for the root of 123456 with remainder 255.

---

[20] The text has *sub* but the procedure requires *supra*, as the diagram in the margin shows.

```
                    (255
            2
            9
    1 2 3 4 5 6
            3 5 1
            7 0
```

Figure 2.9

[14] Again, if you wish to find the root of 987654, find first the root of 9876 which is 99 with a remainder of 75. Put 99 under 65 and 75 above 76 [see Figure 2.10]. Double the root you found to get 198. From this number, put 8 under the 9 in the second place, 9 under the 9 in the third place, and 1 after the last. Join 75 with 54 to make 7554. There are four digits in this number which we must delete gradually by four multiplications by the three digits 1, 9, and 8 with the first digit and finally by the square of [1] itself. Whence the first product must be subtracted only from 7. If we know that the first digit is 3, the product of 3 and 1 in 198 subtracted from 7 leaves 4 above 7 itself. Join it to 5 in the third place to make 45. Now subtract from this the product of 3 and the 9 in 198, to have 18 remain over the fourth and third places. Join 18 to 5 in the second place to make 185. Subtract from this the product of 3 and the 8 in 198. 161 remains above the fourth, third, and second places. Join 161 to 4 in the first place to make 1614. Subtract the square of 3 to give a remainder of 1605. And so we have 993 as the root.

```
                (1605
          1 6
          4 8
          7 5
    9 8 7 6 5 4
              9 9 3
          1 0 8
```

Figure 2.10

[15] Likewise, if you wish to find the root of a seven digit number, find first the root of the last five digits. Join any remainder with the remaining two digits, and proceed as before. For example: we want to find the root of 9876543 [see Figure 2.11]. Since the root of the seven digit number has four digits, we must put the first digit of the root under the fourth place, the following digit under the third, and the next one under the second place. You will find these three digits by finding the root of the last five digits. Their root is 314 with remainder 169 that you place above the fifth, fourth, and third places. Join 169 to 43 the

46   Fibonacci's *De Practica Geometrie*

remaining two digits to make 16943. Then double the 314 that you found and put the 8 under the 4, the 2 under the 1, and the 6 under the 3. You will multiply the first digit by those three digits, and then square the first digit. Consequently there are four products which we must subtract step-by-step from the five digit figure 16943. So the first product to be subtracted is from the last two figures, namely {**p. 23**} from 16. Hence divide 16 by 6 the first product to get 2. Put the 2 in the first place of the root. Then multiply it by the 6 of 628 to get 12 that you subtract from 16 for a remainder of 4. Put this over the 6. Join 4 with 9 to make 49. From this subtract the product of 2 by 2 to leave 45 over the 49. Join this with the 4 in the second place to make 454. Subtract from this the product of 2 by 8 to leave 438. Put the 4 over the fourth place, 3 above the third, and 8 above the second. Join 438 with the 3 in the first place to make 4383. From this subtract the square of 2 for a remainder of 4379. This is less than twice the root you found. Consequently, the root is 3142.

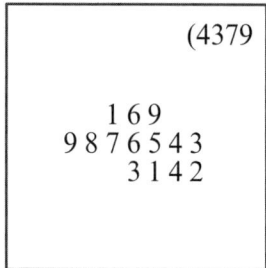

Figure 2.11

[16] Similarly, if you want to find the root of an eight digit number, find the root of the last six digits, join its remainder with the remaining two digits, and proceed as has been said. And so on with finding the roots of numbers with an unlimited number of digits. For if you want to know how many digits are in the root of any number of many digits, give thought to whether the number is odd or even. If it is even, take half its number of digits. The number of digits in the half is the number of roots of the number itself. If however the number of digits is odd, make it even by adding one to it. Consequently the number of digits in the half is the number of digits in the root. Then you begin by putting the first digit under the place where it had fallen.

[17] Now in dealing with the fractions which are left over from finding the whole number root, if you want to get close to the truth, you have to decide whether they are rods for measurements in geometry or degrees for measurements in astronomy. The roots of fractional parts of a rod must be reduced to feet, inches and parts of inches. The roots of fractional parts of a degree are to be reduced to minutes, seconds, and parts of seconds. As noted before, there are 6 feet in a rod, 18 inches in a foot. Further, an inch is $\frac{1}{18}$ of a foot, a foot is $\frac{1}{6}$ of a rod, and one whole rod is 108 inches.

Likewise a degree can be divided into 60 minutes, a minute into 60 seconds, and similarly a second into 60 thirds. Whence a whole degree is 3600 seconds and 21600 thirds, all of which must be memorized. If you wish to be able to find the root of any number of rods, multiply the number of rods by the product of 108 times 108 (11664) and you will find the root of the product. You will have the number of inches that are in the desired root. Inches divided by 18 yield feet for the root, and the number of feet divided by 6 yield rods.

[18] For instance, we want to find the root of 67 rods. Multiply 67 by 11664 to get 781488 the root of which is 884 with a remainder of 32. Divide the 32 by twice 884 or half of 32 (16) by 884 to get nearly $\frac{1}{55}$. Consequently, the root of 67 rods is 884 and $\frac{1}{55}$ inches. Now if the inches are divided by 18, we have 49 feet, 2 and $\frac{1}{55}$ inches. If we divide the 49 feet by 6, we have 8 rods and 1 foot. Consequently, the root of 67 rods is 8 rods, 1 foot, and 2 and $\frac{1}{55}$ inches. Wherefore you must learn how to do this for all similar cases.

[19] If you want to find the root of a number of degrees, square the number of seconds (3600) in one degree to get 12960000. This is the number by which you will multiply the number of degrees whose root you seek. You will find the root in that product. Divide the excess by twice the root you found and you will have the number of seconds that are in the desired root. Dividing the seconds by 60 yields minutes. Minutes divided by 60 produces degrees. And thus you have the degrees, minutes, seconds and parts of one second which are {p. 24} in the root of however so many degrees.

[20.1] If you wish to find step by step the rods, feet, inches, and finally the smallest parts of an inch that are in the root of however so many rods, first find the root of the rods, as has been said. Then change any remaining rods to halves of soldi or feet. Divide them by twice the number of rods you found for the root, and you will have the feet. From what remains from these, change to deniers. From these subtract the product of the square of the number of feet which you found. Triple what remains, that is, change them to inches that you divide again by twice the number of rods and feet that were found in the root. Thus you will have rods, feet, inches, and parts of inches. This is the root of however so many rods.

[20.2] For example: we wish to find the root of 67 rods. First, the integral root of 67 rods is 8 whole rods with a remainder of 3 rods. These are halves of soldi or 18 feet. Divide them by 16 (twice the 8) to get 1 foot with a remainder of 2 feet that make 12 deniers. Subtract from them the square of one foot or 1 denier, because the square of one foot is one denier, as said above. There remain 11 deniers. Multiply these by 3 to get 33 inches. Divide them by twice 8 rods and 1 foot ($\frac{1}{3}$ 16) for 2 inches and a remainder of $\frac{1}{3}$ of one inch. If we subtract the square of two inches from $\frac{1}{3}$, what remains from that third of an inch is about $\frac{1}{18}$ of one inch, because we divided what was left over by twice the root that we had found.

48   Fibonacci's *De Practica Geometrie*

[21] Again, you wish to use this technique for finding the root of 111 rods. Take 10 as the integral root, double it to get 20 by which you are to divide 66 feet (the 11 rods left over). You get 3 feet with a remainder of 6 feet or 36 deniers. From this subtract the 9 deniers that came from squaring 3 feet. 27 deniers remain. Triple these to 81 that you divided by 21 or twice the root you had found to get $\frac{6}{7}$ 3 inches. There is no remainder from which you would subtract the square of the inches. Simply call it a little less than $\frac{6}{7}$ of an inch. Thus you have for the root of 111 rods: 10 rods, 3 feet, and $\frac{5}{6}$ inches. Do remember what we said above about multiplying rods, feet and inches into rods, feet and inches. Keep this in mind as you proceed to find feet and inches in the roots of rods.

[22] Likewise, if you wish to find the root of 1234 rods, take the integral root as 35 rods with a remainder of 9 rods. Now 9 rods make 54 feet. Since these are less than twice the number of integral roots (70), we know that for this root there are no feet. Hence, make 972 inches from the 54 feet. Divide these by 70 to get 13 inches and $\frac{7}{8}$. Again, there is not enough here from which the square of the $\frac{7}{8}$ 13 inches can be subtracted. So, take some fraction less than $\frac{7}{8}$ such as $\frac{6}{7}$ or $\frac{5}{6}$ or $\frac{4}{5}$ or $\frac{3}{4}$, whatever you choose, although by doing so you deviate a bit from absolute accuracy. By this method you can come very close to the number of roots of rods, or rods and feet, or rods, feet, and inches.

Irrational Roots

[23] In my book *On Calculations* we discussed thoroughly how to find minutes and seconds, the fractional parts of degrees.[21] Now the roots of numbers that are not squares can be expressed as lines but not as numbers. The method of expressing the roots as lines is to draw a line representing the length of such a quantity or number, the root of which {**p. 25**} you wish to have, and add an extension of one unit. Find the middle of the extended line to serve as a center about which you construct a circle. Through the point between the original line and its extension erect at right angles another line to the circle. This last line is the true root of the number you seek. For example: we wish to find the root of 67. Draw straight line *ab* of 67 units. Add line *ag*, an extension of 1 unit, so that the full length is 68. Divide line *gb* in two equal parts at point *d*. Using the distance *dg* or *.db*, draw the circle *gebz*. Further, draw the line *ae* at right angles. I say that the line *ae* is the true root of line *ab*, the root of 67.

[24] Draw line *ea* directly toward point *z*. Because line *ba* meets line *ez*, it makes either two right angles or two angles equal to two right angles. Right angle *bae* equals right angle *baz*. And because straight line *ag* through the center cuts a certain straight line *ez* at two right angles, it necessarily cuts that straight line into equal parts [see Figure 2.12]. Therefore line *ea* equals

---

[21] For a table of the 59 sixtieth parts of sixty, see *Liber abaci*, 79, Sigler (2002), 121.

line *az*. Again: because in circle *gebz* the two straight lines *bg* and *ez* intersect one another at point *a*, the product of *ba* by *ag* is as the product of *ea* by *az*. But the product of *ae* by *az* is as the square on *ea*. Whence the product of *ba* by *ag* is as the square on *ae*. But the product of *ba* by *ag* is as the product of 67 by 1. And thus the square on *ae* is the same as 67, which we wanted.

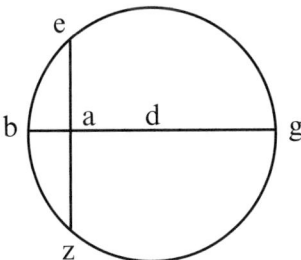

Figure 2.12

2.2. Multiplication of Roots[22]

Now we wish to show how to multiply and add roots of numbers by other roots, how to subtract smaller roots from larger ones, and finally how to divide roots.

[25] Suppose you wish to multiply the root of 16 by the root of 25. Since both 16 and 25 are square numbers, take their roots (4 and 5) and multiply them together to get the desired product, 20. Suppose you wish to multiply the root of 16 by the root of 30. Since the root of 30 is a surd, multiply 16 by 30 to get 480. Then find the root of their product as closely as possible. This gives you the root of your multiplication. Again, if you wish to multiply the root of 20 by the root of 30, multiply 20 by 30 to get 600. Find its root and you have what you wanted to find. For example, let *a* be 20, *b* be 30, *g* the root of 20, and *d* the root of 30. The product of *a* by *b* is *e* or 600. The product of *g* by *d* is *z*. I say that *z* is the root of *e* or 600. First it must be noted that as one number is to another number, so is the product of one to the product of the other as the part of one is to the part of the other. We shall prove this with 12 and 24. The ratio of 12 to 24 is as twice 12 to twice 24, as thrice 12 is to thrice 24. Similarly, the quadruple and the others. Likewise, the same ratio holds for 12 to 24 as for half of 12 to half of 24, the third to the third and so on. With this established, we return to the problem [see Figure 2.13]. Because the number *g* is the root of number *a* or 20, the square of *g* is *a*. And so the product of *g* by *d* is the number *z*. Therefore *a* is a multiple of *g* as is *z* of *d*. Whence *g* is to *d* as *a* is to *z*. Again: because *d* is the root of *b* or 30, then the square of *d* is *b*. But the product of

---

[22] "De multiplicatione radicum" (25.19, *f.* 15v.4).

d by g is z. Consequently z is a multiple of g; so b is a multiple of d. Whence g is to d as z {**p. 26**} is to b. Then as g is to d so a is to z. But *per equale* as a is to z so z is to b. Whence the product of a by b is as the square of z. But the product of a by b is e or 600. And the square of z is similarly e. Therefore z is the root of e, as required. Note that when the roots of numbers are multiplied together, the numbers themselves have among themselves the ratios of squares of the numbers. The product is the square of the number. Hence the product of roots themselves is a rational number.

20	600	30
a	e	b
g	z	d

Figure 2.13

[26] For example: we wish to multiply the root of 8 by the root of 18. Their ratio is as 4 to 9, the square of one number to the square of another. I say that the product of 8 by 18 produces a square number, namely 144. Its root or 12 equals the product of the root of 8 and the root of 18. Likewise if you wish to multiply 10 by the root of 20, multiply 100 the square of 10 by 20 to obtain 2000. Its root is the product of the stated multiplication. Note further that what happens when multiplying 10 by the root of 20. The product equals 10 roots of 20. Whence we can reduce 10 roots of 20 to the root of one number by multiplying the square of 10 by 20 to get 2000. The root of this is found from the 10 roots. The same must be understood for similar cases.

[27] So if you wish to reduce the root of some number to many roots of another, divide the number by some square number. As many units as there are in the root of the squared number, so many roots belong to the number resulting from the division. For example, you wish to reduce the root of 1200 to several roots of another number. Divide 1200 by some square number such as 16 to get 75. Then you have four roots of 75 for the root of 1200. If you were to divide 1200 by 25, then you would have five roots of 48 for the one root of 1200. Thus you can reduce the root of 1200 to several roots by different divisions.

2.3 Addition of Roots[23]

[28] When the roots of numbers are added to the roots of other numbers, their sum is either a rational number or the root of another number. When they cannot be added, then either a number arises from their sum or another root.

---

[23] "De addictione (*sic*) radicum" (26.26, f. 16r.16).

When we wish to join squares, then a number results from their sum. For example, the root of 16 added to the root of 25 unites 4 with 5 (the roots of 16 and 25) make 9 their sum.

[29] When we wish to add roots that have among themselves the ratio of their squares, then the sum is the root of some rational number. We shall demonstrate this in two ways. We wish to add the root of 27 to root of 48. Their squares have the same ratio as 9 and 16. Now the sum of their roots is 7, and the square of 7 is 49. Since 27 and 48 are thrice 9 and 16, triple 49 to get 147. Its root is the sum of the addition of the root of 27 to root of 48 [see Figure 2.14]. This rule can be proved by the ratios of similar triangles thus: let *ab* and *bg* be the parts of the same straight line, with *ab* the root of 27 and *bg* the root of 48. From point *a* create an angle by drawing straight lines *ad* and *de* with *ad* equal to 3 and *de* equal to 4. Join *eg* {**p. 27**} and *db*. Thus the square on line *ab* is to thrice the square on line *ad* as the square on line *bg* is to thrice the square on line *de*. Therefore as line *ab* is to line *ad*, so is line *bg* to *de*. Whence line *db* is equidistant from line *eg*. Hence triangle *adb* is similar to triangle *age*, with angle *a* in common. Angle *adb* equals angle *aeg* and angle *abd* equals angle *age*, the exterior angles being equal to the interior angles. Therefore the square on line *ab* is to the square on line *ad*, as the square on line *ag* is to the square on line *ae*. But the square on line *ab* is thrice the square on line *ad*. Hence, the square on line *ag* is similarly three times the square on line *ae*. But the square on line *ae* is 49. Whence the square on line *ag* is triple this or 147. Therefore, by joining lines *ab* and *bg* the root of 147 appears, as had to be shown.

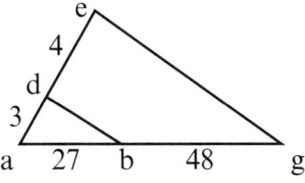

Figure 2.14

[30] Note that all roots of numbers having ratios among themselves can be reduced to the roots of one number, by dividing the antecedent number by the antecedent square, and the consequent number by the consequent square. For example, if you wish to reduce the roots of 27 and 48 to the roots of one number: divide 27 the first number by 9 the first square, and the following number 48 by its accompanying square 16. The root of 3 results from each division [see Figure 2.15]. Consequently, there are as many units in the root of 9 as there are roots of 3 in the root of 27. As many units as there are in the root of 16, so many roots of 3 are there in the root of 48. Thus for the one root of 27 you have three roots of 3. For the one root of 48 you have four roots of 3. If you add them together, you have 7 roots of 3 in sum. If you wish

to reduce them to the root of one number, multiply the square of 7 by 3 and take the root of the product. This gives you the root of 147 for their sum, as above. Another way: first add 27 with 48 to get 75. Multiply 27 by 48. Doubling the root of the product yields 72. Add this to 75 and you have 147 for the square of the stated addition.

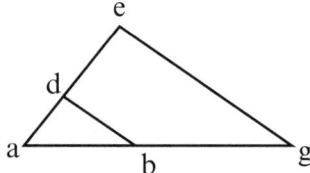

Figure 2.15

[31] We will demonstrate this by an example. Extend straight line *ab* by *bg*. Let *ab* equal the root of 27 and *bg* be the root of 48. We want to know the measure of line *ag*. Because line *ag* is divided in two parts at point *b*, the two squares on parts *ab* and *bg* with twice the product of *ab* by *bg* equals the square on the whole line *ag* [see Figure 2.16]. But the sum of the squares on *ab* and *bg* (27 and 48) is 75. Further, the product of line *ab* and *bg* gives the root of the product of 27 and 48. Twice this plus 75 is 147 for the measure of the square on line *ag*, as we predicted.

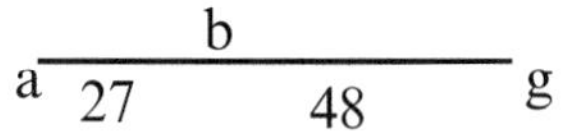

$$\overline{\underset{a\quad 27\qquad\quad 48\qquad}{\quad\quad\quad b\quad\quad\quad\quad\quad}}\;g$$

Figure 2.16

[32] It is not possible to add the root of 20 to the root of 30, because 20 and 30 do not have a ratio between themselves as the square of one number to the square of another. We know this because the product of 20 and 30 is a number that does not have a root. Therefore the sum of their roots does not result in a number nor in the root of a number. For example: [Figure 2.17] let straight line *ba* be the root of 20 and line *ag* be the root of 30. Therefore the square on line *ba* with the square on line *ag* is 50. And the product of *ba* by *ag* is the root of 600. Whence doubling the product of *ba* by *ag* produces the root of four times 600, namely the root of 2400. Since this does not have a root, we know that we cannot have a number or the root of a number from the addition of the foregoing. Rather their addition produces the root of a number and of a root, namely the root of 50 and the root of 2400. So that we can have something, let us find the root of 2400. Its nearest approximation is 49 less $\frac{1}{98}$. After adding 50, we have 99 {p. 28} less $\frac{1}{98}$. We can find the root of this, as we wanted. Or, we may find the individual roots of 20 and 30 as close as possible, add them, and thus have what we wanted.

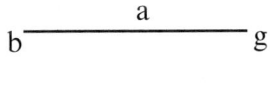

Figure 2.17

2.4 Subtraction of Roots[24]

[33] Suppose you want to subtract one root from another, such as the root of 16 from the root of 49. Then subtract 4 (the root of 16) from 7 (the root of 49) for the remainder 3 the residual of the subtraction. Suppose you wish to subtract the root of 45 from the root of 125. Since their ratio is as the ratio of squares (9 to 25), subtract the root of 9 from the root of 25 for a remainder of 2. Square this to get 4. Then multiply it by 5 to get 20, because 45 and 125 are each five times 9 and 25. The root of 20 remains from the subtraction.

[34] For example: let straight line *ab* be the root of 125. At point *b* attach the straight line *ag* (equal to 5 the root of 25) to make angle *bag*. Take from *ag* line *gd* (3) leaving *da* (2). Through point *d* draw straight line *de* equidistant from side *gb* [see Figure 2.18]. Because line *de* in triangle *agb* was constructed equidistant from line *bg*, sides *ab* and *ag* were cut proportionally at points *d* and *e*. That is, as *ae* is to *eb*, so is *ad* to *dg*. By composition therefore, as *ba* is to *be*, so is *ga* to *gd*. By alternation we have *ba* is to *ag* as *be* is to *gd*. Whence as the square on side *ba* is to the square on side *ag*, so the square on line *bd* is to the square on line *gd*. But the square on line *ba* is five times the square on side *ag*. Whence the square on line *be* is five times the square on line *gd*. Since the square on line *gd* is 9, the square on line *be* is 45. Therefore line *be* is the root of 45. We have taken this from *ab* (the root of 125) to find the remainder *ae*. The ratio of the square [on *ae*] to the square on line *ad* is as the ratio of the square on line *be* to the square on line *gd*. Whence the square on line *ae* is five times the square on line *ad*. But the square on line *ad* is 4. Therefore the square on line *ae* the desired residue is 20. And thus *ae* has been found as the root of 20, as above.

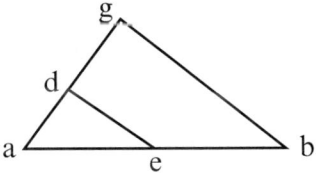

Figure 2.18

---

[24] "De extractione radicum" (28.3, f. 17r.6).

[35] Or, if we reduce the roots of 45 and 125 into [so many] roots of 5, then we have for the root of 45 three roots of 5 and for the root of 125 five roots of 5. So if 3 roots of 5 are removed from 5 roots of 5, then 2 roots of 5 remain. This equals one root of 20, as we found.

[36] Another way: add 45 to 125 to get 170. From this subtract 150 twice the root of the product of 45 and 125 to get 20 for the square of the desired remainder. For example, let line *ab* be the root of 125. To this add line *bg* the root of 45 [see Figure 2.19]. I say that line *ga* is the root of 20. Since line *ab* is divided at point *g*, the two squares on lines *ab* and *gb* are equal to the square on *ag* and twice the product of *gb* and *ab*. For the two squares on *ab* and *gb* are 170. This equals the square on line *ag* and twice the product of *bg* and *ba*. But *bg* by *ba* is the root of the product of 45 and 125 or 75. Twice 75 is 150. Subtract this from 170 to leave 20 for the square on line *ag*, as required.

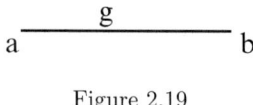

Figure 2.19

[37] If you wish to subtract the root of 20 from the root of 30, even if they are not in the ratio of their squares, find the root of 20 and the root of 30 as accurately as possible. Then subtract the root of 20 from the root of 30 [see Figure 2.20]. While the remainder is not perfect, it is nearly so. Or if you want, add 20 and 30 to get 50. Then multiply them to get 600. From this take the two roots, namely twice its root, that is the root of four times **{p. 29}** 600 which is nearly 49 less $\frac{1}{98}$. Subtract this from 50 and you have $\frac{1}{98}$ 1. Find the root of this and you will have a close approximation of the desired remainder.

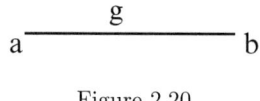

Figure 2.20

2.5 Division of Roots[25]

[38] If you wish to divide the root of 600 by the root of 40, divide 600 by 40 to get 15. The root is the result of the division. For example: let *a* be 40, *b* 600, *g* 15, *d* the root of *a*, and *e* the root of *b*. Divide *e* by *d* (the root of 600 by the root of 40) to get *z*. I say that *z* is the root of *g* or 15. Since the quotient of *e* by *d* is *z*, then the product of *d* by *z* is *e*. Now the square of *d* is *a*. Therefore the product of *a* by *z* equals the product of *e* by *d*.[26] Consequently as *d* is to *z*

---

[25] "De diuisione radicvm" (29.3, f. 17v.13).
[26] Text has "*a* ex *d*...*e* ex *z*"; context requires "*a* ex *z*...*e* ex *d*" (29.9–10).

so is a to e [see Figure 2.21]. Again: z times d makes e. And z squared makes i. Therefore, as d is to z so is e to i. But as d is to z, so a to e is met again. Whence as a is to e, so is e to i. Therefore a, e, and i are in continued proportion. So the product of a by i is as the square of e. But the square of e is b. Therefore the product of a by I is likewise b or 600. But b divided by a yields g. Therefore the product of a by g is b. And the product of a by i is b. So the number i equals the number g. But the square of z is i. Consequently squaring z produces in like fashion g. Therefore z is the root of g, as we had to show.

40	15	600
a	g	b
d	z	e

Figure 2.21

[39] In another way: d times z makes e. The square on e is b. Therefore the product of d by z multiplied by g produces b. Therefore the square on d times g is b. Whence the product of d by z times e equals the square on d times g. By removing the common factor d from both products, the product of z by e remains equal to the product of d by g. Hence, as d is to z, so is e to g. But as d is to z, so is e to I. Therefore e to g and to I is the same ratio. Consequently g equals I. But z is the root of I. Therefore z is the root of g. So if you wish to divide the root of 40 by the root of 600, dividing 40 by 600 gives $\frac{1}{15}$. Find this root and you have what you proposed.

[40] Now we wish to explain how to find as accurately as possible the roots of fractions. First observe that when the roots of squared numbers are divided among themselves or by their ratios, a rational number is always the result. For example: we wish to divide the root of 64 by the root of 16. Sixty-four divided by 16 is 4. Its root is 2, the result of the division. Now 8 (the root of 64) divided by 4 (the root of 16) also gives the root 2. Note that you get the same results from dividing the roots of all the numbers having the same ratio as 16 to 64. If we want to divide the root of 80 by the root of 20, the division produces 2.

[41] If you want to find the root of some fraction or group of fractions, there are two ways to do this. The first way is to take the part or parts of some large number and multiply however many [fractions] you have by that number, and you will find the root of the sum of the products. Then divide by that same number, and you will have what you wanted. For example, you want to find the root of $\frac{2}{3}$, so take $\frac{2}{3}$ of some large number, such as 60. The larger the number you chose, so much closer will you get to the root. Now $\frac{2}{3}$ of 60 is 40. Multiply it by 60 to get 2400 whose root is 49 less **{p. 30}** $\frac{1}{98}$.

56   Fibonacci's *De Practica Geometrie*

Divide this by 60, and you have what you wanted. Now if you want to have this in feet and inches, then take $\frac{2}{3}$ of 108 the number of inches in one rod to get 72. Multiply this by 108 to obtain 7776. The root of this is $\frac{2}{11}$ 88. That is how many inches there are in the root of $\frac{2}{3}$ of one rod. In a similar way, suppose you want the answer in minutes and seconds, say of the root of $\frac{4}{5}$ of one degree. Take $\frac{4}{5}$ of the seconds of one degree, that is, 2880 of 3600. The product of these two numbers is a fourth of 10 368 000,[27] the root of which will give you the number of seconds in the root of $\frac{4}{5}$.

[42] Another way: we want to find the root of $\frac{2}{3}$ of one rod, remembering that multiplying a rod by a rod produces a denier. Thus there are 24 deniers in $\frac{2}{3}$ of one rod. The root of this is 4 feet and 8 inches. These equal 144 eighteenths of one denier. Divide this by 8, twice the root you found, to obtain 16 inches[28] with a remainder of $\frac{16}{18}$. From these subtract the square of 16 inches $\left(\frac{256}{324}\right)$ to leave $\frac{8}{81}$ of one denier. Divide this by twice the root you found to produce about $\frac{2}{11}$ of one inch. Or alternately: take the root of 24 deniers that is 5 feet less one denier. Make 18 eighteenths of this which you will divide by twice the root that you found (10) to yield $\frac{4}{5}$ of an inch. Subtract this from 5 feet and what remains is 4 feet and $\frac{1}{5}$ 16 inches.

---

[27] "2880; que multiplica per 3600 erunt quarta 10368000" (30.6, f. 18r.19). Although the reading is correct, there seems to be too much information here. If there are 2880 seconds in $\frac{1}{5}$ of a degree, then the root of that number is easily found, a bit less than $\frac{2}{3}$ 53 seconds.

[28] Text has "16"; context requires "18" (30.12). The correction vitiates all that follows up to the alternate method. The difference of the two answers makes one wonder if this example [42] were not an addition by someone other than Fibonacci.

3
Measuring All Kinds of Fields

COMMENTARY

The longest of the eight chapters, Chapter 3 focuses on measuring fields of all geometric descriptions. Within its scope are triangles, quadrilaterals, other polygons with straight sides, circles and their parts, and oblique figures with straight and curved sides including fields along the sides of mountains. Fibonacci began the measurement of triangles in a very general way, identifying the three types of triangles and stating what is necessary to find the area of each, enough information to satisfy practitioners, but not so for theoreticians. For the latter and considering the various types of triangles, he discussed all possible ways an altitude may be drawn (he used the word *cathete* for *altitude*). A few remarks on the Pythagorean theorem, the use of Hero's formula for finding the area of a triangle given the lengths of its sides, and the method used by surveyors to measure fields rounds out the section on the measurement of triangular fields. In paragraph [16] we find an unusual and very practical formula for computing the area of an equilateral triangle: take $\frac{13}{30}$ ths of the square on a side.

At its end is a lengthy exposition on ratio and proportion with reference to triangles that offers at the beginning a summary of the basic rules of proportion, including standard conjunction and alternation. Some twenty-four problems hone the skills of the practical geometer, preparatory for new relations in ratios. These are from *composition of ratios* by multiplying together all the antecedent terms and all the consequent terms, *excision of a ratio* by which one ratio is pulled out of another, *conjunction of ratios*, and *disjunction*, the latter two being the usual adding or subtracting of terms in a ratio.

In our text Fibonacci observed that there are eighteen combinations of ratios [79]. This is an uncited reference to Chapter 9 of *Liber abaci* where he leads his reader through the set of propositions.[1] The instruction begins with an example (good teaching!): "Five horses eat 6 sestaria[2] of barley in 9 days. How many days will it take 16 horses to eat 10 sestaria of barley?" In order to solve the problem, he instructs the reader to write the number in reverse order, (see Figure 3.a).

---

[1] For a thorough discussion of this, see Bartolozzi and Franci (1990), 5–9.
[2] One *sestario* is a measure of a pint and a half (Florio [1611], 493).

58 Fibonacci's *De Practica Geometrie*

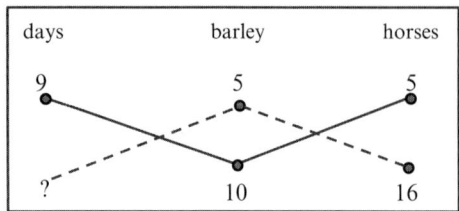

Figure 3a

Then, he multiplies the numbers connected by the solid lines and divides by the product of the numbers connected by (my) dashed lines: $\frac{720}{60} = 12$ days. After showing a different way to solve the problem, Fibonacci faces the reader with the statement, "In this problem we can show 18 combinations of ratios." He begins his explanation thus:

Let the number e be the first line,[3] f the second, d the third, a the fourth, b the fifth, and c the sixth. Further, let the numbers a, e, and c be a certain arrangement (*coniuntio*) called the first set, and the numbers d, b, and f the second set. Now, the ratio of any one of the numbers in the first set to any number in the second set, is composed of the remaining four numbers. Two of these are antecedents and two consequents. Consequently, there are 9 such ratios possible according to one combination of the ratios. ... Similarly, a ratio composed of any number from the second set with any number of the first set is composed of the remaining four numbers, two of which are antecedents and two are consequents. And this makes 9 more combinations.[4]

Noteworthy is his use of letters to represent numbers, an early use of letters as variables. Then he identifies a, e, and c as the first conjunction (they will be multiplied together), and d, b, and f as the second conjunction (again, to be multiplied together). The variables as products are related in the following way,

$$aec = dbf. \tag{3.1}[5]$$

For practical purposes the relationship appears in the diagram shown in Figure 3.b, where the respective paths of multiplication appear as lines.

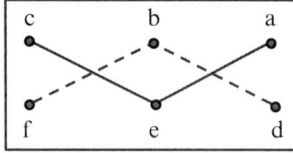

Figure 3b

---

[3] Judging from the marginal illustration, the word *line* refers to the bottom row of numbers.
[4] *Liber abaci* 132.23–30, 36–39; Sigler (2002), 206–207.
[5] Shades of Menelaus' theorem, not to ignore the Chinese *Rule of Five*.

From (3.1) or the diagram we can form nine ratios: take any letter in the top row and combine it with any in the bottom row to make the first ratio; for example, $\frac{c}{f} = \frac{db}{ea}$, and so on, always using 2a letter at the top for the beginning of a new ratio. The next set of nine ratios begins with the bottom row, selecting one of its letters to combine with any one of the three in the top row. Returning to the original problem that is now solved (see Figure 3.c), we can create 18 problems from this one just by blocking out one of the numbers in the diagram and setting up the quotient of products according to (1). In *De practica geometrie* Fibonacci remarked that given the six quantities, it is not possible to form a ratio of two quantities from the same set or conjunction. After this admonition, he moved on to find area of quadrilateral fields.

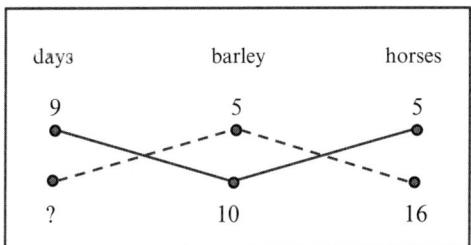

Figure 3c

Part 2, *Measurement of Quadrilaterals*, lists the figures whose areas Fibonacci intended to find: squares, rectangles, rhombi, rhomboids, trapezoids, and generic quadrilaterals lacking equality between any two parts. In this section he would use algebra as a primary tool for solving problems. The theory of algebra that he presents here was offered first by al-Khwarizmi and developed further by Abū Kāmil. Fibonacci begins by defining the three kinds of numbers: square, root, and simple number. Then he lists the three simple and three composite equations, the latter all quadratics. Next he demonstrates how to change the enunciation of a problem into the appropriate standard form of a composite equation. His first examples match the simple equations, then problems expressible in the form of composite equations. By the end of the lesson [94] the reader is well instructed in the three-step technique for solving problems describable in algebraic terms:

1. Establishing an equation from the enunciation terms of algebraic numbers
2. Simplifying the equation to one of the six standard types
3. Applying the proper procedure to reach the answer[6]

Furthermore, as the following exemplifies, his listeners understood more than the simple meaning of some words. From Chapter 3 [95]: "Likewise, the square on the diameter (the area) and the four sides of the given square equal

---

[6] See Oaks (2005), 403 for further development of these strategies.

279 rods." Query: how is uniformity of dimensions maintained? The answer is in the third sentence following: "... the square and four roots equal 279." This suggests an interchange of the words "lines" and "roots" that is carried into [96]. The equivalency of the words is discussed at length below where paragraph [89] of this chapter is analyzed. Finally and by now obviously, the "279 rods" are understood by the listeners as areal, not linear, roods.

Not so simple is a set of problems whose solutions are indeterminate, listed under the title *Multiple Solutions* [139]. Fibonacci began with "find the dimensions of a rectangle of unknown and unequal sides, whose diameter is 20. It is required that the answers be rational." The solution begins by selecting a rectangle of sides 5 and 12 with diameter 13. Because the sides of the unknown rectangle are not known, Fibonacci could declare the two rectangles similar. Thereby he set up the appropriate fourth proportionals and solved the problem. He might just as well have used a rectangle of some other sides. Because of this option, the problem is called indeterminate.[7]

The section on *Other Quadrilaterals* offers rules for finding the dimensions of rhombi, parallelograms, trapezoids, and any quadrilateral not defined by any of the first named. There are, however, polygonal relationships to which he wished to devote some particular attention because of their relationship to inscribed and/or circumscribed circles. He began with a circle inscribed in a regular pentagon that can be divided into five congruent triangles. If the regular pentagon were inscribed in the circle, then the area of the pentagon equals the product of three-fourths of the diameter by five-sixths the length of the chord joining nonadjacent vertices of the pentagon. This procedure leads into a discussion of the binomial and apotame.[8] The section concludes with several examples worked out in panes and rods.

In paragraph [84] the reader is confronted with an underlying extraordinary concept, geometric roots. The Latin is clear "... *embadum eius equatur quatuor radicibus quarum una est quadrilatera ae, secunda* ..." As is obvious from Figure 3.52, the root is the unit rectangle *ae*. If we express this algebraically, then $x^2 = 4x = x + x + x + x$. Now each of these xs is a rectangle, 1 by x. Thus, a geometric root is a rectangle, 1 by x. Hence, it would seem that Fibonacci's concept of a *geometric* root differs from his idea of a *numerical* root.

The problem at [89] is at once the direct descendent of a Babylonian ancestor and an exercise in the development of algebraic terminology. It says, "Given an area and four sides that make 140, and you wish to evaluate the sides from the area."[9] Jens Høyrup (1996₁) ably described the story of the historical descent of our problem in clear detail and diagram. Of more importance, I believe, is the place of [89] as a stage in the development of algebraic

---

[7] "... solutiones non sunt terminate" (216.38).

[8] Fibonacci preferred the word *reciso* to *apotame*.

[9] "Et si embadum et quatuor eius latera faciunt CXL, et uis separare latera ab embado" (59.5–6). Because the problem seeks the value of the length of the side of the square, I chose "evaluate" as the best word for "separare."

3 Measuring All Kinds of Fields   61

terminology. The expression "four sides" in whichever language limps. The four sides of the square? No, not really. What Fibonacci was trying to say is "four unit areas of dimension 1 by the side of the square; that is, $1x + 1x + 1x + 1x = 4x$." This is clear from the fact that moving toward the solution of the problem, he added an area of 4 sides to the square, an area defined by a rectangle. An area was added to an area. This is precisely what modern algebraic terminology in the context of [89] means by $+ 4x$.

The section on measuring circular or round fields is the longest in the chapter. Keeping his fellow Pisans in mind, he noted, "If you wish to do it in the Pisan way: square the diameter, divided by 7, and you will have the area of the circle in panes." He assumed that the reader knew that the diameter also had to be measured in panes. Diameters suggest chords of circles, for which Fibonacci offered an extended treatment.

Strangely and without explanation other than to say the concepts are used in astrology, Fibonacci defined right sine and versed sine in [206]. The brevity of the paragraph is quite unlike Fibonacci. Anyone reading his works knows that whatever he mentioned, he discussed at length. There is mention here but no length. Furthermore, Fibonacci was careful to exemplify whatever tool he discussed. No examples appear in the text. Finally, the Italian translations that I have studied do not mention these sines. Because of these characteristics, may it not be that the paragraph is an addition by a knowledgeable scribe?

At this point I need to mention a major shift that I made in the received text. For reasons stated in the Commentary on Chapter 7, I have moved the entire section relating to Fibonacci's *Table of Chords* to follow paragraph [9] in that chapter. This includes paragraphs [211] through [219] formerly in Chapter 3,[10] which numbers I have kept in Chapter 7 and without any change in paragraph numbering. Hence there is a hiatus between paragraph [210] and paragraph [220] in Chapter 3.

Recognizing again the surveyor's craft, Fibonacci noted that any part of the circumference or an arc of a circle can be measured quite easily: lay a tape of one rod along the arc as many times as is necessary to measure its length. The remainder of the section on circles develops rules for finding the dimensions of half a dozen regular polygons, each inscribed in a circle often of diameter 8. Finally, Fibonacci returned to Ptolemy's theorem on inscribed quadrilaterals and their diagonals, for finding any of the chords. Hence, with these tools for measuring arcs and chord, there was no need in this chapter for *Table of Chords*.

The determination of the area of the last round figure, an oval [225], shows that Fibonacci was familiar with unending processes.[11] Sides abc and dez are curved, the one concave and the other convex.[12] Various straight

---

[10] (95.32 through 100.10).

[11] Fibonacci did not discuss limits; hence, these are not infinite series. See *Elements* XII, 2. He might also have studied its power as used by Archimedes.

[12] Fibonacci used the same phrase "aream uentricis" that I translate as concave or convex area as required by the context.

lines are drawn to create triangles. The triangle leaves some areas between their sides and the figure unmeasured. So smaller triangles are formed and measured, "until nothing appreciable is left of the convex/concave part." This statement surely refers to a geometric technique for exhausting the area desired by smaller and smaller areas of triangles.

The final section of Chapter 3 offers methods for measuring fields located on the sides of hills or mountains. The first technique is simply to lay a tape measure along the descending side, take its length, measure its width, and compute the area according to a method mentioned earlier in the treatise. For an unstated reason Fibonacci spent much space indirectly measuring the length of the base of the hill immediately below the side of the field, bg in the Figure 3.116. He used a wooden instrument called an *archipendulum*.[13] It is shaped like an isosceles triangle with a lead bob attached to a cord hanging from the upper vertex of the triangle, and is used to determine lines parallel to the base of the hill. As the figure suggests, the instrument helps create a series of steps, the sum of which horizontal lengths measures bg, the base of the hill. Fibonacci concluded this section and the chapter with instructions on measuring polygonal fields, some of the sides of which are convex or concave or both. The areas are found by appropriate measurement of rectangular and triangular sections of the fields, some of which mentioned in [246] might involve continuous, if not infinite, processes to fill out the field.

Several algorithms outside the repertoire of ordinary use merit mention; they are expressed in modern symbols.

- To find on the base of a scalene, acute triangle with sides a, b, and c, b being the base and $a > c$, the distance x from a base angle to where the altitude falls:

$$\frac{a^2 - c^2}{2b} + \frac{b}{2} = x.$$

- To find the area within a semicircle of diameter d:

$$\frac{11d^2}{28}.$$

- To find a very approximate area of an equilateral triangle of side s:

$$\frac{13}{30}s^2.$$

- To find the side s and altitude a of an equilateral triangle inscribed within a circle of diameter d:

$$\frac{d\sqrt{3}}{2} = s; \quad \frac{3d}{4} = a.$$

Some appear in more than one section of the chapter as use requires.

---

[13] See Smith II, 358, for an illustration of the archipendulum in use, where the method is called "leveling."

SOURCES

The basic outline for this chapter might have been adapted from *Kitab 'ibm al-misāhāa* (*Book on Mensuration*) by Abū Bekr (*fl.* ix s.),[14] not to rule out al-Karajī's *Kāfī fīl-hisāb* (*Essentials of Reckoning*) with a similar outline.[15] Nor may we rule out the possible existence of other sets of problems that Fibonacci might have gathered in his *iter mathematicum* about the Mediterranean and thereafter. It is possible that Fibonacci produced Chapter 3 from what he knew of current handbooks of land measurers. Here I wish to consider the Latin version of *Kitab 'ibm al-misāha* of Abū Bekr as a source.

The remark by Høyrup, "There can be no doubt that Leonardo had Gerardo's version of *Liber mensurationum* (in full or in excerpt) on his desk while writing parts of the *Practica*,"[16] requires attention. In substantiation, he noted the word-for word correspondence between the solution of a problem in Abū Bekr's tract and Fibonacci's solution. Because it would be unfair to extrapolate from his observation that Fibonacci drew heavily from the translations (which in fact he did not, if at all) for the composition of Chapter 3, a comparison and contrast of the two treatises is appropriate. Both discuss the usual plane and solid objects, although not to the same extent or in such detail. Students of just one tract would become reasonably competent in finding the dimensions of triangles, squares, rhombi, rectangles, and/or parallelograms, trapezoids, cubes, prisms, parallelepiped, circles, arcs, and spheres. Abū Bekr was more interested in trapezoids: 30 paragraphs and problems to Fibonacci's 14. The latter, on the other hand, developed a great interest in triangles: 67 paragraphs and problems to Abū Bekr's 33. As a last contrast, Fibonacci devoted more space to the solids than did Abū Bekr. In general Abū Bekr's treatise contains 158 paragraphs and problems to 225 in Chapter 3 of *De practica geometrie*. Although there are many examples of the extent to which Fibonacci added to the material offered by Abū Bekr, the most prominent difference is his derivation of the value of π. Where the use of π would be necessary, Abū Bekr simply advised the reader to multiply by $3\frac{1}{7}$ with no indication of the origin of the constant; Fibonacci prepared the reader by deriving the value of π by Archimedes' method.[17]

A comparison of individual problems shows a more or less concordance between 31 problems. Twenty-one are classified as similar. That is, the components state the same relationships involving sides, angles, areas, and such parts of the figure but differ in their numerical values. For instance, both texts have the isosceles trapezoidal problem, given the lengths of upper base, side, and distance among the bases, find the diagonal; but the measures are different. In paragraphs <66>[18] of *Liber mensurationum* the values are 4, 6, and 8. The same problem in

---

[14] Busard (1968).
[15] Høyrup (1996$_2$), 9.
[16] Høyrup (1996$_1$), 55.
[17] pp. 88 [194]–91 [200].
[18] Paragraph symbols and numbers are from Busard's edition.

Fibonacci's paragraph [170] shows 8, 13, and 12 for the values. Of the remaining ten problems, the statements of these problems are the same: <2> and [87], <9> and [93], <11> and [88], <25> and [108], <38> and [119], <47> and [127], <48> and [130], <49> and [134], and <52> and [144]. Five of these have the same components but are solved differently; namely, <2> and [87], <9> and [93], <11> and [88], <25> and [108], and <52> and [144]. Four are solved in the same way but differ in wording pair wise; namely, <47> and [127], <48> and [130], <49> and [134], and <81> and [169]. Only one pair has nearly the same wording, <38> and [119]. It is upon this last that Høyrup based his assertion.

My thinking on the remark produced a dilemma. If Fibonacci had a copy of Gerardo's translation of *Liber mensurationum* on his desk and incorporated the section from <38>, a very neat, cleancut proof, into [119], then why did he not incorporate the proofs from <47>, <48>, and <49>, which are more simply stated, into [127], [130], and [134]? If, on the other hand, Fibonacci had an Arabic copy of *Liber mensurationum* that he was translating and from which he prepared the similarly worded proofs in [127], [130], and [134], then how do you explain the congruence of pertinent sections in <38> and [119]?

I lean toward the horn of the Arabic copy of *Liber mensurationum*, for one reason. There is some internal evidence that, at least with respect to the treatises discussed here, Fibonacci was a clearer translator than Gerardo. Compare the following. In each case the first reading in italics is from Gerardo's translation, the second from Fibonacci's text.

1. The section on trapezoids begins
   "*Capitulum quadrati cuius caput est brevius vel lacius.*"
   "Incipit *de*figuris que habent capita abscisa *de*quibus iiiior sunt genera."[19]
2. Irregular trapezoids are introduced with
   "*Capitulum aride diverse vel quadrati diverse latitudinis*"
   "... que diuerse caput abscisa dicitur ..."[20]
3. This type of trapezoid resembles one half of an isosceles trapezoid:
   "*Capitulum dimidii aride id est quadrati cuius cacumen basi equidistat*"
   "que semi caput abscisa dicitur... quorum (laterum) unum eleuatur supra basem secundum rectum angulum..."[21]
4. The upper base leans over an end of the lower base:
   "*Capitulum aryde expanse vel latitudinalis*"
   "... que caput abscisa declinans dicitur ..."[22]

The Latin word *arida* might have had common currency in Gerardo's time, but I think that Fibonacci's descriptions are more easily visualized. His four are described as isosceles trapezoid, irregular trapezoid, right angled trapezoid, and leaning trapezoid. Although I did not find the word *trapezium* in either treatise, I think that the foregoing supports my position that Fibonacci

---

[19] From just before <65> and prior to [169].
[20] From just before <81> and at [177].
[21] From just before <95> and at [174].
[22] From just before <105> and at [180].

adapted and adopted pertinent parts of an Arabic copy of *Liber mensurationum*. The congruence is explained as the best wording that each writer could produce. Nothing rules out the possibility that he had at hand other Arabic and Latin tracts on land measuring.

The *Geometry* by Banū Mūsā provided further direction and assistance to Fibonacci for composing several sections. Proposition VII presents Hero's formula for finding the square of the area of a triangle in paragraphs [31] to [33]. Proposition I and paragraph [185] for finding the area of a regular polygon are allied, as are Proposition IV and paragraph [191] for finding the area of a circle. Proposition VI that represents Archimedes' method for determining the value of π might have assisted Fibonacci in forming paragraphs [194] to [200].[23]

The section on combinations of quantities to form 18 ratios was most probably based on his own work in *Liber abaci*.[24] This in turn was drawn from Ahmed ibn Yussuf's treatise on ratio and proportion, with assistance from Ptolemy's proof of Menelaus' theorem in *Almagest* I.12.

TEXT[25]

There are five parts to this chapter: the measurement of (1) triangles, (2) quadrilaterals, (3) polygons with straight sides, (4) circles and their parts, and (5) oblique figures and those with straight and curved sides. In the fifth kind we measure fields along the sides of mountains.

3.1 Measuring Triangles in General

[1] Triangular fields have three sides. Some are rectangular, others are acute, still others are obtuse. These triangles get their names from their angles. A right triangle has one of its angles at 90°, the other two being equal to a right angle. An oxigonal triangle has three acute angles. An ampligonal triangle has one angle greater than 90°. Some triangles get their names from their sides: isopleural triangles are equilateral; isosceles triangles are equicrural; and diversilateral triangles are also called scalene. Equilateral triangle have three sides each equal to each of the others. Equicrural triangles have only two equal sides. The sides of diversilateral triangles are all unequal to one another. Note that in every triangle three cathetes or perpendicular lines can be drawn. Each of these falls from an angle to its opposite side. In a right triangle only one perpendicular can be drawn within the triangle, and that from the right angle to the side opposite. The other perpendiculars are the sides forming the right angle. In acute triangles, the perpendiculars fall within the triangle. In an oblique triangle two perpendiculars fall outside the triangle, one inside.

---

[23] I am indebted to Clagett (1964), especially p. 224n, for the lead here and in Chapter 6.
[24] *Liber abaci*, 119; Sigler (2002), 180.
[25] "Incipit pars prima tertie distinctionis de mensuratione triangulorum" (30.25, f. 18r.29–30 [de: in]).

66   Fibonacci's *De Practica Geometrie*

[2] *To find the square measure, namely the areas of all triangles* {**p. 31**}, *multiply half the cathete by the whole base or half the base by the whole perpendicular line*. To establish this procedure: if a cathete is drawn from an angle not less in measure than either of the other angles to the side opposite, it falls within the triangle. [That is,] if a cathete is drawn in triangle *abg* from angle *bag* that is not less than either angle *abg* or *bga* to side *bg*, then if it is drawn from vertex *a*, I say that it will fall within triangle *abg*. For if it is not true, then let it fall outside the triangle beyond side [*a*]*b*. Extend side *gb* indefinitely in a straight line beyond point *e* and let the cathete *az* fall on *ge*. Therefore triangle *azb* has a right angle, angle *azb* [see Figure 3.1]. Because in triangle *azb* one side has been extended outside the triangle (*zb* beyond *bg*), the exterior angle *abg* is greater than its opposite and interior angle *azb*. But angle *bag* is not less than angle *abg* that is greater than right angle *azg*.[26] Similarly, angle *bag* is greater than right angle. Hence, in triangle *abg* there are two angles greater than two right angles. But this is impossible because all three angles of a triangle must equal two right angles, as Euclid said in I.32. Therefore a cathete cannot be drawn from vertex *a* to side *bg* outside of point *b*. It can be shown in a similar way that must fall inside of point *g* rather than outside. And this had to be shown.

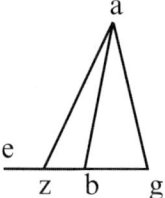

Figure 3.1

[3] From this it can be understood that if a cathete is erected from the side opposite a small angle in an oxigonal triangle, the right angle in a right triangle, the obtuse angle in an ampligonal triangle, it lies within the triangle. We shall show that in a right triangle the sides containing the right angle are cathetes. Consider right triangle *bgd* with right angle *bgd*. I say that straight line *bg* is erected perpendicularly to straight line *gd* and *dg* to *bg* [see Figure 3.2]. Extend straight line *gd* to point *a*. Now *bg* is on *ad* where it makes two right angles or angles equal to two right angles. Now angle *bdg* remains a right angle as does angle *bga*. Hence *bg* is a cathete to *ad*. Similarly, if straight line *bg* is extended, it can be shown that the straight line *dg* is a cathete to *bg*. I say again, from point *b* no other cathete cannot fall to line *ad* except *bg*. For it were possible, let the cathete be *ba* to line *ad*. Then in triangle *bag* there would be two right angles, which is quite inconvenient.[27] Similarly, if we had it fall between *gd* say at point *e*, then in triangle *abe* there would be two right angles, one at *bge*

---

[26] Text has *agb*; context requires *azb* (31.13).
[27] "inconveniens" (31.29).

and the other at *geb*, which is impossible. For no cathete can fall upon line *ad* except *bg*.

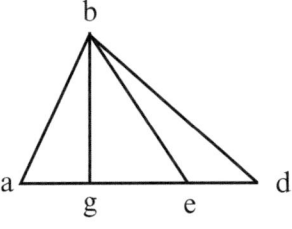

Figure 3.2

[4] Similarly it can be shown that no other cathete except *dg* can be drawn from *d* to *bg*. And this must be shown. If a cathete be drawn in an oxigonal triangle from one of its smaller angles to the side opposite it, it would fall within the triangle. For example: in oxigonal triangle *abg* let the small angle be *bag*. I say that if a cathete be erected from angle *a* to straight line *bg*, it falls within the triangle. For if it were not true, then: if possible let it fall outside [the triangle] on a point *d* [see Figure 3.3]. Now because cathete *ad* falls on line *dbg*, it forms a right angle *adb*. But angle *abg* that is outside triangle *adb* is greater than the interior angle opposite it, which is angle *adb*. But angle *adb* is a right angle. Whence angle *abg* is greater than a right angle. And that is inconvenient for acute triangle *abg* for it cannot have a cathete falling outside of itself; it must fall inside as was shown **{p. 32}**.

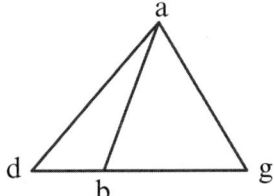

Figure 3.3

[5] In an oblique triangle, if a cathete is drawn from an acute angle to the opposite side, it falls outside the triangle. Let oblique triangle *bdg* have oblique angle *bdg*. I say that if a cathete is drawn from angle *dbg* to side *gd*, it falls outside triangle *bdg*. For were it not true, then if possible let it fall on *dg* at point *a*. Because cathete *ba* is on straight line *gd*, angle *bad* is a right angle. It is larger than right angle *bda*[28] [see Figure 3.4]. Whence in triangle *bda* there are two angles whose sum is greater than two right angles. And this is impossible. For no cathete falls from point *b* onto line *gd* within the triangle.

---

[28] The bottom figure in *f.* 194 is flawed, but corrected here.

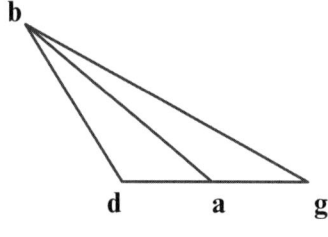

Figure 3.4

[6] Similarly, it is shown that a cathete cannot fall from point *g* onto line *bd* within triangle *bdg*; it must fall outside. To prove this: let a straight line fall on two sides of an angle. Select its midpoint. Draw a straight line from that point to the angle, If this second line is equal to the line lying from the given point to a side of the angle, then the angle is a right angle [see Figure 3.5]. For example: let the sides of angle *abg* be *ab* and *bg*. Let line *de* intersect both sides. From midpoint *z* of line *de* draw line *zb*. I say that if line *zb* equals line *ze* or line *zd*, then angle *abg* is a right angle because the three lines *zb*, *zd*, and *ze* are equal to one another. If a circle is drawn from point *z* with a radius equal to one of those lines, it will pass through points *d, b,* and *e*. The line *de* is within the circle, the center of the circle lies on it, and therefore line *de* is the diameter of the circle. This diameter contains arc *dbe*; therefore it is a semicircle within which is angle *dbe*. Any angle lying on a semicircle is a right angle as Euclid teaches in his third book.[29] From this it is obvious that if straight line *zb* is longer than line *zd* or *ze*, then angle *dbe* is acute. Or if *zb* were less than *zd* or *ze*, angle *abg* would be obtuse.

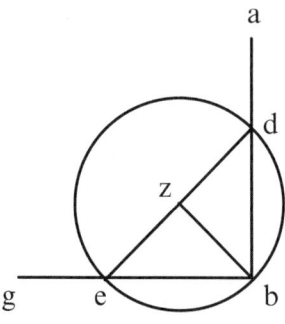

Figure 3.5

[7] *In a right triangle the square on the side opposite the right angle equals the sum of the squares on the other two sides.*[30] Given right triangle *abg* with right angle *agb*, I say that the square on line *ab* is equal to the sum of the squares

---

[29] *Elements*, III.31.
[30] *Elements, I.47.

on lines *ag* and *gb*. At point *g* I draw a cathete *gd* from point *g* to line *ab*, dividing the triangle in two right triangles, *gdb* and *gda*. They are similar to each other and to their origin, as Euclid demonstrated in Book Six.[31] [see Figure 3.6]. Because triangle *gdb* is similar to triangle *agb*, they have proportional sides around their common angle at *b*. In triangle *dbg* side *db* is to *bg*, as side *gb* is a to line *ab* in triangle *bag*. Whence the product of *db* and *ba* equals the square on line *bg*. Again, because triangle *gda* is similar to triangle *agb*, they have proportional sides around their common angle *a*. Therefore as line *da* is to line *ag* in triangle *gda*, so in triangle *agb* line *ga* is to line *ab*. Whence the product of *ad* and *ab* equals the square on line *ag*. Now we demonstrated that the product of *db* by *ba* equals the square on line *gb*. Whence the product of *db* by *ba* and the product of *ad* by *ab* equal the sum of the two squares on lines *bg* and *ga*. But the product of *db* by *ab* with the product of *da* by *ab* equals the square on line *ab*.[32] Therefore the square on line *ab* equals the sum of the squares on lines *bg* and *ga*. And this is what we had to prove.

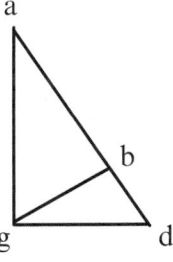

Figure 3.6

[8] Having proved all of this we will show how triangles are measured. But first it must be observed that among right triangles, some are obtuse and others are isosceles **{p. 33}**. Still others are scalene. Among the right oxigonal triangles, some are equilateral, others scalene. Whence, so that we may have a doctrine for measuring perfectly all sorts of triangles, we divide this part of Chapter III into three sections. In the first we shall measure right triangles, in the second acute triangles, and in the third obtuse triangles.

Right Triangles[33]

[9] *The area of all right triangles is found by multiplying a side containing the right angle by half the other side of the right angle.* For example: Consider right isosceles triangle *abc* having sides *ab* and *bc* 10 rods in length and side

---

[31] *Elements*, VI.8.
[32] *Elements*, II.2.
[33] "Incipit differentia prima" (33.6, f. 20r.5).

ac equal to the root of 200 rods. The product of half side *ab* by the whole side *bc* (5 by 10) yields the area of the whole triangle, 50 square rods. This is how it is proved: from point *a* draw at right angles line *ad* equal to line *bc*; then draw line *dc*. You have created rectangle *abcd* of which triangle *abc* is one half, as the attached figure shows [see Figure 3.7]. Therefore the whole square *abcd* is 100 rods found by multiplying 10 by 10, or one side by itself. Whence, half the square or triangle *abc* necessarily contains 50 rods.

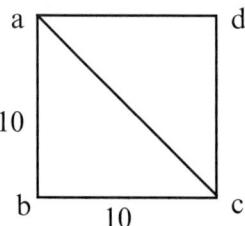

Figure 3.7

[10] Likewise with the right triangle *bcd* with all sides of unequal length: side *bc* is 8 rods long, side *cd* 6 rods long, and side *bd* is 10 rods long. The right angle is at *c*. Whence, the product of half of *bc* by all of *cd* (4 by 6) or half of *cd* by all of *cb* (3 by 8) yields 24 rods for the area of triangle *bc*. This follows from what I taught about triangles [see Figure 3.8]. Or otherwise: divide *bc* into two equal parts at point *e*. From point *e* draw line *ef* equal to and equidistant from line *cd*. Join *df*. Because *cd* is equidistant from and equal to line *ef*, so is line *df* equal to and equidistant from line *ce*, as is obvious from geometry. Whence, line *df* is 4 rods long and equal to line *ef*. Line *bh* is equal to *hd*, and angle *ebh* equals angle *hdf*. Whence lines *eh* and *hf* are equal. Therefore triangle *hfd* equals triangle *beh*. Therefore the whole triangle *bcd* equals the rectangle *ecfd* whose area is the product of lines *ec* and *cd* (4 by 6). Whence the area of triangle *bcd* is the product of half of *bc* (*ec* by *cd*), as we said.

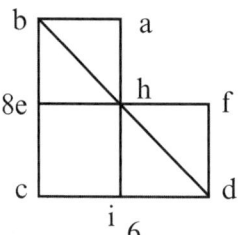

Figure 3.8

[11] In a similar way it can be shown that if from point *i* (the midpoint of *cd*) line *ia* is drawn equidistant from and equal to line *cb* and straight line *ba* is

drawn, then triangle *abh* equals triangle *hid*. If quadrilateral *bcih* is added to both, then the whole quadrilateral *abci* equals triangle *bcd*. The area of the quadrilateral is formed from *ic* and *cb* (3 by 8), as we said above.

[12] Now if you wish to find side *bd* by the other sides, multiply side *bc* by itself (8) to get 64 to which you add the product of side *cd* by itself (36) to get 100. The root of this is 10. And that is the length of the hypotenuse *bd*. With the hypotenuse *bd* 10 rods and the base *cd* 6 rods, the length of the cathete *bc* is sought. Square the hypotenuse (10 by 10) to get 100. Then subtract the square on the base (36) from it and 64 remains. The root of this is 8, the length of cathete *bc*.

[13] Likewise, if the hypotenuse is 10 and the cathete is 8 {**p. 34**} and you want to find the length of the base, subtract the square of *bc* (64) from the square of *bd* (100), to leave 36. Its root is 6, for the length of side *cd*.

Acute Triangles[34]

[14] *The area of an acute triangle equals the product of the cathete by half the length of the base or the product of the base by half the cathete.* To explain: given equilateral triangle *abc* each of whose sides are 10 rods in length, construct cathete *ad* to side *bc*. Since the cathete *ad* to side *bc* creates equal angles at [point] *d*, the triangles *adb* and *adc* are right triangles [see Figure 3.9]. And since lines *ab* and *ac* are congruent,[35] and line *ad* is common to both triangles, the bases *bd* and *bc* are congruent. Consequently, triangles *abd* and *adc* are congruent. And because triangle *adb* is a right triangle, its area is found by multiplying the cathete *ad* by half the base *bd* ($\frac{1}{2}$ 2). Similarly with acute triangle *adc*: its area is the product of cathete *ad* and half the base, *dc*. Hence, the area of the entire triangle *abc* is found by multiplying the cathete *ad* by half the base *bc*, as we said above.

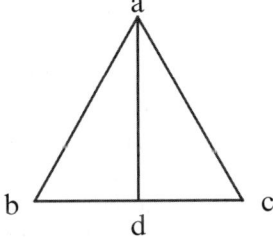

Figure 3.9

---

[34] "Incipit differentia secunda" (34.3, *f.* 20v.8).
[35] Wherever I use the word *congruent*, the text has *equalis*.

[15] Given the same information, we show that the area of triangle *abc* is the product of the length of base *bc* by half the length of cathete *ad*. If you want to know the length of cathete *ad*, then subtract the square[36] of line *bd* (25) from the square on line *ad* (100) to get 75. The root of this number that is a little less than $\frac{2}{3}$ 8 rods is the length of cathete *ad*. The product of $\frac{2}{3}$ 8 rods by 5 or half the base *bc* is a little less than $\frac{1}{3}$ 43 rods for the area of the entire triangle *abc*. Similarly, the product of the whole base (10) by half the cathete that is a little less than $\frac{1}{3}$ 4 is almost $\frac{1}{2}$ 43. Or, multiply *bd* by *ad* (5 times root of 75)[37], and you have the root of 1875 for the area of triangle *abc*. The root is approximately $\frac{1}{3}$ 43 less $\frac{1}{36}$.

[16] You can find the area of the same triangle in another way, namely by taking a third and a tenth of the square on one side (100).[38] What you get is quite close to the area, because the ratio of the area of any equilateral triangle to the square on its side is a little less than the ratio of 13 to 30 [see Figure 3.10]. If you want to find the area of an acute isosceles triangle *def* with equal sides *de* and *df* of length 10 rods and side *ef* of 12 rods, draw cathete *dg* between the equal sides to base *ef*. Because it falls on half of *ef*, you can find the length of cathete. Subtract the square on *eg* from the square on *de* (36 from 100) to get 64 that is the square on cathete *dg*. Hence, *dg* is 8 rods long, the root of 64.

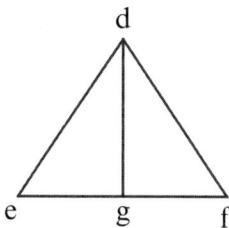

Figure 3.10

[17] Similarly, 8 multiplied by half the base *ef* (6), or the whole base by half the cathete (12 by 4), produces 48 rods for the area of triangle *def*. Multiply the cathete by half the base; then form from the triangle a rectangle with length 8 rods (the same as the length of the cathete) and width 6 rods (half the base) [see Figure 3.11]. For example, draw triangle *def* again, and from point *d* draw line *dh* equidistant from and equal to line *gf* that equals line *ge*. Draw line *hf* and it equals cathete *dg*. Whence quadrilateral *dgfh* equals triangle *def*. For the area of quadrilateral *dgfh* {**p. 35**} is found by multiplying *dg* by *gf* (half of *ef*).

---

[36] Leonardo uses the phrase *potentia totius lateris* for square on the line, here (34.19) and often below.

[37] His method is to put the 5 within the radical, after the Arabic custom.

[38] A later hand drew a line though "tertiam et decimam partem accipe" and wrote above "multiplica per 13 et divide per 30" that anticipates the sentence after next (f. 20r).

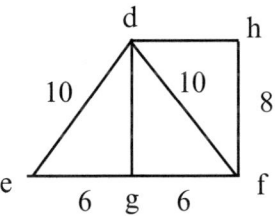

Figure 3.11

[18] Likewise, given acute scalene triangle *abc* with side *ab* 13 rods, side *ac* 15 rods, and base *bc* 14 rods. In this triangle the cathete cannot be found until first the place on the base is found where the perpendicular or cathete would fall. There are three ways to find that place.[39] The first way is to add the square on one side to the square on the base, and to subtract the square on the other side from their sum. Divide half the remainder by the length of the base. The quotient will be the segment for that part from which was added the square on the side with the square on the base. For instance in the attached figure, the square on base of length 14 is 196. Add this to the square of 13 or 169. Subtract from their sum the square on the remaining side *ca* (225) to get 140. Divide half of this or 70 by the base or 14 to get 5. This is the length of segment *bd*. What remains is 9 rods, the length of *dc*, the difference between 14 and 5 [see Figure 3.12]. The second way: the square on side *ac* (225) added to the square on the base *cb* (196) is 421. If from this the square on side *ab* (169) is subtracted, the remainder is 252. Half of this is 126 that divided by the base leaves 9 for [the length of the segment]. There is another way: add the lengths of the two hypotenuses, 13 and 15, to get 28. Take half of this or 14 and multiply it by one, the difference with one of the hypotenuses, to get 14. Divide this by half the base (7) to get 2. Add this to the half base to get 9 that is the major part lying under hypotenuse *ac*. In a similar manner subtract 2 from 7 to find the lesser distance *bd* of length 5 rods as was found by the first method.

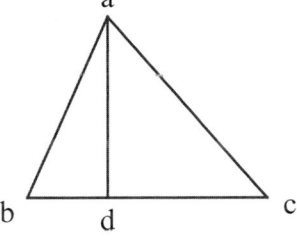

Figure 3.12

---

[39] The two ways are based on *Elements*, II.12, 13.

74   Fibonacci's *De Practica Geometrie*

[19] A third way is to subtract the square on the smaller hypotenuse from the square on the greater (225 less 169). Divide the remainder (56) by the base (14) to get 4 that added to the base becomes 18. Half of this is 9, the longer segment. Or take 4 from the base to leave 10 whose half is 5, the shorter segment. This way seems to me to be more practical than the others. Once the place has been found, if you want to find the perpendicular, subtract the square (25) of the shorter segment from the square on side *ab* (169) to get 144. Its root is 12, the [length of the] cathete *ad*. Or, subtract the square (81) of the longer segment from the square (225) of *ac*, and what remains is likewise 144 for the square on the cathete *ad*. Whence the length of the cathete is 12, as we said. The multiplication of the cathete by half the base or conversely yields an area of 84 rods for the entire triangle *abc*.

[20] In the isosceles triangle above we showed that the area of the triangle equals the area of the rectangle whose length equals that of the cathete and whose width is half the base [see Figure 3.13]. What we wish to show now is that the area of triangles can be equated with the area of rectangles whose lengths equals the bases of the triangle but widths are half those of the cathetes. Divide cathete *ad* in two equal parts at point *e* as shown elsewhere. Through point *e* draw line *fg* equal to and equidistant from line *bc*. Complete lines *fb* and *gc* that are coequal and equidistant from each other. Because line *fg* is equidistant from and equal to line *bc*, and because cathete *ad* was divided into two equal parts at point e^{40} {p. 36}, line *fg* cuts lines *ab* and *ac* in two equal parts at points *h* and *i*, as shown in geometry.[41] The angle at *e* is a right angle as are the angles at *d*. Whence the angles at *f* and *g* are also right angles. Therefore if line *ae* is cut from *ad* at *e* and line *eh* cut from *fe* at *h*, and if triangle *aeh* is placed on triangle *bfh*, then line *ae* falls on line *fb*. Consequently because line *fb* equals line *ed* that equals line *ea*, and line *eh* is on line *hf*, and line *ah* is on line *hb*, angle *f* equals angle *aef*. Whence angle *f* is a right angle. For the same reason therefore the angle at *g* is a right angle, and triangle *cig* equals triangle *aei*. Therefore the area of triangle *abc* equals the area of quadrilateral *fbcg* because they share the same base and [the height of the quadrilateral] is half the cathete. And this had to be proved. Nor must it be overlooked that in quadrilateral *fbcg* angle *bcg* equals the angle at f. Because they are opposite angles, angle *fbc* equals the angle at *g*. All of this is clearly explained in Euclid's book where it is shown that in every figure whose opposite sides are equals, its opposite angles are also equal.[42] Therefore angles *fbc* and *bcg* are right angles. Therefore quadrilateral *fbcg* is a rectangle,

---

[40] An impossible clause is omitted here: "et per punctum e protracta est basis bc equidistans recte" (35.43–36.1)
[41] *Elements*, I.45.
[42] *Elements*, I.34.

as demanded. In Euclid's Second Book it is shown how to find a perpendicular segment in an acute triangle.[43]

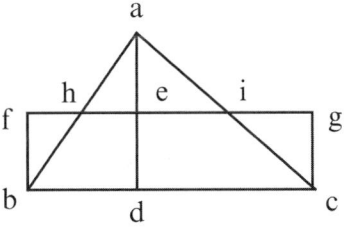

Figure 3.13

[21] We wish to demonstrate geometrically how to find the same segment by a second and a third method. Given again triangle *abc*: draw cathete *ad*. At points *b* and *c* draw two right angles with sides *eb* and *fc*. Let line *eb* equal line *ac* (15), and line *fc* equal line *ab* (13). Draw lines *fe*, *fd*, and *de*. Bisect line *ef*, the point of bisection being *g*. From point *g* draw line *gh* equidistant from both *fc* and *eb*. Through point *f* draw *fik* equidistant from *bc*. Again through point *g* draw line *gl* equidistant from lines *cb* and *fik*. Because triangles *adc* and *adb* have right angles at *adc* and *adb*, the square on side *ac* equals the sum of the squares on the two lines *ad* and *dc*. The square on line *ab* equals the sum of the squares on lines *ad* and *db*. Whence if the square on line *ad* is taken from both [groups], then the square on the larger segment *dc* plus the square on the smaller *db* is as much as the square on line *ac* plus the square on line *ab*.

[22] Whence the squares on lines *ab* and *dc* equal the squares on lines *ac* and *db*. But line *fc* equals line *ab*, and line *be* equals line *ac*. Whence the squares on lines *fc* and *cd* equal the squares on lines *eb* and *bd*. But the squares on *fc* and *cd* equal the square on line *fd*, since angle *fcd* is a right angle. Similarly, the square on line *de* equals the squares on lines *eb* and *bd*. Whence lines *fd* and *de* are mutually equal. Therefore triangle *fde* is isosceles [see Figure 3.14]. And because the base has been bisected at point *g*, line *dg* is a cathete to line *ef*. Whence both angles *dge* and *dgf* are right angles. Again, because line *gh* is equidistant from line *fc*, and line *cb* falls on them, angles *fch* and *ghe* are also right angles. But angle *fch* is a right angle. Whence angle *ghe* is also a right angle. Exterior angle *ghd* is a right angle because it equals its interior and opposite angle (angle *fch*). The cathete therefore is line *gh* falling on line *bc*. Likewise, because line *fik* has been extended through point *f*, it is equidistant from line *cb*. Line *be* {p. 37} is equidistant from line *fc*. Therefore quadrilateral *kbcf* is a parallelogram. Consequently, the opposite sides are mutually equal. Therefore side *fk* equals side *bc*, and side *bk* equals side *cf*. Therefore 13 measures *bk* and 2 measures *ke*. For line *ih* is equal to both *fc* and *kb*. Quadrilaterals *kbhi* and *ihcf* are both parallelograms and 13 measures line *hi*.

---

[43] *Elements*, II.13.

76 Fibonacci's *De Practica Geometrie*

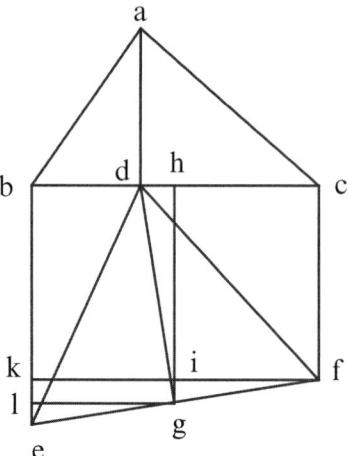

Figure 3.14

[23] Likewise because line *ef* falls on the two equidistant lines *eb* and *gh*, exterior angle *fgi* equals its opposite interior angle *gel*. And angle *fig* equals angle[44] *gle* both of which are right angles. The remaining angle *gfi* equals the other remaining angle *egl*, and line *fg* equals line *ge*. Whence the remaining sides are equals to the other sides that subtend equal angles; namely, side *gi* equals side *el*, and side *fi* is similarly equal to side *gl*. But side *gl* equals side *ik*, because quadrilateral *lkig* is a parallelogram. Therefore side *fi* equals side *ik*. Whence side *ch* equals side *hb*. Therefore the base *bc* is divided into two at point *h*, and *ch* measures 7.

[24] Likewise, quadrilateral *lkig* is a parallelogram, side lk equals side *ig*. But it has been shown that *ig* equals side *le*. Whence *kl* also equals *le*. But the whole line *ke* measures 2. Therefore, 1 measures each of these lines, *kl*, *le*, and *ig*. Whence 14 measures the whole line *hg*. Again, because angle *dgf* is a right angle, the two angles *dgh* and *hgf* equal one right angle. Similarly because triangle *gif* is a right triangle with right angle *gif*, the remaining two angles, *igf* and *gfi*, equal one right angle. Therefore angles *dgh* and *hgf* equal angles *hgf* and *gfi*. Whence, if angle *hgf* is subtracted from both, what remains is angle *dgh* equal to angle *gfi*. So angles *gif* and *ghd* are equal. The remaining angles *igf* and *hdg* are also equal. Therefore triangle *fig* is similar to triangle *ghd*. Whence as line *fi* is to *ig* (7 to 1), so is *gh* (14) to *hd*. Whence by dividing the product of *gi* and *gh* by *fi*, we have *hd*. And this is the second way [of which we spoke]. Namely, we added side *ab* to side *ac*, that is *fc* to *eb*, and we have 28. Half of this is 14 or line *gh*. And, *ig* is at once the excess of *gh* over *fc* or *ab* and the excess of *ac* or *eb* over *gh*. Multiplying *gh* by *ig* we have 14.

---

[44] Text has *ei* (a line); manuscript shows *ei* (a pronoun) (37.8, f. 22r.32).

Dividing this by 7 (equal to *ch* or *fi* that is half base *bc*), we have 2 for the measure of *bd*. After adding this to 7 the same half base, we have 9 for line *cd* that is the longer segment. Or, we subtract *hd* from *hb* (7 less 2) to get 5 for the shorter segment. And this is what we had to show.[45]

[25] Consider again triangle *abg* with side *ab* equal to 13, side *ag* being 15 and the base *bg* 14 and draw cathete *ac* to line *bg*. Because side *ag* is longer than side *ab*, the longer segment is *gc* rather than *cb*. Whence by subtracting *cd* equal to *cb* from *cg*, line *bd* has been divided in two equal parts at point *c* to which line *dg* is added. Whence the product of *dg* by *bg* with the square on line *cd* or *cb* equals the square on line *cg*.[46] Whence the square on line *cg* (the longer segment) exceeds the square on line *bc* (the short segment) {p. 38} as measured by the product of line *dg* by line *bg* [see Figure 3.15]. Now it was shown above in the other figure that the excess of the square on segment *ac* over the square on segment *cb* is as the excess of the square on side *ag* to the square on side *ab*. Whence the excess of the square on side *ag* to the square on side *ab* is as the product of line *dg* by line *bg*. But 225 the square on side *ag* exceeds 169 the square on side *ab* by 56. Whence the product of *dg* by *bg* is 56. Now *bg* is 14 that divided into 56 yields 4 for the measure of line *dg*. Subtracting 4 from 14 (base *bg*) leaves line *bd* equal to 10. Halving this leaves 5 for the smaller segment *bc* that we were to show.

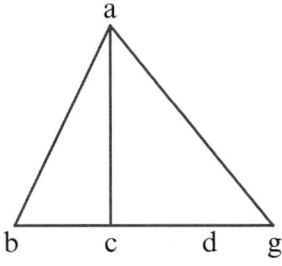

Figure 3.15

Oblique Triangles[47]

[26] If an oblique triangle is also isosceles, the cathete is drawn to the longer side, and you proceed as was described above for the acute isosceles triangle. But if the oblique triangle is scalene, such as triangle *abg* with sides *ab*

---

[45] Boncompagni's text identifies the following as *modus tertius* (37.36, *f.* 22r.23).
[46] See *Elements*, II.5.
[47] "Incipit differentia tertia" (38.10, *f.* 23r.3).

78 Fibonacci's *De Practica Geometrie*

measuring 13 rods, side *bg* 4 rods, and side *ag* 15 rods, and if you wish to construct the cathete from angle *b* to the longer side (*ag*), it falls within the triangle. You can find the segment or cathete and also the area of the triangle by the methods that we explained for the acute scale triangle. But if you want to draw cathetes from angle *a* or *g*, they will fall outside the triangle. The reason for their falling outside the base needs attention.

[27] From 225 the square on the longest side subtract 169 and 16, the squares on the other two sides, *ab* and *bg*, which leaves 40. Half of this is 20 that you divide by 4 to get 5 the length of the segment *bd* where you erect cathete *ad*. And if you divide that same 20 by 13 base *ab* you get $\frac{7}{13}$ 1 rods for segment *be* upon which you erect cathete *ge* [see Figure 3.16]. Then, subtracting 25 the square on *db* from 169 the square on *ab* produces the same result as taking 81 the square on *dg* from the square on *ag*. What remains is 144 for the square on the cathete *ad*. Its root is 12, the length on cathete *ad*. Multiplying this by 2 half its base gives 24 rods for the area of triangle *abg*. For example, let triangle *adg* be a right triangle. Compute its area by multiplying 6 by 9, half the cathete *ad* by the whole base *dg*. Or, you can compute the area by multiplying *ad* by *dc* half the base. Either way, the area of triangle *adg* is 54 rods. From this subtract 30, the area of right triangle *adb*. What remains is 24 rods for the area of triangle *abg*. Compute again by multiplying half the cathete *ad* by the whole base *bg* or the cathete by half the base, as we said before. Similarly, if you multiply the cathete *ge* by half the base *ba*, you have the same area. You can find cathete *ge* if you subtract the square on line *eb* by the square on line *bg* or the square on line *ea* by the square on line *ag*. The cathete measures $\frac{9}{13}$ 3. Take half of this to get $\frac{11}{13}$ 1. Multiplying this by 13 or base *ab*, we find again the area of triangle *abg* to be 24 rods.

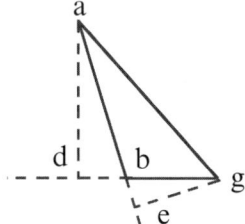

Figure 3.16

[28] It is clearly demonstrated in the Second Book of Euclid, where is shown the above method for finding the segments for the perpendicular from the obtuse angle that falls outside an obtuse triangle.[48] We can indeed find the segments by two other methods {**p. 39**} namely by what we demonstrated

---

[48] *Elements*, II.12

above. The first of these follows: add 13 to 15, that is, side *ab* to side *ag* to get 28. Multiply its half 14 by the difference between it and either side, to get 14. If you divide this by 2, half the base, you obtain 7 for *df*. Take 2 or *bf* from this and 5 remains for segment *bd*. If you add 2 or line *fg*, you have 9 for whole line *dg*. Another way is to subtract the square on line *ab* from the square on line *ag*, 225 less 169. Divide 56 the remainder by 4 base *bg* to get 14. Again subtract the base and 10 remains whose half is 5 for segment *bd*.

[29] This is very clearly shown if we extend line *gd* to point *h*, so that line *dh* equals line *db*. Join *ah*, as seen in another figure [see Figure 3.17]. Now cathete *ad* in triangle *ahg* falls within the triangle. And because line *hd* equals line *db*, and cathete *ad* falls between them, therefore line *ah* measures 13, line *ag* 15, line *bg* 4, and the measure of line *bh* is unknown. Now line *bh* is divided in two parts at point *d* that is on the line extended from *bg*. Line *dg* is the longer segment and line *dh* is the shorter segment in triangle *ahg*. Whence the square on line *dg* exceeds the square on line *dh* by as much as the square on line *ag* exceeds the square on line *ah* or 56. But the square on line *dg* exceeds the square on line *dh* by the product of lines *bg* and *gh*. Whence *bg* times *gh* equals 56. Whence the quotient of 56 and *bg* is 14 for the whole line *hg*. Subtracting 4 line *bg* from the whole line leaves 10 for line *bh*. Its half gives 5 for segment *bd*, as found above. Let us draw lines *gc* and *hi* at right angles on *hg*, so that *gc* equals both lines *ab* and *ah*, line *hi* equals line *ag*, and *c* and *i* are joined to complete the figure [see Figure 3.18]. As we described for the acute triangle, you will find that *kl* is half the sides *ah* and *ag*, and *hl* is half the base *hg*. Whence *dl* is 2 and *lg* remains unknown. Now the difference between line *kl* and either line *cg* or *hi* is *mk* equal to 1. Triangle *dlk* is similar to triangle *kmc*. Whence as *dl* is to *lk* so is *km* to *mc* Whence by dividing the product of *kl* and *km* by 2 or *dl*, 7 is obtained for the measure of line *mc* or *lg*. By adding 2 or *ld* you have 9 for line *gd*. From this subtract *gb* or 4 to get 5 for segment *bd*, as required.

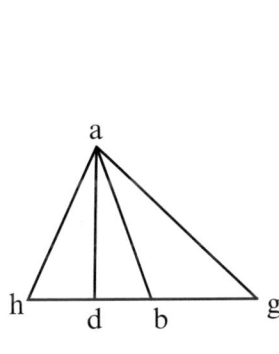

Figure 3.17

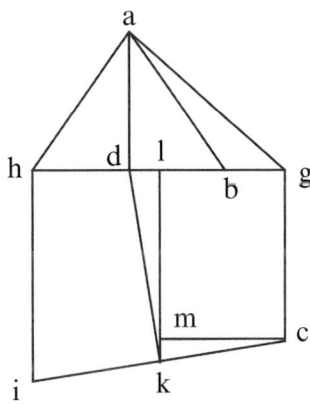

Figure 3.18

80   Fibonacci's *De Practica Geometrie*

[30] Another way: again we use the obtuse triangle *acb* with angle *acb* obtuse, and the measure of *ac* 13, of *ab* 20, and of *bc* 11. Draw the cathete outside the triangle to line *bg*. From points *c* and *b* draw *bd* and *cf*, of which *bd* equals *ac* and *cf* equals *ab*. Set *bd* equal to 13 and *cf* equal to 20. Join the points *d* and *f*, *f* and *g*, and *d* and *g*. Then from point *e*, the midpoint of *df*, draw cathete *eh* to line *gb*. And join *e* and *g*. Through point *d* draw *dk* equidistant from line *bh* [see Figure 3.19]. Because triangles *agc* and *agb* are right triangles, the square on line *ac* equals the [sum of the] two squares on lines *ag* and *gc*, as does the square on *ab* equal the sum of the squares on *ag* and *gb*. Whence the squares on lines *ac* and *gb* equal the squares on lines *ab* and *gc*. That is, the squares on lines *fc* and *gc* equal the squares on lines *db* and *bg*. Whence lines *dg* and *gf* equal each other. And line *ge* is the cathete to *df*. As is shown by the foregoing, *hk* is 13 {**p. 40**} the equal of line *db*, *ke* is the excess of line *he* over *bd*, and triangle *ghe* is similar to triangle *ekd*. Whence as *dk* is to *ek*, so is *eh* to *hg*. If the product of *ek* and *eh* is divided by *dk* or *bh*, that is, half of line *bc*, then line *hg* emerges as $\frac{1}{2}$ 10. Subtract $\frac{1}{2}$ 5 line *ch* from this, and what remains is 5 for segment *cg*, as was necessary to show. Whence cathete *ag* will be 12. If we multiply half of this by the base *cg*, 6 times 11, we obtain an area of 66 rods for the triangle *acb*.

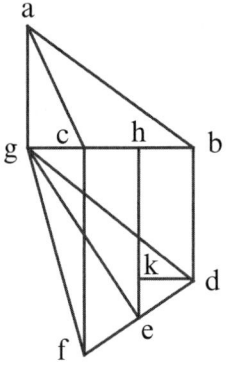

Figure 3.19

Hero's Theorem

[31] In order to present in this book a perfect program for mensuration, we shall instruct on how any triangle can be measured without finding its cathete. Add the sides of the triangle together and take half the sum. Subtract in order the sides of the triangle from the half. Then multiply the remainder from one side by the remainder of another. Then multiply the product by what remains from the third side. Then multiply the final product by half the sum of the

three sides. Find the root of the product for the area of the whole triangle. For instance: The sum of the sides 13, 11, and 20 of the triangle described above is 44, half of which is 22. The difference between this and the longest side is 2 rods, and the next side is 9 rods, and the third side is 11 rods. Now multiply 2, the difference from the first side, by 9, the difference from the next side, and then by 11, the difference from the third side, to reach 198. Multiply this number by 22, half the sum of the sides, to get 4356, the square of the area of the triangle. Its root is 66, as we found above for the area of this triangle.

[32] To prove this: in triangle *abg* bisect the two equal angles, *abg* and *agb* by straight lines *bt* and *tg*. From point *t* draw cathetes *th*, *te*, and *tz*. Join *a* and *t*. Because the lines opposite angles *thg* and *tzg* are equal, angle *thg* equals angle *tzg*. Also angle *tgh* equals angle *tgz*. Whence the remaining angle *gth* equals angle *gtz*. Because triangles *thg* and *tgz* are equiangular, and since they have side *gt* in common, the other sides subtending the equal angles are equal. Also side *th* equals side *tz* and *gh* equals *gz* [see Figure 3.20]. It can be shown in similar fashion that line *hb* equals line *be*, that line *th* equals *te*, and triangle *thb* equals triangle *teb*. And since both lines *te* and *tz* equal line *th* and quantities equal to the same quantity are equal to one another, therefore line *te* equals line *tz*. Now line *ta* is a common line. Therefore the two lines *te* and *ta* are equal to the two lines *tz* and *at*. And angle *aet* equals angle *azt*, and side *at* is held in common. Whence triangles *aet* and *azt* are equiangular and equilateral.[49] Whence side *az* equals side *ae*. And because lines *az* and *ae* are equal, if the common line *eb* is added to them, line *ab* equals the two lines *az* and *eb*, that is, *az* and *bh*. Again, because line *zg* equals line *gh*, then the two lines *ag* and *hb* equal the two lines *ab* and *gh*. Therefore *ag* and *hb* (or *eb*) are half the sides of triangle *abg*. Whence *eb* is the excess of half the sides of triangle *abg* over side *ag*.

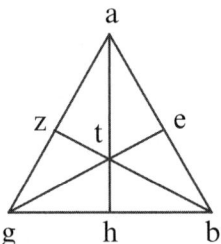

Figure 3.20

[33] It can be shown in similar fashion that line *ae* is that by which half the sides of triangle *abg* exceed side *bg*, and that *hg* or *gz* is the excess over side *ab*. Whence sides *ab* and *hg* are half of the sides of triangle *abg*. Similarly, sides *ag* and *hb* are half of the sides of the same triangle. Therefore, extend lines *ab*

---

[49] By equilateral Fibonacci must mean that their respective sides are equal.

and *ag* directly to points *l* and *m*. Let line *bl* equal line *hg*, and line *gm* equal line *hb*. Therefore each {**p. 41**} of the lines *al* and *am* will be equal to half the sides of triangle *abg*. Then extend line *at* to point *k* and draw lines *lk* and *km*, and let a straight line be under angle *alk*. Whence a straight line is also under angle *amk*. Because the two lines *al* and *ak* equal the lines *ak* and *am*, and angle *lak* equals angle *kam*, and because line *lk* equals line *mk*, all the other angles equal the angles subtended by the equal sides. Angle *akl* equals angle *akm*, and angle *alk* equals angle *amk*. But the straight line under right angle *alk* equals the line under right angle *amk*, as we had said. Now cut from line *bg* line *bn* equal to *bl*[50] [see Figure 3.21]. Draw lines *nk*, *kg*, and *kb*. Because *gh* is the excess of half the sides of triangle *abg* over side *ab*, it equals line *bl*; this is line *bn*. Whence *ng* equals *gm* since it is the excess of half the sides on side *ag*. Because triangles *gmk* and *klb* are right triangles, the square on line *gk* equals the [sum of the] squares on lines *gm* and *mk*; this is *gn* and *mk*. And the square on line *bk* equals the two squares on lines *kl* and *bl*, that is, *kl* and *bn*. But the square on line *lk* equals the square on line *km*. Whence as much as the square on line *kg* exceeds the square on line *kb*, by so much does the square on line *ng* exceed the square on *nb*. Whence the line *kn* is the cathete to line *bg* and it equals line *kl*. And because angles *knb* and *blk* are right angles, the remaining angles *nbl* and *lkn* equal two right angles. But the angles *ebn* and *nbl* likewise equal two right angles. Whence angle *ebn* equals angle *lkn*, and angle *lkb* is half of angle *lkn*. It is therefore equal to angle *ebt* that is half of angle *ebh*. And the angle at *l* equals the one at *e* since both are right angles. There remains angle *etb* equal to angle *kbl*. Therefore triangle *kbl* is similar to triangle *ebt*. Therefore the ratio of *kl* to *lb* is as the ratio of *be* to *et*. Therefore the product of *kl* and *et* equals the product of *lb* and *be*. But the ratio of the square on *et* to the product of *et* and *lk* is as the ratio of *et* to *lk*. And the ratio of *et* to *lk* is as the ratio of *ae* to *al*. Since *et* is equidistant from line *lk*, the ratio therefore of *ae* to *al* is as the ratio of the square *et* to the product of *et* and *lk* or the product of *eb* and *bl*. Therefore the ratio of *ae* to *al* is as the ratio of the square on *et* to the product of *eb* and *bl*. Therefore the product of square *et* and the square on *al* is as the product of *ae* to the product of *eb* and *bl*. And the product of the square on *et* to the square on *al* is as the product of *ae* to the product of *eb* and *bl* multiplied by *al*.[51] But the product of the square on *et* by the square *al* equals the square of the area of triangle *abg*, as we shall demonstrate in what follows. Whence the product of *ae* (the excess of half the sides of triangle *abg* over the side *bg*) multiplied by *eb* (the excess of side *ag* multiplied by *bl* that is the excess of side *ab*) and *al* (as half the sides of triangle *abg*) returns the square of the area of triangle *abg*, as we had to demonstrate.

---

[50] In the figure, the text has *br*; the manuscript has *bk* (41, f. 24v). Regardless, Fibonacci did not locate *n* for the cut.

[51] Text has *producti al*; manuscript has *producti ad al* (41.30, f. 25r.11).

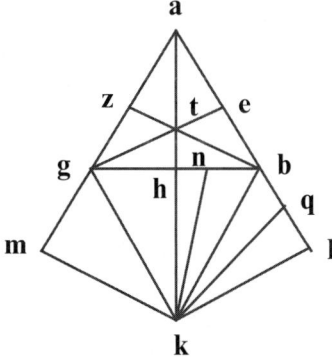

Figure 3.21

[34] But it has to be shown how the product of the square of square on *et* and the square on *al* is as the square of the area of triangle *abg* [see Figure 3.22]. Because triangle *abg* can be dissected at point *t* into three triangles *atb*, *btg*, and *gta*, each of the cathetes *te*, *th*, and *tz* equals any of the others. Therefore the product of *et* by half the base *ab* yields the area of triangle *atb*. Similarly the product of *th* or *te* by half *bg* produces the area of triangle *btg*. And for the same reason, the product of *tz* or *te* by half of *ag* returns the area of triangle *atg*. Whence the product of *et* by *al*, namely by half the sides of triangle *abg*, yields the area of triangle *abg*. Whence the product of the square on *et* to the square on *al* is as the square of the area {**p. 42**} triangle *abg*, as had to be shown.[52]

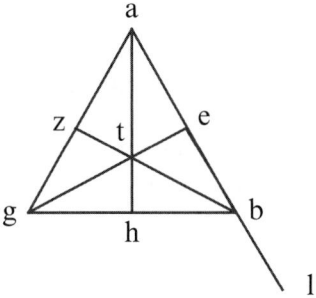

Figure 3.22

[35] If in another triangle the measures of only two sides are given, and you want to know its area and the length of the other side, as in triangle *abg* with sides *ab* and *bg* known, then consider first whether the given sides are the legs of the same angle such as right angle *abg* or of an angle greater or less than a right

---

[52] Clagett (1964), 636, thought that Fibonacci took his proof from the Banū Mūsā (ibid), 278–289. This opinion may need a rethinking. Compare for instance the figures around which the proofs rotate in the two texts.

angle [see Figure 3.23]. First as to the right angle: Because side *ab* is the cathete to side *bg*, the product of *ab* by half of *bg* gives the area of triangle *abg*. And if we sum the squares on the given sides *ab* and *bg*, then we know the square on the line *ag*, the root of which is the measure of the length of side *ag*.

Figure 3.23

[36] But if angle *abg* is less than a right angle, then select on line *ab* some point *d* from which you can draw a cathete *de* and you will measure the sides of triangle *deb*. If the ratio of *be* to line *bg* equals the ratio of *bd* to *ba*, the angle at *g* is a right angle because line *ad* is equidistant from side *ag*. Whence, if the square on the side *bg* is subtracted from the square on the side *ab*, the square on the side *ag* becomes known. Or, because the line *de* is equidistant from line *ag*, it is proportional to the ratio of *bd* to *ba* as *ed* is to *ga*. Whence if we multiply side *ba* by side *ed*, and divide the product by *db*, side *ag* is known [see Figure 3.24]. For example: let side *ab* be 20, *bg* 12, and angle *abg* less than a right angle. Further, let the measure of *bd* be 5 rods, of *de* 4 rods, and of *eb* 3. Therefore as *bd* is to *ba*, so is 5 to 20 and 3 to 12 (this is *be* to *bg*). Whence the line *de* is equidistant from line *ag*. Whence angle *agb* is a right angle since angle *deb* is a right angle. So if the square on *bg* is subtracted from the square on side *ab* (400 less 144), what remains is 256 for the square on the side *ag*. Its root is 16 for the length of side *ag*. Or, if you multiply *ba* by *ed* (20 by 4) and divide by *db* (5), you reach 16 for side *ag*. Its half multiplied by *gb* (8 by 12) yields 96 for the area of the triangle.

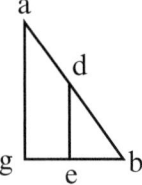

Figure 3.24

[37] And if the ratio of *be* to *bg* were less than the ratio of *bd* to *ba* as can be seen in another triangle, then angle *agb* will be less than a right angle. Whence triangle *abg* is an acute triangle. Draw a cathete from point *a* to line *bg*. Whence in order to find the segments, let *bd* be to *ba* as *be* is to *bf*. Draw the line *af* because that is the cathete to line *bg* [see Figure 3.25].

For example, let ab be 20, bg 17, and bf 12. Since then as bd is to ba, so is be to bf. Whence using what has been said you can find that the measure of cathete af is 16. If we add its square 256 to 25 the square on line fg, we would have 281 for the square on line ag. And if we will multiply half the cathete af by the whole line gb, we will have 136 for the area measure of triangle abg.

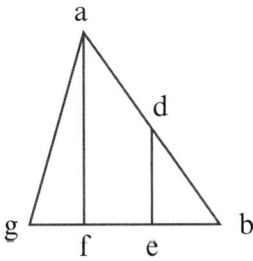

Figure 3.25

[38] If the ratio of be to bg were greater than the ratio of bd to ba, the angle agb would be greater than a right angle. Whence the cathete from point a would fall outside the triangle abg. Whence let line bg be drawn to point h outside triangle abg. And let the ratio of be to bh be as the ratio of bd to ba. Draw the cathete ah to line bh. For example: Set side ab equal to 20, bg to 7, bd to 5, be to 3, and de to 4. Whence, bh will be 12 [see Figure 3.26]. This you will find if you divide the product of be and ba by db since be is to bh as bd is to ba, and again ed is to ah as bd is to ba. Whence dividing the product of de and ba by db, 16 results for the cathete ah. And if bg (7) is subtracted from bh (12), 5 remains for the segment gh. If its square (25) is added to the square on line ah, you have 281 for the square on line ag. Therefore the measure of side ag is the root of 281. Multiplying half of ah by gb yields 56 for the area of triangle {p. 43} abg.

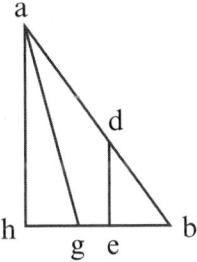

Figure 3.26

[39] If only one side of a triangle is known, you want to use it to find the other sides and its area. For instance in triangle dez let side dz be known. I would

take a part, say az, of the line dz, and I would draw a line ab from point a on line az that is equidistant from line de. I would measure the lines ab and bz, so that the parts of triangle abz are known. Because line ab is equidistant from line de, they are in the same ratio as za to zd and zb to ze and also ab to de. But the ratio of za to zd is known [see Figure 3.27]. Whence sides de and ez are known. For example: let dz be 18, az 6, ab 7, and bz 5. Thus the ratio of zd to za is a triple [ratio]. Because de is thrice ab and ez is similarly thrice bz, therefore side de is 21 and side ez is 15. And since the sides of triangle dez are known, therefore the area of the triangle can be found by any method explained above. So that you may find the area of triangle abz by the method explained above, by taking half a side (9), we take the difference of the sides that sum to 9 (2, 3, and 4), then multiply them together (2 by 3 by 4) and the product by 9 to get 216. The root of this is the area of triangle abz.[53]

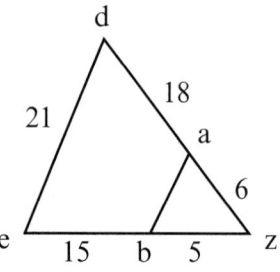

Figure 3.27

[40] Since triangle abz is similar to triangle dez, the ratio of the area of triangle abz to the area of triangle dez is as the square on side za to the square on side zd, as Euclid showed.[54] Because side dz is thrice side az, the square on line dz is nine times the square on line az. Whence the area of triangle dez is nine times the area of triangle abz. Because, if we multiply 216 by 81 the square of nine, we have 17496 for the square of the area of triangle dez. Hence, its root, a bit more than $\frac{1}{4}$ 132, is the required area.

[41] We did say that triangle abz is similar to triangle dez because their respective angles have the same measure. [The reasons are these.] Line ab is equidistant from line de. Whence angle zab equals angle zde, exterior angle to interior angle. For the same reason angle zba equals angle zed. Further, they have angle dze in common. Whence the triangles abz and dez are equiangular. Therefore the triangles are similar.

---

[53] An obvious application of Hero's formula.
[54] *Elements*, VI.19.

The Surveyors' Method[55]

[42] Let the surveyor stand at the larger angle of the triangle and look at the longer side on which the cathete would fall. If he cannot see this perfectly, then he will use a tape and fix one end of it at the vertex of the angle. He will move the tape along the longer side on which the cathete would seem to fall. Thereby he will extend the tape a bit beyond the longer side. And then he will pull the tape about until he can touch both ends of the longer side with his hand. He marks each point of contact to enable him to divide the distance in between in two equal parts. And there is the place for the cathete. Then he can measure the cathete in rods, multiplying it by half the length of the base or half the longer side to produce the area of the triangle. In proof of this method, let side *bc* of triangle *abc* be longer than side *ab* or *ac*. Whence the cathete will be drawn from angle *a* to side *bc*. If sides *ac* and *ab* are equal, then the cathete will be drawn to the middle of side *bc*. If one side is less than the other, the cathete will be drawn toward the shorter side.

[43] In practice, the surveyor stands at point *a* and considers where the cathete must fall on line *bc*. Having decided on this, he puts one end of the tape at point *a* and moves it over line *bc* toward side *ab* that is shorter than side *ac*. [In the figure] let the tape be *ad* {p. 44} that he holds at point *d* somewhat outside the triangle and drags it toward point *c* until point *d* touches line *bc*. Let the point of contact be e[56] so that tape *ad* is on line *ac* [see Figure 3.28]. Then he drags the tape toward *b* until point *d* touches line *bc*, where the partner has chosen; call the point *f*. So now tape *ad* is on line *af*. Divide line *ef* in two equal parts at *h*. Draw a line *ah* from *a* to *h*. I say that *ah* is the cathete to line *bc*. Therefore because line *af* equals line ae[57] and line *fh* equals line *he*, he can measure in rods both cathete *ah* and side *bc*. By multiplying half the cathete by the entire base *bc* or conversely, the area of triangle *abc* is found. Now if triangle *abc* were too large for a tape shorter than the cathete, or the area of the triangle were to be measured in ells, or it were an orchard, or ready for harvest so that these directions cannot be followed to find the cathete, then as many rods as there are in side *ab*, so many feet are there along line *ai*. Similarly, as many rods as there are on side *ac*, so many feet are on side *ak* [see Figure 3.29]. Now draw line *ik* and find the cathete in triangle *aik* with the tape and the foregoing directions. Let it be *al*. Drag the cathete *al* directly to point *m*. Line *am* then is the cathete for triangle *abc*, as is clear from geometry.[58]

---

[55] "Modus uulgaris quo uti debent agrimensores, et est sufficiens in mensuratione omnium trigonorum" (43.27–28, f. 26r.24).
[56] Text has *c*; context requires *e* (44.3).
[57] idem (44.7).
[58] *Elements*, I. Postulate 2.

88   Fibonacci's *De Practica Geometrie*

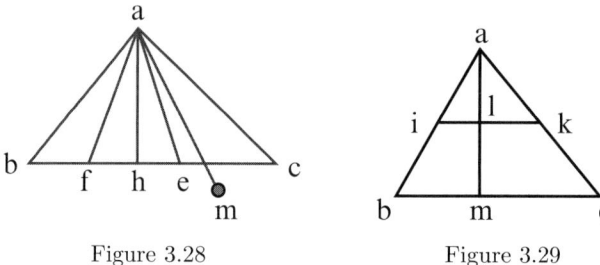

Figure 3.28   Figure 3.29

3.2 Ratios and Properties of Triangles and Their Lines[59]

Lines Within a Single Triangle

[44] When two sides of a triangle are known together with a line drawn through them equidistant from the remaining line, then we know the parts of one line, the sections of another, and the length of the drawn line. Given triangle *abg*: construct line *de* equidistant from the base *bg*, cutting the sides *ab* and *ag*. Let *ad* be 4, *bd* 2, *ag* 15, and *bg* 14. I say that the lengths *ae* and *eg* are known [see Figure 3.30]. Because line *de* is equidistant from line *bg*, the segments of the sides of the triangle are proportional, as Euclid shows in Book Six.[60] For as *ad* is to *db*, so is *ae* to *eg*. Therefore as 4 is to 9, so is *ae* to *eg*. Similarly by conjunction: *ad* is to *ab* as *ae* is to *ag*. So as 4 is to 13, so is *ae* to *ag* (15). Hence, the product of 4 and 15 equals the product of 13 and line *ae*. Therefore, by dividing 60 by 13, we have $\frac{8}{13}$ 4 for line *ae*. Subtracting this from 15 leaves $\frac{5}{13}$ 10 for line *eg*.

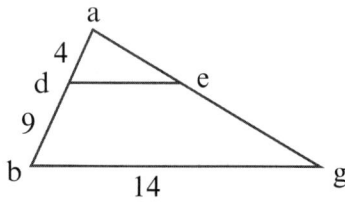

Figure 3.30

[45] Or in another way by alternation: as *ab* (13) is to *db* (9), so is *ag* (15) to *eg*. Whence if we divide 135 the product of 9 and 15 by 13, we get the same $\frac{5}{13}$ 10 for line *eg*. Or, because 4 is to 9 so *ae* is to *eg*. Divide line *ag* in 4 and

---

[59] "De proportionibus et accidentijs, que fiunt in trigonis per protractionem linearum in ipsis" (44.17–18, f. 26v.21–22).

[60] *Elements*, VI.2.

3 Measuring All Kinds of Fields   89

9 (the [proportional] parts of 13). Then line ae is $\frac{4}{13}$ parts of line ag, and line eg is likewise $\frac{9}{13}$ parts of line ag. Whence taking $\frac{4}{13}$ and $\frac{9}{13}$ of 15, we have the aforementioned $\frac{8}{13}$ 4 and $\frac{5}{13}$ 10. But let ad be 4, db 9, ae $\frac{8}{13}$ 4, and eg unknown. So, as 4 is to 9, so is $\frac{8}{13}$ 4 to eg. If you multiply 9 by $\frac{8}{13}$ 4 and divide the product by 4, you will get $\frac{5}{13}$ 10 for line eg. And if we do not know the length of ae only, then divide the product of 4 and $\frac{5}{13}$ 10 by 9, because as bd is to da, so is ge to ea.

[46] We can show in a similar fashion that line de is known. Since it is equidistant from line bg, then triangle ade is similar to triangle abg. Because the sides opposite equal angles are proportional, then as ad is to de, so is ab to bg. So if the product of ad and bg (4 by 14) is divided by ab, $\frac{4}{13}$ 4 is found for the length of line de. And this is what we had to show {p. 45}.

[47] We want to find the length of a line that has been drawn between two sides of a triangle, is not equidistant from the third side, and its end points are known. Referring to the same triangle bag draw the line ez. Let eb be two thirds of line ba and z be the midpoint of bg. Hence, be is $\frac{2}{3}$ 8 and ea is $\frac{1}{3}$ 4. We want to know the length of ez [see Figure 3.31]. Draw the cathete that measures 12 as shown above. The shorter segment ad is 5 and the longer dg is 9. Through point e draw line ei equidistant from the base ag. Through point z draw ztk equidistant from the cathete bd. Because line zk is equidistant from line bd, zg is to gb as gk is to gd and as zk is to bd. But gz is half of gb. Therefore gk is half of gd and zk is half of bd. Therefore gk is $\frac{1}{2}$ 4, kd is the same, and zk is 6. Again, draw ef equidistant from cathete bd. Then as ae[61] is to ab, so is af to ad and ef to bd. Because af is $\frac{2}{3}$ 1, fd is $\frac{1}{3}$ 3, and ef is 4 that is a third part of bd. Because the lines ef and tk are equidistant from the cathete bd and from one another, the lines et and fk are a matched pair. Whence tk equals ef and et equals fk. But fk equals the two lines kd and df ($\frac{1}{2}$ 4 and $\frac{1}{3}$ 3)[62]. Hence the total length of line fk is $\frac{5}{6}$ 7. Therefore te is $\frac{5}{6}$ 7. Similarly, tk is 4 since it equals ef, and tz is 2. And because line zk intercepts the two lines ei and ag, exterior angle zte equals its opposite and interior angle zkf. But since angle zkf is a right angle because of right angle bdk, so also is angle zte. But triangle zte is a right triangle, the two sides of which containing the right angle are known, as we said. Whence the third side ez is also known because its square equals [the sum of] the squares on the two lines zt and te. Now the square of zt is 4 and of te is 61 and a third and a thirty-sixth part of one. Adding[63] these to 4 yields $\frac{13}{36}$ 65 for the square on line ez. Its root is a little more than $\frac{1}{12}$ 8 for the length of line ez. And this we had to show.

---

[61] Text has ac; context requires ae (45.11).
[62] Text has $\frac{1}{3}$ 4; manuscript has $\frac{1}{2}$ 4 (45.15, f. 27r.25).
[63] Text and manuscript have $\frac{13}{49}$ 65; context requires $\frac{13}{36}$ 65 (45.23, f. 27r.34).

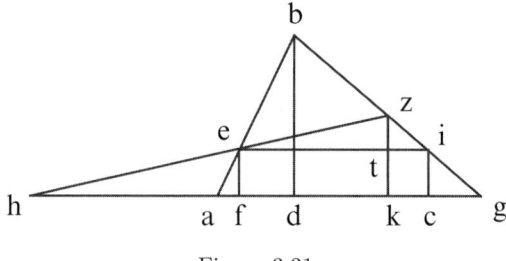

Figure 3.31

[48] We can find the measures of line zt and te differently. Triangle zti is similar to triangle bdg because side ti is equidistant from side dg and side zt is equidistant from side bd. For side zi is similar to side bg and side it is similar to side gd. Consequently the remaining lines zt and bd are similar. Therefore, as zi is to bg, so is it to gd and zt is to bd. Now gi is a third part of gb, and gz is half of gb. Therefore iz is $\frac{1}{2}$ 1, a sixth part of gb. Therefore zt is 2, a sixth of bd, and it is $\frac{1}{2}$ 1, a sixth of gd. Subtracting this from ie that is $\frac{1}{2}$ 9 leaves te equal to $\frac{5}{6}$ 7, as we had said.

[49] Another way: draw ic equidistant from line bd. Then triangle icg is similar to triangles {p. 46} bdg and zti. Therefore as gi is to gb so is gc to gd and ic to bd. Now gi is a third part of gb, so is gc a third of gd and ic a third of bd. Therefore gi is 5, gc is 3, and ic is 4. Because triangle zti is similar to triangle icg, so zi is to ig as ti is to cg. Now zi[64] is half of gi. Whence it is half of gc and zt is half of ic, as required.

Lines Extended Outside a Single Triangle

[50] Suppose we extend lines ze and ga to point h outside the triangle and we want to know the length of lines ah and eh. Then we multiply lines te and ef, divide the product by zt, and we will have the line fh. And we will do this, because triangle zte is similar to triangle efh. For te is equidistant from line gh. And line ze intersects them. Exterior angle zet equals its opposite and interior angle ehf. And angle zte equals angle efh because both are right angles. The remaining angle tze equals the other remaining angle feh. Whence as zt is to te, so is ef to fh. Consequently the product of te and ef divided by zt equals line fh. Or, alternately: as zt is to ef, so te is to fh. For zt is half of ef. Whence, fe is half of gh; that is to say, fh is twice te. But te is $\frac{5}{6}$ 7; hence, fh is $\frac{2}{3}$ 15. From this subtract fa and ah remains equal to 14. Likewise, as zt is to ef, so ze is to eh. Hence it is twice ez. But because ez is irrational, we take the ratio of squares, namely, as the square on line z is to the square on line ef (4 to 16), so is to the square on ze to the square on eh. Whence the square on eh is four times the square on ez. Or, add together the squares on line af and ef together with the square on line eh.

---

[64] Text has gi, manuscript has zi (46.4, f. 27.v12).

[51.1] Another way is to extend line hz outside the triangle until it intersects line bl at point l. Let line bl be equidistant from lines hg. Because line hl intersects the equidistant lines bl and hg, angle elb equals angle zhg. Further, angle lbz equals angle zgh. The other [angles] at z are mutually equal since it is a vertex [see Figure 3.32]. Therefore triangle lbz is similar to triangle hgz. Consequently, as gz is to zb, so is hg to bl. Now gz equals zb and hg equals bl. Again, because triangles leb and eha are similar, then ae is to eb as ah is to bl, that is to hg. Since ae is a third of ab, ae is half of eb, and ha is half of bl (hg), then ag is half of hg. Whence ha equals ag. But ag is 14. In similar fashion ah is 14. So the two lines hg and bt equal 28. Now suppose we want to join lines ze and eh. Because zt is to ef, then ze is to eh. But zt is to ef as half of zt is to half of ef (1 to 2). Add a certain line mno {**p. 47**}. Let mn be 1 and no 2[65] [see Figure 3.33]. And because mn is to no as ze is to eh, and they are proportional conjointly, so mn is to mo and ez is to zh. Whence as the square on line mn is to the square on line mo, so is 1 to 9; thus the square on line ze is to the square on line zh. Whence the square on line zh is nine times the square on line ze. So if we multiply $\frac{13}{36}$ 65 (the square on ze) by 9, we have $\frac{1}{4}$ 588 for the square on line zh.

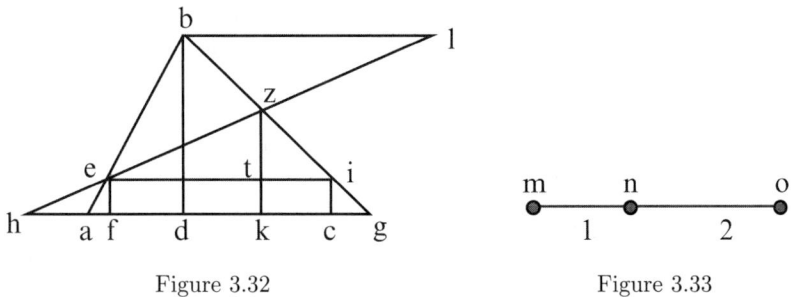

Figure 3.32   Figure 3.33

[51.2] In another way: because triangle zkh has a right angle at k, let the squares on lines zk and kh be added to produce the square on line zh. For kd was found above to be $\frac{1}{2}$ 4, da 5, and ah 14. Whence all of line kh is $\frac{1}{2}$ 23. The square on this is $\frac{1}{4}$ 552. The square on line zk is 36. And thus we have for the square on line zh $\frac{1}{4}$ 588, as we said.

[52] Again, let triangle bdg with side hg 13 as we said, gd 14, and bd 15. Extend gd to point a so that ga is 10. Select point z on gb, so that gz is 5. Query: if az is extended to e, what is the length of segments de and eb[66] [see Figure 3.34]? Draw from point b a line bi equidistant from line gd. Extend line ae to i. Consequently the two triangles bzi and azg are similar. Whence as ag is to gz, so is ib to bz. But ag is twice gz because ib is 16 or twice bz.

---

[65] Text has *sit mni et non sit 2*; manuscript has *sit mn 1 et non sit 2* (47.1, f. 28r.10).
[66] Figure 3.34 in the text does not meet the instructions; hence, a new figure was drawn in its place.

Likewise, because triangle *aed* is similar to triangle *eib*, as *ad* is to *bi*, so is *de* to *eb*. Now *ad* is to *bi* as an eighth part of *ad* is to an eighth part of *bi*. But 3 is an eighth of *ad* and 2 is an eighth of *bi*. Therefore, as 3 is to 2, so is *de* to *eb*. And they are proportional conjointly. Whence as 3 is to 5 so *de* is to *eb*. Whence *de* is $\frac{3}{5}$ of *db* or 9, and *eb* is 6. This had to be shown.

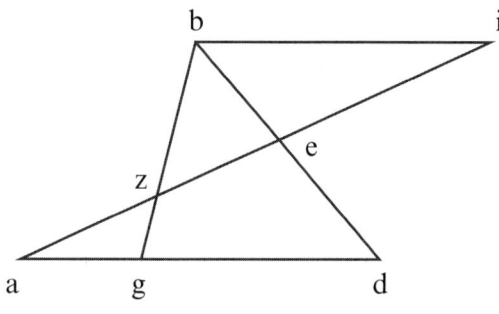

Figure 3.34

[53] Let these segments be known: *bz, zg, be,* and *ed,* as before, and let the triangle be the same. The ratio of *ag* to *bi* is not known. First, because *de* is to *eb* (3 to 2), so is *ad* to *bi*. Therefore *ad* is $\frac{1}{2}$ 1 of *bi*. Again, because *gz* is to *zb* (5 to 8), so is *ag* to *bi*. Consequently, *ag* is $\frac{5}{8}$ of *bi*. All of *ad* was found to be $\frac{12}{8}$ of *bi*, since all of *ad* is $\frac{1}{2}$ of it.[67] Subtracting $\frac{5}{8}$ from $\frac{12}{8}$ leaves *gd* with $\frac{7}{8}$ *bi*. Therefore as 7 is to 8, so is *gd* to *bi*. Whence multiplying 8 by 14 (by *gd*) and dividing the product by 7, or [taking] a seventh of 14 (2), and multiplying by 8, we get 16 for line *bi*. Take $\frac{5}{8}$ of 16 to get 10 for line *ag*.

[54] Likewise, let the same triangle be *abg* and select a point *d* outside it that is not on the extension of line *bg*.[68] Through point *d* draw line *de* {p. 48} equidistant from the base *bg*. Let *de* and *bg* be known as will be *ea*, because all of *ab* is known. Select point *z* on *ab* and connect points *d* and *z*, and extend the line to point *i*. I say that the ratio of *gi* to *ia* is known. Now complete the figure[69] [see Figure 3.35]. Because triangle *dze* is similar to triangle *zat*, the ratio of *ez* to *za* exists and is known. So also known is *de* to *at*;[70] whence *at* is known. Again, because triangles *hzb* and *zat* are similar, so is the ratio *az* to *zb* known. Likewise known is *at* to *bh*. Because *bh* is known, so *gb* is known. Therefore, all of *gh* is known. And because triangles *hig* and *iat* are similar, so also is the ratio of the known *hg* to the known *at*, and *gi* to *ia*. Therefore the ratio of *gi* to *ia* is known. And this we had to prove.

---

[67] Text has *ab* and $\frac{1}{2}$ 61. There is no *ab* nor 61. Context requires these corrections (47.38).
[68] There is no figure available that fits the instructions, nor does one seem possible.
[69] Only from Figure 3.35 do we learn that lines *zd* and *gb* must be extended to meet at *h*, to prepare for triangle *hzb* in what follows.
[70] Text has *ad*; manuscript has *at* (48.5, f. 28v.14).

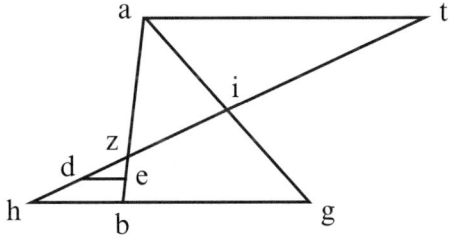

Figure 3.35

[55] Again, let triangle gab be known with cathete gd. Select a given point e on the cathete. Through points a and e construct line aez. I say that the ratio of bz to zg is known [see Figure 3.36]. Draw line gi equidistant from line ab and extend az to i to produce the similar triangles aed and ieg. Because de is to eg, so is ad to gi. Therefore gi is known together with ad. Further, the ratio of gi to ab is as the ratio of bz to zg. Therefore the ratio of bz to zg is known, as we had to prove.

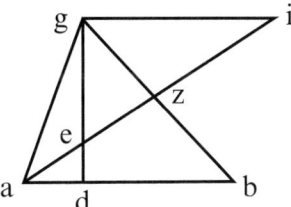

Figure 3.36

[56] In triangle dez let there be a given point a on line dg that is not the cathete [see Figure 3.37]. Through that point line eab passes. The ratio ga to ad is known; likewise ge is known. I say that given the foregoing the ratio of zb to bd is known. Complete the triangle.[71] Because the terms in the ratio ga to ad are known as is the term eg in the ratio of eg to di, therefore di is known. Because ez is to di (a known ratio) as zb to bd, therefore the ratio of zb to bd is known, as required.

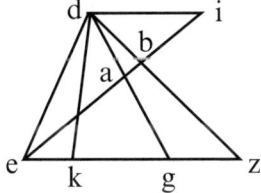

Figure 3.37

---

[71] Leonardo uses this phrase several times as though his reader is expected to know that additional points and lines are necessary. (48.19–20).

[57] And if the ratio of the segments of the line going through point *a* are equidistant from the cathete and given, then also known are its end point *g* on the base. Also known is the other end point *t* on the line *di*, as can be seen in this figure [see Figure 3.38]. I say again, that the ratio of *zb* to *bd* is known. Because triangles *eag* and *ait* are similar, so *ga* is to *at* as *eg* is to *ti*. In triangle *dez* drop the cathete *dk*. Thus *dt* equals *gk* because of quadrilateral *dg* is a parallelogram. Now if a line equal to *gk* (*td*) is added to line *ti*, then all of *di* is known. And because *ez* is to *di*, so is *zb* to *bd*, because the ratio *zb* to *bd* is known, as we had said.

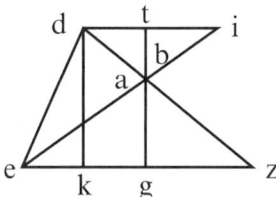

Figure 3.38

[58] And if a point is given within the triangle and between the cathete and an angle from which a line is extended through a point to the side subtending that angle, then similarly the segments of that side are known. Let the point *e* be in triangle *abg* whose cathete *ad* is given. Draw line *bez* from point *b* through point *e* [see Figure 3.39]. I say again that segments *gz* and *za* are known. Draw line *it* through point *a* equidistant from line *bg*. Complete line *bt*. Through point *e* draw line *hi* equidistant from the cathete *ad*. Let the ratio *he* to *ei* be known, as well as the sides of triangle *abg* and the line *bh*. Because the sides of triangle *abg* are known, the segment *bd* will be known. If from this the known *bh* is subtracted, the remaining *hd* is also known. Hence *ia* is known, since it equals *hd* on account of parallelogram *id*. Because triangle *beh* and *eit* are similar, as *he* is to *ei* so is *bh* to *it*. Therefore *it* is known. If from this you remove *ia*, the remaining *at* is known. And because *bg* is to *at*, so is *gz* to *za*. Therefore the sections in the ratio *gz* to *za* are known, as we said.

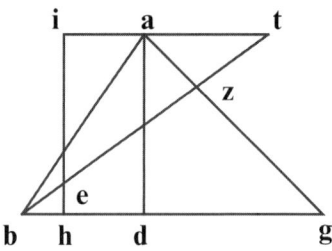

Figure 3.39

[59] Similarly, if through a given point within the triangle a line is drawn from a given point on one side to another side {p. 49}, the ratio of the segments can be found. For example: in triangle bdg let point e be given on the base gd, and within the triangle let point a be given [see Figure 3.40]. Similarly known is line zk equidistant from the cathete bi. Through the points e and a draw line et. I say that the ratio of dt to tb is known. Complete the figure. Because line zk is equidistant from the cathete bi, bk is equidistant from iz. Therefore bk equals line iz and zk equals the cathete bi. Since this is known, also known are zk and za. The remaining ak is known. Because za is to ak, so is ez to kh. Therefore kh is known. To this add kb or zi that was given as known, and bh is known. And because the triangles etd and tbh are similar, as ed is to bh so is dt to tb. And this we had to prove.

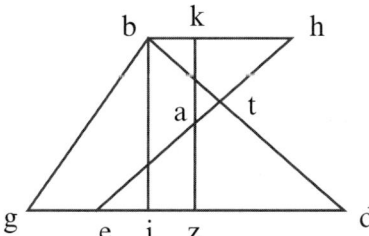

Figure 3.40

[60] Likewise: let the triangle abg have sides ab equal to 13, bg to 14, ag to 15, and cathete ad to 12. Select point e[72] outside the triangle, from which line ez[73] is drawn equidistant from cathete ad. And let ez be 2 and zb be 1; whence zd[74] equals 4. Further, select point i[75] within the triangle so that it is 3 and equidistant from cathete ad. Let tz be 9. Through points e and i draw line eik. I say that the ratio of gk to ka is known [see Figure 3.41]. Draw line al equidistant from the base bg, and extend line ek[76] to point l. Connect points h and m through point i. Let tm equal ez. Connect e and m. Because ze and it are equidistant from cathete ad, so is line tm equidistant from line ze, and they are mutually equal. Line em will be equal to and equidistant from line zt. Therefore em is 9 and equidistant from line al, and th is 12 because it equals ad. Therefore the entire line mh is 14. For mi is 5 and ih is 9. Therefore as mi is to ih so is em to hl. That is, as 5 is to 9, so is 9 to hl. Therefore hl equals $\frac{1}{5}$ 16. Whence the entire line al is $\frac{1}{5}$ 21. Likewise: because ez is equidistant from ti, triangles enz and $in\,t$ are similar. Whence as ti is to ez, so is tn to nz. Whence tn equals $\frac{2}{5}$ 5.

---

[72] Text has d; but the line is drawn from e to z on the base, not from d to z (49.11).
[73] Text has az; manuscript has ez (49.12, f. 29r.31).
[74] Text has ad; manuscript has zd (49.12, f. 29r.31).
[75] Text has 1; manuscript has i (49.13, f. 29r.32).
[76] Text has eh; manuscript has ek (49.15–16, f. 29r.35).

96 Fibonacci's *De Practica Geometrie*

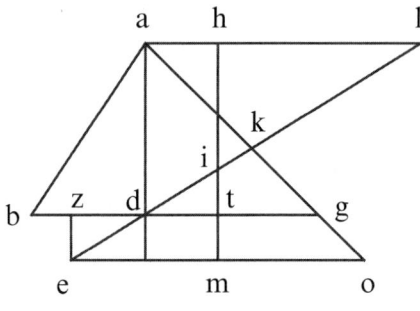

Figure 3.41

[61] In another way: because triangles *nit* and *lih* are similar to one another, as *ti* is to *ih* or as 3 is to 9, so *nt* is to *hl*. Therefore *nt* is $\frac{2}{5}$ 5, a third of *hl*. To this add *tg* or 4 so that *ng* becomes $\frac{2}{5}$ 9. As *ng* is to *al*, so is *gk* to *ka* [see Figure 3.42]. Now *al* is $\frac{1}{5}$ 21 (equal to 106 fifths) and *gn* equals 47 fifths. Therefore as 47 is to 153 the sum of 47 and 106, so is *gk* to *ga* equal to 15. Whence as 47 is to 51 a third of 153, so *gk* is to 5 a third of 15.[77] Whence if we multiply 5 by 47 and divide the product by 51, we get $\frac{1}{3}\frac{13}{17}$ 4 for line *gk*. What remains for *ka* is the remainder from 15, namely $\frac{2}{3}\frac{6}{17}$ 10. In another way: let lines *em* and *ag* intersect at *o*. Then as *eo* is to *al*, so is *ek* to *kl*.[78] By this ratio we shall find *ak* and then we have *kg*, as required.

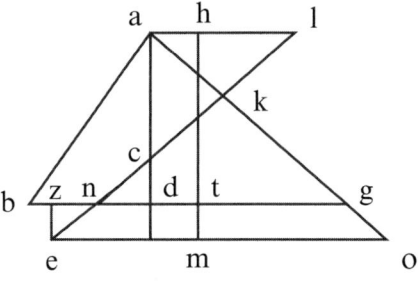

Figure 3.42

[62] Let *go* and *oe* be given and triangle *agd* be similar to triangle: if we want information about point *c* through which line *el* passes, we use cathete *ad*. Then because triangle *ncd* is similar to triangle *lca*, *nd* is to *al* as or as two fifths is to 106 fifths, namely [by conjunction] as 2 to 108, so *dc* is to *da*. Since *dc* is $\frac{2}{9}$, *ca* equals $\frac{7}{9}$ 11. Or,[79] if we extend the cathete *nx* it will be *ex* to *xn* as *nd* is to *dc*. For *ef* equals *dz* and *dn* equals *fx*. Whence *ex* is known.

---

[77] The ratio of proportionality is 1 : 3.
[78] Text has *ka*; context requires *kl* (49.32).
[79] This addition becomes obtuse because the cathete in [62] is *ad*. The figure does not locate *x*.

[63] Again, let *ge* be 2 in place of *ez* in another figure [see Figure 3.43]. The sides of the triangle and its cathete and also the line *ti* remain as we stated above. Let angle *egt* be a right angle because lines *it* and *ad* are equidistant. Draw line *ef* through points *e* and *i*. We want therefore {**p. 50**} to know the lengths of the sections of lines *gb, ad,* and *ab*. Extend line *ef* to point *c*, and line *ti* to point *l*. Join points *c, a,* and *l* so that the line is equidistant from line *bg*. Then as *eg* is to *ti*, so is *gz* to *zt*. Because *eg* and *ti* are to *ti* (that is, as 5 is to 3), so is *gz* to *zt*. Therefore *zt* is $\frac{2}{5}$ 2, since gz[80] is 4. Likewise, as *ti* is to *il*, so *zt* is to *tc*. Because *lc* is $\frac{1}{5}$ 7, subtracting *la* from this leaves *ac* equal to $\frac{1}{5}$ 2. Likewise, as *td* is to itself and *ac*, so is *dh* to *da*. Because *dh* equals $\frac{1}{4}$ 9, *ha* remains as $\frac{3}{4}$ 2. Again, because as *zb* is to itself and *ac*, so *bf* is to *ba*. Therefore *bf* equals $\frac{3}{73}$ 11. The remaining *fa* equals $\frac{70}{73}$ 1.

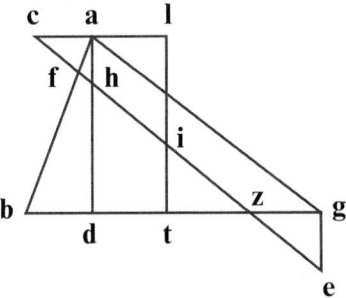

Figure 3.43

[64] In a right triangle of known sides, let the side subtending the right angle be extended outside the triangle for a known distance, and from the end of this line let a known line be produced at the right angle [see Figure 3.44]. For example, let the right triangle be *abc* with known sides and right angle *abc*. Extend side *ac* outside the triangle to point *d*, and let the entire line *ad* be known. From point *d* to point *b* draw line *bd*. I say that line *db* is known. The proof follows. Extend line *ab* indefinitely through point *c*. At point *d* draw line *de* equidistant from line *bc*. Angle *aed* will be a right angle because it is equal to angle *abc*. Since a straight line falls on two equidistant lines, each interior angle equals the opposite exterior angle. Now line *ae* falls on the equidistant lines *bc* and *ed*. Whence angle *aed* equals angle *abc*. For the same reason therefore angle *ade* equals angle *acb*, and they have angle *a* in common. Whence triangles *abc* and *aed* are equiangular and similar to one another.

---

[80] Text and manuscript have *gt*; context requires *gz* (50.4, f. 29v.33).

98   Fibonacci's *De Practica Geometrie*

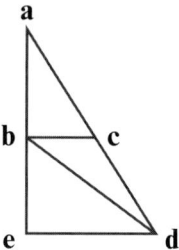

Figure 3.44

[65] Similar triangles have proportional sides about corresponding angles. Whence, *ac* is to *cb* and *ad* is to *de*. Alternately, as *ad* known is to *ac* known, so *ed* is to *bc* known. Whence *ed* will be known. Similarly, as *ad* is to *ac*, so is *ae* to *ab* known. Whence line *ae* is known. If the known line *ab* is removed from it, line *be* is known. If its square is added to the square on line *de*, the square on line *bd* becomes known. and thus it is shown that line *bd* is known, as I said.

[66] The above figure leads to the solution of the following problem proposed to me by a Veronese: "A certain tree is standing near the bank of a certain river. The height of the tree is 40 feet that I represent by the line *bg*. The distance from the foot of the tree to the river is 5 feet, represented by line *bc*. Now there is a certain point *a* on the tree (*bg*) such that *ba* is 10 feet. At point *a* the tree is cut through. The upper part *ag* falls over a distance of 30 feet on a line *ad* touching the point *c*, and line *ad* is 30 feet. He asked, 'What is the length of the line *db* coming from the tip of the tree to the point at its foot?'" Wishing to solve the problem I constructed the figure above and proceeded as follows: the sum of the squares on lines *ba* and *bc* (100 and 25) is 125. This represents the square on line *ac*. Now, because *ad* is to *ac*, so *ed* is to *bc*. Thus the square on line *ad* is to the square on line *ac*, 900 to 125. So the square on line *ed* is to the square on line *bc* or 25. But the ratio of 900 to 125 in reduced form is 36 to 5. Therefore as 36 is to 5 so the square on line *ed* is to 25. Alternately, 36 is to the square on line *ed* as 5 is to 25. But the ratio of 5 to 25 is one fifth. Whence 36 is one fifth of the square on line *ed*. So, I multiplied 36 by 5 and got 180 for the square on line *ed*. Then I subtracted this square from the square on line *ad* {p. 51} or 900. 720 remained for the square on line *ae*. Or in another way: because *ad* is to *ac*, so is *ae* to *ab*. Therefore 36 is to 5 as the square on line *ea* is to the square on line *ba* or 100. Alternately, as 5 is to 100, so is 36 to the square on line *ae*. So, multiply 36 by 20 because 5 is $\frac{1}{20}$ of 100, and you have again 720 for the square on line *ae*. Therefore *ae* is the root of 720. If I took from this line *ab* equal to 10, what remained for me is the root of 720 less 10 for the line *eb*. I squared this and got 820 less the root of 288000 for the square on line *eb*. To this I added the square on line *ed* that is 100 to get 1000 less the root of 288000 for the square on line *bd*. Whence *bd* is 1000 less the root of 288000. And thus

3 Measuring All Kinds of Fields   99

in order to reduce this to a rational number I took the root of 288000 that is $\frac{2}{3}$ 536 less $\frac{1}{96}$. Subtracting this from 1000 leaves $\frac{11}{32}$ 463, the root of which is $\frac{1}{2}$ 21 and one fortieth of one foot for the length of line *bd*.

[67] Not to be overlooked is to show how to find the square on line *eb* called the residue, recisum, or abscissa. It is the difference between two lines commensurable only in their squares, such as between lines *ae* and *ab*. For example, let *ae* be the root of the rational number 720 and *ab* the number 10. Because line *ae* was divided in two parts at point *b*, the squares on lines *ae* and *ab* equal twice the product of *ab* by *ae* and the square on line *eb*, as was shown above. Therefore subtract twice the product of *ab* and *ae* from the squares on lines *ae* and *ab*; that is, subtract 20 times the root of 720 from 820. Now 20 roots of 720 equal the root of 288000, the number arising from the product of 400 the square of 20 and 720. The residue then is 820 less the root of 288000, as I demonstrated above.

Composition of Ratios

[68] Again, given triangle *abg*: from angles *b* and *g*, construct lines *be* and *dg* intersecting each other at point *z*. Given further the ratio of *ge* to *ea* and of *bd* to *da*, then these ratios are known: *bz* to *ze* and *gz* to *zd*. But before we can demonstrate this, we must consider the composition of ratios from two or more ratios.[81] *Composition of ratio results in a composed ratio; that is, the ratio of the product of all the antecedent terms to the product of all the consequent terms* [see Figure 3.45]. For example: given the ratios of 2 to 3 and 4 to 5: by multiplying 2 times 4 and 3 times 5 we compose the ratio of 8 to 15. Likewise from the ratios of 2 to 3 and 4 to 5 and 6 to 7, we compose the ratio of 48 to 105 by multiplying together the antecedents to get 48 and the consequents to obtain 105. Now the ratio of 48 to 105 can be reduced to the ratio of 16 to 35. Understanding this kind of composition means that between 16 and 35 fall three numbers in the afore-stated ratios, namely as 2 is to 3 as 16 is to 24; and 4 is to 5 as 24 is to 30; and 6 is to 7 as 30 is to 35. From these proceeds the composition of the ratio of two quantities or numbers. Now between these quantities some other quantity can be selected.[82] Given the ratio of two numbers, we can discover its composition to be the ratio of the first term to a selected number as the ratio of the selected number is to the second term. An example in numbers may help: since 7 lies between 3 and 10, I say that the ratio of 3 to 10 is composed of the ratio between 3 and 7 and {p. 52} 7 to 10. The ratio of the product of the antecedents to the product of the consequents remains 3 to 10. For if we multiply the antecedents 3 and 7 to get 21 and at

---

[81] *Elements*, V, *passim*.
[82] Fibonacci uses appropriate forms of the verb *eicere* here (51.39) and frequently below, that I translate as *select*.

the same time multiply 7 by 10 the consequents to get 70, the ratio of 21 to 70 is the ratio of the seventh part of 21 to the seventh part of 70. And that is the ratio of 3 to 10, as we said. Similarly, two or more quantities can be selected, and the ratio will be composed of two quantities from those ratios. For instance, if between 3 and 10 you select 5, 6, and 7, then the ratio of 3 to 10 is composed of the ratios of 3 to 5, 5 to 6, 6 to 7, and 7 to 10.

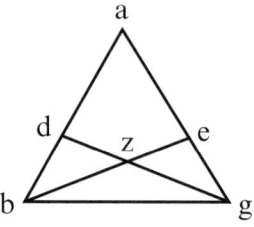

Figure 3.45

Excision of Ratios

[69] With all of this understood, it is important for us to show how some ratio is drawn from a given ratio. For instance, if you wished to remove the ratio of 6 to 5 from a ratio of 7 to 8, multiply the antecedent of the ratio from which the ratio will be taken by the consequent of the ratio that is being removed, namely 7 by 5. This gives 35 for the antecedent of the residual ratio. Next, multiply the antecedent of the ratio that is being removed by the consequent of the ratio that is being taken, namely 6 by 8, for the consequent of the desired ratio. Therefore if you take the ratio of 6 to 5 from the ratio of 7 to 8, what remains is the ratio of 35 to 48. Whence, if we add the ratio of 6 to 5 to the ratio of 35 to 48, we obtain the ratio of 210 to 240. Now the ratio of the thirtieth part of 210 to the thirtieth part of 240 is the ratio of 7 to 8.

Conjunction of Ratios

[70] With this understood and before we return to the proposal, we need to show how the ratio of sides ga and ea in triangle abg is composed of the ratio that the whole line[83] gd has to its part zd and of the ratio that bz has to be. The demonstration is called "by conjunction" [see Figure 3.46]. Draw the line ei from point e equidistant from line gd as shown in the figure.[84] Triangle aie will be similar to

---

[83] Leonardo calls this line *reflexa*, a term used throughout this section wherever the line in question is one of the three lines drawn within the triangle.
[84] The theorem and figure (without line-segment bg) appears in Ptolemy, *Almagest*, I.13, which is considerably easier to follow.

triangle *adg*. Whence, as *ga* is to *ae*, so is *gd* to *ei*.[85] Select line *zd* from between *gd* and *ei*. Thereby, as we said above,[86] the ratio of *gd* to *ei* is composed of the ratios of *gd* to *dz* and of *dz* to *ei*. But the ratio of *zd* to *ei* is as the ratio of *bz* to *be*, because triangle *bzd* is similar to triangle *bei*, and line *zd* is equidistant from line *ei*. Therefore the ratio of *gd* to *ei* is composed of the ratios of *gd* to *dz* and of *bz* to *be*. But the ratio of *gd* to *ei* is as the ratio of *ga* to *ae*. Therefore the ratio of *ga* to *ae* is composed, as we said, from the ratios of *gd* to *dz* and of *bz* to *be*. I repeat that the ratio of *ga* to *ae* is composed of the ratios of line *gd* to line *be* and of the ratio of *bz* to *zd*, because the ratio of *gd* to *ei* equals the ratio of *ga* to *ae*. If the line *be* is selected from within the ratio of *gd* to *ei*, then the ratio of *ga* to *ae* will be composed of the ratios of *gd* to *be* and of *be* to *ei*. But the ratio of *be* to *ei* is as the ratio of *bz* to *zd*. Therefore the ratio of *ga* to *ae* is composed of the ratios of *gd* to *be* and of *bz* to *zd*. And that is what had to be shown.

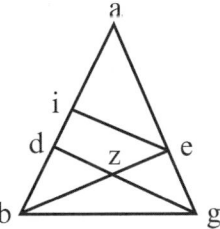

Figure 3.46

[71] For it always happens that when some ratio is composed of two ratios, that the same ratio is composed of the ratio of the antecedents and the permuted consequents. For example: the ratio of 8 to 15 is composed of the ratios of 2 to 3 and of 4 to 5. Similarly, the ratio of 8 to 15 is composed of the permuted ratios of 2 to 5 and of {p. 53} 4 to 3, because the product of 3 by 5 equals the product of 5 by 3. In each composition 3 and 5 are consequents, and 2 and 4 are antecedents. In a similar way it can be shown that the ratio of *ba* to *ad* is composed of the ratio of *be* to *ez* and of *gz* to *gd*. If we draw a line equidistant from line *be* from point *d* within the triangle, then a combination of composite ratios can be created from the ratio of *ga* to *ae*.

[72] Again I say that the ratio of *bd* to *da* is composed of the ratio of *bz* to *ze* and of the ratio of *ge* to *ga*. The proof follows: FROM a certain point *a* draw line *at* equidistant from line *be*, as shown in another figure [see Figure 3.47]. This is called "by disjunction." Extend line *gd* to point *t*. Because lines *at* and *zb* are equidistant from each other and line *ab* intersects both lines, angles *tad* and *dbz* are equal. For the same reason angle *atd* equals angle *dzb*. And the opposite angles at point *d* are equal. Therefore triangles *atd* and *dbz*

---

[85] *Elements*, VI.4.
[86] [69].

are equiangular. Further, the sides about the equal angles are proportional. Whence, as *db* is to *bz*, so is *da* to *at*. Alternately therefore as *bd* is to *da*, so is *bz* to *at*. Therefore let us select the line *ze* from between *zb* and *at*. Then the ratio of *bz* to *at* is formed. Thus the ratio of *bd* to *da* is composed of the ratios of *bz* to *ze* and of *ze* to *at*, that is, of the ratio of *ge* to *ga*. Since the triangles *gez* and *gat* are similar, the ratio of *bd* to *da* is composed of the ratios of *bz* to *ze* and of *ge* to *ga*.

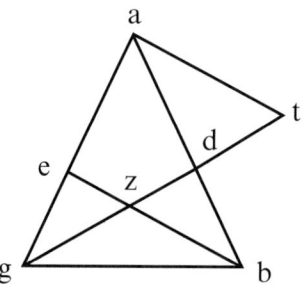

Figure 3.47

[73] Or, the ratio of *bd* to *da* is composed of the ratios of *bz* to *ga* and of *ge* to *ez*, which we now show. Select line *ga* from between lines *bz* and *at*. Thus the ratio of *bz* to *at* will be composed from the ratios of *bz* to *ga* and of *ga* to *at*. But the ratio of *ga* to *at* is as the ratio of *ge* to *ez*. Therefore the ratio of *bz* to *at* (that is, the ratio of *bd* to *da*[87]) is composed of the ratios of *bz* to *ga* and of *ge* to *ez*. This is what we had to show. And thus one combination of the composition of the ratio of *bd* to *da* has been shown. In like manner the ratio of *ge* to *ea* is shown to be composed of the ratios of *gz* to *zd* and of *bd* to *ba*, if we draw a line from point *a* equidistant from line *gd* and connect it with the extension of line *be*.

[74] Now let us return to the proposition. I will show that if ratios of *ge* to *ea* and of *bd* to *da* are known, then the ratios of *bz* to *ze* and of *gz* to *zd* are also known, because the ratio between *ge* and *ea* is known to be composed of the ratios between *gz* and *zd* and between *bd* and *ba*. If the known ratio of *bd* to *ba* is selected from the ratio of *ge* to *ea*, what remains known is the ratio of *gz* to *zd*. Similarly, if the ratio of *ge* to *ga* is selected from the ratio of *bd* to *da*, the ratio of *bz* to *ze* remains known [see Figure 3.48]. Let me express this more clearly. Let line *ag* be 16, line *ab* be 15, *ge* a third of line *ea*, and line *bd* half of *da*. Therefore the ratio of *ge* to *ea* is 1 to 3. Now if the ratio of *bd* to *ba* is withdrawn from the ratio of 1 to 3, what remains is the ratio of 3 to 3 or the ratio of *gz* to *zd*. Whence lines *gz* and *zd* are equal. Thus the ratio of *gz* to *zd* is known. Similarly, if we subtract the ratio of 1 to 4 (*ge* to *ga*) from the ratio of 1 to 2

---

[87] Text has *db*; context requires *da* (53.23).

(bd to ba),[88] what remains is the ratio of 2 to 1 by the ratio of bz to ze. Howsoever the ratios of the sections of the slant lines may be known, we cannot know the lengths of these lines except by knowing (in advance) the root of 97. Let us study {p. 54} side ag to find where the perpendicular would fall upon it from angle b. By what we demonstrated previously[89] for finding such places, we find it to be point e. Whence line be is the cathete to line ag. Therefore triangles beg and bea are right triangles. Whence, if the square on the side eg that measures 16 is selected from the square on the line bg that measures 97, the square on line be remains with a measure of 81. Whence line be is 9. And because bz is to ze as 2 is to 1, if we join the ratio of bz to ze, we will find it to be 3 to 1. Whence be is thrice the length of ze. Consequently, since be is 9, ze is 3 and zb is 6.

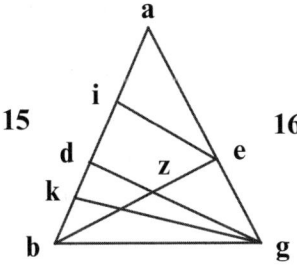

Figure 3.48

[75] Likewise in order to find the length of line gd, we drop the cathete gk from point g onto line ab, and the segment bk is known by the method that we demonstrated for finding such segments.[90] Whence kd becomes known as $\frac{4}{5}$ 2. Consequently if we subtract the square on line bk ($\frac{1}{5}$ 2) from the square on line gb (97), what remains for the square on line gk is $\frac{4}{25}$ 92. Adding to this the square on line dk (8 less $\frac{4}{25}$) produces 100 for the square on line gd. Therefore line gd measures 10. Whence each of the lines gz and zd measures 5 since both had been found equal to one another.

[76] Again reproducing the figure of the triangle above but without its base, I say that the ratio of gd to dz is composed of the two ratios of the lines ga to ae and of be to zb. We prove this now [see Figure 3.49]. From point z I draw line zi equidistant from line ga. Thus triangle dzi is similar to triangle dga. Whence gd is to dz as ga is to zi. By removing ae from between ga and zi, the ratio of ga to zi is composed from the ratios of ga to ae and of ae to iz. But the ratio of ea to zi is as the ratio of eb to bz because triangles bzi and bea are similar. Therefore the ratio of ga to zi (that is the ratio of gd to dz) is composed from the ratios of ga to ae and of eb to bz. And this we had to prove.

---

[88] Text has da; context requires ba (53.40).
[89] See page 86 [69] above.
[90] See page 60 [18] above; viz. $\frac{1}{5}$ 2.

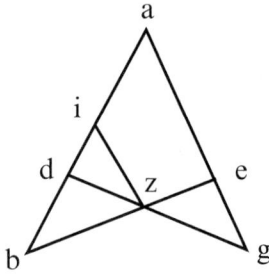

Figure 3.49

[77] From this it is obvious that when the ratio of two quantities is composed from the ratio of the third quantity to the fourth and of the ratio of the fifth to the sixth, then the ratio of the third to the fourth will be composed of the ratios of the first to the second and of the sixth to the fifth. For the ratio of the first quantity ag to the second ae is composed from two ratios, namely from that of the third quantity gd to the fourth zd and of fifth bz to the sixth be. Now we find the ratio of the third quantity gd to fourth zd to be composed of the ratios of the first ga to the second ae and of the sixth be to the fifth bz.

[78] I will show again from the above figure that the ratio of gd to dz is composed of the ratios of ga to zb and of be to ae, as follows. For as gd is to dz so is ga to zi. Attach line zb between ea and zi. Then the ratio of ga to zi is composed from the ratios of ga to zb and of zb to zi. If the ratio of bz to zi is as the ratio of be to ea, then the ratio of ga to zi (namely the ratio of gd to dz) is composed of the ratios of ga to bz and of be to ea, as we predicted. And thus in this figure we have shown another combination of ratios.

Several Combinations of Ratios

[79] There are eighteen combinations of ratios that can be shown in the figure. I have discussed all of these in my book on *The Rule of Barter*.[91] From this, I have selected only the information necessary for finding some quantity given five other quantities, and of these six quantities the ratio of any one to another is composed from two ratios made from the remaining four quantities. I am not concerned about showing the remaining combinations {p. 55}. As we must come to an understanding of unknown quantities by means of the five known quantities, so am I concerned about showing this in an orderly fashion with the above-mentioned numbers. So let the ratio of the first quantity a to the second b be composed from the ratios of the third g to the fourth d and of the fifth quantity e to the sixth z, and let the quantity a be unknown. Now because the ratio of a to b comes from the ratio of the products of the antecedents of the two remaining ratios to the product of the two consequents

---

[91] *Liber abaci*, Ch. 9, p. 119; Sigler (2002), 180. Fibonacci refers to Ptolemy and Ahmed ibn Yussuf.

3 Measuring All Kinds of Fields   105

of the same, so the product formed by multiplying the antecedents by the quantity b is equal to the product of multiplying the consequents by the quantity a. Whence, if we multiply e by g that are the antecedents by the quantity b, we then divide the product by the product of d by z that are consequents. Thus the quantity a becomes known.

[80] For example, let a be 16, b 12, g 10, d 5, e 6, and z 9. Now multiply 6 by 10 (e by g) to get 60. Afterwards[92] multiply the product by 12 to get 720. Likewise if we multiply 9 by 5 and the total by 16, we get 720 [see Figure 3.50]. Therefore the product of e by g multiplied by b equals the product of z and d multiplied by a. If we divide the product of e, g, and b or 720 by 45 the product of z and d, we will obtain 16 as the quantity of a. But if the quantity b is unknown and the other five are known, then divide 720 (9 times 5 times 16) by the product of g and e (60) to get 12 for b. If g[93] is unknown, then divide 720 (the product of a, d, and z) by the product of b and e (72) to get 10 for g. And if only d is unknown, then again divide 720 (the product of e, g, and b) by the product of a and z (144) to obtain 5 for the quantity d. Again, if only the quantity e is unknown, then we divide again 720 (the product of z, d, and a) by the product of b and g (120), to find 6[94] for the quantity e. And finally, if z is unknown while the others are known, then again we divide 720 (the product of e, g, and b) by the product of a and d (80), to realize 9 for the value of z.

prima	tertia	quinta
16	10	6
a	g	e
secta	quarta	sexta
12	5	9
b	d	z

Figure 3.50

[81] Be it noticed that the quantities a, d, and z are called the primary group, and the other three, namely b, g, and e are designated the secondary group. The ratio of any one of the first group to any of the second group, such as the ratio of a from the first group to b in the second group, is composed from two ratios of the remaining four quantities, namely from the ratio of g to d and of e to z, as had been demonstrated. Or, the ratio of a to b is composed from the ratio of g to z and of e to d. Likewise the ratio of a to g[95] is composed from the ratios of b to d and of e to z, or from the ratios of b to z and of e to d. Indeed the ratio a

---

[92] The text reads *quibus ductis in 6* signifying that 10 was multiplied by 6 (55.13); I omitted this as redundant.
[93] The text has 9; manuscript has g (55.20, f. 33r.27).
[94] The text has b; manuscript has 6 (55.26, f. 33r.29).
[95] The text has z; manuscript has g (55.38, f. 33v.7).

to e is composed of the ratios of b to z and of g to d, or from the ratios of b to d and of g to z. In like manner the ratio of the quantities d and z that are from the first group to three quantities remaining from the ratios of the four other quantities; and thus there are nine {p. 56} ratios composed from three quantities of the first group by three quantities of the second group. In the same way another nine compositions of ratios from quantities of the second group to three groups of the first quantities, and any of these composites produces a combination. And thus among them all there are eighteen combinations of ratios for the same six quantities. And however the ratios are changed, they no longer belong to the aforementioned groups. Whence the product of one of whichever group will be 720, as we saw above, arising from the composite antecedent of the ratio to the consequents of the composites. For it is not possible for any term of the six quantities to have a composite ratio to another from its own group [formed] out of the ratio of the remaining four quantities, namely the ratio of quantity a to quantity d or to quantity z cannot be composed from the ratios of the remaining four quantities. Nor can the ratio of quantity d to quantity a[96] or to quantity z be composed. Nor can the ratio of quantity z to quantity d or a be composed from the remaining four quantities.

[82] Similarly, the ratio of quantities b, g, and e are not composed among themselves, since they are in the same group. And thus there are 12 ratios in these six quantities that are not composed from the two ratios of the remaining four quantities. Having explained all of this, we now come to the matter of quadrilateral fields.

3.3 Measurement of Quadrilaterals[97]

[83] The area of quadrilateral fields with right angles was discussed above in another part.[98] If the sides are equal, simply multiply one side by itself; if they are unequal, then multiply the length by the width for the area. The remaining quadrilaterals are divided into four groups: rhombi, rhomboids, trapezoids, and those with unequal, equidistant sides. We shall treat the measurement of each of these in its own place.

Algebraic/Geometric Models[99]

[84] Before we concern ourselves with the dimensions of these quadrilaterals, I want to discuss certain aids for measuring quadrilaterals, that assist our understanding of this craft. They are six rules derived from three kinds of numbers, as follows.

---

[96] The text has z; manuscript has a (56.11, f. 33v.26).
[97] "Incipit pars secunda tertie distinctionis de mensuratione quadrilaterorum" (56.19–20, f. 33v.32).
[98] Chapter 1.
[99] This section parallels the division of numbers and their relationships as found in al-Khwārizmī's *al-jabr* rather than in the corresponding section of Abū Kāmil's *al-jabr*.

Whole numbers and fractions are roots of squares, squares themselves, or simple numbers. Numbers multiplied by themselves are called roots of the square product or *census*.[100] Numbers without any relationship to squares or roots are simply called numbers. Consequently according to this way of thinking, every number is a root, a square, or a simple number. These are the three kinds of number for which there are three simple rules and three composite rules. The simple rules consider numbers as roots equated to squares, or a number to a square, or parts of square to roots or a number, or when a number is equated to roots or squares or conversely, as found in arithmetic or geometric problems. The composite rules consider roots equated to squares and number, or squares to roots and number, or number to roots and squares. For an example of a square equal to roots, consider a square equal to four roots. The root of the square {**p. 57**}, therefore, is 4 and the square is 16. In other words, the side of an equilateral and equiangular square surface is 4 and its area is 16. As many units as there are in each side, so many roots are in its area. This is seen in quadrilateral *abcd*, each of whose sides is 4 rods long[101] [see Figure 3.51]. Whence the area equals four roots of which the first is quadrilateral *ae*, the second *zt*, the third *ik*, and the fourth quadrilateral is *lkcd*. If the side of the square were 5, let it equal roots; then its square would be 25. When four squares equal 24 roots, then one square equals 6 roots, and each root is 6 for a square of 36. When half a square or a *census* equals 4 roots, then a *census* equals 8 roots, and the *census* is 64 with 8 for its root. Likewise if the fifth part of a square equals three roots, the square or area equals 15 roots. So, the area is 225 and the side of the square is 15. Similarly, whenever the number of squares is more or less than unity, it must be brought to unity.

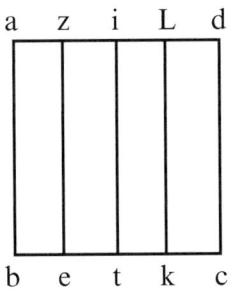

Figure 3.51

[85] In the same way when roots equal squares, the roots must be found that equal one square. Likewise, were a square to equal a number such as 36 rods, then the area would be 36 with side 6. If five squares equal 125, then

---

[100] In Fibonacci's day *census* was a common expression for a square number and was the usual Latin word for the Arabic *mal*. For a thorough discussion of the transition of the word *mal* into *census*, see Oaks and Alkhateeb (2005).

[101] For the remainder of the paragraph, keep in mind that Fibonacci is referring to a square equal to four or five roots.

108   Fibonacci's *De Practica Geometrie*

the square is 25 and its root is 5.[102] If the fourth part of a square equals 16 dragmas, as required by arithmetic, then the *census* or square is 64 with a root of 8. Similarly, all larger or smaller number of squares must be brought to one *census*. It is the same way when the number equals squares. An example of roots equaling a number is, a root equals 4. Therefore the root is 4 and the *census* is 16. If you were to say that six roots equal 30, then one root equals 5. Similarly if you were to say that half a root equals 9, then the root is 18 and the area is 324.

[86] Now for the case where roots equal squares and a number: if you were to say 36 roots equal three *census* and 105 dragmas, this becomes 12 roots equal to one *census* and 35 dragmas. Or, if you were to say 5 roots equal half a *census* and 12 dragmas, this becomes 10 roots equal to one *census* and 24 dragmas. For the case where the squares equal roots and a number: If you were to say three squares equal 12 roots and 36 dragmas, then you have one *census* equal to 4 roots and 12 dragmas. And if you were to say a fourth of a *census* equals two roots and 12 dragmas, then you have one square equal to 8 roots and 48 dragmas. For the case where a number equals squares and roots: if you were to say 78 dragmas equal two squares and 10 roots, this becomes one square and 5 roots equal to 39. If you were to say 32 equals half a square and six roots, this becomes one {p. 58} square and 12 roots equal to 64 dragmas. Thus we must always change the problem to one square. By however so much the large number of squares or parts thereof are changed to one square, by that much are the roots and number changed so that we may evaluate the squares and their roots, as to be seen in what follows.

Squares

[87] If you wish to know the diameter *ag* or *bd* of square *abdg* the sides of which are 10 rods, double its area of 100 to 200, take the root, and you have the length of the diameter. For example: because angle *abg* in triangle *abg* is a right angle, the square on side *ag* opposite the right angle equals the sum of the squares on lines *ab* and *bg* [see Figure 3.52]. But the square on side *bg* is the area of square *abdg*. And the square on side *ab* equals the square on side *bg*. Whence the two squares on lines *ab* and *bg* are twice the square on side *bg*. But the square on the diameter *ag* equals the squares on the lines *ab* and *bg*. Therefore the square on the diameter *ag* is twice the square on side *bg*. But the square on side *bg* is the area of square *abgd*. Therefore the square on side *ag* is twice the area of square *abgd*. And this we had to show. And that is the way it is where the *census* equals a number, such as an area equals 100. Whence, twice the area or the square on the diameter is 200.

---

[102] The text has 25; manuscript has 5 (57.18, *f.* 34v.5).

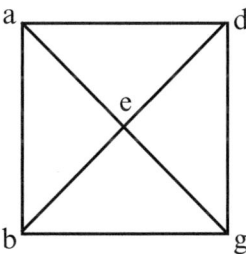

Figure 3.52

[88] Let me say it again: the diameter *ag* equals the diameter *bd* because side *bg* equals side *ad*. If side *ad* is added to both, then the two lines *ab* and *bg* equal the two lines *ba* and *ad*. Now angle *abg* equals angle *bad*. Whence diameter *ag* equals diameter *bd*. I say it again, the diameters *ag* and *bd* bisect one another equally at point *e*. Because lines *ad* and *bg* are equal to each other, and the lines at their endpoints *ab* and *gd* are equal to each other, therefore *ab* and *dg* are equidistant. Because lines *bd* and *ag* fall upon them, so are angles *adb* and *dbg* equal, and angle *dag* equals angle *agb*. And the remaining angles *aed* and *beg* are equal. Therefore triangles *aed* and *beg* are equal as are lines *be* and *ed*. Similarly, lines *ge* and *ae* are equal. Therefore the diameters *ag* and *bd* bisect one another. And this we had to prove. For if the diameter of a given square is the root of 200, and you do not know the area nor its side, then take half of the 200 or 100 and you have the area. And its root 10 is the {p. 59} side. Since two squares equal 200, one square (the area of the square) is 100. And if the square on the diameter with the area of the square would be 300, then three squares equal 300. Whence a third of that or 100 is the area. The 200 that is left is for the square on the diameter. The root of the area is the side of the square.

[89] Given an area and four sides that make 140; and you wish to evaluate[103] the sides from the area. Add the rectangular surface *ae* to a square *ezit*. Let *ai* be extended to line *it* and *be* extended to *ez*. Let each of the lines *be* and *ai* be 4 to represent the number of sides of the square. Whence the surface *ae* equals four sides of the square *et* since the side *ei* is one of the sides of the surface *ae*. And the surface *et* contains the area of the square *zi* and its four sides. Therefore the surface *za* is 140. And that is what we had said, namely a *census* with four roots equals 140. Let the square *et* be the *census* and the surface *ae* be the four roots [see Figure 3.53]. Divide the line *ai* in two equal parts at point *g*. Because line *ti* was added to line *ai*, the surface of rectangle *it* by *at* with the square on line *gi* equals square on line *gt*.[104] But the surface *it*

---

[103] Fibonacci wrote "uis separare" (59.5). Looking ahead, however, we know that he wanted to find the value of each side; hence, my "evaluate".

[104] *Elements*, II.5.

110   Fibonacci's *De Practica Geometrie*

by *at* equals the surface of *zt* by *at*, since *it* equals *tz*. Therefore the surface *zt* by *at* with the square on line *gi* equals the square on line *gt*. But *zt* by *at* is the surface *za* that is 140. To this add 4 the square on line *gi* to get 144 for the square on line *gt*. Whence *gt* is 12, the root of 144, as required. Whence if *gi* (2) is taken from *gt*, what remains is *it* (10), the side of square *et*. Its area is 100. If the four sides are added, then there is the required 140.

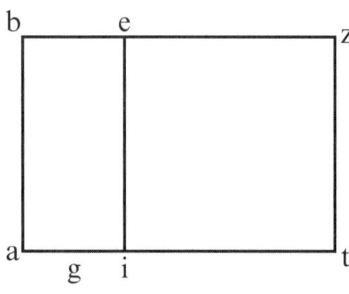

Figure 3.53

[90] Thus in all problems in which a number is equal to one square and roots, add the square of half the roots to the number and find the root of the sum. From this take half the given number of roots. What remains is the root of the *census*. The root being multiplied by itself makes the *census*.[105] For example, let 133 dragmas equal one *census* and twelve roots. So, add 36 the square of half the roots to 133 to get 169. Is root is 13. Take 6 (half the roots) from this to get 7 for the root you wanted of the *census*. And the *census* is 49.

[91] Likewise, let there be a quadrilateral from the area of which the four sides are removed; what remains is 77. Consider quadrilateral *bd*, and let point *a* be on line *gd* with *gb* equal to four rods. Through point *a* draw line *az* everywhere equidistant from lines *bg* and *de*. Because *ga* is 4, the surface of *ba* must necessarily contain the four sides (4 roots) of the quadrilateral *bd*. Whence if from the quadrilateral *bd* the quadrilateral *ba* (the four sides) are removed, what remains is the surface *zd* (77). Because the two surfaces *ba* and *zd* are equal to the quadrilateral *bd*, the *census* equals roots and number [see Figure 3.54]. That is, the square *bd* equals 4 of its roots and 77. Divide *ga* in two equal parts. And because line *ga* is divided in two equal parts at point *i*, and line *ad* lies in the same straight line, the product of *ad* by *gd* with the square on line *ai* is equal to the square on line *di*.[106] But the product of *ad* by *dg* is as the product of *da* by *de*, since *de* is equal to line *dg*. But *da* by *de* is the surface *zd* that is 77. Therefore *da* by *de* makes 77. Add to this the square on

---

[105] This and the next examples are based on *Elements*, II.5 and 6. (59.24).
[106] *Elements*, II.6.

line ai (4) to get 81 for the square on line di, the root of which is 9 equal to line di. If to this is added line ig, then side dg is 11. Whence the area of the square bd is eleven times eleven {**p. 60**} (121) and the four sides (surface ba) is 44. The surface zd remains 77, as required.

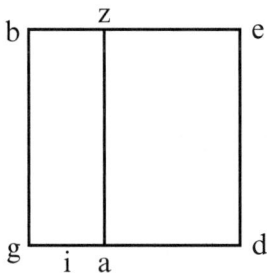

Figure 3.54

[92] That is how it is to be done in all problems in which the square equals roots and number; that is add the number to the square of half the roots from which you take the root of this sum, to which root you add half the roots. And thus you have the root of the square that multiplied by itself makes the desired square. Thus we describe the case where the *census* equals ten roots and 39 dragmas. To this is added the square of five roots (25 added to 39) to make 64. The root is 8 to which is added half the roots (5) to make 13 for the root of the *census* and the *census* is 169. And its ten roots are 130[107] that is produced by multiplying 10 by 13.

[93] Likewise, if you subtract the area of a square from the sum of its four sides or four roots, 3 rods remain. Add to it square ge with each side measuring less than 4 rods. Let there be added line de to line ad so that the whole length ae is 4. Divide line ae in two equal parts at point b. Draw line az equal to and equidistant from line dg. Extend line fg to point z [see Figure 3.55]. Because line ae is 4 and ef is the side of square ge, then the product of fe by ea (the surface ze) is equal to the four roots (the sides of square ge). If you take square ge from these, what remains is surface ga of 3 rods. But square ge and surface ga equal surface ze. Therefore the four roots equals the *census* and the three rods. And so it is necessary for us to find the *census* and its root. Because line ae equal to 4 was divided into two equal parts at point b and in two unequal parts at point d, the product of ed by da with the square on line bd equals the square on line be described as half the line ae.[108] But the product of ed by da produces the surface ga which is 3. Since dg equals de, therefore surface gd by da with the square on line bd equals the square on line be that is 4. Therefore the

---

[107] Text has 150; manscript has 130 (60.9, f. 36r.2).
[108] *Elements* II.5.

112   Fibonacci's *De Practica Geometrie*

square *bd* is 1 with root 1. If it taken from line *be* what remains from 1 is the side of square *ge*, the area of which is the desired *census* or 1. And if half the line *ae*[109] were between *d* and *e* at point *b*, as seen in this other figure [see Figure 3.56], add line *bd* to line *eb* (2) to make *de* equal 3, as we had foretold.

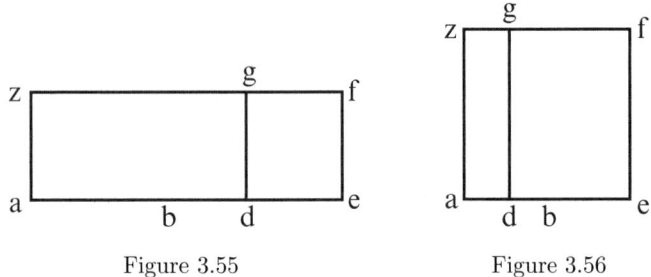

Figure 3.55   Figure 3.56

[94] Such is the procedure for all such problems in which roots equal a square and a number: subtract the number from the square of half the roots; then add and subtract the remainder from the aforementioned half. Thus you have the root of the desired square. For example: 12 roots equal one square and 27 dragmas. Half the roots is 6 the square of which is 36. Subtract 27 from this to get 9 the root of which is 3. If this is subtracted from half the roots, the 3 that remains is the desired root and the *census* is 9. Or, if the 3 is added to the 6 you have 9 as the required root and the square is 81.[110] Thus always, whenever roots equal a *census* and number, there are two solutions. Whence in certain problems sometimes one answer fits, sometimes the other.

Squares

[95] Likewise, the square on the diameter (the area) and the four sides of the given square equal 279 rods. The size on the side on the square is required. Because the square on the diameter is twice the area of the square, therefore the square on the diameter with the square is triple the square. Then three times the square and four roots equal 279. In order to reduce this to one *census*, take a third {p. 61} from all the foregoing, and you will find that the *census* and $\frac{1}{3}$ 1 roots equal 93 rods. Take half the roots or $\frac{2}{3}$ and square them to get $\frac{4}{9}$. Add this to 93 to reach $\frac{4}{9}$ 93. From the root of this subtract $\frac{2}{3}$ or half the roots, and 9 remains for the side of the square. Therefore its area is 82 and the square on the diameter is 162.

---

[109] Text and manuscript have *ag*; context requires *ae* (60.27, *f.* 36r.20).
[110] Note the equivalence of *census* and *quadratum* throughout [94].

[96] Again, four sides of a square equal $\frac{2}{9}$ of the whole square. Consider square *abgd*. Select points *z* and *e* on sides *ad* and *bg*, so that each length *az* and *be* is 4 rods [see Figure 3.57]. Connect *e* and *z*. Parallelograms *ae* and *zg* lie between equidistant lines *ad* and *bg*. Whence as parallelogram *ae* is to parallelogram *zg*, so is base *be* to base *eg*. But parallelogram *ae* equals four roots of square *ag*. Therefore parallelogram *ae* is $\frac{2}{9}$ of square *ag*, and parallelogram *zg* is $\frac{7}{9}$ of[111] square *ag*. Therefore surface *ae* is to surface *ze* as 2 is to 7. Whence as 2 is to 7, so is *be* or 4 to *eg*. So, divide the product of 4 and 7 by 2 to get 14 for line *eg*. Or, in another way, because 4 is twice 2, so 14 is twice 7 or *eg*. Add 4 or *eb* to produce 18 or *bg*, namely the side on square *ag*. This is what we had to show.

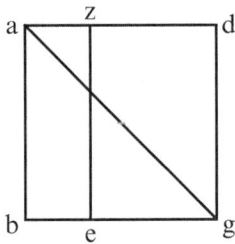

Figure 3.57

Algebraic Methods

[97] In another way and algebraically:[112] because surface *ae* equals four roots or $\frac{2}{9}$ on square *ag*, the four roots equal $\frac{2}{9}$ of the *census*. Whence, in order to make the census[113] one, divide the product of 9 and 4 by 2, or take half of 9 and multiply by 4. For as many two ninths are in nine ninths (of the *census*), so many fours are in the root of the square *ag*. Whence the square *ag* equals 18 roots, as we said, and the area contains 324 rods.

[98] Again, 4 sides and $\frac{3}{8}$ of the area[114] of a square equals $\frac{1}{2}$ 77. Whence change the $\frac{3}{8}$ of a *census* to one *census*; and you will have a *census* and $\frac{2}{3}$10 roots equal to $\frac{2}{3}$ 206. We will find this by multiplying the 4 roots and $\frac{1}{2}$ 77 by 8 and dividing each product by 3. And so half the roots is $\frac{1}{3}$ 5, the square of which is $\frac{4}{9}$ 28. Add this to the constant $\frac{2}{3}$ 206 to get $\frac{1}{9}$ 235 the root of which is $\frac{1}{3}$ 15. From this take half[115] the roots to get 10 for the side of the square, and the area is 100.

---

[111] Text has $\frac{1}{9}$; manuscript has $\frac{7}{9}$ (61.11, f. 36v.13).

[112] Fibonacci's introduction of an algebraic method at this point indicates a clear distinction in his mind between geometric techniques and algebraic methods. The latter is a numerical method, distinct from geometry. A modern reader would see algebra is most of what preceded.

[113] Fibonacci used the verb *reintegrare* for the process of changing the coefficient of the squared term to unity (61.17).

[114] Text has $\frac{3}{3}$; manuscript has $\frac{3}{8}$ (61.21–22. f. 36v.23).

[115] Text has *dietatem*; manuscript has *medietatem* (61.26, f. 36v.28).

[99] If four sides equal the area of the square, then four roots equal a *census*. So one side is 4 and the *census* is 16. And if the sides are twice the area, then four roots equal two squares. Whence one *census* equals two roots.[116] Therefore the side of the square would be 2 and the area 4. Likewise if three sides or 3 roots are taken from the area of the square, what remains is 40. Therefore 3 roots and 40 equal a *census*. So add $\frac{1}{4}$ 2 the square of half the roots to 40 to get $\frac{1}{4}$ 42. Add half the roots to $\frac{1}{2}$ 6 to reach 8, the side of the square.

[100] If you divide the area of a square by its diameter to get 10, what are its diameter and side? Because the quotient of the area by the diameter is 10, therefore the product of the diameter and 10 is the area. Further, the product of the diameter and twice 10 is twice the area. But twice the area equals the square on the diameter. Therefore by multiplying the diameter by 20 you obtain the square on the diameter. So multiply the diameter by itself to obtain that square. Hence the diameter is 20 and its square 400. Now 200 is half of this. The side therefore is the root of 200, which is a bit less than $\frac{1}{7}$ 14.

[101] Likewise if you take four sides from the area, four remains; what therefore is the length of the side? Consider square *abgd* above [see Figure 3.58]. Let each of the segments *az* and *de* be 4. Join *ze* {**p. 62**} as before. And so surface *dz* equals four sides of square *db*, and surface *eb* stays as 4. Therefore the four roots and 4 equals square *db*. So divide line *az* in two equal parts at point *i*.[117] And the product of *zb* and *ab* with the square on line *iz* will be equal to square on line *ib*. But the product of *zb* by *ab* equals the product of *zb* by *ze*. But *zb* by *ze* makes the surface *eb* or 4. Therefore the product of *zb* by *ab* is 4. This added to the square on line *ia* makes 8 for the square on line *ib*. Therefore line *ib* is the root of 8.[118] If we add line *ia* or 2, we have 2 and the root of 8 for the whole line *ab*. And that is the side of square *db*.

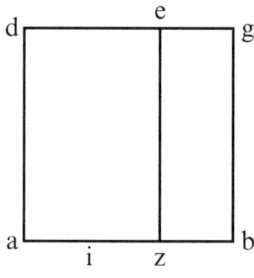

Figure 3.58

[102] If 6 the diameter of a square is added to one of the sides of the same quadrilateral, how long is the side? Double the square of 6 to get 72 and take its root. Add 6 and you have the side. Therefore the side is 6 and the root of 72. For

---

[116] Text omits what manuscript supplies: *duabus censibus. Quare census equatur* (61.29, f. 36v.30).
[117] Text has 1; context requires *i* (62.3).
[118] Text has 5; manuscript has 8 (62.6, f. 37r.19).

example, consider a certain line ab that equals the diameter of a given quadrilateral. Its side is bg, and ga is the 6 by which the diameter exceeds the side [see Figure 3.59]. Construct square ad on line ab, in which eb is the diameter. At point g draw line gz equidistant from lines ae and bd. Through point i draw line tk equidistant from lines ed and ab. Then select point L on line bg so that gL equals ga. Construct the same figure in square gh. Now because quadrilateral ad is a square, the squares about the diameter are tz and gk. Because the side of square tz is line ti equal to ag, tz therefore equals 6, and square tz is 36.[119]

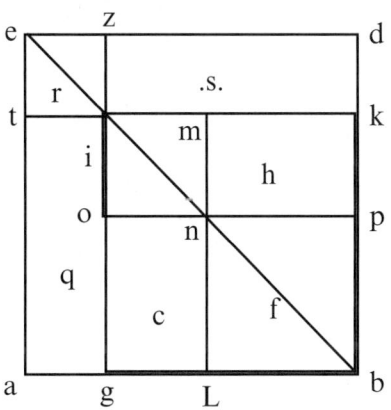

Figure 3.59

[103] Likewise let squares om and lp straddle the diameter of square gk. Square om equals square tz because line on equals line gl, line gl equals line ga, and line ga equals ti. And because line bg is the side of the square with diameter ba, the square described on line ba is twice that which is described on line bg. Therefore square ad is twice square gk. Because gnomon qrs equals square gk, take square tz from gnomon qrs, square om (both being equal) from square gk, and what remains are the supplementary parallelograms ai and id equal to gnomon cfh. But the supplement ai equals the surface il because they have a common side ig and lines lg and ga are equal. Similarly, supplement id equals surface pi. Therefore the two surfaces il and ip equal gnomon cfh. But if we take away from each the surfaces gn and nk, what remains is square lp equal to twice square om. But square om is 36. Whence square lp is 72 whose side bL is the root of 72. If we add 6 or lg to this, we have for the whole side 6 and the root of 72, as required. Now if to this is added ga, then the whole line ab[120] or diameter of the given quadrilateral is 12 and the root of 72.

[104] Or, in another way: line ag is 6 and gi is the side of the square whose diameter equals line ab. Because of this quadrilateral ae equals 6 roots of square gk, and quadrilateral id equals quadrilateral ia. Therefore quadrilateral

---

[119] *Elements*, II.8.

[120] Text and manuscrript have ag; context requires ab (62.35, f. 37v.15).

116   Fibonacci's *De Practica Geometrie*

id is 6 roots of square *gk*, and square *tz* is 36. Hence the whole gnomon *qrs* is 36 and 12 roots of square *gk*. And gnomon *qrs* equals square *gk*, as was shown. Whence if we make line *bg* thing,[121] square *gk* will be a *census*. Hence, a *census* equals 12 roots and 36 dragmas. Then we must proceed according to what was said about a *census* equal to roots and a number **{p. 63}**.

[105] Again, the product of the diameter and a side is 100. Find the length of the diameter and the side of the square. Because the product of the diameter and the side is 100, if we multiply the square on the diameter by the square on the side (the area of the square) we have 10000 the square of 100. Whence if we multiply the square on the diameter by twice the area (the square on the diameter) what we have is its double, namely 20000. Therefore the diameter is the root of the root of 20000. Since the area is half the square on the diameter, it is the root of 5000.

[106] Likewise, I multiply the diameter by the square of the square and obtain 500. How much is the diameter and its side? Because the product of the diameter and the area is 500, then the product of the diameter by twice the area (by the square on the diameter) is 1000. Therefore the diameter is the cube root of 1000. For the cube root of 1000 is 10. Therefore the diameter is 10 and its square is 100. The area is half of this, or 50. And the side is the root of 50.

Rectangles[122]

[107] Consider rectangle *abcd* with shorter sides *ab* and *cd*, each 6 rods in length. The longer sides *ad* and *bc* are 8 rods long. Its area is the product of the sides *ab* and *bc*; hence the area is 48 [see Figure 3.60]. Now if you wish to know its diameter *ac*, add the squares on lines *ab* and *bc* to get 100. Its root is 10 for the diameter *ac*. Likewise: if the diameter is 10 and the side *cb* is 8, what is the side *ab*? From the square on line *ac* subtract the square on line *bc*. What remains is the square on line *ab*. And conversely, as we discussed above on the right triangle.

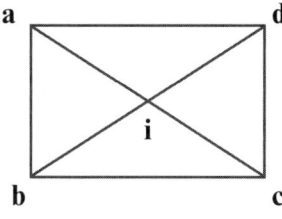

Figure 3.60

[108.1] Similarly, we can find the diameter to be 10 because of the equal triangles *abc* and *bad*. I say that the diameters are bisected at point *e*, for lines *ad* and

---
[121] Fibonacci introduced algebraic terminology in line with his earlier decision to show how to solve the problems algebraically.
[122] "Explicit de quadrilatero; incipit de parte altera longiori" (63.14, f. 38r.2).

bc are equidistant. Whence angle *ade* equals angle *ebc*. Consequently angle *dae* equals angle *ecb*. The remaining angles *aed* and *bec* are equal, for the triangles *aed* and *bec* are equiangular with sides *ad* and *bc* mutually equal. Whence the remaining sides that subtend equal angles are respectively equal: sides *be* and *ed* are equal as are sides *ce* and *ea*. Because the diameter *ac* equals diameter *bd*, they were bisected into equal parts at point *e*. Whence all these lines are equal: *ea, eb, ec*, and *ed*.[123] Each of them measures 5, for each is half the diameter, as had to be shown. Suppose you want to know the lengths of the sides if the area were 48 and sum of a short and a long side were 14. Then subtract the area (48) from the square (49) of half of 14 (7), to get 1 whose root added to 7 yields 8 for the long side. Subtract this from 14 for 6 the short side.

[108.2] For example, consider the rectangle *bgde* with short side *bg* and long side *gd* [see Figure 3.61]. Extend line *bg* to point *a*. And let *ga* equal line *gd*. Divide line *ab* in two equal parts at point *c*. Therefore the whole line *ba* is 14. Whence *bc* or *ac* is 7. And because line *ba* has been divided into two equal and unequal parts at points *g* and *c*, the product of *bg* and *gd* with the square on line *gc* equals the square on line *ca*. But the product of *bg* by *ga* equals the product of *bg* by *gd*. And the product of *bg* by *gd* is the area. Therefore the product of *bg* by *ga* is 48. If to this is added the square on line *ge* {**p. 64**} we have 49. Therefore the square on line *gc* is 1. Since its root 1 is line *gc*, the square on *ac* is 8 for all of *ag* or *gd* the long side. What remains is *gb* or 6, as we said. Again, let the area be 48 and add the measures of the longer and shorter sides. Let half the sum of the two be 1. Add its square to 48 to make 49. The root here is 7. Again, add this to 1, the half of the two lines added together, to make 8 for the long side. Take 2 from this, the amount by which the long side exceeds the short side, to make 6 for the short side. The figure above helps with understanding this.

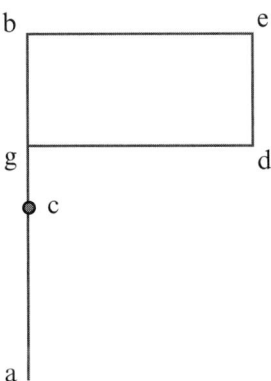

Figure 3.61

[109] Let the long side be *gd* as before, and equal to line *ga*. Subtract 2 for line *gf* from *ga*. Then line *af* equals line *gb*. Divide line *gf* in two equal parts at

---

[123] Text has four *i*; context requires four *e* (63.31, *f*. 38r.12).

point c. Then line ac equals line cb. Therefore line ab is divided in two equal parts at point c and in two unequal parts at point g. So line gc equal to 1 lies between the two sections. Therefore the product of ag and gb with the square on line ac equals the square on line ac. But the product of bg and ga is 48, and the square on cg is 1. Therefore the product of bg and ga with the square on line gc is 49. Its root or 7 is the line ac. If cg or 1 is added to this, the whole line ag is 8 for the long side. Likewise if cg is taken from cb or 7 equal to ca, what remains is gb or 6, as we said.

[110] Or in another way: let the short side be the root; then the long side is the root and 2. And because the product of the short side by the long side is 48, it follows that the product of the root by the root and 2 is likewise 48. Now the product of the root by itself is a *census*, and the product of the root by 2 is 2 roots. Therefore a *census* and two roots equal 48. Proceed as we said above regarding a *census* and roots equal to a number; and you will have the answer.

[111] Likewise, let the diameter be 10 and the area 48. How much is each side? Add twice the area to the square on the diameter to get 196. Its root is 14, the length of both sides. For example: Consider the quadrilateral $abgd$ with diameter ag 10. Extend line ab to point e so that line be equals line bg [see Figure 3.62]. Because line ae is divided wherever in two parts at point b, then the two squares on the parts ab and be with twice the product ab and be are equal to the square on the whole line ae. But the part be equals the side bg. Therefore the squares on lines ab and bg with twice the product ab and bg equal the square on line ae. But the squares on the sides ab and bg equal the square on the diameter ag, and twice the product of ab and bg is twice the area or 96. If this is added to 100, we have 196 for the square on line ae. Therefore the root of ae is 14, as we said. After this, in order to separate ab from bc (or bg), proceed as we said above[124] where we said that the area 48 and the two sides 14 are joined.

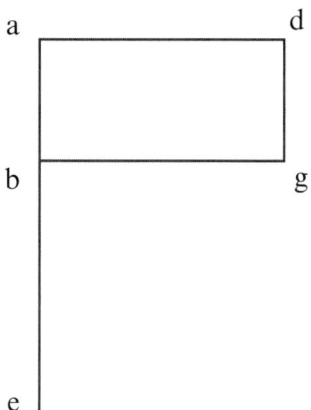

Figure 3.62

---

[124] [108].

[112] Again, let the diameter be 10 and the two sides together be 14. I want to know the area and the length of each side. So, multiply 14 by itself to get 196. Subtract 100 the square on the diameter to leave 96. Half of this is 48 for the area. Afterwards by what was said before you can sort out the sides. These are found in the figure above in this way: the square on the diameter *ag* equals the two squares on the sides *ab* and *bg* that are equal to the squares on *ab* and *be*. But the squares on *ab* and *be* with twice the product of *ab* and *be* equal the square on line *ae*. Therefore the square on the diameter with twice the product of *ab* and *be* equals the square on line *ae*. Whence if you subtract 100 the square on the diameter from 196 the square on line *ae*, what remains {**p. 65**} is 96 for twice *ab* by *be*, that is, twice *ab* by *bg*. Therefore *ab* by *bg* makes 48, half of 96. But *ab* by *bg* is the area. Therefore the area is 48, as we said. Using the aforesaid method, you will find that the short side is 6 and the long side 8.

[113] Again, I have joined the diameter with one side and found the sum to be 16, and the other side is 8. What is the diameter and what is the side joined to it? Multiply 16 by itself to get 256. From this subtract 64 the square on the given side to leave 192. Divide this by twice 16 to reach 6 which is the side joined to the diameter. Subtract this side from 16 and 10 remains for the diameter [see Figure 3.63]. For example: consider the adjacent quadrilateral *bgde*. Let the sum of the diameter *bd* and one side *bg* be 16 rods. Let side *dg* be 8. Make side *bg* a root so that the diameter *db* is 16 less the root. Squaring this produces 256 and one square less 32 roots for the square on the diameter. But the square on diameter *bg* equals the two squares on sides *bg* and *gd*. Therefore the squares on sides *bg* and *gd* with 32 roots equal 256 and the square on side *bg* less 32 roots. Restore the roots. And the squares on lines *bg* and *gd* with 32 roots equal 256 and the square on side *bg*. After subtracting the square on *dg* from both parts, what remains is the square on side gd and 32 roots equal to 256. Since *gd* equals 64, subtract it from both sides to leave 32 roots equal to 192. So we simplify the solution of this problem to one of the six paradigms mentioned above, namely that of roots equal to a number. Whence divide 192 by 32 to get 6 for the root equal to side *bg*. Subtract this from 16 to get 10 for the diameter *bd*, as I said.

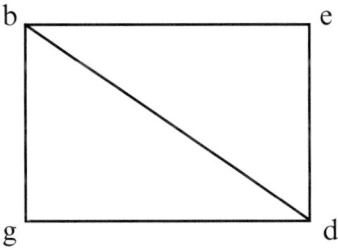

Figure 3.63

[114] If the diameter is 4 longer than the short side and the long side is 8, how long is the diameter? Square 8 to get 64. Add this to the square of four to get 80. Divide 80 by twice 4 to reach 10 for the length of the diameter. Take from this 4 by which the diameter exceeds the short side to get 6 for that side. And so the [solution of this] problem is changed to one of the six paradigms in this way: First it is obvious that the sum of the two squares of the sides is equal to the square on the diameter. Hence, let the diameter be the root, square it, for the square on the diameter. Likewise, let the root of the diameter less 4 be the small side, which squared makes the square on the diameter and 16 less 8 roots of the diameter for the square on the short side. If the square on the long side is added to this, you have the square on the diameter and 80 less 8 roots of the diameter. This equals the square on the diameter. Restore therefore the roots from both side, and the square on the diameter and 80 remains. This equals the square on the diameter and 8 roots. Subtract from both sides the square on the diameter and what remains is 80 equal to 8 roots of the diameter. Divide 80 by 8 to get 10 for the diameter, as I said.

[115] If the product of the long side by the area is 384 and the short side is 6, you want to know the area. Because the product of the short side by the long side produces the area, and the product of the area by the long side produces 384, then the product of the short side by the square on the long side produces the same quantity [384]. Whence if we divide 384 by 6 or the short side, what results is 64 for the square on the long side. Its root 8 is the long side.

[116] And if it were proposed that the area is 48 and the quotient of the long side by the short side is $\frac{1}{3}$ 1, you want to know one of the sides. Let the short side be 3 because of the $\frac{1}{3}$ which is denominated by it. Multiply 3 by $\frac{1}{3}$ 1 to get 4 which you put for the long {p. 66} side. Therefore the ratio of 3 to four is the ratio of the small side to the large side. So the product of 3 and 4 is 12 that divided into 48 yields 4 whose root is 2. This multiplied by each of 3 and 4 yields 6 for the short side and 8 for the long side. Another way: multiply 3 by 48 and divide by 4 to leave 36 whose root is the short side. Likewise, multiply 4 by 48 and divided by 3 to get 64 whose root is the long side. And if the short side is divided by the long side, the result is $\frac{3}{4}$. Put 4 for the long side because it lies in the denominator of the $\frac{3}{4}$. Now take $\frac{3}{4}$ of 4 to get 3 which you put for the short side. Proceed as above and you have 6 for the short side and 8 for the long side.

[117] Likewise, let the short side and the diameter be 16 with the long side 2 more than the short side. You wish to know the measures of the diameter and one of the sides. Because the long side exceeds the short by 2, if we were to add the long side to the diameter, we would have 18. So square 18 and square 16. Add the products to get 580. Subtract from this the square of the aforementioned 2 to obtain 576. The root of this is 24 for the sum of the two sides and the diameter. Subtract the diameter and 16 short side from this, and the 8 that

remains is the long side. Subtract 2 to get 6 for the short side. The long side is 8. Or subtract 18 (the sum of the diameter and the long side) from 24.

[118] If you wish to compute by algebra, make the short side the thing, so the diameter is 16 less the thing. Square the thing to get a *census* which you add to the square on the long side which you find thus: because the long side exceeds the short side by 2, put for the long side thing and 2 dragmas; the square of this is a *census* and 4 roots and 4 dragmas for the square on the long side. Add this to the square on the short side (*census*) to obtain two *census* (two squares of the short side) and 4 roots and 4 that equal the square on the diameter, namely the square of 16 less thing. The square is 256 and a *census* less 32 roots. Restore the roots leaving 256 and a *census* equals to two *census* and 26 roots and 4 dragmas. Subtract from both parts a *census* and 4 dragmas to leave a *census* and 36 roots equal to 252. Halve the roots according to what we said above where the *census* and roots equal number. And if the long side with the diameter is 18 and the long side exceeds the short by 2, then the short side with the diameter would be 16, and you proceed as above.

[119] Again: let two sides with the area be 62, and the long side exceeds the short by 2. What is the measure of each side? To find this, take 2 from 62 leaving 60. Add the 2 to half the sides to get 4. Add this to 60 to obtain 64 whose root is 8. Take this as the long side. If you want the short side, take 2 from 8 to obtain 6 for the short side.[125] For example: let the short side be thing, then the long side is thing and 2 dragmas. Their product is the area. Whence multiply thing (the short side) by thing and two dragmas to obtain a *census* and 2 roots for the area. Add the two sides to this, namely the 2 root and 2 dragmas, to obtain a *census* and 4 roots and 2 dragmas which equals 62 dragmas. Take from both sides 2 dragmas and what remains is a *census* and 4 roots which equals 60. Et cetera.[126]

[120] And if the product of the long side by the diameter is 80 and the short side 6, square 80 and 6 to obtain 6400 and 36. Square half of {p. 67} 36 and add it to 6400 to get 6724 whose root is 82. Add half of 36 to reach 100. Its root is 10 for the diameter. Use this to divide 80 for the long side. Or, take 16[127] from 80 to reach 64 whose root is the long side. If you wish to compute all of this by algebra, then because the product of the long side and the diameter is 80, then the product of their squares is the square of 80 or 6400. But the square on the diameter is the sum of two squares, that of the long side and of the short side. And the square on the short side is 36. Therefore the product of the square and the square on the long side and 36 is 6400. Whence make

---

[125] Fibonacci's text (66.32–39) from my "To find...6 for the short side" is word for word from *Liber mensurationum*, 94 <38>.

[126] Fibonacci seemed confident that his reader would know how to complete the square on this problem.

[127] Text and manuscript have 18; context demands 16 (67.3, f. 40r.33).

the square on the long side thing. Thus the product of thing and itself and 36 makes a *census* and 36 roots.[128] This equals 6400, et cetera. Similarly if you proceed according to this rule, the product of the short side and the diameter is 60, and the long side is 8. Then a *census* and 64 roots equals 3600.

[121] Likewise: the short side with the area makes 54; the long side is 2 more than the short side. How much therefore is each side? Because the product of the short and long sides is the area, if we make the short side thing, then the long side is thing and 2. By multiplying this by thing what results is a *census* and 2 roots for the area. If we add the short side or root to this, then we have a *census* and 3 roots equal to 54. Therefore square $\frac{1}{2}$ 1 or half the roots to obtain $\frac{1}{4}$ 2, and add it to 54 to obtain $\frac{1}{4}$ 56. From its root or $\frac{1}{2}$ 7 subtract $\frac{1}{2}$ 1 to reach 6 for the short side; the long side consequently is 8. If the long side and area were 56, then let the short side be 2 less than the long side. Call the long side thing and the short side thing less 2. Multiply the long side by the short side to obtain a *census* less 2 roots. Add one root to make a *census* less a root equal to 56. Restore therefore the root to leave a *census* equal to one root and 56. Take half of one, and add its square to 56. Add the root of $\frac{1}{4}$ 56 or $\frac{1}{2}$ 7 to half of one[129] to find 8 for the long side.

[122] If the sum of the four sides and the area of a rectangle is 76, and the long side is two more than the short side, then put thing for the short side and thing and 2 for the long side. Multiply thing by thing and two to obtain a *census* and 2 roots equal to the area. Add to this 2 roots for the 2 short sides and two more and 4 (by which the long sides exceed the short sides) for the two long sides. Then you have a *census* and 6 roots and 4 dragmas equal to 76. Take 4 from both sides to leave a *census* and 6 roots equal to 72. Add the square of half the roots to 72 to yield 81. From its root or 9 take half the roots to leave 6 for the short side.

[123] If the short side is subtracted from the area, 42 remains. Let the long side be 2 more than the short side. Make the short side thing that you multiply by thing and 2 or the long side to obtain a *census* and 2 roots equal to the area. Take from this one root or the short side and what remains is a *census* and root equal to 42. Add $\frac{1}{4}$ to 42 the square of half the root to give $\frac{1}{4}$ 42. From its root or $\frac{1}{2}$ 6 subtract $\frac{1}{2}$ to leave 6 that is the short side.

[124] If you subtract the long side from the area and 40 remains and the long side is 2 more than the short side, then it is obvious that if the short side is

---

[128] In other words, $x(x + 36) = x^2 + 36x$.

[129] In this sentence and the previous, where the word "one" appears, the text has "root": "accipe quidem medietatem radicis, et multiplica eam in se, erit $\frac{1}{4}$. Quod adde super 56, erunt $\frac{1}{4}$ 56. Super quorum radicem, scilicet $\frac{1}{2}$ 7, adde medietatem radicis, erunt 8, quod est latus longius" (67.25-27). This is another stage in the development of algebraic terminology where the word "coefficient" would have been most helpful, because it is the number of roots (in this case, one) that is halved and operated upon.

subtracted from {p. 68} the area, there remains two more than when the long side is subtracted from the area. Therefore with the small side removed from the area, 42 remains. So proceed as above and you will have what you want, God willing.

[125] Or by another way: let the long side be thing. Multiply it by the short side or thing less 2, and you will have a *census* less two thing*s* equal to the area. Subtract one of the roots or the long side, and what remains is a *census* less three roots equal to 40. Restore the roots, and you will have a *census* equal to three roots and 40. Et cetera.

[126] Likewise, subtract 4 roots from the area and 20 remains; the long side is 2 more than the short side. How much is each side? Let the short side be thing and the long side thing and two. Whence 4 sides are 4 things and 4 dragmas. The product of thing and things and 2 is a *census* and two things equal to area. From this take the 4 sides or 4 things and 4 dragmas to leave a *census* less two thing*s* and 4 sides equal to 20. Restore the two things (2 things and 4 dragmas) to leave a *census* equal to 2 thing*s* and 24 dragmas. Half the roots will be 1 whose square added to 24 yields 25. Add its root to 1 to make 6 for the short side. Et cetera.

[127] Likewise, given an area of 48: if the long side and the short side are added to the diameter, their sum equals half the area. Find the measures of the diameter and each side. Square 24 half the area to get 576. Subtract twice the area from this to leave 480. Divide its half by 24 or the sum of the sides and the diameter to leave 10 for the diameter.

[128] Or another way: double the area to get 96; then divided it by 24 to have 4. Subtract this from 24 to leave 20. Half of this is the diameter. Subtract this from 24 to get 14 for the sum of the two sides. This leads to the problem: the sum of the two sides is 14 and the area is 48. Proceed as said above and you have the solution. Now if you do not know what to do, let the line *ab* be 24 ells for the sum of the sides and the diameter. Let *ac* equal the long side of the given rectangle and *cd* the short side. What remains is *bd* equal to the diameter. Now construct on line *ab* the square *ae* and draw the diameter *fb*. Through points *c* and *d* draw lines *cpig* and *dkmh* equidistant from each other and lines *af* and *be*. Through points *i* and *k* draw lines *lim* and *nko*. Because *ae* is a square, then the figures about its diameter are squares. Therefore *kdbo*[130] and *knfh* are squares. Again, because {p. 69} quadrilateral *nh* is a square, so are *pimk* and *ilfg* squares. Further, line *db* is the side of square *do*. Whence square *do* equals the square on the diameter. *kp* the side of square *pm* equals line *cd*, and *cd* equals the short side. Therefore square *pm* equals the square on the short side. Similarly it can be shown that square *lg* equals the

---

[130] Text has *c*; context requires *o* (68.34).

124   Fibonacci's *De Practica Geometrie*

square on the long side because of line *fg* and *li* which equal line *ac*[131] [see Figure 3.64]. And *ac* equals the long side of the given triangle. Because side *kp* of square *pm* equals line *pi*, *pi* equals the short side and *il* equals the long side. Therefore the supplement[132] *ni* equals the area of the given quadrilateral, and supplement *ih* is equal to supplement *ni*, as Euclid showed.[133] Therefore supplements *ni* and *ih* are 96, twice the area of the given rectangle. If this is subtracted from 576 the area of *ae*, 480 remains for the area of squares *lg* and *pm* and gnomon *rst*. But square *do* and gnomon *rst* are half the surface *ao*. Therefore the surface *ao* is 240, the product of *ab* by *bo*. Therefore if we divide 240 by *ab* or 24, what results is 10 for line *ob* or line *bd*, which equals the diameter. Therefore the diameter is 10.[134]

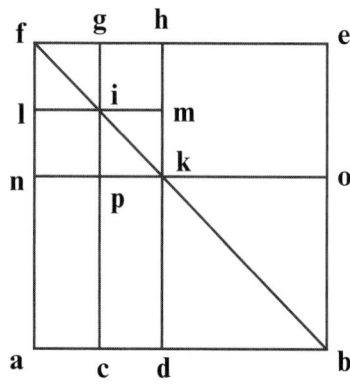

Figure 3.64

[129] Or, according to another rule noted above:[135] if twice the area or the supplements *ni* and *ih* are subtracted from the square on line *ab* (namely 24 times 24), and the area is twice the measure of line *ab*, then twice the area is four times 24. Therefore if four times 24 is subtracted from the square of 24, what remains is 20 times 24 for the area of squares *lg* and *pm* and the gnomon *rst* whose surface *ao* is one half [of *ae*], as was shown. Whence the surface *ao* is 10 times 24. Therefore if *ab* is 24, then it necessarily follows that *bo* is 10, as required.

[130] Likewise, let the sum of the two sides and the diameter be 24 and the long side 2 more than the short side. What therefore is the measure of each side? Double 576 the square of 24 to 1152. Add this to 16 the square of 4,

---

[131] In Figure 3.64 (p. 68), replace *t* with *f* and *r* with *t* as seen on *f.* 41r.
[132] That is, the adjacent square.
[133] *Elements*, I.43.
[134] The diameter of what figure? (69.17, *f.* 41v.24).
[135] [112].

the excess of the long side over the short side, to read 1156. From the root of this or 34 subtract the given 24. What remains is 10 for the diameter. Fourteen, the difference with 24, is the sum of the two sides. Take 2 from 14 and halve the 12 that remains for 6, the measure of the short side. To demonstrate this, consider square *abcd* whose sides and diameter are the given measures [see Figure 3.65]. Let *be* equal the long side and *ef* the short side and the diameter equal to *fc*. Extend diameter *ac*. Through point *f* drawn line *fgh* equidistant from both lines *ba* and *cd*. Through point *g* draw line *igk*. Let line *ig* equal line *bf*. Take from line *gi* line *gl* which equals line *fe*. Therefore line *li* remains equal to line *eb*. Therefore *gl* equals the short side and *li* is the long side. Take from *li* line *im* (the excess of the long side over the short side), and what remains is *ml* equal to *lg* the short side.[136] Extend line *ad* to point *n* and let *dn* equal *fc* the diameter. Construct the square *nopa* on *an*. Because squares *bd* and *pn* have angle *a* in common together with the same diameter, then if diameter *ac* is extended to point *o*, line *ao* is also the diameter. Extend line *dc* to *q* and line *bc* to *r*. Then square *qr* contains in itself the square on the diameter of the given rectangle. Therefore the whole square *pn* equals the square *bd* and the gnomon *stu*.[137] {p. 70} I shall show that the gnomon *stu* is equal to the square *bd* and the square on line *im* by which long side exceeds the short side. For the supplements *pc* and *cn* are equal to the surfaces *bk* and *fd*. But the surfaces *bk* and *fd* equal the gnomon *xyz* and the square *fk*. to which is added the equal of square *qr*. What is equal to square *fk* is twice square *fk*[138] since gnomon *xyz* is equal to square *stu*.

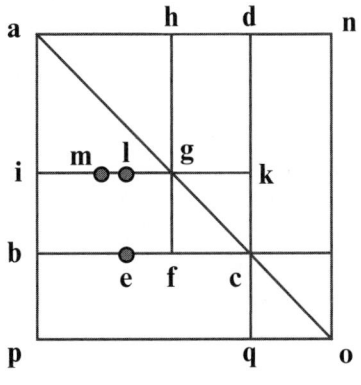

Figure 3.65

[131] It remains to be shown that twice square *fk* equals squares *ih* and *ei* described on line *im*. For line *gm* is divided in two parts by point *l* that adjoins

---

[136] Text and manuscript have *maiori*; context requires *minori* (69.37, f. 42r.12).
[137] Text has *stu*; context requires *xyz* here (69.43) and at (70.1).
[138] Text and manuscript have *fk* in both places (70.5, f. 42r.30).

straight on line *mi*. And so the square that is described by line *gi* (square *ih*) with that described on line *im* is the double of those described on line *gl* and *li*. But the squares described by lines *gl* and *li* equal square *fk*. Whence twice square *fk* equals square *ih* and *ei* described on line *im*. Therefore the gnomon *xyz* equals square *bd* and the square of the excess of the long side, as we said.

[132] Now we come to the cause.[139] We multiplied 24 by 24 above for the square on *bd*. Then we doubled it by adding it to itself. Next we added 4. That is, we added the gnomon *xyz* to the square on *bd*.[140] Thus we had 1156 for the square *pn* one of whose sides is the root. Therefore *po* is 34. Taking *pq* or *bc* from this, what remains for *qo* is 10 the equal of line *cf*. Therefore *cf* is 10, as I had said.

[133] We can find the sides in another way, namely by calling the short side thing. Then the long side is thing and two. Subtracting these from 24 what remains for the diameter is 22 less two thing*s*. Square the thing to get a *census*. Then square thing and two to get a *census*, 4 things, and 4 dragmas. Putting all this together we have two *census*, 4 thing*s* and 4 dragmas. These equal the square on the diameter or the square of the 22 less two thing*s* squared. The product is 4 *census* and 484 less 88 roots. Restore the roots and collect the terms in two *census* and 4 dragmas and you will have 92 roots equal to 2 *census* and 480. Reduce this to one *census* and you have a *census* and 240 that equal 46 roots. Et cetera. Or, call the long side thing. Then the short side is a thing less 2. Subtracting this from 24 leaves 26 less two things. Proceed as before and above, and you will have a *census* and 336 equal to 50 roots.

[134] Let the diameter be 2 greater than the long side that is longer than the short side by the same amount. You wish to know the measure of the diameter and the sides. When the excess of the sides are equal,[141] always multiply the excess by 5 to find the diameter, by 4 for the long side, and multiply by 3 for the short side. For example: let the excess of the sides be 2. So multiply 2 by 5 to get 10 for the diameter, by 4 to get 8 for the long side, and by 3 to obtain 6 for the short side. This happens because of the rectangle whose long side is 4, short side is 3, and diameter is 5. The excess of these sides is 1. Whence as 1 is to the excess in the problem, so are the three sides to the sides of the rectangle whose excess was given. For example: if the excess were 3, then triple the given measures of the sides; namely the diameter is 15, the long side 12, and the short side is 9.

[135] Or if it were proposed that the diameter is 20 and you wish to know the excess of the sides: then divide the 20 by 5 to get 4 that is the excess of the sides. Multiply this {p. 71} by 4 and 3 to obtain 16 and 12 that measure the sides.

---

[139] Figure 3.65 helps here.
[140] Again the gnomon must be *xyz*, because its area can hardly be 4.
[141] Fibonacci is working from an $n-1, n, n+1$ triangular paradigm.

[136] Likewise, let the long side be 20. Divide it by 4 because four is similar to the long sides and 3 to the small side and 5 to the diameter. Dividing 20 by 4 leaves 5 for the excess. Multiply this by 3 and then by 5 to obtain 15 for the short side and 25 for the diameter. Again, the short side is 18. Divide it by 3 to get 6 the difference between the sides. Multiply this by 4 then by 5 to obtain 24 for the long side and 30 for the diameter.

[137] If the excesses were unequal as in a square: let us propose that the diameter is 1 more than the long side which is 7 more than the short side. Then we solve the problem by algebra. Let the short side be thing, the long side thing and 7, and the diameter thing and 8. Multiply thing by thing to obtain a *census*. Square thing and 7 to obtain a *census* and 14 thing*s* and 49 dragmas. Combining all of these, we have two *census* and 14 thing*s* and 49. They equal the square on the diameter or the square of one thing and 8 that is a *census* and 16 thing*s* and 64. Subtract a *census* and 14 thing*s* and 49 from both parts. What remains is a *census* equal to 2 roots and 15. Whence halve [the number of] roots to get 1 whose square is 1. Add this to 15 to get 16. To its root or 4 add half the roots to get 5 for the short side. Add 7 to get 12 the long side. Add 1 for the diameter, 13.

[138] Again: let the long side be 7 more than the short side and the diameter be 13. How much therefore is each side? Subtract the square of the excess from the square on the diameter, that is 49 from 169, to leave 120. Half of this is 60 for the area. For example: let the line ab equal the two sides, with bg the small side and ga the long side. Further let ac be the 7 by which the long side ag exceeds the small side bg. Consequently gc equals gb. On line ab construct square ad as shown in the figure [see Figure 3.66]. The side of square gk is line gb or gi. The side of square hf is line hi or if that is equal to line ag. The supplement ai, therefore, is equal to the given rectangle whose dimensions are ag by gi. Now these are equal to the two sides of the given rectangle, and the supplement id equals the supplement ai. Therefore the supplements ai and id are twice the area of the given rectangle [ai]. Therefore twice the area with the squares hf and gk equals the square ad. But the two hf and gk are equal to the square on the diameter. Therefore the supplements ai and id with the square on the diameter of the given rectangle equal the square ad. But square ad with the square on line ac is described to be twice the squares which are described by lines ag and gb. But those which are described by the squares on lines ag and gb are equal to the square on the diameter. Therefore twice the square on the diameter is equal to square ad and the square on line ac. But one square on the diameter is equal to the squares hf and gk. Therefore the square on the diameter equals the supplements ai and id together with the square on the line ac. Whence if we subtract 49 the square on line ac from the square on the diameter, what remains for us is 120 for the supplements ia and id. Their half is 60 for the area ai of the quadrilateral equal to the given quadrilateral. Therefore the area of the given quadrilateral is 60, as I said. In order to find the sides, let

128   Fibonacci's *De Practica Geometrie*

the area be 60 and the long side 7 more than the short side. Then do as we taught you before. Or, let the short side be thing and the long side be thing and 7. Add together the squares of thing and of thing and 7. The sum is two *census*, 14 roots, and 49 dragmas. Collect terms with the square on the diameter.[142] And you have what you were looking for {**p. 72**}.

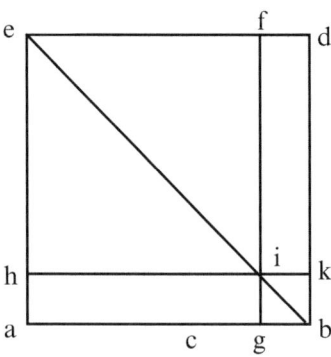

Figure 3.66

Multiple Solutions

[139] Likewise:[143] consider a rectangle whose diameter is 20. Further, the difference between the diameter and the long side is not the same as the difference between the long side and the short side. What are the measures of the sides? First, create a quadrilateral whose sides and diameter are rational, and the differences of diameter and sides are unequal, say with short side equal to 5, long side 12, and diameter 13. Let the 13 be the antecedent in a ratio with the diameter of 20. Whence multiply 20 by 12 and 20 by 5, and divide both products by 13. Thus you have $\frac{6}{13}$ 18 for the long side and $\frac{9}{13}$ 7 for the short side.

[140] If the long side were 20 and you wish to find the short side or diameter, then multiply 20 by 5 and 20 by 13, and divide each product by 12. Thus you have $\frac{1}{3}$ 8 for the short side and $\frac{2}{3}$ 21 for the diameter.[144] Again: let the short side be 20. Then multiply 20 by 12 and 20 by 13. Then divide each product by 5 (or, take a fifth of 20) to get 4. Then multiply by 12 and 13, to find the long side to be 48 and the diameter to be 52.

---

[142] "que oppone quadrato dyametri" is out of place (71.42). The task is to reintegrate the *census*.
[143] A section on indeterminate equations and a set of problems solved by similar quadrilaterals begins here.
[144] Text has $\frac{2}{2}$ 21; manuscript has $\frac{2}{3}$ 21 (72.10, f. 43v.15).

[141] So that we can proceed safely in similar problems, certain things have to be established. We begin by considering how to add a square number to a given number to produce a square number[145] [see Figure 3.67]. Let the number *a* be given. You add a square number so that their sum is a square number; that is, the square number has a root. [This is how it is done.] Take two unequal factors of *a*, and let them be *bg* and *gd*. Divide *bd* in two equal parts at point *e* [see Figure 3.68]. Because the number *bd* has been divided in two equal parts at point *e* and in unequal parts at point *g*, then the product of *dg* by *gb* with the square of the number *ge* equals the square *de*. But the product of *dg* and *gb* equals the number *a*. Therefore the number *a* with the square of the number *ge* equals the square of the number *de*, as required.

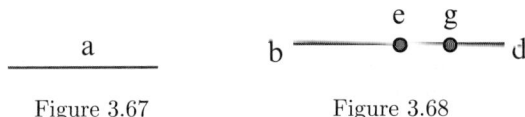

Figure 3.67   Figure 3.68

[142] If however a given side, say 13, of a quadrilateral contains a right angle, and you want to measure the other side and the diameter, then square 13 to get 169 for line *ab*. Then find two numbers whose product is also *ab*. Let them be *ab* and *bg* equal to unity. Now any number multiplied by unity remains the same number. Therefore the product of *ba* and *bg* is *ab* or 169 [see Figure 3.69]. Divide the number *ag* in two equal parts at point *d*. Let *gd* be half of *ag*, namely 85; and this is the diameter. What remains is 84 or *bd* for the other side. Because we can find a limitless set of pairs of unequal numbers with fractions whose products[146] are *ab*, there is a limitless choice of measures for the other side and diameter that measure the given side.

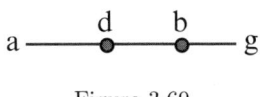

Figure 3.69

[143] Now suppose the diameter is given at 34, what are the sides? Double 34 to get 68, the number for *ef* [see Figure 3.70]. Divide *ef* in two equal parts at point *g*, with *gf* equal to 34. Let *d* be a point on the number *ef* so that the ratio of *fd* to *de* is as a square number to another square number. So let *fd* be 4 which makes *de* be 64. And because *fd* and *de* are in the same ratio as the squares on *fd* and *de*,[147] there is a square number,

---

[145] Compare with Appendix [1].
[146] Fibonacci is thinking geometrically because he uses the word *superficies*.
[147] *Elements*, VI.20.

256, whose root is 16 for one of the sides. The other side dg is 30 and the diameter is 34.

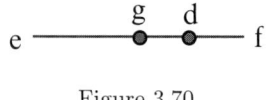

Figure 3.70

[144] Likewise: assume that the difference of the diameter over the long side equals the difference of the long side over the short side and that the product of the difference by the diameter is 20, for the reason given above. Divide 20 by 5 to get 4 whose root 2 is the difference. So multiply the difference by 5 and 4 and 3. Thus you have a diameter of 10 with sides 8 and 6 **{p. 73}**.

[145] Suppose that the ratio of the diameter to the length of a quadrilateral is the same as the ratio of its length to the width, and that the diameter measures 10. In this situation divide 10 in mean and extreme proportion[148] and let the smaller side be the mean term. Multiply this by 10, and the root of the product is the long side. For example: [see Figure 3.71] consider the adjoining right triangle abg which is half of a rectangle of diameter ag, long side ab, and short side bg. Draw cathete bd to side ag. Then as ga is to ab so is ab to bg. And because line bd is the cathete to side ag, triangles bdg and bda are similar to each other and to the whole triangle abg. Whence as ga is to ab so is ba to ad. But as ga is to ab, so is ab to bg. Therefore the ratios of line ab to lines bg and ad are the same. Whence line ad and line bg are equal. Likewise because triangles abg and bdg are similar to each other, so the ratio of ag to bg is as the ratio of bg to gd because bg equals line ad. Therefore as ag is to ad, so is ad to dg. Therefore ag equal to 10 has been divided in mean and extreme proportion. And the mean is the long part ad and the short part is dg.[149]

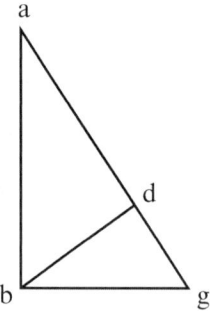

Figure 3.71

---

[148] *Elements*, II.6.
[149] Text has dg; context requires dg (73.16).

[146] If you want to find its measure according to the rule of Euclid,[150] add the square of 5 or half of the 10 to 100, the square on line ag, to make 125. From its root take 5, half of line ag. And thus you have the short side or line gb equal to the root of 125 less 5. In order to find the long side ab: because the ratio of ga to ab is as the ratio of ab to bg, so is the surface bg by ga equal to the square on line ab. Whence multiplying ag by gb (that is, 10 by the root of 125 less 5), what results is the root of 12500 less 50 for side ab. d There are many different problems about sides and diameters and areas of two quadrilaterals whose solutions you can find by what has been said here.

Other Quadrilaterals[151]

[147] The remaining quadrilaterals are divided in four parts. (1) rhombi all of whose sides are equal but lack right angles; (2) rhomboids with opposite sides equal and equidistant with opposite angles equal; (3) fields with two equidistant but unequal sides of which there are four kinds as will be explained below; and (4) fields with diverse sides no two of which are equidistant.

Rhombus[152]

[148] Let rhombus $abcd$ have equal sides of 13 rods. Since we wish to use this measure, it is important that we know one of the diameters. Let the short diameter be bd, 10 rods. Therefore the diameter cuts rhombus $abcd$ into two equal triangles, each being isosceles [see Figure 3.72]. Now the two sides ab and ad are equal to the two sides cb and cd, and they share line bd. Whence the two angles bad and bcd are equal. And thus triangle abd equals triangle cbd. Therefore if we want to know the area of this rhombus, we have only to double the area of either triangle abd or triangle bcd, and thus we have what we wanted. The area of triangle abd is found by multiplying cathete ae by half the base bd as explained {p. 74} in the section on triangles. So if we multiply cathete ae by the whole line bd, we get twice the area of triangle abd or the area of rhombus $abcd$. Cathete ae falls in the middle of line bd because triangle abd is isosceles. To understand this, subtract the square on line eb from the square on line ab (169 less 25) to get 144. Its root 12 is the cathete ae. For the same reason cathete ce is also 12, ce being in a straight line with

---

[150] *$Elements$, II.11.
[151] "Incipit pars secunda tertie differentie" (73.26, f. 44v.6).
[152] "Incipit de rumbo" (73.34, f. 44v.13).

ea since angle *aed* and *dec* are right angles. Whence the diameter *ac* is 24. Therefore the area of rhombus *abcd* is found by multiply half the diameter *ac* by the whole diameter *bd* to get 120.

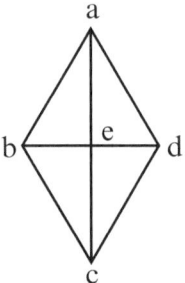

Figure 3.72

[149] In similar fashion: if the measure of diameter *ac* is given as 24 rods, we can use the foregoing method to find the diameter *bd*. Because triangles *bac* and *dac* are isosceles, they are equal. Whence if we subtract the square on side *ae* from the square on *ba* (169 less 144), what remains is 25 for the square on cathete *be*. Therefore *be* is 5. Whence the entire diameter measures 10. By multiplying cathete *be* by half the base *ac*, we get the area of triangle *abc*. By multiplying cathete *be* or half the diameter *bd* by the whole diameter *ac*, what remains is the area of entire rhombus *abcd*. That is, the product of *be* and *ac* (5 and 24) is 120, the area of rhombus *abcd*. Therefore the area of any rhombus can be found by multiplying one diameter by half the other. This is a universal rule. Suppose that the major diameter were 24 and the minor 10, and you want to know the measure of the sides of the rhombus. Square each diameter (*ac* and *eb*) and add the products. The root of the sum or 13 is the same as one of the sides.

[150] Now, we can propose many problems about diameters, areas, and sides of rhombi. All of them can be reduced to problems of rectangles whose long side is half the length of the major diameter of the rhombus and the short side is half its minor diameter. We now prove that the quadrilateral contains half the area of the rhombus. As seen in the figure, draw *af* equal to and equidistant from line *eb* [Figure 3.73]. Join points *f* and *b*. I say that the quadrilateral *ef* is half of rhombus *abcd* and that its sides are equal to half the diameters *ac* and *bd*. Now *ae* is half of *ac* and *be* is half of *bd* because triangle *abd* is half of rhombus *abcd*. But triangle *abd* is equal to quadrilateral *ef*, for both areas are found by multiplying *ae* by *eb*. Therefore quadrilateral *ef* is half of rhombus *abcd*, as I said. Of the many problems regarding the relationship of rhombi to quadrilaterals, I wish to propose these.

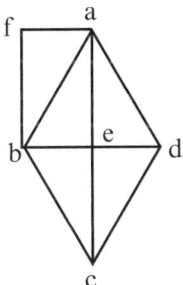

Figure 3.73

[151] Suppose the sum of two diameters of a rhombus were 34, and its area 120. What would be the measure of the diameter? Since the sum of the diameters is 34, their half or *ae* and *eb* is 17, and the area of quadrilateral *ef* is 60. Therefore I have reduced this problem to a quadrilateral problem. For example: the area of a rhombus is 60 and the sum of its sides is 17. Subtracting 60 from $\frac{1}{4}$ 72, the square of half of 17, leaves $\frac{1}{4}$ 12. Subtracting its root from half of 17 leaves 5 for the measure of line *be*. What remains from the 17 is 12 for line *ae*. Therefore twice these numbers are the diameters, 24 and 10.

[152] Likewise: let the sum of the diameters be 34[153] and the long side 14 more than the short side. What is the area? Take 14 from[154] 34 to leave 20. Its half is 10 for the minor diameter. The remainder of 24 measures the major diameter. Finally, multiply half of one diameter by the other, and you have the area {p. 75}.

[153] Again, I added the measures of the two diameters with the area of a rhombus and got 154; the major diameter is 14 more than the minor. Because the two diameters of the rhombus are equal to the four sides of rectangle *ef*, let the short side be thing and the long side thing and 7. Multiply thing by thing and 7 to reach a *census* and 7 roots which is the area of quadrilateral *ef*. Because quadrilateral *ef* is half the rhombus, double the *census* and 7 roots to two *census* and 14 roots to equal the area of the rhombus. To this add 4 thing*s* and 14 (the four sides) to produce two *census* and 18 roots and 14 to equal 154. Take 14 from both parts and reduce it to one *census* to obtain a *census* and 9 roots equal to 70. Square half the roots or $\frac{1}{2}$ 4 to get $\frac{1}{4}$ 20. Add this to 70 to reach $\frac{1}{4}$ 90. Subtract $\frac{1}{2}$ 4 from its root to leave 5 for *be*. Double this to 10 for diameter *bd*. Add 14 to get 24 for the longer diameter. Subtract it from 154 to obtain 120 for the area.

---

[153] Text has 24; manuscript has 34 (74.39, *f.* 45r.28).
[154] Text has *et*; manuscript has *ex* (74.40, *f.*, 45r.29).

[154] Again: I added the short diameter and a side of the rhombus and got 23. The major diameter is 14 more than the short diameter. What is the side and each diameter? Because the major diameter is 14 more than the minor, half the long diameter adds 7 to half the short diameter, or side *ae* is 7 more than side *eb* [see Figure 3.74]. Therefore *be* with *ae* adds 7 to diameter *bd*. But *bd* with side *ab* is 23. Therefore *be* and *ea* with *ab* are 30.

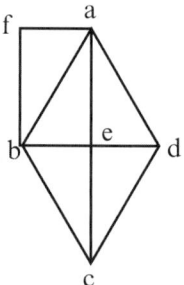

Figure 3.74

[155] I added two sides of the quadrilateral with its diameter and got 30, and the long side is 7 more than the short side. Do as instructed above. Et cetera.

[156] Likewise: square each diameter, add them, and you have 676, and the area is 120. How much is each diameter? Take a fourth of 676 because the squares on lines *ae* and *eb* are a fourth of the squares on the diameters. Since their sides are half of them, you have 169 whose root is 13 for *ef* the diameter of the quadrilateral and side of the rhombus. Because it has been shown above that the square on the diameter of the rectangle adds the square of the excess of the sides to double the square of the area, consequently if we subtract the area of rhombus *abcd* (which is twice the area of quadrilateral *ef*) from the square on the diameter itself (take 120 from 169), what remains is 49. Its root 7 is the difference between the sides. Therefore the area of the quadrilateral is 60, and the long side is 7 more than the short side. How large are the sides? Et cetera.

[157] Likewise, I multiplied the major diameter of the rhombus by the minor and got 240. The major diameter is 14 more than the minor. What is the measure of each diameter? By multiplying half of one diameter by all the other diameter you have the area of the rhombus. Therefore 240 or the product of one diameter by the other is twice the area of the rhombus. Therefore the area of the rhombus is 120 which is twice the area of the quadrilateral *ef*. Therefore the area of quadrilateral *ef* is 60.

[158] The foregoing suggested this problem: there is a rectangle whose area is 60 and whose long side is 7 more than the short side or half of 14. The major diameter exceeds the minor diameter by the same amount. Et cetera.

[159] Again: I added the two diameters to get 34, and their product is 240. How much is each diameter. To find them, consider the line ab of measure 34 rods [see Figure 3.75]. Divide it in equal and unequal parts at points d and g. Let ad equal the shorter diameter. The remaining db is the major diameter. Now the product of ad and db with the square on line dg equals the square on line gb; that is, 17 squared is 289. If from this we subtract {p. 76} the product of ad by db, 49 remains for the square on dg. Therefore dg is 7. Add bg to this to make line bd or the major diameter 24. The da that remains is 10 for the minor diameter.

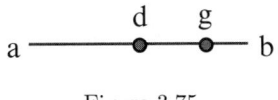

Figure 3.75

[160] Again: I divided the major diameter by the minor to get $\frac{2}{5}$ 2, and the area of the rhombus is 120. What is the measure of each diameter? Because the ratio of whole to whole is the same as part to the same part, so the ratio of the major diameter to the minor is as half the major to half the minor. And half the major is ef, the long side of quadrilateral (or line ae), and half the minor is line eb, the short side [see Figure 3.76]. Now all numbers which have one and the same ratio, if the larger are divided by the smaller, always come out the same in the division process. Therefore, if we divide the long side of the quadrilateral ef by the short side, we still get $\frac{2}{5}$ 2. Therefore this is the problem: the area of the quadrilateral is 60, half the area of the rhombus. I divided the long side by the short side, and got $\frac{2}{5}$ 2. So multiply 1 by $\frac{2}{5}$ 2, then by 60 to reach 144. Its root is 12. Divide it by the numbers in the ratio, that is, by 1 and next by $\frac{2}{5}$ 2, to get first 12 and then 5 for the sides of the quadrilateral or halves of the diameters. Therefore the major diameter is 24 and the minor is 10.

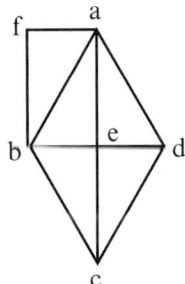

Figure 3.76

[161] To help you understand the procedure: let a be the unit and b $\frac{2}{5}$ 2 [see Figure 3.77]. Multiply a by b to get g. Let the short side of the quadrilateral

136   Fibonacci's *De Practica Geometrie*

be d, the long side e, and the area f. Because the quotient of the long side by the short side is $\frac{2}{5}2$, so the ratio of 1 to $\frac{2}{5}2$ equals the ratio of the short side to the long side; or, as a is to b so is d to e. Whence the product of a and e is the same as the product of b and d. Therefore let either product be the number h. Square h and get 1. The product of d and f is k. I say that the number k equals the number 1.[155] The product of a and b produced g, of d and e gave f, and of g and f came k. Therefore the number k is the result of multiplying a by b, then by d, and finally by e [see Figure 3.78]. Likewise, you multiplied a by e to yield h. Similarly, the product of b and d produced h. Further, the product of h by h made 1. Therefore the product 1 is the result of multiplying a by e, then by b, and finally by d. But the product of a and b multiplied by d and then by e equals the product of a and e multiplied by b and then by d. Therefore k equals 1, as we said.

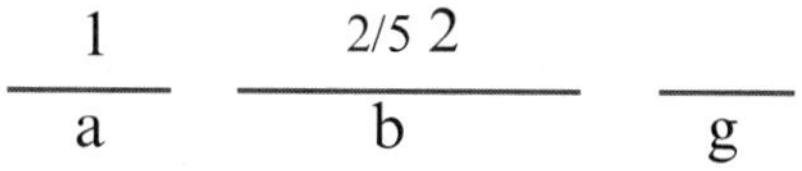

```
      1              2/5 2
 ―――――――――    ―――――――――――――  ――――――         d│ e│ f│h│ i │ k│
      a              b              g
```

        Figure 3.77    Figure 3.78

[162] For we did indeed above multiply a by b to get $\frac{2}{5}2$, which we multiplied by the area (namely g by f), and we had the number k which is the number i or 144. We took the root of this for h because the square of h produced i.[156] Therefore h is 12. And because the product of a and e is h or 12 and a is 1, we therefore divided 12 by 1 (or h by a) and got 12 which is e, the long side. Likewise, the product of b and d is h or 12, and b is $\frac{2}{5}2$. Whence we divide 12 by $\frac{2}{5}2$ or h by b, and we have 5 for d, the short side. For if the area f were 100, then k (or 1) would be 240 whose root is h. But 240 does not have a root.[157] Whence since we cannot divide h by a or b, we took their squares and divided them by 1.

[163] Another way: we took the ratio in pure numbers which the unit a has to b, and they are 5 and 12. Therefore let a be 5, b 12, and g 60. Whence k or i[158] will be 6000. This we divide by the squares of the numbers a and b (25 and 144) to get the root of 240 for the long side. The short side is the root of $\frac{2}{3}41$.

[164] Or another way: Let the short side be thing, the long side two thing$s$ and $\frac{2}{5}$ thing. Multiply thing by two thing$s$ and $\frac{2}{5}$ to get **{p. 77}** two *census* and two fifths of a *census* which equal the given area. Therefore let the given area be 100. Divide this by $\frac{2}{5}2$ to get $\frac{2}{3}41$ whose root is d, the short side. And because as a is to b, so is d to e. Therefore the square of number a is to the

---
[155] Text has i; context requires 1 (76.22).
[156] See next note.
[157] Fibonacci means that there is no rational root.
[158] Text has *1*; manuscript has i (76.40, f. 46v.12).

square of number *b* as the square of quantity *d* is to the square of quantity *e*. Whence: multiply the square of number *b* by the square of quantity *d* (144 by $\frac{2}{3}$ 41), and divide the product by the square of number *a* or 25 to obtain 24[159] whose root is the long side.

Rhomboids[160]

[165] Rhomboids are parallelograms having only equal opposite sides and angles. In order to measure these, we draw a diameter that divides the figure in two equal triangles. If we multiply one cathete by the diameter to which it was drawn, we have the area of the parallelogram. To make this clear, let there be parallelogram *abcd* with opposite and equidistant sides *ab* and *cd* each equal to 30 rods [see Figure 3.79]. The other two sides, *ac* and *bd*, are also equal and equidistant from each other. One of these sides measures 13 rods and the diameter is 37 rods. The diameter divides the parallelogram in two equal triangles *abc* and *dbc*. Each of these is obtuse because the square on the diameter *bc* is larger than the sum of the squares on the two sides *ba* and *ac*, or *bd* and *dc*. Draw cathete *ae* from point *a* in triangle *abc* to its base *bc*. Multiply cathete *ae* by base *cb*, and you have the area of the entire parallelogram *abcd*. Or, in triangle *bcd* draw the cathete *df* to base *bc*, multiply cathete *df* by the base *bc*, and you have again the area of parallelogram *abcd*.

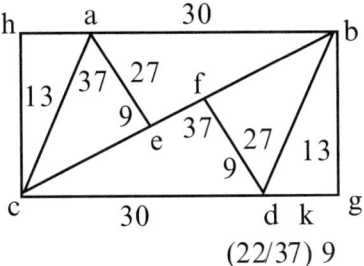

Figure 3.79

[166] For example: Parallelogram *abcd* is twice triangle *bcd* whose area was found by multiplying cathete *df* by half of the bases *bc*. Whence to multiply cathete *df* by all of base *bc* is to double the area of triangle *bdc*. Therefore this procedure produces the area of the whole parallelogram that is twice [the area of the] triangle *bcd*. You will find that both cathetes, *ae* and *df*, measure $\frac{27}{37}$ 9. Multiplying $\frac{27}{37}$ 9 by 37, the length of diameter *df*, produces 360, the area of the parallelogram. Similarly, Draw cathete *bg* to the base *cg* but outside of

---

[159] Text has 250; context requires 24 (77.7).

[160] "Incipit de rumboide differentia secunda" (77.8, f. 46v.23). After the first sentence, I chose to use the word *parallelogram* for Fibonacci's *rhombodes*.

138   Fibonacci's *De Practica Geometrie*

triangle *bcd*. If we multiply cathete *bg* by *cd*, the base of the triangle, we will have the area of the entire parallelogram *abcd*. Likewise: to multiply cathete *ch* by base *ab* produces the area of the same parallelogram.

[167] We can find each cathete by the rule described above for the obtuse triangle. For example: subtract the squares on the sides *ca* and *ab*, or *bd* and *dc* (169 and 900) from the square on side *bc* (1369) leaving 300 {p. 78}. Dividing half of this by the base *cd* or 30 yields 5 for the measure of *dg* or *ah*. Its square is 25. Subtract this from 169 the square on *bd* to leave 144. Its root is 12 for the cathete *bg*. Multiplying this by the base *cd* (12 by 30) yields 360 rods for the area of the parallelogram, as we found above. Further, let us consider the area of a parallelogram by its two cathetes, *di* and *ak*, which are measured by the diameter *ad* and sides of the parallelogram. Because the diameter and cathetes create within the parallelogram two equal right triangles, *acd* and *abd*, as seen in the figure [Figure 3.80], each cathete *id* or *ak* measures 12 rods. Be it noted that from wherever between [points] *a* and *i* the cathete is drawn, it falls between points *k* and *d* on line *kd*. You can find the length of the cathete from what we have said above about right triangles or what we taught about using a tape in triangles.

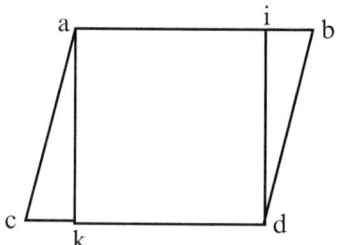

Figure 3.80

[168] If the area of a parallelogram is cut in two by the short diameter to make two right triangles, as in parallelogram *bcde*, its sides *bc* and *de* measure 35 rods, sides *bd* and *ce* are 37 rods, and the short diameter is 12 [see Figure 3.81]. I say that parallelogram *bcde* is divided in two right triangles because the square on line *ec* equals the sum of the squares on the lines *bc* and *be*.[161] Whence angle *cbe* is a right angle, as is angle *bed*. Further triangle *cbe* is congruent with triangle *bed*. And because the product of cathete *be* by half the base *ed* produces the area of the triangle *bed*, if we multiply the diameter *be* by the side *ed*, we will have 420 rods for the area of all of parallelogram *bcde*. The same holds for any parallelogram which is divided by whichever diameter into two obtuse triangles.

---

[161] Figure 3.81 as shown represents what appears in the margin of Boncompagni's text; but it certainly does not represent what is given in the text (78, *f.* 47r).

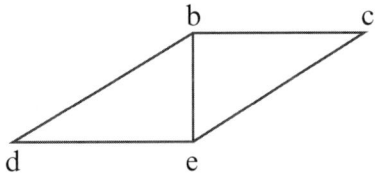

Figure 3.81

Trapezoids[162]

[169] There are three kinds of quadrilateral fields with two sides unequal and parallel. The first is the figure whose head, as is said, was cut off, leaving the two sides equal to one another, as in quadrilateral *abcd*. Side *ab* measures 8 rods and is equidistant from side *cd* that measures 18 rods.[163] Each of the other sides *ac* and *bd* measures 13 rods [see Figure 3.82]. In this figure[164] side *ab* is where "the head was cut off" and side *cd* is the base under the cut. Having drawn the cathete from the cut to the base, the area of this figure is the product of the cathete by half the sum of sides *ab* and *cd*. Now, if you wish to draw a cathete from either point *a* or point *b* to the base *cd*, subtract 8 the length of the cut from 18 the base leaving 10. Its half is 5 for either length, *ce* or *df*. The cathete will fall from point *a* to point *e* or from point *b* to point *d*. So, if you subtract 25 the square on line *ce* from 169 the square on line *ae* (or the square on line *df* from the square on line *bd*), what remains is 144. Its root or 12 is the length of the perpendicular *ae* or *bf*. The product of 12 by half the sum of sides *ab* and *cd* (half of 26 or 13) leaves 156 for the area of quadrilateral *abcd*. For example: the two cathetes themselves, *ae* and *bf*, create rectangle *aefb*. Its area is found by multiplying cathete *ae* by side *ef* equal to side *ab*. Now 8 rods measures *ef*. Hence the product of 12 rods by cathete *ae* results in 96 rods for the area of quadrilateral *aefb*. By subtracting this rectangle from quadrilateral *abcd*, what remains are two equal right triangles, *aec* and *bfd* {**p. 79**}. Now the product of cathete *ae* and half of *ec* produces the area of triangle *ace*. Whence, the product of line *ae* by all of *ec* yields 60, the area of the two triangles *aec* and *bfd*. Adding this to 96 the area of the quadrilateral *aefb* produces 156, the area of quadrilateral *abcd*, as we said above.

---

[162]"Incipit de figuris que habent capita abscisa de quibus iiiior sunt genera, in differentia prima" (78.23–24, f. 47v.3–5).

[163]These measurements are presumed in several problems that follow.

[164] As shown in Boncompagni's edition, Figure 3.82 is clearly incorrect. Point *f* should be to the right of point *d* in order to fit the definition above as well as the mathematics that follows, requiring the 12 and 13 on the right side to exchange their positions, as seen in manuscript f. 47v.

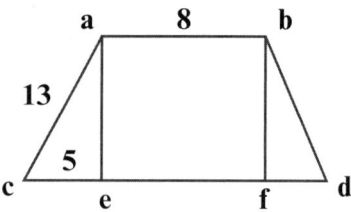

Figure 3.82

[170] If you wish to find diameter da or cb, then you add 169 the square on the line de (or cf) to 144 the square on the cathete ae (or bf) to find 313 whose root is the measure of diameter da or bc. And if you wish to find the point of intersection of the two diameters, add the bases 26 and square this for 676. Then square the upper base for 64; square the lower for 324. Multiply 64 by 313 the square on one of the diameters, and divide the product by 676. Or, take 16 a fourth of 64, multiply it by 313, and divide by 169 (a fourth of 676) to reach $\frac{3}{13}\frac{8}{13}$ 29, because the root is line ag or bg. Similarly, multiply 81 (a fourth of 321) by 313, divide the product by 169 a fourth of 676, and you will have the square on line gc or gd.

[171] Note that when we took the squares on those lines, one of them, namely 313 the square on one diameter, does not have a root. For if the diameter were rational, we would have multiplied it by 8 and 18, and divided the products by 26. And thus we would have had the abscissas of the diameters, all of which we wish to demonstrate geometrically. Because line ab is equidistant from line ce, triangles agb and dgc are similar. Hence, angles abg and gcd are equal, as are angles bag and gdc. Whence, as ab is to bg, so is dc to cg. And by alternation as ad is to cd, so is bg to gc. Likewise again, as ab is to cd so is ag to gd. Because ab is $\frac{4}{9}$ of cd, whence bg is likewise $\frac{4}{9}$ of gc or ag is $\frac{4}{9}$ of gd. Because terms that are proportional by disjunction are proportional by conjunction, therefore as the square on ab is to cd (or as 8 is to 26, or 4 to 13 being the least ratio), so bg is to bc as ag is to ad. Therefore bg is $\frac{4}{13}$ of bc; similarly ag is $\frac{4}{13}$ of ad. Whence if we take the rational part $\frac{4}{13}$ of the diameter bc, we would have the line bg or ag. The remainder gc or gd would be $\frac{9}{13}$ of the whole diameter. Now because 313 the square on the diameter does not have a root, we take the squares of 4 and 13 or 16 and 169. Then we multiply 16 by 313 and divide by 169. Whence because the square on ab is to cd, so is bg to bc. Therefore as the square on line ab is to the square on the sum of the lines ab and cd; that is, as 64 is to 676, or as a fourth of 64 is to 16 a fourth of 676 to 169, so is the square on line bg to 313 the square on the diameter bc. And the square on line gc has the same ratio to the square on line ad. Whence line ag equals line bg, for they have the same ratio of equals to equals.[165] Because if equals subtracted from equals leave equals, so are lines gc and gd equal. And

---

[165] "ad equalia" (79.35).

they have a ratio to the root of 313 the square on the whole diameter, as the ratio of *cd* has to itself and line *ad*. And this ratio is 9 to 13. Whence the ratio of the square on line *gc* or *gd* is to 313 the square on the diameter as the square of 9 has to the square of 13 or 81 to 169. Whence by multiplying 81 by 313 and dividing by 169, we have the square on line *gc* or *gd*, as we wanted to show.

[172] If lines *ca* and *db* are extended[166] to point *h* **{p. 80}** as shown in this other figure [Figure 3.83], quadrilateral *abcd* is transformed into triangle *hcd*. If you want to know the measure of line *ah* or *bh*, then take half the width from half the base of the part cut off (4 from 9). Divide the remaining 5 by the product of half the width by line *ca* (4 by 13) to get $\frac{2}{5}$ 10 for the measure of line *ah* or *bh*. If you multiply the same 4 by 12 the measure of cathete *ae* and divide by 5, you will get[167] $\frac{3}{5}$ 9 for cathete line *ih* of triangle *hab*. If this is extended to point *k*, then the entire line *hk* is the cathete of triangle *hcd*. Because a certain line *ab* in triangle *hcd* has been drawn equidistant from the base *cd*, you have triangle *hab*[168] similar[169] to *hcd* with equal adjacent angles; namely, angle *hab* equals angle *hcd*, and exterior angle equals interior angle. Similarly angle *hba* equals the angle at *d*. The common angle is at *ahb*.

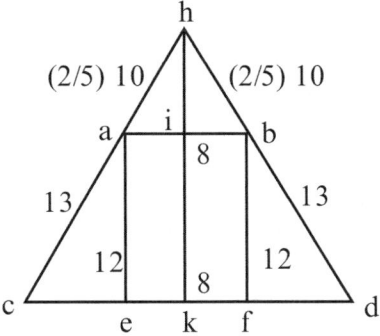

Figure 3.83

[173] Now, as is well known in geometry similar triangles have equal angles and proportional sides. Whence, as side *ha* is to *ab*, so is *hc* to *cd*. And as *hb* is to *ba*, so is *hd* to *dc*. Alternately, as *ha* is to *hc*, so is *hb* to *hd*, and *ab* to *cd*. Again, as *ab* the base of triangle *hab* is to the base *cd*, so side *ha* is to side *hc* and *hb* to *hd*, and also cathete *ih* to cathete *hk*. Therefore *ab*, a part of *cd* (8 of 18), is the same part *ha* of *ac* ($\frac{2}{5}$ 10 of $\frac{2}{5}$ 23), and *hb* of *hd*, and cathete *ih* of cathete *hk*. Now 8 is $\frac{4}{9}$ of 18. Whence *ha* of *hc*, *hb* of *hd*, and also *ih* of *hk* are four ninths, the one of the other. For in this figure triangle *cea* is similar

---

[166] The phrase *in partes ab* is omitted here (79.43) as a locator for the endpoints of the two lines.
[167] Text has $\frac{2}{5}$ 9; context requires $\frac{3}{5}$ 9 (80.6).
[168] Text has *hai*; context requires *hab* (80.10).
[169] Text has *scilicet*; context requires *simile* (80.10).

to triangle *aih* because they have equal angles. Angle *hia* equals angle *aec* because they are both right angles. And the angle at *c* equals angle *iah* because line *ab* is equidistant from line *cd*. What remains under *ahi* equals what remains under *cae*, because the three angles of each triangle equal two right angles. For as *ce* is to *ea* (5 to 12), so is 4 or *ai* to *h*. Multiplying 4 by 12 and dividing by 5 we have $\frac{3}{5}$ 9 for cathete *hi*.[170] Likewise, as *ec* is to *ca* or 5 to 13, so is *ia* or 4 to *ah*. So by multiplying 4 by 13 and dividing by 5 we have $\frac{2}{5}$ 10 for line *ha*. By these proportions we can investigate the heights and lengths and depths of things, as we shall show in their own place.[171]

[174] The second kind of figure in the third category is called "half-headed".[172] It has two unequal sides equidistant from one another. The other two sides are also unequal: [see Figure 3.84] one of them making right angles with the lower and upper bases, the other making an acute angle with the lower base. For instance, quadrilateral *abcd* with upper base *ad* measures 18 rods and is equidistant from the lower base *bc* whose length is 30 rods. The cathete *ab* is 16, and side *dc* is 20. To find the area of the whole quadrilateral, add 18 to 30 (sum the two bases) to get 48. Take half of this or 24 and multiply it by 16 the perpendicular line *ab*. Therefore the area of quadrilateral *abcd* is 384 rods.[173] For example: upon line *bc* at point *d* draw cathete *de* which divides quadrilateral *abcd* in two parts, rectangle *abed* and right triangle *dec* Now *be* equals *ad* and *ad* is 18. Similarly *be* is 18, and cathete *de* equals cathete *ab*, both measuring 16. Therefore the area of quadrilateral *abed* is 288 which was found by {p. 81} multiplying *de* by *eb*, or 16 by 18. The area of triangle *dec* is found by multiplying cathete *de* by half of *ec*, or 16 by 6, to make 96. Adding this to 288, the area of quadrilateral *abed*, yields 384 for the area of quadrilateral *abcd*, as I said.

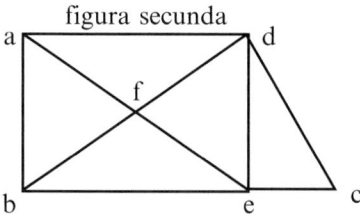

Figure 3.84

---

[170] Computation corrected from $\frac{2}{5}$ 9.
[171] This looks forward to Chapter 7.
[172] "semicaput abscisa dicitur" (80.30).
[173] Text has 484; context requires 384 (80.38).

[175] Suppose you wish to know the measure of diameter *ac*. Because triangle *abc* is orthogonal, add the squares on the lines *ab* and *bc* (259 and 900) to obtain 1156. Its root is 34 the length of diameter *ac*. Likewise, if you want the measure of diameter *bd*, then add the square on cathete *de* to the square on base *eb* (256 and 354) to get 580. Its root is a surd, the length of diameter *bd*. Whence we say that the measure of diameter *bd* is the root of 580, or the square on diameter *bd* is 580. In order to know where the diameters intersect, we proceed as above by adding the width to the base (18 and 30) to obtain 48. As 18 is to 48, so *af* is to the whole diameter *ac*. Now 18 to 48 is as 3 to 8. So as 3 is to 8 so is *af* to *ac*. Whence multiply 3 by 34 and divide by 8 (3 times 17 and divided by 4) to obtain $\frac{3}{4}$ 12 for line *af*. The $\frac{1}{4}$ 21 that remains from the 34 is the length of line *fc*. Similarly, because triangles *afd* and *bfc* are similar, *af* is to *ac* (3 to 8), as *df* is to *db*. Therefore *df* is $\frac{3}{8}$ of *db* and *fb* remains as $\frac{5}{8}$ of *db*. Because line *db* is surd, we take their ratio in squares. Therefore the square of 3 is to the square of 8 (9 to 64) as the square on line *df* is to the square on the diameter *bd* (580). Whence we multiply 9 by 580 and divide by 64 (or multiply 9 by 145, a fourth of 580, and divide by 16, a fourth of 64). We must always use this method of resolving a proportion[174] which we taught in *Liber abaci*,[175] namely, to multiply or divide by the smallest numbers in the same ratio. The ratio of 580 to 64 is the same as the ratio of a fourth of 580 to a fourth of 64. What results is $\frac{9}{16}$ 81 for the square on line *df*. Again, because line *fb* is $\frac{5}{8}$ of diameter *bd*, multiply 25 the square of 5 by 145 and divide the product by 16 to get $\frac{9}{16}$ 226 for the square on line *fb*.

[176] If line *ba* and *cd* are extended from endpoints *a* and *d* until they meet at point *g*, as shown in the figure [Figure 3.85], and you want to know the measure of line *ag*, then multiply *ed* by *da* (16 by 18) and divide by *ce* (12) to get 24. Since triangles *dec* and *gad* are similar, then as *ce* is to *ed*, so is *da* to *ag*. *Ex equali* as *ec* is to *cd*, so is *ad* to *dg*. Whence the product of *cd* and *da* (20 by 18) divided by *ec* returns 30 for the measure of *dg*, as shown in the above figure.

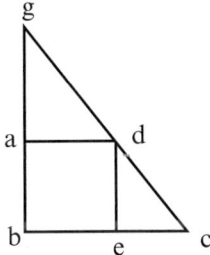

Figure 3.85

---

[174] "modum evitationis" (81.22).
[175] Op.cit.

[177] The third kind of quadrilateral in the third category is the trapezoid.[176] Its bases are parallel but unequal in measure, and the other two sides make unequal acute angles with one base. For example, in quadrilateral *abcd* [see Figure 3.86], the upper base *ab* is 10 rods long and equidistant from the lower base *cd* that is 24 rods long. One side *ac* is 13, and the other *bd* is 15. The area of this figure is found by multiplying the cathete by half the sum of the two bases. Whether you wish to draw the cathete from point *a* or point *b* to the base *cd*, it is first necessary to locate endpoints of the cathetes on the lower base. The procedure is this: subtract the upper base from the lower (10 from 24) to get 14 {p. 82}. And then take 169 from 225, the square on side *ac* from the square on side *bd*. What remains is 56. Divide this by the remainder 14 already computed to reach 4. Add this to 14 to get 18. Half of this is 9 for *de* the long part of side *bd*. The 5 that remains from the 14 is *fc* the short part of side *ac*, as we said about the acute triangle. Now subtracting the square on the short part *cf* from the square on side *ac* (169 less 25) leaves 144. Its root is 12 for cathete *af*. Similarly, taking the square on *ed* from the square on *bd* leaves 144, the square on *be*. Whence *be* is 12 as is *af*. Adding the two bases, 10 and 24, produces 34. Multiply its half by 12 the cathete *af* or *be* to get 204 for the area of quadrilateral *abcd*.

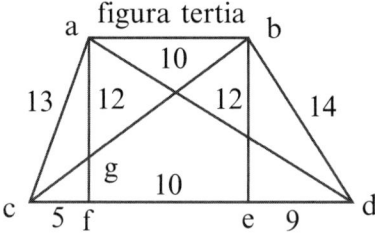

Figure 3.86

[178] You can find the measure of the diameter *cb* by taking the root of the square on *cb* which you obtain by adding the squares on line *ce* and cathete *eb*. If you want to know where the diameter *cb* intersects cathete *af*, you have two methods. First, because line *ab* is equidistant from line *cf*, *ab* is to itself and *cf* as *ag* is to *af*. Therefore *ag* is $\frac{2}{3}$ of *af* or 8, leaving 4 for *gf*. Because triangles *agb* and *cgf* are similar, *bg* is similarly $\frac{2}{3}$ of *bc*. Likewise its square is $\frac{4}{9}$ of the square on its diameter. Hence multiply 4 by 369 and divide by 9, or take a ninth of 369 and multiply by 4, to get 164 for the square on line *bg*. Because *bg* is $\frac{2}{3}$ of *bc*, *gc* remains as $\frac{1}{3}$ of *bc*; whence its square is 41, $\frac{1}{9}$ of 369. The other way is based on similar triangles *ceb* and *cfg*; *cf* is to *ce*, so is *fg* to *eb*, a third part. Therefore *fg* is 4, as I said. Similarly and for the same reasons, *cg* is $\frac{1}{3}$ of *cb*. Et cetera.

---

[176] "que diuerse caput abscisa dicitur" (81.35).

[179] According to what we have said in this part, we can do the following: find the diameter *da* in triangles, know where it intersects cathete *be*, where the former intersects diameter *bc*, even if a line were drawn upon the given parts of sides *db* or *ca* from the angles on points *c* and *d* or from some given point on line *cd*, whether inside or outside the figure. And if it were extended outside the figure equally with line *ab* until both met, we can know the point of intersection and the measure of the extended lines. For instance: if you wish to extend sides *ca* and *db* outside the figure until they meet at point *h* above the upper base 10 [see Figure 3.87], you can find the measure of line *ah* to be $\frac{2}{7}$ 9 by dividing the product of base *ab* and line *ac* (10 and 13) by 14, the difference of the two bases. Similarly, we can find the length of line *bh* to be $\frac{5}{7}$ 10 by dividing the product of *ab* and *bd* (10 and 15) by 14.

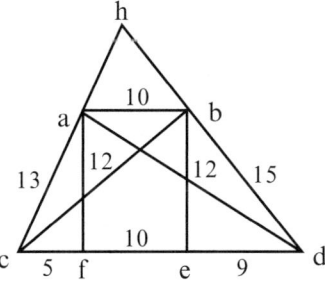

Figure 3.87

[180] The fourth type of figure, called a leaning trapezoid,[177] has unequal and equidistant bases. The remaining two sides are drawn from an acute angle and an obtuse angle on the lower base [see Figure 3.88]. Consider quadrilateral *abcd*. Its upper base *ad* measures 12 rods, lower base *bc* measures 16 rods, side *ab* is 15 rods, and side *dc* is 13 rods. The area of this figure is found by multiplying the cathete by half the sum of the bases. The cathete falls within the figure *abcd* from point *a* to base *bc*, or if it falls from point *d*, it falls outside. Whence if we want to find the measure of either the interior or the exterior cathete, it is necessary to find first the measure of the segment of the base on whose end-points they fall. The way to do this is to subtract the upper base from the lower base (16 less 12) to get 4. Its square or 16 added to 169 the square on side *cd* yields 185. Subtract this from 225 (the square on *ab*) to leave {p. 83} 40. Divide half of this by 4 to get 5 for the far side of the segment *ce*, as we showed in the section on obtuse triangles. Add the 5 to the 16 that measures the base *cb* to obtain 21 for the extended line *cb*. Take from this 12 the measure of the upper base, and 9 remains for the measure of the interior

---

[177] *caput abscisa declinans* (82.33). In *Liber mensurationum* this type is treated under the title *Capitulum aryde expanse vel latitudinis*, op. cit. 109.

segment *fb*. Whence by subtracting 81 from 225, the square on *fb* from the square on *ba*, what remains is 144. Its root is 12, the measure of cathete *af* or *de*. Now multiply the 12 by half the sum of the bases to obtain 168 rods for the total area of quadrilateral *abcd*.

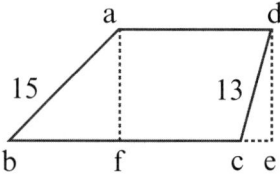

Figure 3.88

[181] In similar figures also called leaning trapezoids, one of the cathetes drawn from the upper base to the lower base falls within the figure, the other falls outside as seen in the attached figure [Figure 3.89]. When one falls inside upon an angle of the base, the cathete is opposite the angle from which it was drawn. And the other falls outside. When both fall outside, as seen in the figure below,[178] consult what was shown above to find the measures of their cathetes. The square on *be* is 441 and the square on *ed* is 144. Hence, their sum is 585, the square on diameter *bd*. Likewise, by adding 49 to 144, the square on line *ef* to the square on cathete *fa*, the sum is the square on diameter *ac*. You can find the sections of the diameter by the method discussed above. Now it is true that if sides *ab* and *dc* are extended to meet at point *k*, then *ka* is to *kb*, and *kb* and *kd* to *kc* as *ad* is to *bc*; that is, $\frac{3}{4}$. Whence *ab* of *kb* and *dc* of *kc* are fourth parts. Therefore *ka* is three times *ab* and *kd* is three times *dc*. Whence *ka* is 45 and *kd* is 39.[179]

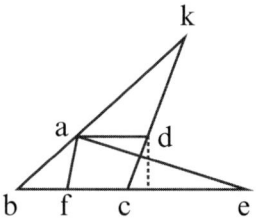

Figure 3.89

---

[178] There is no figure for this in Boncompagni's edition nor did I see one in Urbino 292 or in other manuscripts, the case in which the upper base is longer than the lower.

[179] Text and manuscript have 29; context requires 39 (83.21, f., 50v.17).

3 Measuring All Kinds of Fields   147

[182] If none of the sides of the quadrilateral is equidistant from another side, as in quadrilateral *abcd*, let side *ab* be 13 rods, side *bc* 15 rods, side *dc* 17 rods, and side *da* 16 rods <see Figure 3.90]. In order to measure the quadrilateral and assuming you know the measure of one diameter, use that diameter to divide the figure into two triangles, find the area of each, then add them together for the area of the quadrilateral. For example, let diameter *ac* be 14 rods, whereby the area of triangle *abc* is 84 rods and the area of triangle *acd* is $\frac{1}{3}$ 104 rods. The sum of these areas is $\frac{1}{3}$ 188 for the area of the whole quadrilateral *abcd*. This is the universal method for all quadrilaterals. Another way: [extend side *cd* to point *e*][180] from point *a* draw line *ae*.[181] Add the area of triangle *ade* to the area of quadrilateral *abcd*, which we found by the method described above. Thus you have the area of quadrilateral *abce*. For a concave quadrilateral[182] such as *defg* with unequal sides [see Figure 3.91], one diameter *eg* falls within the figure [from point *e* to point *g*], and the other *df* falls outside from point *d* to point *q*. You can find the area of this figure by adding the areas of triangles *deg* and *feg* or by subtracting the area of triangle *dgf* from the area of triangle *def*.

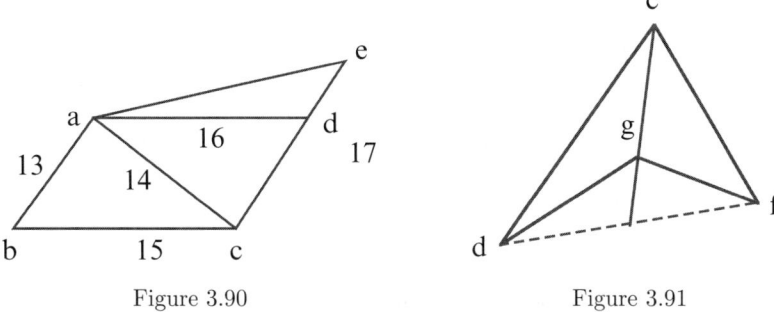

Figure 3.90   Figure 3.91

3.4 Measurement of Multisided Fields[183]

[183] The method for measuring a field with many sides is to divide the field into triangles and add the areas of the triangles. Thus you have the area of the multisided figure. It must be noted that if the multisided figure has five sides, you solve it by adding three triangles. If it has six sides, then add four.

---

[180] This addition is necessary to justify what follows for finding the area of the enlarged figure.
[181] The text has a line drawn from point *a* parallel to line *bc* that is impossible.
[182] "figura que barbata dicitur" (83.32).
[183] "Incipit pars tertia in dimensione camporum plura latera quam quatuor habentium" (83.37–38, f. 50v.33–34).

148   Fibonacci's *De Practica Geometrie*

Thus and always: any multisided figure is dissected into two triangles less than the number of sides. Although {p. 84} multisided figures can be measured by dissecting them into triangles, when we wish to proceed somewhat more subtly in certain matters, namely with pentagons (they have five sides), we can dissect them into two pieces, of which one is a triangle and the other is a quadrilateral with two equidistant sides. For instance, pentagon *abcde* is cut into triangle *abe* and trapezoid *ebcd*, side *be* being equidistant from side *cd*. Then add the area of triangle *abe* to the area of trapezoid *bcde*, and you have the area of pentagon *abcde*. Similarly, it is possible to make two trapezoids out of a hexagon (a figure with six sides), each of which has two equidistant sides. Or, one trapezoid can be made with two equidistant sides and two triangles. And thus you can learn how to work with the other multisided figures.

[184] Now if the multisided figure you wish to measure is both equiangular and equilateral and different from what has been described, you can still find its area because a circle falls within it that contains the midpoint of each side. Multiply the semidiameter of the circle by half the number of sides of the figure and you will have its area [see Figure 3.92]. To understand this somewhat better consider the equilateral and equiangular pentagon *abcde* within which we wish to inscribe a circle tangent to its sides. The method is this: divide angles *eab* and *abc* equally by two lines *af* and *fb* and extend lines *fc, fd,* and *fe*. Name the midpoints of each side as *g, h, i, k,* and *l*. Join lines *fg, fh, fi, fk,* and *fl*. I will show you that these lines are equal. Because pentagon *abcde* is equiangular, angle *fab* equals angle *fba*. Further, they are half the angles of the pentagon. Whence triangle *fab* is an equilateral triangle with equal sides subtending equal angles. Further line *fa* equals line *fb*. *fg* is the cathete to the middle of line *ab*. Since *la* is half of line *ae*, it equals line *ag*. Since line *fa* is attached to both lines *la* and *ae*, the two lines *ga* and *af* equal the two lines *fa* and *al*. And angle *gaf* equals angle *fal*. Whence the base *fl* equals base *fg*, angle *afl* equals angles *afg*, and angle *agf* is a right angle as is angle *alf*. Whence line *fl* is the cathete {p. 85} on line *ae*. And because *al* equals line *el*, and if line *fl* is added to both, the two lines *fl* and *la* are equal to the two lines *fl* and *le*. And the angles at *l* are equal since each of them is a right angle. Whence line *fe* equals line *fa*, triangle *afl* equals triangle *lfe*, and the whole triangle *bfa* equals the whole triangle *afe*. It can be shown in a similar way that any of the lines *fh, fi,* and *fk* is equal to either of the lines *fg* and *fl*. Whence the circle *ghikl* is described about the center *f* by one of the lines *fg* or *fh* [see Figure 3.93]. Further, pentagon *abcde* is divided into five equal triangles: *fab, fbc, fcd, fde,* and *fea*. And their respective cathetes, *fg, fh, fi, fk,* and *fl*, are all equal to one another. Now by multiplying *fg* by half of *ab*, the area of triangle *fab* is found. If we multiply *fg*, the semidiameter of the circle within the pentagon, by five times half of *ab* which would be half of the sides of pentagon *abcde*, we obtain five times the area of triangle *fab*. That is the area of pentagon *abcde*, as we said. We use the same procedure for

any equilateral and equiangular figure that contains such a circle. It should be obvious from this that if the semidiameter is multiplied by a number that is more than half the measure of the circumference, the area will be greater than the area of the circle.

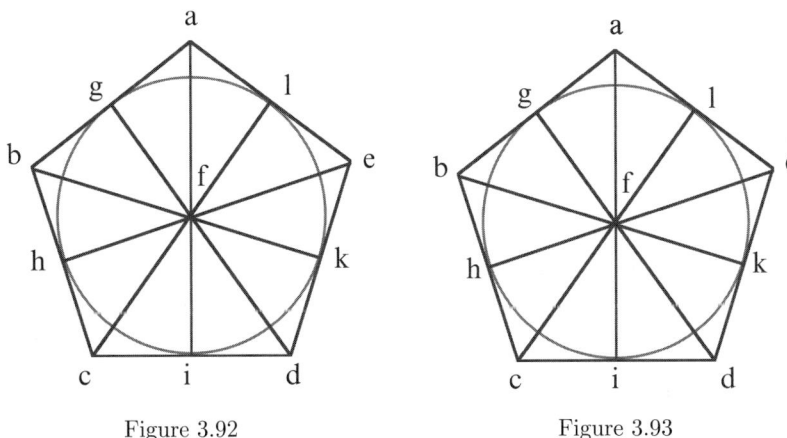

Figure 3.92   Figure 3.93

[185] There is another way to measure an equilateral and equiangular pentagon: inscribe it within a circle that touches all the vertices. Multiply the dodrans[184] of the diameter by the distance of the chord of the angle of the pentagon, and you will have its area. To make this clear consider pentagon *abgde* inscribed in circle *abgde*, with diameter *az* and center *c* [see Figure 3.94]. Draw line *be* which is the chord for pentagonal angle *bae*. Take *ci* as half of the semidiameter *cz*. The whole line *ai* is a half and a quarter of the diameter *az*. Now as *ai* is to *ac*, so is *te* to *tk*. For as *ac* is two thirds of *ai*, so is *tk* two thirds of *te* or *bt* since *bt* equals *te*. Hence *tk* is as third of the whole chord *be*. Whence *bk* is a half and third of chord *be*. Therefore I say that the area of the pentagon *abgde* equals the product of *ai* by *bk*. The proof follows. Since *ia* is to *ac* as *te* is to *tk*, the product of *ca* and *te* (or *tb*) equals the product of *ia* and *tk*. But the product of *ca* and *bt* is twice triangle *cba*. Therefore the product of *ai* and *tk* is twice triangle *cba*, because *tk* is twice that from *ek*. If we multiply *ia* by *ek* we obtain an area equal to the area of triangle *cba* which is a fifth of the area of pentagon *abgde*. Whence if we multiply *ai* by *bk* (that is, five times *bk*), we obtain a number that is five times the area of triangle *cba*. But five times triangle *cba* equals the area of pentagon *abgde*. Therefore by multiplying *ai* by *bk* we find the area of pentagon *abgde*, as I said.

---

[184] A Roman term for three-fourths; also equal to 9 inches. A *distunce* is five-sixths.

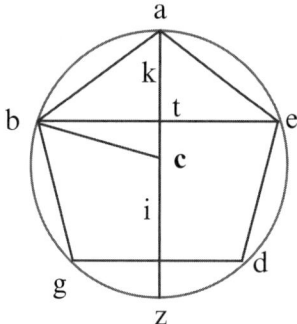

Figure 3.94

[186] It must be noted that if the measure of the diameter of the circle is rational, then the side of the inscribed pentagon is a minor line. That is, it is the root of the fourth recised or fourth cut off length[185] which consists of a number less a root. The greater of the two terms is more than the smaller. Further, the chord under the pentagonal angle is called the major line, is incommensurable in length with it, and is the root of the fourth binomium. It consists of a number and a root in which the larger term can be greater than the lesser. The side of the pentagon and the chord below the pentagonal angle are incommensurable with each other in length. For example: the pentagonal side ab is the root of 40 less the root of 320, and chord be is the root of 40 and the root of 320, together with diameter az equal to 8, as will be shown in its place {p. 86}.

[187] If all the boundaries of the field are not straight lines, as in quadrilateral $abcdez$ that has two straight sides az and cd with the remaining sides abc and dez being curved, then you must measure them as I shall instruct you [see Figure 3.95]. Since you can draw straight lines ac and dz, you can find the area of the resulting quadrilateral $acdz$ according to previous instructions. Add to this the area of the convex part[186] zed, subtract the area of the concave part[187] abc, and you have the area of the field. This is how you measure the area of the convex part zed. Make e the midpoint of arc zed, and create triangle ezd by drawing lines ez and ed. What remains from the whole convex part zed are sections zge and egd.[188] By considering each of these as rectilinear triangles, you can find the areas of the remaining four parts in the usual manner. Always do the same with whatever parts that remain until nothing

---

[185] See *Liber abaci*, 356–358; (Sigler [2002], 494–497).
[186] "aream uentris" The figure requires a convex part (86.5).
[187] "aream uentris" The figure requires a concave part (86.6).
[188] Text and manuscript have *ead*; context requires *egd* (86.9, f. 52r.23). Note that the letter g is missing in Figure 3.95.

appreciable is left of the convex part *zed*. Hence if all the areas of the triangles within the convex part are added together, you will indeed have its area. Proceed in a similar way to find the area of the concave part *abc*. Thus you will have the area of the whole field.

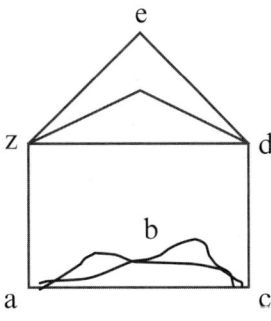

Figure 3.95

3.5 Measuring the Circle and its Parts[189]

[188] Assuming a field you wish to measure is circular, you must know its diameter. Multiply this by $\frac{1}{7}$ 3; or multiply it by 22 and divide by 7. Either way you have the circumference that contains the circle. To find the area, multiply half the circumference by half the diameter; or take eleven fourteenths of the square on the diameter. Either way you have the area of the circle. Or, if you wish to do it in the Pisan way, then square the diameter, divided by 7, and you will have the area of the circle in panes.

[189] So that all of this may be clearly understood, consider the adjacent circle *abdg* in which are two points *b* and *d*. Join them by a straight line. Find its midpoint *e* and draw line *ag* through it, making right angles with line *bd* [see Figure 3.96]. Further, let line *ag* be the diameter of the circle, midpoint *z* being the center of the circle. Let the measure of the diameter be 14 rods. Multiplying this by $\frac{1}{7}$ 3 produces 44 rods for the circumference *abgd*, also called "the periphery". Or, multiply 14 by 22 and divide by 7 to obtain the same 44 for the curve *abgd*. If we multiply its half or 22 by half the diameter, we obtain 154 for the area of circle *abgd*. Or, if we take $\frac{11}{14}$ of 196 (by multiplying 196 the square on the diameter by 11 and dividing by 14), we get the same 154 for the area of the circle. Similarly, if we square the diameter to get 196 and divide by 7, we find the area of circle *abgd* to be 28 panes. This equals the aforesaid 154 rods, because there are $\frac{1}{2}$ 5 rods in each panis.

---

[189] "Incipit pars quarta in dimensione circulorum et eorum partium" (86.15, *f.* 52r.29).

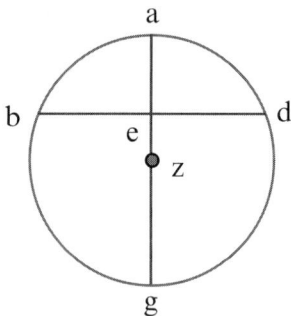

Figure 3.96

[190] Knowing the measure of the circumference of the circle and wishing to find its diameter, you divide the circumference by $\frac{1}{7}$ 3 or multiply by 7 and divide by 22. Or, multiply the twenty-second part of the diameter by 7 to get 14 for the diameter. If you want to know the area given the circumference, square its half, multiply by 7, and then divide by 22. For example, half the measure of curve *abgd* is 22. Its square is 484 that multiplied by 7 gives 3388 **{p. 87}**. Divide this by 22 to reach 154, as we said above. Or, if we divide 484 by 22 we get 22 that multiplied by 7 produces 154 again. If the diameter of the circle were 10, then the circumference would be $\frac{3}{7}$ 31, that is, the product of 10 and $\frac{1}{7}$ 3. So if we multiply 5 (half the diameter) by $\frac{5}{7}$ 15 (half the circumference), we get $\frac{1}{7}$ 78. Or, if we multiply 100 the square on the diameter by 11 and divide the product by 14, or if we multiply 11 by half of 100 and divide the product by half of 14 or 7, we also get $\frac{1}{7}$ 78 for the area of the given circle. Or, if we divide 100 by 7, we get $\frac{2}{7}$ 14 panes that equal the $\frac{1}{7}$ 78 rods. If $\frac{2}{7}$ of one panis is reduced to the usual parts of soldi and deniers, multiply the 2 above the fraction bar by 33 the number of soldi in one panis, then divide by 7 to get 4 soldi and $\frac{4}{7}$ 8 deniers of measure. Therefore, for the area of the given circle we have one starium, 2 panes, 4 soldi, and $\frac{4}{7}$ 8 deniers. This is the procedure for similar problems.

[191] Now if you wish to know why the area of a circle results from multiplying half its circumference by half the diameter of the circumference, we return to circle *abgd* with center *e* and construct within it a rectilinear figure of so many sides [see Figure 3.97]. Let this be quadrilateral *abgd* in which I create four triangles around center *e*: triangles *eab, ebg, egd*, and *eda*, each of which has equal corresponding sides. Lines *ea, eb, eg*, and *ed* are equal to one another because they are drawn from the center to the periphery. Whence, if cathetes are drawn from the center, each falls upon the midpoint of the base of its triangle. So from these midpoints *z, i, t*, and *k* on the bases, we draw lines from the center *e* to the periphery, namely, *el, em, en*, and *eo*. And then we draw these lines: *al, lb, bm, mg, gn, nd, do*, and *oa*. Thus four triangles are created on the bases *ab, bg, gd*, and *da*. Because cathete *ez* is a straight line falling upon *ab* another straight line, if we multiply *ez* by half of *ab*, we

obtain the area of triangle *eab*. Similarly, because *lz* is the cathete for triangle *lab*, the area of triangle *lab* is found by multiplying the cathete by half of *ab*. Similarly, if we multiply all of *el* (the semidiameter of the circle) by half of *ab*, we have the area of quadrilateral *ealb*. Similarly, if we multiply *em* or *el* by half of line *bg*, we obtain the area of quadrilateral *ebmg*. In the same way if we multiply *en* by half of *gd*, and *eo* by half of *la*, we obtain the areas of quadrilaterals *egnd* and *edoa*. That is, if we multiply *el* (the semidiameter) by half the sides of quadrilateral *abgd*, we obtain the area of the many-sided figure inscribed in the circle. But the area of the multisided figure *albmgndo* is less than the area of the circle. Therefore, by multiplying the semidiameter of the circle by half the measure of sides *ab*, *bg*, *gd*, and *da*, we obtain a measure less than the area of the circle. But half the sum of the sides is less than half the circumference *abgd*. Therefore by multiplying the semidiameter of the circle by less than half the circumference of the circle we obtain a number less than the area of the circle. We demonstrated above in the measurement of multi-sided figures containing a circle, that the product of the semidiameter and a number more than half the circumference will produce an area more than the area of the circle. Whence we concluded that the product of the semidiameter of the circle and half its circumference equals its area.

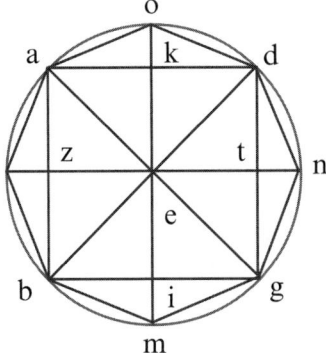

Figure 3.97

[192] But you may ask again: how does one find the area of the circle by {p. 88} taking eleven fourteenths of the square on its diameter? Because the area of any circle is to the area of any other circle as the ratio of the squares on their diameters, so are the ratios of their diameters to the ratios of their areas, as Euclid demonstrated in Book XII.2. Then, alternately as the square on the diameter of one circle is to its area, so are all the squares on the diameters of all circles to the areas of all those circles. Whence if the ratio of the square on the diameter of one circle to its area has been found, then the ratio of the square on the diameter of any circle to its area is known. Let the square on the diameter of the given circle be 196 and its area 154; then, in lowest terms

154 Fibonacci's *De Practica Geometrie*

their ratio is 14 to 11. Whence as 14 is to the square on the diameter, so is 11 to its area. Whence if we multiply 11 by the square on the diameter of some circle and divide the product by 14, we obtain the area of that circle. And this was what we had to show. Likewise, if we reduce 154 rods to panes, we obtain 28 panes. Therefore as 196 is to 28 panes, so the square on the diameter of a circle is to the panes measuring that circle. But the ratio of 196 to 28 is 7 to 1. Whence as 7 is to 1, so is the square on the diameter of the circle to the panes of its area. Whence if you take a seventh of the square on the diameter of any circle, you obtain the panes measuring the area of the circle.

[193] Be it noted that whatever is the ratio of one quantity to another, the ratio remains the same if a multiple is taken of the ratio. So, whatever is the ratio of the diameter of one circle to the diameter of another circle, the ratio of the circumference of the first circle to the circumference of the second circle is the same, as is the ratio of their semi circumferences. Similarly, the ratio of the squares on the respective diameters is the same as the ratio of the semi circumferences of the two circles. Because the ratio of the squares on the respective diameters is the same as the ratio of the measures of the areas of the two circles, so does the ratio of the squares of their semi circumferences equal the ratio of their areas. Alternately, therefore, as the ratio of the square on the semi circumference of one circle to its area, so is the ratio of the square on the semi circumference of the other circle to is area. For example: let the measure of the semi circumference of the given circle be 22. Its square is 484 and its area is 154. In reduced form, the ratio is 22 to 7. Therefore as 22 is to 7, so is the square of the semi circumference of one circle to its area. Whence if we multiply 7 by the square of the semi circumference of the one circle and divide the product by 22, we obtain the area of that circle.

[194] It remains to be shown how Archimedes the Philosopher found that the ratio of the circumference of a circle to its diameter is three and a seventh. I am not going to repeat the beautiful demonstration with his numbers since smaller numbers can show the same thing quite well [see Figure 3.98]. Consider circle *abgd* with diameter *ag*, center at *c*, and line *ez* tangent to the circle at *a*. Thus diameter *ag* is a cathete to line *ez*. Then on line *ac* from point *c* construct angle *ace* {p. 89} to be a third of a right angle.[190] Because angle *eac* is a right angle, angle *aec* is two thirds of a right angle. For two right angles measure the sum of the angles of every triangle. Line *az* equals line *ae* which joined to line *ez* makes triangle *caz* equal to triangle *cae*. Angle *ezc* is equal to angle *cea*, for each of them equals $\frac{2}{3}$ of a right angle. Likewise, if angle *cez* is twice angle *eca*, angle *ecz* is similarly equal to $\frac{2}{3}$ of a right angle. Therefore triangle *cez* is equiangular and equilateral. Whence side *ez* is the side of an equilateral and equiangular hexagon containing circle *abgd*.[191] Following the

---

[190] The distance $\overline{ea}$ on the tangent line equals the radius times $\sqrt{3}$.
[191] Printed between lines (89.6–7) and over this statement is "idest circa circulum *abgd*" (f. 54v.19).

usual practice let *ce* be 30; whence *ae* is 15. And because triangle *cae* is a right triangle: if 225 the square on side *ae* is taken from 900 the square on side *ce*, what remains is 675 for the square on side *ca*. Therefore side *ca* is the root of 675. If we compute carefully,[192] we will find that the number[193] is very close to 26 rods less $\frac{1}{13}$ 2 inches, a rods being 108 inches. Then divide angle *eca* in two halves by line *cf* that divides arc *ab* at point *y*. Now we have it from Euclid[194] that equal central angles intercept equal arcs on the periphery. Therefore arc *ay* equals arc *by*, and *ae* is half a side of the hexagon. Consequently, *af* is half the side of the dodecagon containing circle *abdg*.[195]

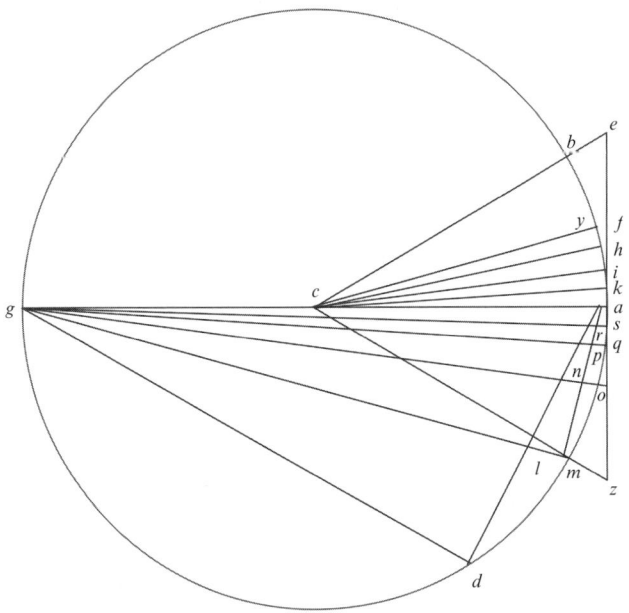

Figure 3.98

[195] And because angle *eca* was divided in two by line *cf*, *ec* is to *ca* as *ef* is to *fa*, as Euclid declared in Book VI.[196] Whence as the sum of *ec* and *ca* is to *ea* (that is, 56 less $\frac{1}{3}$ 2 inches to *ea*), so is 15 the sum of *ef* and *fa* to line *fa*. Alternately then, as the sum of *ec* and *ca* is to *ea* (that is, as 56 less $\frac{1}{13}$ 2 inches is to 15), so is *ca* to *af*. Whence let *ea* be 56 less $\frac{1}{13}$ 2 inches and *af*

---

[192] "si subtiliter ceperemus" (89.10).
[193] $\sqrt{675}$.
[194] *Elements*, III.26.
[195] Printed between lines (89,15–16) and over this statement is "idest circa circulum *abgd*" (f. 54v. 29).
[196] *Elements*, VI.1.

156   Fibonacci's *De Practica Geometrie*

be 15. So if we join the squares on lines *ca* and *af*, we will have 3359 less $\frac{2}{3}$ 16 inches for the square on line *cf*. The root of this is 58 less $\frac{4}{5}$ 4 inches, equal to the measure of side *cf*. Then divide angle *fc* in two equal parts by line *ch* to make *ah* a half side of the equilateral figure having 24 sides that is circumscribed about circle *abgd*. And because angle *fca* was bisected by line *ch*, the ratio of *fc* and *ca* to *ca* is as the ratio of *fa* to *ah*. Alternately therefore, as the sum of *fc* and *ca* is to *fa* (that is, as 114 less $\frac{7}{8}$ 6 inches is to 15) so is the ratio of *ca* to *ah*. So let *ca* be 114 less $\frac{7}{8}$ 6 inches and *ah* be 15. Consequently, if we take the root of the sum of those squares, we have 115 less $\frac{16}{23}$ 8 inches for line *ch*. Again, bisect angle *hca* by line *ci*. Then line *ae* is the half side of the equilateral figure having 48 sides that is circumscribed about circle *abgd*. From this, the ratio of *ai* to *ac* is as the ratio of 15 to the sum of *ac* and *ch* that is very nearly 229 less $\frac{41}{72}$ 15. Greater precision is not possible because irrational numbers lack rational roots.[197] Therefore let *ca* be 229 less $\frac{41}{72}$ 15 and *ae* be 15.

[196] Again, divide angle *ica* in two equal parts by line *ck*. Line *ak* becomes the half side of a 96 sided figure circumscribed about circle *abgd*. Add again the squares on line *ca* and *ai* to obtain the square on side *ci*. Its root is 229 and a bit less than $\frac{1}{23}$ 7 inches. But the ratio of *ca* to *ak* is as the ratio of the sum of *ic* and *ca* to *ai*. Therefore the ratio of *ca* to *ak* is almost $\frac{1}{5}$ 458 to 15. But the ratio of *ca* to *ak* is as the ratio of diameter *ga* to twice **{p. 90}** *ai*. But twice *ai* is the side of the equilateral figure of 96 sides circumscribed about circle *abgd*. Thus, as $\frac{1}{5}$ 458 is to 15, so is diameter *ga* to one of the sides of the figure of 96 sides. Consequently, if we multiply 15 by 96, we obtain 1140 for the measure of the sum of the sides of that figure. Therefore the ratio of all of the sides of the figure to the diameter of the circle within it is 1440 to $\frac{1}{5}$ 458.[198]

[197] Again, I look for the ratio of the circle to its diameter by the side of the figure of 96 sides lying within it in this way. I place in the same circle *abgd* the hexagonal side equal to the semidiameter *ca*. Then I join *g* and *d* to form right triangle *gda*, because it lies in the semicircle. Now according to Euclid Book III[199] every angle lying in a semicircle is a right angle. And because line *ad* is the side of the hexagon, its perimeter *ad* is a third of the perimeter of *dag*. Whence the perimeter *gd* is twice the perimeter *da*. Whence angle *gad* is twice angle *agd*, and both equal one right angle. Therefore angle *agd* is the third part of a right angle. I place in the aforementioned order the diameter *ag* of measure 30. Whence straight line *ad* is 15, and line *gd* is 26 less $\frac{1}{13}$ 2 inches, as we showed above. Then I divide angle *agd* in two equal parts by line *gm*, and I draw line *am*.

---

[197] It needs to be remembered that Fibonacci used the simple word number in various unqualified senses. Here (89.34–36) "rational roots" translates "*in numeris*."

[198] Text has $\frac{1}{3}$ 458; manuscript has $\frac{1}{5}$ 458 (90.6, *f.* 55r.4).

[199] *Elements*, III.31.

Then the ratio of line *al* to *ld* is as *ag* is to line *gd*. And after we have conjoined them, the ratio of line *ad* to *ld* is as the sum of lines *ag* and *gd* to line *gd*. Alternately, as *ag* and *gd* {**p. 91**} is to *ad* (that is, 50 less $\frac{1}{13}$ 2 inches to 15), so is *gd* to *dl*. Because angle *agd* was divided in two equal parts by line *gm*, angle *agm* is equal to angle *dgm*, and angle *gdl* is equal to angle *gma*. Both of these are right angles because they are in semicircle *gdma*. The other angle under *gld* is equal to the angle under *gam*. Therefore equiangular triangle *gdl* is equal to triangle *gma*. Whence as line *gd* is to *dl*, so is line *gm* to *ma*. Consequently, I find line *gm* to be 56 less $\frac{1}{13}$ 2 inches, and line *ma* to be 15. Line *am* is the side of a dodecagon, since its perimeter *am* is half the perimeter of *amd*.

[198] Again I divide angle *agm* in two equal parts by line *gno* and join points *a* and *o*. I find the root of the sum of the squares on lines *gm* and *ma* to be 58 less $\frac{4}{5}$ 4 inches. And as the sum of the lines *ag* and *gm* is to line *ma* (that is, as 114 less $\frac{7}{8}$ 6 inches is to 15), so is *gm* to *mn*. But as *gm* is to *mn*, so is *go* to *oa*, for triangle *gmn* and *goa* are similar right triangles. Therefore as 114 less $\frac{7}{8}$ 6 is to 15 so is *go* to *oa*. Whence I find *go* to be 114 less $\frac{7}{8}$ 6 inches and *oa* to be 15. Again I take the root of the squares on lines *go* and *oa*, and have for line *ga* 115 less $\frac{16}{23}$ 8 inches. Whence line *oa* is the side of a figure of 24 sides inscribed in circle *abdg*.

[199] Again I divide angle *ago* in two equal parts by line *gq* and join points *q* and *a*. As *ga* and *go* is *oa*, so is *go* to *op*. But as *go* is to *op*, so is *gq* to *qa*. Therefore as 229 less $\frac{41}{72}$ 15 inches is to 15, so is *gq* to *qa*. When I make *gq* to be 229 less $\frac{41}{72}$ 15 inches and *qa* 15. I add their squares, take the root, and get 229 and a bit less than $\frac{1}{2}$ 37 inches. Line *aq* is the side of a figure with 48 sides. Again I divide angle *agq* in two equal parts by line *grs* and join points *s* and *a* to become the side of a figure with 96 sides within circle *abgd*. Because angle *agq* was divided in two equal parts by line *gs*, the ratio of *gq* to *qr* is as the ratio of *gs* to *sa*. Since triangles *gqr* and *gsa* are similar, $\frac{1}{5}$ 458 is to 15 as *gs* is to *sa*. Joining the squares on lines *gs* and *sa*, I take the root of the sum and have $\frac{4}{9}$ 458 for the diameter *ga*. Therefore I multiply line *sa* by 96 to get 1440 for the sum of all sides of the figure inscribed within circle *abgd*. Whence as 1440 is to $\frac{4}{9}$ 458, so are all the sides of the figure inscribed in circle *abgd* to the diameter of circle *ga*.

[200] We have found by the investigation of the side of the exterior figure that the ratio of all its sides to the diameter of the circle is 1440 is to $\frac{1}{5}$ 458. And the circumference is less than all the sides of the figure containing the circle, and more than all the sides of the figure inscribed within the circle. Therefore the ratio of the circle to its diameter is as 1440 is to $\frac{1}{3}$ 458 which lies between $\frac{4}{9}$ 458 and $\frac{1}{5}$ 458. But the ratio of 1440 to $\frac{1}{3}$ 458 is as thrice one number to thrice the other, that is as 4320 is to 1375, which is reduced to 864 to 275. But the ratio of 864 to 275 less $\frac{1}{11}$ is as $\frac{1}{7}$ 3 to 1. Because there is less difference between these ratios than between [the ratio of the circumference of] a circle and its diameter, the ratio they have is $\frac{1}{7}$ 3 to 1. The wise men of

antiquity held that the [circumference of a] circle is thrice and a seventh of its diameter. And this is what I wanted to show.

[201] If you want to measure a semicircular field, measure the circle by one of the ways that I showed and then take half of that area for the area of the semicircle {p. 92}. To make this clear, consider the semicircle *abg* with diameter *ag* of 24 panes. Complete the circle *abgd* so that the addition *agd* is also a semicircle [see Figure 3.99]. Whence if we take half the area of circle *abgd*, we will have the area of the given semicircle *abg*. Or, take half the diameter (12) and multiply it by $\frac{1}{7}$ 3, and you will have $\frac{5}{7}$ 37 for arc *abg*. If you will have multiplied half the diameter by half the arc *abg*, or a fourth of the diameter by all of arc *abg*, then you will have $\frac{4}{7}$ 226 for the area of semicircle *abg*. Or, if you take a twenty-eighth part of the square on the diameter and multiply it by 11, you will find the same area. And if you take a fourteenth of the same square on the diameter, you will have $\frac{1}{7}$ 41 for the area in panes.

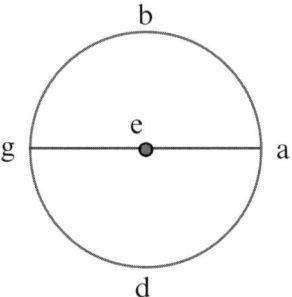

Figure 3.99

[202] Now if you want a different approach to an understanding of arc *abg* the semicircumference of the circle, then erect at the center of the diameter *ag* perpendicular line *eb* as the semidiameter of circle *abgd*. Join lines *ab* and *bg* [see Figure 3.100]. The angle under *abg* is a right angle since it is in semicircle *abg*, or because *be* equals lines *ea* and *eg*, and both triangles *aeb* and *beg* are isosceles. Whence angles *eba*, *eab*, *ebg*, and *bge* are all equal to one another, each being half of a right angle. Whence the two angles *abe* and *ebg* equal one right angle. Therefore angle *abg* is a right angle. So the two lines *ae* and *eb* equal the two lines *be* and *eg*. And angles *aeb* and *beg* are right angles, the lines *ab* and *bg* being equal one another. Whence the square on diameter *ag* is twice the square on each of the lines *bg* and *ba*. Also each of the squares on lines *bg* and *ba* equals each of the squares on lines *ae* and *be*. And again the sum of the squares on lines *be* and *eg* is twice the square on line *bg*. Whence as *ag* is to *gb* so is *gb* to *be* or to *eg*. Whence if we multiply *ag* by *be* (24 by 12), we have 288 for the square on each of the lines *ab* and *bg*. Or, if we take half the square on the diameter *ag* (or if we double the square on the semidiameter *ge* or *ea*), we have again 288 for the

square on each of the lines *gb* and *ba*. And *gb*[200] is the arc of half the semicircle *abg*. Then if we bisect chord *bg* at point *c* and draw line *df* through points *c* and *e*, line *df* becomes the diameter of circle *abgd*, makes right angles at point *c* with chord *bg*, and divides arc *bfg* in two equal parts. Whence if from point *f* we draw lines *fb* and *fg*, each will be a chord of the fourth part of semicircle *gba*.

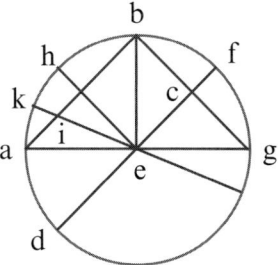

Figure 3.100

[203] We come to an understanding in this way. Because angle *bce* is a right angle: if we subtract the square on line *bc* or 72 (the fourth part of the square on chord *bg*) from the square on line *be* (opposite right angle *bce*), 72 remains for the square on line *ce*. Whence the entire line *dc* measures 12 plus the root of 72. (This is called a binomial since it cannot be expressed in rational numbers.) Whence if we take *ec* from *ef*, 12 less the root of 72 remains for line *cf*. (This is called the abscissa or recise or apotame since it consists of a number less a root.) Then we take the squares on lines *fc* and *cb*, and you have the square on chord *bf* or *fg*. In a moment I will explain how to get the square on the abscissa *cf*. Arrange the terms 12 and 72 as I show in the margin [see Figure 3.101]. Multiply 12 by 12 to get 144 and the root of 72 by the root of 72 to get 72. The sum of the numbers is 216. From this take twice 12 multiplied by the root of 72 {**p. 93**}; you obtain 24 roots of 72. And thus you have for the square on line *cf* 216 less 24 roots of 72 which is one root of 41472. Or, because line *ef* was divided deliberately at point *c*, there are two squares on lines *ef* and *ec* equal to the square on *cf* and twice the rectangular surface *ec* by *ef*. For the squares on lines *ef* and *ec* are 216. If we take from this twice the product of *ef* and *ec* (24 by the root of 72), we obtain indeed 216 less the root of 41472. (This called the first recise, as will be shown in its own place.) If we add to this 72 the square on line *bc*, we have 288 less the root of 41472 for the square on chord *bf*. (This is called the fourth recise whose root is the line called minor.) Since you want a close approximation of the measure of line *bf*, subtract the root of 41472 (which is a little less than $\frac{2}{3}$ 203) from 288 to leave somewhat more than $\frac{1}{3}$ 84 for the square one of those four chords *gf*, *gb*, *bh*, or *ha*.

---

[200] Text and manuscript have *ge*; context requires *gb* (92.27, f. 56r.22).

Figure 3.101

[204] Or, you can find the root in another way by the root of the square on line ce which is a little less than $\frac{1}{2}$ 8. Subtract this from 12 line ef, and a little more than $\frac{1}{3}$ 3 remains for line cf. Now adding its square (a little more than $\frac{1}{3}$ 12) to the square on line bc gives us likewise a bit more than $\frac{1}{3}$ 84 for the square on one of the four chords. If we multiply this by 16 the square of 4, we will have 1350 for the square of the sum of the four chords. The root of this is about $\frac{3}{4}$ 36. But arc abg is $\frac{5}{7}$ 37. Finding the four chords left us still somewhat far from knowing their lengths. So again divide one of those four in two equal parts, say, chord ah at point i. Draw through points i and e diameter kl to divide arc akh in two equal parts at point k. Then draw chord ak. It will be the side of the figure falling within circle abgd, a figure of 16 sides. We understand this by what was demonstrated, namely by taking the square on line ai (about $\frac{1}{11}$ 22, the fourth part of the square on line ah) from the square on line ae (144). What remains is $\frac{10}{11}$ 121 for the square on line ei. Whence that line is about $\frac{1}{11}$ 11. And if we take it from line ek, what remains is about $\frac{10}{11}$ of one rods for ak. If we add the square of this (about $\frac{9}{11}$) to the square on line ai, we will have $\frac{10}{11}$ 21 for the square on chord ak. This is the chord of an eighth part of arc abg. So if we multiply $\frac{10}{11}$ 21 by 64 the square of eight, we will have about 1402 for the square on the eight equal chords falling within semicircle abg. The root is a bit less than $\frac{1}{2}$ 37. But arc abg is more, about $\frac{5}{7}$ 37. Whence if we use the same technique to find the chord of the semi arc ak, we will be closer to the length of arc abg. And thus by continuing to divide the arc, we will come to know the measure of the chord whose difference with the measure of the arc is negligible. In this way you can find the measure of any chord you wish.

[205] In order to understand the foregoing more clearly, consider circle abcd whose diameter ac is 10, and let the measure of chord bd be given as 8. We wish to measure an arc by knowing the length of its chord. First, I will show how to use the chord to find the length of each arrow.[201] By knowing the arrow you can find the length of the chord. This is how it is done. Because line ac was divided in two equal parts at point f and in two unequal parts at point e, the product of ae by ec together with the square on line ef is equal to the square on line fa. But fa is equal to fb, since both of them were drawn from the center f to terminate on the periphery of the circle. Next, the product of ae and ec with the square {p.94} on line ef is equal to the square on line fb [see Figure

---

[201] These are segments ae and ec of diameter ac in Figure 3.101; see 93.38.

3.102]. But the square on line fb is equal to the two squares on lines be and ef. Therefore the product of ec and ae with the square on ef is equal to the two squares on lines be and ef. Whence if we take the square on line ef from both, what remains is the product of ae by ec equal to the square on line be. Whence if we take the square on line ef from the square on line bf (that is, subtract 16 from 25), what remains is 9 for the square on line ef. Therefore ef is 3. If cf is added to this, then the arrow ce becomes 8. Similarly, after taking fe from af what remains is 2 for line ae. Now the product of ae by ec or 2 by 8 equals the square on line be. Whence the lines ae, eb, and ec are in continued proportion: as ae is to eb, so is eb to ec. Therefore, by knowing the measure of chord bd, we know the measures of the arrows ae and ec. Now ae and ec are known arrows; whence ae is 2 and ec is 8. We want to find [the measure of] chord bd. Because the product of ae and ec equals the square on line be: if we multiply ae by ec (2 by 8), we have 16 for the square on line be. Doubling its root 4 yields 8 for chord bd. Now chord bd is known as is the arrow ea; but the diameter ac is unknown. So you square half the chord bd to get 16, divide this by 2 the arrow ae, to find 8 for the arrow ec. Whence the diameter ac is 10. Similarly, if we divide the square on line be by the arrow ec, we obtain the arrow ae.

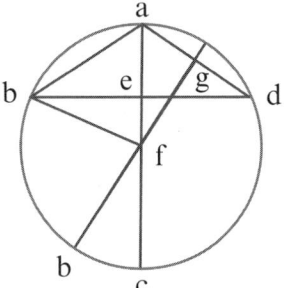

Figure 3.101

[206] It must be noted that in semicircle abc, straight line be is called the right sine of both arcs ab and bc, straight line ae is called the versed sine of arc ab, and straight line ec is the versed sine of arc bc, as they are used in the art of astrology. With these concepts explained, let us return to our subject.[202]

[207] Let us draw the chords ba and ad for the two arcs ba and ad; these are known. If we join the squares on lines ae and ed or ae and eb, then the square on each of the lines ab and ad is the root of 20. Whence if we divide chord ad at point g and draw the diameter hi through the points g and f, we will find the measures of the arrows ig and gh. Since we joined the squares on lines ag and gi, we have the square on chord ai that is the chord of arc ai. That in turn is the fourth part of the whole arc bad. Now, as we often do, we can get

---

[202] Fibonacci seemed to have assumed that his readers would know that the versed sines in [206] are the arrows in [205] and elsewhere, "arrow" being from the Latin "*sagitta*" from the Arabic "*sahem*".

a close idea of the measure of the whole arc *bad*. If we subtract arc length $\frac{3}{7}$ 31 from the circumference *abdg*, we find the measure of arc *bcd*.

[208] There is another way to find the chords of semi arcs that leads to knowing [their measures]. Ptolemy mentioned it in his *Almagest*.[203] Let the diameter *bd* and chord *ad* be known in circle *abgd*, and we wish to find the chord of the half arc *ad*. Draw chord *ab* which with the given chord *ad* makes angle *dab* a right angle. Whence the square on the diameter *bd* equals the squares on the two chords *da* and *ab*. Construct line *ba* equal to line *be*. Bisect angle *abd* by line *bz*. Join lines *ad*, *ze*, and *za* [see Figure 3.103]. From point *z* above diameter *bd* draw cathete *zi*. Now line *ab* equals line *be*. If line *bz* is added to both, then the two lines *ab* and *bz* equal the two lines *zb* and *be*. And angles *abz* and *zbe* are equal. Whence base *az* equals base *ze*. But line *az* equals lines *zd*, since they are both equal to the arc lengths *az* and *zd*. Because equal angles have equal arc lengths, whether the angles are measured at the center [of the circle] or on the arc lengths. Angles at *b* are equal to one another {**p. 95**}. And they are on the peripheries *az* and *zd*. Whence straight line *ze* equals straight line *zd*, and therefore triangle *zed* is equilateral. Whence point *i* that locates cathete *zi* falls in the middle of *ed*. And because triangle *bzd* is in the semicircle, it is a right triangle. On its base a cathete has been drawn from a right angle. Then triangle *bzd* can indeed be divided into two triangle similar to each other. Each has a right angle and one angle in common with the whole triangle *bzd*, as Euclid shows in VI.8. Whence as *bd* is to *dz*, so is *zd* to *di*. Whence the product of *di* and *bd* equals the square on line *zd*. For *id* is known since it is half of *ed* which is known because of *be* that is equal to the known chord *ba*. Whence if chord *ba* (that is *be*) is taken from the diameter *bd*, what remains known is *ed*. So *id* its half is known. Whence if we multiply the known *di* by the known *bd*, we get the square of the chord *zd* as known. Therefore straight line *zd* is known, as we said.

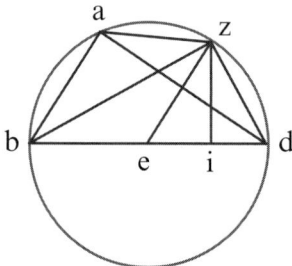

Figure 3.103

[209] This can be shown in numbers. Let the diameter *bd* be 10, chord *da* 8, whence chord *ab* is 6. Since straight line *be* equals chord *ab*, it is also 6. Subtract this from the diameter to leave 4 for line *ed*. Its half or 2 is line *id*. The product of *id* and *bd* produces 20 that is equal to the square on chord *zd*. Whence [the

---
[203] *Almagest*, I.10.

measure of] chord *zd* is the root of 20, as we had to show. Likewise, if we take the square on line *zd* from the square on the diameter *bd*, 80 remains for the square on chord *bz*. Its root is 9 less $\frac{1}{18}$. If we subtract this from the diameter *bd*, what remains is $\frac{1}{18}$ 1. If we multiply its half by the diameter *bd* or if we multiply $\frac{1}{18}$ by half the diameter or 5, we get $\frac{5}{18}$ 5 for the square on the chord of the semi arc *dz*. In this way we can find the chords of the halves of any number of given arcs.

[210] This method is not used by surveyors who prefer a more common way of proceeding. The common procedure for finding the measure of an arc is to set the length of some line as one foot that can be curved and extended. They use this to measure whatever arc they choose. Or they might use a rope of one or more rods in length with which they try to measure an arc of a circle, imagining that they were fitting a measuring rod around a circle so that the rope would not deviate from the circumference of the circle. Thus they can find the measure of any arc of a circle.

[220][204] Since you know how to find arcs by their chords and chords by their arcs by what we have said and you want to find the area of a sector of a circle, then find its arc, multiply its half by the semidiameter of the circle and the result is the area of the sector.[205] For example: let sector *abgd* lie between straight lines *ab* and *ad* and arc *bgd*. Since the sector is figure *abgd*, each of the straight lines *ab* and *ad* is a semidiameter of the circle. Whence point *a* is the center of the circle from which sector *abgd* was selected. Now complete the circle from which sector *abgd* was selected, and the circle is *gbed* [see Figure 3.104]. Let arcs *be* and *ez* equal arc *bgd*. Complete straight lines *ae* and *az*. And so each sector *abe* and *aez* equals sector *abgd*. Whence as arc *db* is to arc *be*, so is sector *abgd* to sector *abe*. Similarly, sector *aez* equals sector *adb*. Whence as arc *db* is to arc *ez*, so is sector *abgd* to sector *aez*. And so three sectors, *abgd*, *abe*, and *aez*, are mutually equal.

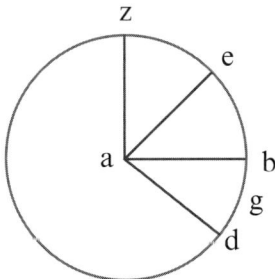

Figure 3.104

[221] Since those two sectors are twice the third, it follows that sector *adbe* is twice sector *aez*, and arc *dbe* is twice arc *ez*. Whence as arc *dbe* is to arc *ez*,

---

[204] The jump in paragraph and page numbers is due to the transfer to Chapter 7 of paragraphs [211] to [219] on Fibonacci's *Table of Chords*.

[205] Text and manuscript have *sectionis*; context requires *sectoris* (100.14, f. 61r.41).

so is sector *adbe* to sector *aez*. Whence, if we divide the whole circle into sectors, we will find that the ratio of one of them to the whole circle is as its arc to the circumference of the circle. But the ratio of an arc of any sector to the whole circumference is as half the arc to half the circumference. For the ratio of the product of the semidiameter of a circle by half the arc of a sector is to the product of the semidiameter by half the circumference, as half the arc of the sector is to half the circumference. But the area of the circle comes from the product of the semidiameter and half the circumference. Therefore as the product of the semidiameter of the circle and half the arc of the sector is to the area of the sector, so is the sector to the area of the circle. Therefore, the areas of all the sectors is the product of the semidiameter of their circle by half their arcs. And this is what I wanted to demonstrate.

[222] This can be made clear with numbers. Let two straight lines *ab* and *ad* equal 5 rods, arc *bgd* be 8 rods, and the diameter 10 rods. Multiply the semidiameter *ad* by half the arc (5 by 4), and you get 20 for the area of the sector *abgd*. If you want to know the area of sector *abez*, multiply the semidiameter *ae* by half the arc *bez* to get 40 for the area of sector *abg*. If you want only the area of some small sector of a semicircle, such as the area of sector *bez* with chord *ag* equal to 16 rods and arrow *bd* equal to 4 rods, you can find the diameter of the circle in this way: extend line *bd* straight on to point {p. 101} *e*. Let the ratio of straight line *de* to straight line *da* be as *da* is to *db*. Whence *ed* will be 16, namely twice line *da* since *da* is twice line *db*. In another way: multiply *ad* by *dg* (8 by 8) to get 64, and divide by 4 the straight line *bd* to get 16 for line *de*, as we said. Consequently, the whole diameter *be* is 20 [see Figure 3.105]. Take *f* to be the center of the circle, draw lines *fa* and *fg* to locate sector *fabg*. So, if we multiply *fb* by half of arc *abg* or by arc *bg*, we obtain the area of sector *fabg*. If we remove from this the area of the rectilinear triangle *fag* which is found by multiplying *fd* by *dg*, what remains is the area of the sector *abd* bounded by straight line *ag* and arc *abd*. And if you want to know the area of the rest of the circle contained by straight line *ag* and arc *aeg*, multiply the semidiameter *fe* by half of arc *aeg*, and add the product to the area of triangle *fag*. Thus you will have the area of segment *aeg*. Follow this procedure in similar situations.

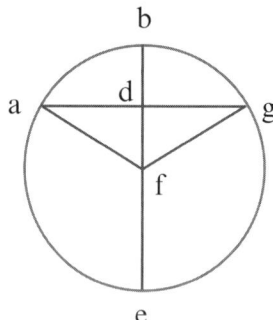

Figure 3.105

[223] Should it happen that you have to measure a field in the shape of a sector, as field *abgd* that is composed of triangle *abd* and circular segment *bgd* [see Figure 3.106], then add the areas of triangle *abd* and circular segment *bgd*. Understand that the figure is not a sector if the straight lines *ab* and *ad* are not equal to each other, or if any line *ag* drawn below the figure is not equal to one of the straight lines *ab* or *ad*.

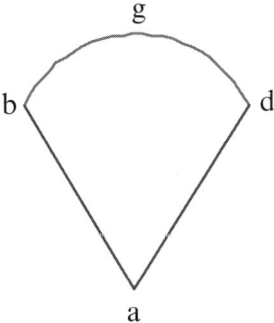

Figure 3.106

[224] Likewise: if you want to find the area of a field that has a figure like a fish that is composed of two circular segments *ezi* and *eti* [see Figure 3.107], take the area of one of the segments and add it to the other for the area of figure *ezit*.

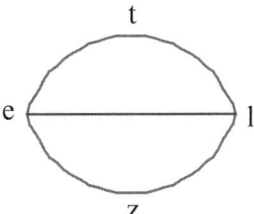

Figure 3.107

[225] Again: suppose there is a field described as *elana* or oblique which is circumscribed by only one line. It resembles a circle but with unequal diameters *ag* and *bd* that intersect each other at right angles [see Figure 3.108]. In order to measure this field, try to use straight lines. First, the larger quadrilateral is bounded by straight lines *ba*, *bg*, *da*, and *dg*. What remain are four segments of the circle:[206] the first contained by straight line *ab* and

---

[206] "et remanebunt ex ipsa figura pectora quatuor" (101.29, f. 62r.6) and elsewhere below *segment* for *pectus*.

166   Fibonacci's *De Practica Geometrie*

curved line *aeb*, the second by straight line *bg* and curved line *bzg*, the third by straight line *gd* and curved line *gid*, and the fourth by straight line *da* and curved line *dta*. Now if we removed the rectilinear triangles *eab*, *zbg*, *igd*, and *tad* from the four segments, not much remains in the figure apart from the little bit contained by the 8 segments. And if in any of them {**p. 102**} a triangle is drawn, try to proceed in the same way for the remaining segments. Thus the entire oblique figure can be resolved into rectilinear figures, so that nothing perceptible remains. If we can possibly find the area of just one of all the rectilinear figures, then we undoubtedly have the area of the whole figure.

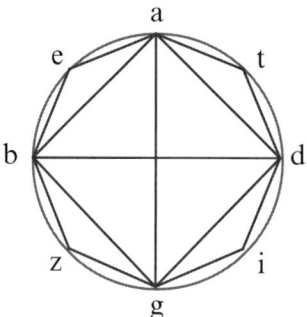

Figure 3.108

[226] Otherwise: join the two diameters *ag* and *bd* as one, multiply their half by $\frac{1}{7}$ 3, and whatever results is the measure of the curved line *abgd*. If its half were multiplied by half the mean of the two diameters, you would have the area of the figure. You can see this in numbers: let diameter *db* be 16, diameter *ag* 12, and their sum 28. Multiply half of this by $\frac{1}{7}$ 3 and you will have 44 for the length of curved line *abgd*. Now multiply its half or 22 by 7 half the mean of the diameters, and you obtain 154 for the area of the figure.

[227] *If a triangle is inscribed in a circle so that its vertices touch the circumference and the lengths of the sides of the circle are known, then the diameter of the circle can be found.* To clarify this consider triangle *abg* inscribed in circle *abgd* with vertices *a*, *b*, and *g* on the circumference. From point *a* draw a diameter that cuts side *bg* of the triangle at point *e*. I say that knowing the lengths of the sides, I can determine the length of the diameter *ad* [see Figure 3.109]. First, let *ab* and *ag* the two sides of the triangle equal one another; then draw lines *bd* and *gd*. Thus each of the triangles *abd* and *agd* is a right triangle because each is drawn on the diameter of the circle. Further, because side *ab* equals side *ag*, sides *bd* and *dg* are equal. Whence arc *bd* equals arc *dg*. Now equal arcs have equal central angles. Whence angle *bad* equals angle *dag*. Whence base *be* equals base *eg*. Therefore diameter *ad* cuts straight line

bg^{207} in two equal parts as well as the angles, as Euclid demonstrated in Book III.[208] Triangles *aeb* and *aeg* are equal right triangles and similar because the angles of one equal the angles of the other. And because triangle *adb* has angle *bad* in common with triangle *abe*, and angle *aeb* equals angle *abd* since both are right angles, therefore triangles *abd* and *aeb* are equiangular triangles. In a similar way it can be shown that triangle *agd* is equiangular with triangle *aeg*. Therefore the four triangles *aeb*, *aeg*, *abd*, and *agd* are all similar to one another. Now similar triangles have proportional sides about the equal angles. Whence as *da* subtends right angle *abd* that contains line *ab*, so *ab* or *ag* subtends the right angles at line *ae*. Whence the product of *ad* by its cathete *ae* equals each of the squares on lines *ab* and *ag* or the product of *ab* and *ag*. So, if we were to multiply *ab* by *ag* or take the square on line *ab* or on line *ag*, and divide the result by cathete *ae*, the outcome is the measure of the diameter *ad*. The cathete *ae* is known because the sides of triangle *abg* are known. Whence the diameter *ad* is known.

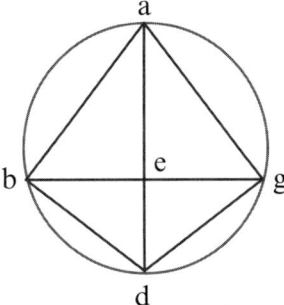

Figure 3.109

[228] To show this in numbers: let each of the lines *ab* and *ag* be 10 rods, and straight line *bg* be 12. Whence cathete *ae* is 8. Indeed, from the product of *ab* and *ag*, or from the square on line *ab* or *ag*, 100 arises. Dividing this by *ae* or 8, we obtain $\frac{1}{2}$ 12 for the diameter *ad*. Or in another way: because the two straight lines *ad* and *bg* intersect one another in circle *abgd*, the product of *ae* and *ed* equals {**p. 103**} the product of *be* and *eg*. So, if we multiply *be* by *eg* and divide by *ae* (36 by 8), then we have $\frac{1}{2}$ 4 for line *ed*. Therefore the whole diameter *ad* is $\frac{1}{2}$ 12, as I said.

[229] Now, suppose one of the lines *ab* or *ag* together with the diameter *ad* is known while straight line *bg* is unknown: because triangle *abd* is a right triangle and the line *be* drawn from the right angle to the base *ad* creates right triangles *abe* and *bed* similar to each other and to triangle *abd*, it follows that *da* is to *ab* so is *ab* to *ae*. So, if we divide the square on line *ab* (100) by

---

[207] Text and manuscript have *eg*; context requires *bg* (102.23, f. 62v.2).
[208] *Elements*, III.3.

the diameter *ad* we obtain 8 for the cathete *ae*. If we take its square from the square on side *ab*, what remains is 36 for the square on line *be*. Or, if we multiply *ae* by *ed* (8 by $\frac{1}{2}$ 4), we obtain the same 36 for the square on line *be*. Therefore 6 measures *be*, and the whole line *bg* is 12. Or in another way: because the triangle *aeb* is similar to triangle *adb*, then as *ad* is to *db*, so is *ab* to *be*, because sides opposite equal angles are proportional, the equal angles being *abe* and *adb*. Whence if we multiply *db* by *ba* ($\frac{1}{2}$ 7 by 10) and divide by *ad*, we obtain 6 for line *be*. This is half of line *bg*.

[230] Now, let lines *ab* and *ag* be unequal with *ab* the shorter as shown in this other figure[209] [Figure 3.110]. Draw cathete *az* in triangle *abg*. Because there are two angles in segment *bdga*, namely angles *bga* and *bda* both equal to one another, [and because] angle *azg* equals angle *abd* since both are right angles, angle *zag* remains equal to angle *bad*, for the triangles *azg* and *abd* are equiangular. Likewise it may be shown that triangle *azb* is similar to triangle *agd* for they are in the segment bounded by line *ga* and arc *abdg* of angle *abg* and *adg*. Whence the angles are mutually equal, and angles *azb* and *agd* are right angles. Whence the remaining angle *azb* equals the other remaining angle *gad*. Therefore triangle *azb* is similar to triangle *agd*. And because triangle *abd* and *azg* are similar, then as *da* is to *ab*, so is *ga* to *az*. Whence if we multiply *ab* by *ag* and divide by *az*, we obtain the diameter *ad*. For a numerical example: let *ab* be 13, *ag* 15, and *bg* 14. Whence *bz* is 5, *zg* is 9, and *az* 12. From [the product of] *ab* and *ag* we obtain 195 that divided by *az* or 12 leaves $\frac{1}{4}$ 16 for the diameter *ad*. Let the diameter *ad* be known with either straight line *ab* or *ag* and *bg* the chord of arc *bag* unknown: because triangles *adg* and *azb* are similar with proportional sides opposite equal angles, as *ad* is to *dg* so is *ab* to *bz*. Whence the product of *ab* and *dg* equals the product of *ad* and *bz*.

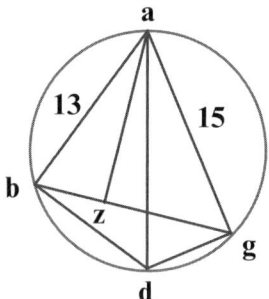

Figure 3.110

---

[209] I supplied the figure missing in the margin of Boncompagni (103).

[231] Again: because triangles *abd* and *azg* are similar, *ad* is to *db* as *ag* is to *gz*. Whence the product of *ag* and *db* equals the product of *ad* and *zg*. But the product of *ab* and *gd* was equal to the product of *ad* and *bz*. Therefore the product of *ab* and *gd* with the product of *ag* and *bd* equals the two products *ad* by *bz* and *ad* by *zg*, the latter two products being equal to the product of *ad* and *bg*. Therefore the product of *ab* and *dg* with the product of *ag* and *db* equals the product of *ad* and *bg* [see Figure 3.111]. Therefore if we multiply *ab* by *dg*, add this product to the product of *ag* and *bd*, and divide all of this by *ad*, we obtain the chord *bg*, as we said. This can be shown in numbers: from the square on diameter *ad* take the squares on lines *ab* and *ag*. What remains known are the squares on lines *bd* and *dg* because triangles *abd* and *agd* are right triangles. So $\frac{3}{4}$ 9 measures line *bd* and $\frac{1}{4}$ 6 measures line *gd*. Whence if **{p. 104}** we multiply $\frac{1}{4}$ 6 by 13 which is *gd* by *ab*, and $\frac{3}{4}$ 9 by 15 that is *bd* by *ag*, we have $\frac{1}{2}$ 227. Divide this by $\frac{1}{4}$ 16 the diameter *ad* to return 14 for the chord *bg*. This is very useful for finding the chord of any arc formed by two arcs whose chords are known [see Figure 3.112]. Indeed, the chords *ab* and *ag* for arcs *ba* and *ag* were known, and by knowing these we found chord *bg* of arc *abg* which is the sum of arcs *ba* and *ag*. Ptolemy demonstrated this employing a different method to construct the table of arcs and chords for *The Almagest*.[210]

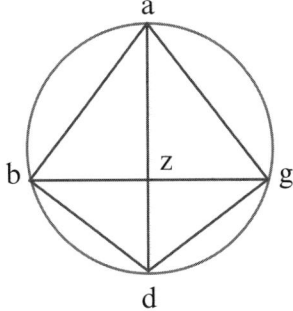
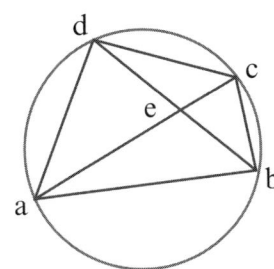

Figure 3.111   Figure 3.112

[232] Ptolemy used a similar figure to demonstrate that for any quadrilateral inscribed in a circle, the product of its two diameters equals the sum of the products of the opposite sides, which is proved thus [211] Given quadrilateral *abcd* inscribed in circle *abcd*, in which *ac* and *bd* the diameters of the quadrilateral intersect each other at point *e*. I say that the product of *ac* and *bd* equals the sum of the two products *ab* by *cd* and *ad* by *bc* [see Figure 3.113]. Now angles *bae* and *ead* are either equal to each other or unequal. First, let them be equal. It will be shown that triangle *aed* is similar to triangle *abc*. Because angles *abd* and *acd* are equal to each other, they are on the same segment. Angle *aed* is similarly

---

[210] Op. cit., I.10.
[211] Op. cit., I.10.3. Fibonacci's proof differs from that of Ptolemy.

equal to angle *abc*. Whence as *ac* is to *cb*, so is *ad* to *de*. Therefore the product of *ad* and *cb* equals the product of *ac* and *de*. Let it be similarly shown that triangle *aeb* is similar to triangle *adc*. Whence as *ac* is to *cd* so is *ab* to *be*. Therefore the product of *ac* and *eb* equals the product of *ab* and *cd*. Consequently the sum of the products of *ad* and *bc* and of *ab* and *cd* equals the product of *ac* and *bd*, as I said. But if angle *bae* is greater than angle *ead*, let angle *bai* equal angle *ead*. To both of these add angle *iae*. Thus angle *bae* becomes equal to angle *iad*. Whence it is shown that triangle *iad* is similar to triangle *abc*, and triangle *aib* is similar to triangle *adc*. Thus we are where we were above. And it is shown that the product of the diameters *ac* and *bd* is equal to the sum of the products of the opposite sides. Whence if one of the chords is unknown, you can find it provided you know the others. For example: let the measure of one of the diameters be unknown, with the other diameter and sides given. Multiply *ab* by *cd* and add the outcome to the product of *ad* and *bc*. Divide the sum by the given diameter. If one of the sides were unknown such as side *ab*, then subtract the product of *ad* and *bc* from the product of *ac* and *bd*. Divide the remainder by *cd* to obtain *ab*. Understand it thus for finding the other lengths.

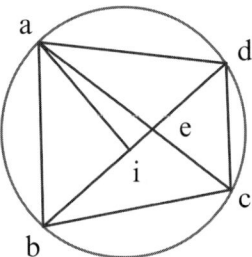

Figure 3.113

[233] Suppose that in a circle of diameter 8 you wish to find the side of an inscribed equilateral triangle. Remove a fourth part of the square on the whole diameter[212] or triple the square on the semidiameter, and you will have 48 for the square on one of the sides of the inscribed triangle. Multiply this by half the cathete or 3 and you will have three times the root of 48 which is a root of 432 for the area of the triangle, that is, 20 rods, 4 feet, and $\frac{3}{4}$ 12 inches. Whence the ratio of the area of each triangle inscribed in the circle to the square on its diameter is 20 rods, 4 feet, and $\frac{3}{4}$ 12 inches to 64. So, if we multiply it by the square of the diameter and divide by 64, we will have the area of the triangle inscribed in the circle. Because the triangle inscribed in the circle is half the hexagon inscribed in the circle {p. 105} the ratio of the area of the hexagon to the square on the diameter of its circle will be as 41 rods, 3 feet, and $\frac{1}{2}$ 7 inches to 64. Hence, the area of any hexagon inscribed in a circle equals the product of the square on the diameter and 41 rods, 3 feet, and $\frac{1}{2}$ 7 inches divided by 64.

---

[212] Fibonacci means that three-fourths of the diameter is left with which to compute.

[234] Further, if you want to find the side of a pentagon or decagon inscribed in circle *abgd* with diameter *bd* measuring 8, draw cathete *ea* to the diameter at its center *e*. Then divide *ed* in two equal parts at point *z*, and draw straight line *az* [see Figure 3.114]. Lay out line *zi*[213] equal to line *az*, and draw line *ai*. I say that line *ai* is the side of the pentagon and that *ie* is the side of the decagon, which I prove thus. Because line *ed* was divided in two equal parts at point *z*, and line *ei* was added in a straight line, the product of *ei* and *id* together with the square on line *ez* equals the square on line *zi*. But line *zi* equals line *az*. Therefore, the product of *ei* and *di* with the square on line *ez* equals the square on line *az*. But the squares on lines *ae* and *ez* equal the square on line *az*. Therefore the product of *ie* and *id* with the square on line *ze* equals the two squares on lines *ae* and *ez*. After removing from both [sides of the equality] the square on line *ez*, what remains is the product of *ei* and *id* equal to the square on the semidiameter *ae*, that is, the square on semidiameter *de*, since line *de* equals line *ae*. Therefore line *di* has been divided in mean and extreme ratio. For as *id* is to *de* so is *de* to *ei*, and the side of the hexagon is line *de*. Since therefore a straight line has been joined to the hexagonal side that has been divided in mean and extreme ratio at the point of juncture, that line joined to the hexagonal side is the side of the decagon, as Euclid demonstrated in Book XIII.[214] Whence line *ei* is the side of the decagon. And because, as Euclid showed in the same book[215] that a pentagonal side can be over a hexagonal side and a decagonal side, straight line *ai* is the pentagonal side since [the square on][216] side *ai* can be the equal to [the sum of] the squares on lines *ae* and *ei*, namely the squares on the hexagonal and decagonal sides.

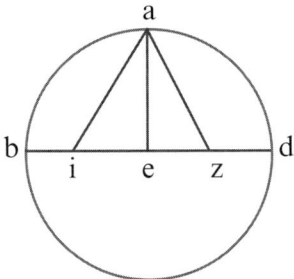

Figure 3.114

[235] I shall show this in numbers, using the minor line, namely the root of the fourth recised. For example: let diameter *bd* be 8, whence the two lines *ae* and *ed* are each 4, and half or *ze* is 2. Whence if we join the squares on lines *ae* and *ez*, we have 20 for the square on each of the lines *az* and *zi*. Therefore line *ei* consists of a root less a number and the root is 20 less 2. If we square it,

---

[213] Text has *z*; manuscript has *zi* (105.7, f. 64v.19).
[214] *Elements*, XIII.9.
[215] Ibid., XIII.10.
[216] Only the added bracketed phrase makes the clause correct.

we obtain 24 less 4 roots of 20. The four roots are one root of sixteen times 20 or 320. If we add this to the square on line ae or 16, we have 40 less the root of 320 for the square on line ai. Its root is line ai. And because it is the root of a number less a root, we say that it is a minor line. Since the difference between the square of 40 or 1600 and 320 is not a square number, the approximate root of the recised is accepted. The root of 320 is about 18 less a ninth. Subtract this from 40 leaving $\frac{1}{9}$ 22 whose root is 4 rods, 4 feet, 3 inches, and 17 points. This is the best approximation for line ae, the side of the pentagon inscribed in the circle. From this it is easy to find all the sides of a pentagon inscribed in any circle. For as 4 rods, 2 feet, 3 inches, and 17 points are to 8, so is the pentagonal side you want to the diameter of its circle. Whence if you multiply the diameter of whatever circle by 4 rods, 2 feet, 3 inches, and 17 points {p. 106} and divide by 8, you will have the side of the pentagon[217] inscribed in the circle.

[236] Now, if you want to find be the chord of pentagonal angle bae in pentagon $abgde$[218] inscribed in circle $abgde$, draw the diameter bz of the circle and let its measure be 8. Join z and e for the side of a decagon. Because angle zeb is in semicircle aeb, it is a right angle [see Figure 3.115]. So, if we subtract the square on the decagonal side ze from 64 the square on the diameter, we will find 24 less the root of 320, as above. The root of this is a major line since it is the root of a fourth binomial. What remains is 40 and the root of 320, the length of chord be. Therefore, whatever the size of the square of a pentagon by that same amount is the chord of an angle of the pentagon.[219] Now the pentagonal side is the root of the fourth recised or number less the root, and the square of the chord is the fourth binomial. Because it is minor to its recised binomial, therefore the root of the fourth recised is called the minor line, and the root of the fourth binomial is called the major.

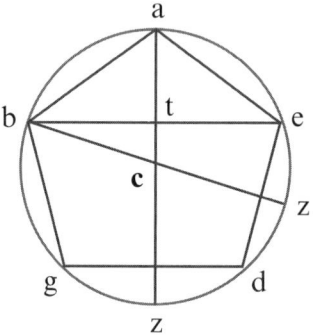

Figure 3.115

---

[217] Text has *latus decagonicum*; manuscript has *latus penthagonicum* (106.1, f. 64v.19).
[218] Text has *acde*; context requires *abgde* (106.2).
[219] "Vnde potest haberi, quod quorumcumque nominum fuerit quadratum penthagonici, eorumdem nominum erit corda anguli penthagonici" (106.8–9, f. 64v.32–33).

[237] In another way: construct the diameter *ati* cutting the chord *be* at point *t*. Draw line *ib*, and triangle *iba* is a right triangle. Further, it is divided in two triangles by cathete *bz*, both similar to one another and to the whole triangle. Whence, as *ia* is to *ab* so is *ab* to *at*. So, if we divide 40 less the root of 320 the square on line *ab* by 8 the diameter *ai*, we have 5 less the root of 5 for line *at*. And if we take this from *ai*, what remains is 3 less the root of 5 for line *ti*. And if we multiply *ti* by *ta* and quadruple the product, we find again 40 less the root of 320 for the square on line *be*. For the root of 320 is 18 less $\frac{1}{9}$. Adding this to 40 makes 58 less a ninth for the square on line *be*. The root of this is 7 [rods], 3 feet, 11 inches, and 14 points. By this chord you can find the measure of similar chords lying in any circle, if you multiply it by the diameter of its circle and divide the product by 8.

[238] To find the area of the aforementioned pentagon, multiply $\frac{3}{4}$ of the diameter or 6 by $\frac{5}{6}$ of chord *be*. Or, multiply $\frac{5}{6}$ of the diameter six or 5 by the whole chord *be*, to get nearly 38 rods and $\frac{1}{2}$ 1 deniers for the measure of the area of pentagon *abgde*. In fact we can find the area of any pentagon inscribed in any circle, if we multiply 38 rods and $\frac{1}{2}$ 1 deniers by the square on the diameter of the circle and divide by 64 the square of the diameter of the model circle. The reason is that Euclid at the beginning of the twelfth book showed that similar polygons in circles are to each other as the squares on their diameters. Conversely therefore as the square on the diameter of one circle is to a multilateral inscribed figure, so is the square of another circle to a similar multilateral inscribed figure. We can find chord *be* in another way as proved in the fourteenth book of Euclid,[220] namely that the chord of a pentagonal arc with the pentagonal side is five times the square on the semidiameter of the circle. So, if we take five times the square on half the diameter, we have 80. If we subtract from this the square on the side of the pentagon, namely 40 less the root of 320, what remains is 40 and the root of 320 for the square on chord *be*, as we said. Or if we were to multiply the diameter by $\frac{5}{81}$ which is 4096, we would have 320 whose root is a minor term. You can do this for other circles.

[239] We found that the area of an equilateral, equiangular triangle inscribed in a circle of diameter {p. 107} 8 is 20 square rods, 4 feet, and $\frac{3}{4}$ 12 inches. The area of the square is 32 or half of 64 from the square on the diameter. The area of the pentagon is $\frac{1}{24}$ 38. That of the hexagon is 41 rods, 3 feet, and $\frac{1}{2}$ 7 inches. The area of the octagon is 45 rods, 1 foot, and $\frac{1}{2}$ 9 inches. It was found by multiplying twice the diameter by half the side of the square, namely 16 by the root of 8. The area of the decagon is 47 rods and $\frac{1}{2}$ 2

---

[220] "In circulo tetragonus lateris pentagoni simul cum tetragono corde anguli pentagonici quincuplum est tetragono lateris exagonici" ibid., XIV.3 (Busard [1987], 400).

174 Fibonacci's *De Practica Geometrie*

inches. It was found by multiplying five times the fourth part of the diameter by the side of the pentagon, or 10 by 4 rods, 4 feet, 3 inches, and 17 points. By multiplying half the diameter by half the sides of the inscribed hexagon, the area of the dodecagon is found to be 48 rods. Now that you know all of this, you can easily find the area of similar figures inscribed in other circles, provided you do not forget it. So much for demonstrations about the circle. Now we move on to measuring[221] fields that lie along the slopes of mountains.

3.6 Measuring Fields Along Slopes of Mountains[222]

[240] When you wish to measure a field that lies up or down the side of a mountain, inquire diligently about the sides and area of a surface lying in a plane below the mountain. It will be the area of the fourth kind of a field, for mountains are not measured according to the apparent surfaces on them because house, building, trees and even seeds are not raised at right angles to the surfaces. Whence the area of the planes are sought over the apparent surfaces of the mountains on which they lie. On these known planes everything can be erected at right angles. I shall presently explain just how using the sides of the inclined[223] surfaces you can find the sides of superficial planes lying under them [see Figure 3.116]. Let line ab represent the side of an inclined surface lying on a mountain side, line bg represent the side of a plane lying under side ab, and a perpendicular falling from a to g makes a right angle at g. It is necessary to find most correctly the hidden side bg and the perpendicular ag by means of the visible side ab. This can be done in either of two ways. The first way is the one used by experienced surveyors: place and hold firm one end of a measuring tape[224] at point a and extend the tape out over line ab toward point b until the tape is equidistant from line gb. (This you can do with an instrument called an *archipendulum*, which I shall explain below.) Then holding the other end of the tape firmly, drop a pebble on line ab. Counting from where the stone first landed, repeat the procedure measuring the number of rods correctly along the tape for the entire length of ab.

---

[221] The text and manuscript have "diminutionem camporum"; context requires "dimensionem camporum" that appears in the title of the section (107.13, f. 65v.12).

[222] "Incipit pars quinta in dimensione camporum qui in montibus iacent" (107.15, f. 65v.22).

[223] Text has "decliuarum"; context requires "declinarum" (107.24).

[224] The phrase "caput metientis pertice" names the tape by the unit it measures, a rods. (107.30).

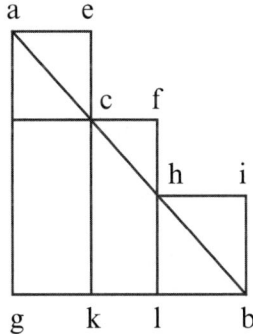

Figure 3.116

[241] For example: let the first position of the tape be line *ae* equidistant from *bg*. At point *e* drop a rock to point *c*. Again, put the tape at point *c* and using the archipendulum fix line *cf* equidistant from *bg*. Again, from point *f* drop a rock to point *h*. Again, for line *hi* to be equidistant from *bg* hold the tape out to point *i*. Finally, at point *i* drop the rock on point *b*. Now, as many times at the tape fell over line *ab*, so many rods {p. 108} will be on side *gb*. For example: because rods *ae* is equidistant from side *gb*, angle *eag* is a right angle since there is a right angle at *g*. And because *ec* locates the position of the little rock, if we draw line *ek* through *c* from point *e* above line *gb*, straight line *ek* will be the cathete to line *gb*. Whence straight line *ek* is equidistant and equal to line *ag*. Whence *gk* equals the length of rods *ae*. Similarly: if we draw line *fl* through *h*, we will find under the same condition that line *kl* is equal to the rods in *cf*[225]. Again, if we draw line *ib* from *i* to point *b* though where the little rock fell, it will be equidistant from and equal to line *hl*. Whence *lb* equals the rods in *hi*. Therefore as many rods as there are in the lines above *ab* and equidistant from *bg*, that many rods are in the length of line *gb*, as we said.

[242] The archipendulum is a wooden instrument in the form of an isosceles triangle. A lead weight on a string hangs from one angle. You can put the archipendulum atop the tape.[226] And the weight and string will fall from the upper angle over the middle of the base. Therefore the tape stands equidistant to the plane you wish to measure. In the attached figure [Figure 3.117] line *po* represents the tape on which is the archipendulum *abg*. The lead weight and string *ad* fall from point *a* to the middle of

---

[225] Text has *ef*; context requires *cf* (108.6; *f.* 66r.16).
[226] Literally "rods" (pertice); but the reference is to a tape measured in rods.

line *bg*. And if in the preceding figure the line attached to side *ab* descends correctly, it is not necessary to apply the tape with the archipendulum more than once. Because when you applied tape *ae* equidistant from side *gb*, you erected at location *ec* a rod equal to line *ec*. Then triangle *aec* will be similar to triangle *abg*. Whence, since line *ab* intersects the equidistant lines *ae* and *bg*, angle *ace* will be equal to angle *abg*. Similarly: because line *ab* intersects the equidistant lines *ag* and *ec*, then angle *gab* will be equal to angle *eca*. Whence the remaining angle *aec* will be equal to the other remaining angle *agb*. Therefore as *ac* is proportional to the whole line *ab*, so is the tape *ae* to the whole line *gb*. If we multiply *ab* by *ae* and divide by *ac*, the measure of line *gb* results. That is, if you make the tape equal to the line *ac* and use it to measure the line *ab*, you will have in similar fashion the measure of line *bg*. Whence as so many lines equal to *ac* are in *ab*, so many rods *ae* are in side *gb*. Similarly: as many lines equal to *ac* are in line *ab*, that many equal lines *ec* are in the altitude *ag*. So, if we multiply *ab* by *ec* and divide by *ac*, the results is the height of cathete *ag*. This way of finding the cathete is quite useful for measuring the altitudes of pyramids.

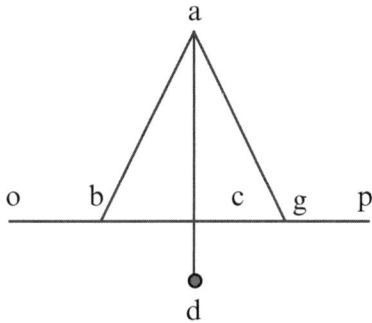

Figure 3.117

[243] Our wise predecessors spoke about rods and similar triangles in this way. Consider the given figure *agb* with side *ab* lying along the side of a mountain for which you wish to know the length of the base *bg* as well as the altitude *ag* [see Figure 3.118]. Erect a perpendicular rod *bc* of whatever length at the foot of the mountain. Attach another rod to it at *c* that makes a right angle, its other end lying on *ab*, and let this rod be *ce*. Thus triangle *cbe* is similar to triangle *abg*. Whence as *be* is to *ba*, so is *ec* to *bg*. So multiply *ab* by *ce* and divide by *be* to find side *bg*. Likewise: because *be* is to *ba*, so is *cb* to *ag*. So multiply *ab* by *bc* and divide by *be*, and thus you will have the height *ag*.

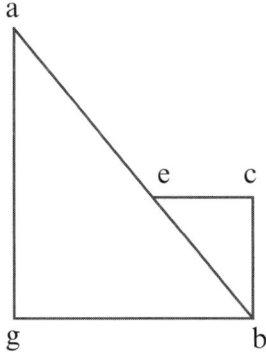

Figure 3.118

[244] Suppose there is a field on the mountain side that looks like the surface of some part of a column, so that the surface appears pushed in as in surface *abde* with sides *ab* and *de* as arcs {p. 109} of a circle. The remaining sides *ad* and *be* are straight lines, and side *be* is at the top of the mountain. Also side *ad* is inverted at the base[227] [see Figure 3.119]. Understand first of all that the sides of the existing plane below the surfaces *agdz* and *gz* as well as the altitudes *bg* and *ez* fall orthogonally on the plane *adzg* and at points *g* and *z*.[228] Whence angles *agb* and *dze* are right angles. I measure arcs *ba* and *ed* with tape and archipendulum, dropping little rocks on those arcs, beginning at points *b* and *e* which are at one end of the field and going down to points *a* and *d*. And thus I have the lengths *ag* and *dz*. And because side *ad* is a straight line, also in the plane of the surface *gadz*, I measure it as I measured the sides of the plane. Then I will measure side *be*, if it were equidistant from and equal to side *gz*. And that is known if the cathetes *bg* and *ez* are equal. And thus I know side *gz*. Now however so much is known about the equality or lack between cathetes *bg* and *ez*, will be shown in the attached figure. Or it can appear to the eye, if points *b* and *e* are in the same plane. If one of them is higher, let line *be* ascend or descend to point *e*. Let point *e* be higher than point *b*. Whence I will measure side *eb* with tape and archipendulum in the order described above. And thus I will know side *zg*. And if plane *adzg* is orthogonal, I will multiply *ga* by *ad* to find the area of quadrilateral *az* which is the area of the apparent concave surface. And if plane *az* is not orthogonal, I will try to find the diameter *gd*. Knowing this I can measure arch *bd* with tape and archipendulum. Then I can join the areas of triangle *gad* and *dzg* as one, and have what I wanted.

---

[227] "sit inversus radicem." (109.2, *f.* 66v.22).
[228] The locations of the letters in Figure 3.119 are taken from *f.* 66v.

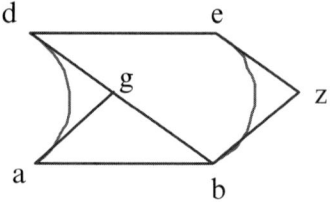

Figure 3.119

[245] And suppose that the protuberance bulges toward either side as does arc *cfh* whose higher part is at point *f* [see Figure 3.120]. Then I understand that the side of plane *ca* can be joined orthogonally with the altitude *af*. And if arc *fe* equals arc *fh*, then the chord of arc *cfh* is twice line *ca*, namely the length of the side of the whole plane contained by arc *cfh* as long as arc *cfh* is no greater than a semicircle. Because if it were greater than a semicircle, then the diameter of the circle would be the length of the apparent surface. But let arc *fh* be less than arc *fc*. I understand that line *hi* is joined orthogonally to line *fi*. Further I understand that line *cak* is equidistant from line *ih* that equals line *ak*. So, if *hk* is drawn, it will be equal to and equidistant from line *ia*.[229] Whence the angle at *k* is a right angle. Whence the whole line *ck* is the side of the plane contained by arc *cfh*, and *hk* is the altitude from *a* to *i*. With all of this understood and using tape and archipendulum I will measure from *f*[230] to *c* and from *f* to *h* to give me the length of side *ck*.

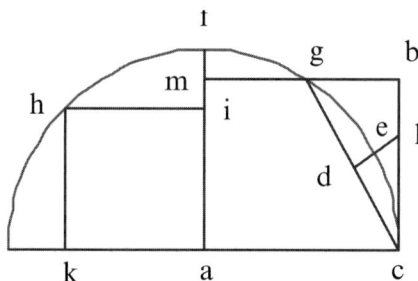

Figure 3.120

[246] Or in another way: I will erect rod *cb* orthogonally at point *c*. And I will fit it to another rod of sufficient length that when making a right angle at point *b*, it becomes another side touching arc *fc*. This rod is line *bg*. So I will investigate as carefully as possible the lengths of rods *cb* and *bg*. And I will add their squares and take the root of their sum for the length of chord *cg*. I

---

[229] The second *hk* in "line *ia* et *hk*" is redundant.
[230] Text has *abf*; context requires "ab *f*" (109.33).

shall divide this in two equal parts at point d. From point d I will erect cathete de to chord cg. Consequently point e divides arc cg in half. And because it says in the eleventh book[231] that every triangle is in one plane: if I draw line de from point e through surface bcg, it will intersect one of the sides cb or bg or angle b contained by them. When cb and bg are equal, then it will intersect {p. 110} at point b. And if one of them is longer, then it falls on the longer of them. So let cb be longer than bg. Mentally draw line de until it meets line cb at point l. Consequently triangles cbg and cdl[232] are right triangles and similar to each other because they have angle lcd in common. Whence as bc is to bg, so is cd to dl. And so by multiplying bg by cd and dividing by bc, I know dl. Likewise, as bc is to cg, so is cd to cl.[233] And so, by multiplying cg by cd and dividing by bc, I know the length of cl. This I take carefully, go, and take a straight rod that falls from l to e. I take its length from ld [which is already] found, and what remains known is the arrow ed. Whence, if I multiply cd by dg, divide by de, then add ed to the quotient, I will have the diameter of the circle. Its arc is cfh. And so having found the diameter you can find the chord of twice arc gf from our tables. If you take half the chord, you will clearly know line gam. If you add to this the length of rod bg, you will know the whole line bm, that is line ce.

[247] Similarly: if with that same diameter you find the chord of twice arc fh, its half is line hi equal to line ak. And thus you know all of line ck, that is, the length of the plane lying under the protuberant surface. And since you want to know the sides of the planes apparently lying along the sides of the mountains, you can carefully find their areas by those same sides, whether the plane has three or four or more sides, or is round, or just part of a circle, or has an oblique form deviating from true roundness. And since you know the form of the plane by those things spoken of in this the third chapter, you can find its area. And so we bring this chapter to an end and start the fourth chapter in which we will teach you how to divide a field of whatever shape.

---

[231] *Elements*, XI.2: "Si due recte secuerint se invicem, in uno sunt epipedo. Et omne trigonum in uno est epipedo" (Busard [1987], 311). Is Fibonacci's use of "superficie" instead of "epipedo" an editorial change on his part for the sake of his readers; or, is it an indication of another translation of Euclid's *Elements*?

[232] Text has edl; context requires cdl (110.3).

[233] Text has el; context requires cl (110.6).

4
Dividing Fields Among Partners

COMMENTARY

The title of the chapter may suggest more than it offers. Except in two problems, there is no discussion of the partners; they are in the background, taken for granted. The emphasis is on the kinds of fields and how to divide them, given a variety of conditions. Fibonacci offered a simple introduction dividing the chapter in four parts, each named after the shape of the field to be partitioned: triangular, quadrilateral, multilateral, and circular. The *Table of Contents* displays the complexity as the chapter develops, the figures ranging from triangles through quadrilaterals to pentagons, circles, and parts of circles. The conditions for dividing the figures are these: in two equal parts, several equal parts, two parts in a given ratio, and several parts in a given ratio. The divisions contain additional conditions. They require drawing a line that passes through a point situated at a vertex of the figure, at a vertex or a point on a side, on a side not produced, on one of two parallel sides, on the middle of the arc of a circle, on the circumference or outside of the circle, inside the figure, outside the figure, either inside or outside the figure, either inside or outside or on a side of the figure, and in a certain part of the plane of the figure. Furthermore, the line may be drawn parallel to the base of the proposed figure or parallel to the diameter of the circle. It may also be required to draw more than one transversal, such as through one point, through two points, or lines parallel to one another in which case the problem is indeterminate. Finally, circles and their parts are also divided.

Two further remarks about the propositions: I have reworded almost all of them after a format whereby each of the fifty-nine propositions (my enumeration in bold type) is easily recognized; namely, "*To (do something)*." Lest the reader be apprehensive about missing something Fibonacci said or implied in his statements, I added the original statements in footnotes. Some propositions were not easily reworded. So I abstracted the essence that I formulated as above and placed the whole within brackets.

A true mathematician, Fibonacci saw in some of the rules more than the statement reflected. For example, following the first rule, *To divide a triangle in two equal parts from one of the angles*,[1] he placed two corollaries:

---

[1] Translation Ch. 4, [1].

182   Fibonacci's *De Practica Geometrie*

C1. "Triangles with one angle of one equal to one angle of another have a ratio composed of the sides containing the equal angles."[2]
C2. "If a line is drawn in a triangle, cutting two sides of the triangle and making with those two sides a triangle having one angle in common with the original triangle, then the ratio of one triangle to another is as the product of the sides containing that angle."[3]

He supplied proofs for each of these.

To prepare his reader for dividing triangles from a point given within a triangle, Fibonacci laid down several principles. They are

P1. "If two lines are drawn from two angles of at triangle to the midpoints of the opposite sides, they will intersect proportionally, and each part of the line lying between the angle and the point of intersection is twice the other part of the line."[4]
P2. *A converse of P1*: "A line drawn from an angle through the point of intersection divides the side opposite the angle in two equal parts."[5]
P3. *From P1*: "A line drawn from the remaining angle to the side opposite will pass through the point of intersection."[6] Fibonacci's proof for this principle is a lengthy *reductio ad impossibile*, or as he put it, *quod est inconueniens*.

Two additional items bear mention. In the proof for the method separating a quadrilateral into two equal parts from a point outside the figure, Fibonacci set the solution by moving a pair of parallel lines until one of them is in line with the point.[7] This may be something he learned from Archimedes who employed verging in some of the proofs in *On Spirals*.[8] Secondly, throughout Chapter 4, not to ignore the other chapters, he demonstrated his expertise in theoretical geometry: his problems are carefully solved and his proofs are Euclidean based.

There is a particular method of proof that Fibonacci makes good use of, in some twenty propositions in this chapter. It appears for the first time in [4]. The task is to divide a triangle into two equal areas by a line drawn from a point d on a side. The point is not the midpoint of the side. So Fibonacci set the point closer to vertex b than to g. After finding midpoint e on bg, he constructed the crosshatch seen in Figure 4.5. This crosshatch appears time and again in many subsequent propositions. Furthermore, the same stages in the proof are followed there as here; namely,

---

[2] Ibid. [2].
[3] Ibid. [3].
[4] Ibid. [6].
[5] Ibid. [7].
[6] Ibid. [8].
[7] [41] to [43].
[8] Heath (1921), II, 65.

(1) $\triangle\ ade = \triangle\ adz$ because of common base and equal altitudes.

(2) Add $\triangle\ abd$ to each triangle in (1).

(3) Quadrilateral $abdz = \triangle\ abd + \triangle\ ade = \triangle\ abe = \frac{1}{2} \triangle\ abg$.

(4) $\therefore \triangle\ adg$ has been cut in half by line dz. Q.E.D.

The same procedure is followed in [20] where a third of a triangle is to be cut off by a line drawn from a point that is not a point of trisection of the side.

However, there is a solution to problem [68] for which a proof [69] is provided, the first apparently incomplete and the second faulty. What seems incomplete is an unannounced line-segment required for an operation. The fault lies in concluding that because corresponding sides of two triangles are proportional, the triangles are congruent, as the Latin text states.[9] In fact they are only similar, as I have corrected the text. Such an obvious error suggests to me that someone else added these two paragraphs. Because both are at the end of a major section, I wonder if they were not slipped in by an over-eager instructor or copyist who thought he had captured Fibonacci's method successfully.

Two research projects provided assistance for the preparation of this chapter, "An Historical-Critical Note on the Division of Areas," by Antonio Favaro (1882/3) and *Euclid's Book on Division of Figures* by Raymond Clare Archibald (1915). Both authors focused on Euclid's tract, beginning with an overview of available manuscripts and studies. Archibald modified some of Favaro's views, particularly that John Dee did not make a copy or a Latin translation of Euclid's treatise. Although each scholar made significant contributions to the study of *De divisione*, their work carried them over into *De practica geometrie* of Fibonacci and, consequently, influenced my work.

Favaro listed fifty-seven rules for dividing areas, by page and line numbers in Boncompagni's edition, a count that is three less than mine. It is important to keep in mind that there is no numbering of the rules in the text available to Favaro or me. The numbering arises out of a sense of ordering from one's study of the treatise. Hence, differences of opinion are to be expected.[10] For instance, where Favaro decided on two or more distinct and separate *regulae*, I saw the others as an extension of, and therefore not separate from, his first. Thus I opined that his 7 and 8 are extensions of 6. Similarly, in my view 12 and 13 are part of 11 as is 46 of 45. On the other hand, he saw one where I saw three; I had divided his 15 into my 13, 14, and 15. Four of his rules seem to be more descriptions or conclusions than methods, namely, 27, 32, 38, and 48. Finally, Favaro did not consider seven statements that I judge to be rules. These are recorded by zeros in Table 4.1 which aligns my numeration with his, my numbers appearing first.

---

[9] The same error occurs below in [74].

[10] Favaro drew my attention to a rule **49** that I had neglected to number in my list.

Table 4.1 Table of Hughes/Favaro enumerations of Fibonacci's rules for division of fields

1 = 1, 2 = 2, 3 = 3, 4 = 4, 5 = 5, 6 = 0, 7 = 0, 8 = 6, 9 = 9, 10 = 10, 11 = 11, 12 = 14, (13, 14, 15) = 15, 16 = 16, 17 = 17, 18 = 0, 19 = 18, 20 = 19, 21 = 20, 22 = 21, 23 = 22, 24 = 23, 25 = 24, 26 = 25, 27 = 0, 28 = 26, 29 = 28, 30 = 0, 31 = 29, 32 = 30, 33 = 31, 34 = 33, 35 = 34, 36 = 35, 37 = 36, 38 = 37, 39 = 39, 40 = 0, 41 = 0, 42 = 40, 43 = 41, 44 = 0, 45 = 42, 46 = 0, 47 = 43, 48 = 0, 49 = 44, 50 = 45, 51 = 47, 52 = 49, 53 = 50, 54 = 51, 55 = 52, 56 = 53, 57 = 54, 58 = 55, 59 = 56, 60 = 57.

Apart from his studied analysis of the history of Euclid's treatise up to his day, Archibald incorporated Fibonacci's proofs from *De practica geometrie* that fit the corresponding proposition in *De divisione*. Each of these is carefully annotated, with references to Euclid's *Elements*, whether or not such citations appear in Fibonacci's tract. Furthermore, he offered the reader a commentary on Heath's explanation of applications of areas that sharpens one's understanding of an important Euclidean tool used often by Fibonacci. Occasionally Archibald introduced words or phrases, although not in Boncompagni's transcription, that clarify meanings. For instance, in Proposition 10, he added the bold italicized words, "Join *eg* **meeting bc in k** and produce it **to meet ad** in *f*."[11] Furthermore, the figures in Archibald's treatise are far more useful than those drawn in the margins of Boncompagni's work, many of which are incorrect or composites of several cases.

SOURCE

As the *Table of Correspondences* (Table 4.2) following the end of *Sources* indicates, Fibonacci probably developed much of this chapter from an Arabic copy of Euclid's *On Divisions*.[12] He might have had at hand a copy of Thābit ibn Qurra's corrected text. At least an early copy of the abstract composed by the tenth century Persian geometer al-Sijzī[13] was available because of certain proofs. Noteworthy is the near identity of the proofs for Propositions 3, 10, 53, and 59 in Fibonacci's text with the proofs for Theorems 18, 19, 28, and 27 in al-Sijzī's tract. If the other proofs in Chapter 4 of *De practica geometrie* are original with Fibonacci, then he probably did not have access to ibn Qurra's corrected edition. Furthermore, he nearly doubled the number of propositions. In his proofs we can find 76 uses of 41 propositions from ten books of the *Elements*.[14] Here, if nowhere else, we see Fibonacci as a world-

---

[11] Archibald (1915), 42–43; "et copulabo rectam *eg* et protraham eam in puncto *f*" (124.8-9).
[12] Archibald (1915).
[13] Hogendijk, J. (1993).
[14] These numbers were reached from an inspection of the proofs in Archibald (1915), the list in Folkerts (2002), 100, and Boncompagni's text.

Table 4.2 Table of Correspondences

F:	1	2	3	4	5	6	7	8	10	11	12	13
E:		3	18	26			1		19	27	2	
P:	1	4	7		8	8	3	5,6				11
G:							159	160			162	

F:	16	20	21	24	25	26	27	30.1	29	31	32	34
E:	6	10		4	8	29		7	12	32	13	5
P:	12	19	2									
G:							163					

F:	35	36	37	38	39	40	41	42	43	53	59	60
E:		33	14	16	17	15		34,35	36	29	27	
P:	20		15	16,17		21,22	23				27,28	
G:			166									

class Euclidean geometer. In putting to rest any consideration of dependence on Abraham's text, Archibald lauded Fibonacci correctly: "Compared with Leonardo's treatment of divisions Savasorda's seems rather trivial. But however great Leonardo's obligations to other writers, his originality and power sufficed to make a comprehensive and unified treatise."[15]

Depending on how the propositions are identified and counted, relationships among the propositions in Fibonacci's chapter with the Arabic–Euclid, *On Divisions*, and Plato of Tivoli's translation can be described as follows. Fibonacci added 32 propositions to the 28 that he adopted from the Arabic–Euclid, *On Divisions*. Fibonacci's tract and Plato of Tivoli's translation have just 12 propositions in common; nearly all the wording is different. In short, the number of propositions are: for Abraham 30, for Euclid 35, and for Fibonacci 60.

The *Table of Correspondences* (Table 4.2) collates identical propositions from four sources: my translation (F with proposition numbers), Archibald's premier edition of Euclid's *On Divisions* (E with proposition numbers), Curtze's transcription of Plato of Tivoli's chapter on division (P with paragraph numbers), and Grant's translation of portions of Fibonacci's Chapter IV (G with page numbers). Some propositions are combined in one or separated into two.

TEXT[16]

{p. 110} We divide Chapter 4 in four parts. In the first we instruct on how to divide triangles, in the second quadrilaterals, in the third polygons, and in the fourth circles and their parts.

---

[15] Archibald (1915), 19.

[16] "Explicit distinctio tertia, incipit quarta de diuisione camporum inter consortes" (110.25–26, f. 67v.25–26).

4.1 Division of Triangles[17]

[1] **1** *To divide a triangle in two equals parts from one of the angles.*[18] Draw a line from the angle to the middle of the opposite side, and you have what you wanted. For example: we wish to divide triangle *abg* in two parts from point *a* [see Figure 4.1]. So divide side *bg* in two equal parts at point *d* to which you draw line *ad*. Therefore triangle *abg* has been divided in two equal parts: triangles *abd* and *adg*. Because they are on equal bases and have the same altitude, namely the cathete drawn from *a* to line *bg*, they are equal to one another. Now triangles with the same altitude are to one another as their bases, as stated in the beginning of the sixth book.[19] Hence as *bd* is to *dg*, so is triangle *abd* to triangle *adg*, for base *bd* equals base *dg*. Whence triangles *abd* and *adg* are equal to one another, as we said. Or if we had drawn the cathete from *a* to line *bg*, then it is the cathete for both triangles *abd* and *adg*.[20] Multiplying half the cathete by the bases *bd* and *dg* is the same as multiplying half the cathete by the base *bg*. For by multiplying {p. 111} half the cathete by the bases *bd* and *dg*, the areas of triangles *abd* and *adg* are found. Whence we have proved that triangle *abd* equals triangle *adg*.

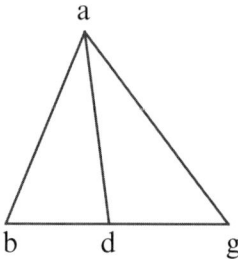

Figure 4.1

[2] Triangles with one angle [of one] equal to one angle [of another] have a ratio composed of the sides containing the equal angles. The proof follows. Let triangles *abg* and *gez* have equal angles at *g* [see Figure 4.2]. I say therefore that triangles *abg* and *gez* have a ratio composed of the ratios of the sides containing the equal angles, such as side *bg* to *ge* and as side *ag* to side *gz*. The proof follows. Draw line *ae* and form triangle *age* between triangles *abg* and *gez*. Therefore the ratio of triangle *abg* to triangle *gez* is composed from two ratios, namely from the ratio triangle *abg* has to triangle *age* and from the ratio triangle *age* has to triangle *gez*. But the ratio of triangle *abg* to

---

[17] "Incipit pars prima de diuisione triangulorum" (110.30, f. 67v.29).
[18] "Cum itaque triangulum aliquem in duas equas partes ab uno angulorum diuidere uis..." (110.31–32. f. 67v.30).
[19] *Elements*, VI.1.
[20] Fibonacci envisions an oblique triangle with altitude outside the triangle.

triangle *age* is as base *bg* is to base *ge* since they share one altitude. Also, the ratio of triangle *eag* to triangle *egz* is as side *ag* is to side *gz*. Therefore the ratio of triangle *abg* to triangle *gez* is composed of the ratios of sides *bg* to *ge* and of *ag* to *gz* that contain equal angles, as required. Now the ratio of triangle *abg* to triangle *gez* is composed of the ratios of *bg* to *gz* and *ag* to *ge* with sides *ag* and *bg* as antecedents and sides *ge* and *bz* as consequents. And as we said in the tract on angles,[21] because the ratio as composed follows from the products of the antecedents to the products of the consequents, then the ratio of triangle *abg* to triangle *gez* will be as the products of *bg* and *ag* to the products of *eg* and *gz*. Whence if the product of side *eg* by side *gz* equals the product of side *bg* by side *ga*, then triangle *egz* will equal triangle *abg*. And if less, then less, and if greater, then greater. And this is what I wanted to prove.

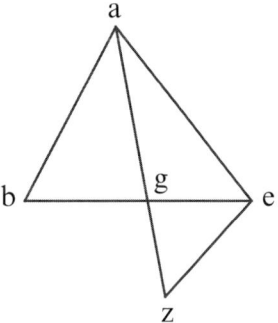

Figure 4.2

[3] If a line is drawn in a triangle, cutting two sides of the triangle and making with those two sides a triangle having one angle in common with the original triangle, then the ratio of one triangle to another is as the product of the sides containing that angle[22] [see Figure 4.3]. For the proof: let triangle *abc* begin with line *de* cutting sides *ca* and *cb* at points *e* and *d*. I say that triangle *abc* has the ratio to triangle *dec* as the product of *ac* and *cb* has to the product of *dc* and *ce*. The proof follows. I will apply on side *ac* triangle *acf* equal to triangle *dec*.[23] Because triangles *abc* and *afc* share the same altitude, as *bc* is to *fc*, so is triangle *abc* to triangle *afc*. But the ratio of *ac* to *fc* is as the product of *ac* and *ab* to the product of *ac* and *cf*. Therefore the ratio of triangle *abc* to triangle *afc* is as the product of *ac* and *bc* to the product of *ac* and *cf*. And because triangle *dec* equals triangle *acf*, the ratio of triangle *acb* to triangle *dce* is as the product of *acb* and *acf*.[24] Because

---

[21] Ch. 3 [44].
[22] Archibald 31 n. 88.
[23] *Elements*, I.44.
[24] This is a strange notation. *acb* represents $(ac)(cb)$; similarly, *acf* (111.35, f. 68v.12).

triangles *acf* and *dce* equal one another and have an angle in common, they have sides in common about this angle. And this is a mutual ratio, as stated in Euclid VI.15. Therefore, as *ac* is to *dc*, so is *ce* to *cf*. Whence the product of *dc* and *ce* equals the product of *ac* and *cf*. Therefore the ratio of triangle *abc* to triangle *dec* is as the product of *ac* and *cb* to the product of *dc* and *ce*. And this is what had to be proved.

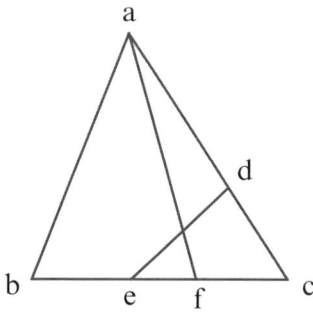

Figure 4.3

[4] **2** *To divide a triangle in two equal parts by a line drawn from a given point on one side.*[25] Given triangle *bdg* in which point *a* on line *gd* is given, first let {**p. 112**} point *a* be in the middle of line *gd*. Join point *a* with point *b*. Then triangle *bgd* will be divided by line *ba* in two equal triangles, *bga* and *bad*, with equal bases and one altitude [see Figure 4.4]. Or since the product of *bd* and *dg* is twice the product of *bd* and *da*, triangle *bgd* is twice triangle *bad* because they have one angle in common. But let no point be given as the midpoint of any side, as in this other triangle *abg* in which point *d* is given as close to point *b*. I will divide side *bg* in two equal parts at point *e*, and I will join lines *ad* and *ae*. And through point *e* I will draw line *ez* equidistant from line *ad*, and I will draw line *dz* [see Figure 4.5]. I say therefore that triangle *abg* is divided in two by line *dz*. The proof follows. Two equal triangles *ade* and *adz* sitting on base *ad* with sides *ad* and *ez* are equidistant from one another. To both triangles add triangle *abd*. The two triangles *abd* and *adz* become quadrilateral *abdz* equal to two triangles *abd* and *ade* that is triangle *abe*. But triangle *abe* is half of triangle *abg*. Whence quadrilateral *abdz* is half of triangle *abg*. What is left, namely triangle *adg*, is the other half of triangle *abg*. Therefore triangle *abg* was divided into equal parts at point *d* by line *dz*, as had to be proved.

---

[25] "Et si in aliquo laterum sit punctus datus a quo lineam rectam protrahere uis diuidentem triangulum in duo equa..." (111.41–43, *f*. 68v.18–19). Archibald 24 (The numerals following Archibald refer to sections).

4 Dividing Fields Among Partners   189

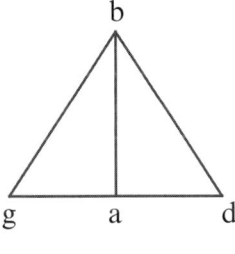

Figure 4.4

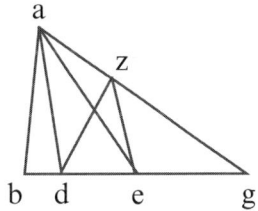
Figure 4.5

[5] If you wish to find point z numerically, you know that the part ed of dg is the same as part az is of ag. Since ez in triangle adg was drawn equidistant to side ad, if you multiply ed by ga and divide the product by gd, what arises is az. For instance, let side ab be 13, bg 14, and ga 15. Whence ge or eb is 7. Let de be two rods; whence gd will be 9. Therefore if you multiply ed by ga (2 by 15) you get 30. Dividing this by gd gives $\frac{1}{3}$ 3 for az. Subtracting this from ag leaves $\frac{2}{3}$ 11 for line gz. So, if we multiply it by gd (9) we get 105 or half of the product of bg and ga. Whence it was proved that triangle zgd is half of triangle abg because both have the angle at g in common. Now the ratio of triangle abg to triangle adg is composed of the two ratios of the sides containing the common angle; namely, the ratio of bg to gd and of ag to gz. Whence as the product of the antecedents bg and ga is to the product of the consequents dg and gz, so is triangle abg to triangle zdg. But the product of bg and ga is twice the product of dg and gz. Therefore triangle abg is twice triangle zdg. Whence, if we would have multiplied ag by half of gb (15 by 7) and divided the product by gd (9), we would have gotten $\frac{2}{3}$ 11 for line gz, as we found above.

Preparation for Dividing Triangles by a Point Given Within the Triangle[26]

[6] If two lines are drawn from two angles of a triangle to the midpoints of the opposite sides, they will intersect proportionally. Further, each part of the line lying between the angle and the point of intersection is twice the other part of the line. And if a line is drawn from the remaining angle through the point of intersection to the remaining side, it divides that side in two equal segments. In triangle abg, if lines ae and bz are drawn from angle abg and bag to the midpoints of sides ag and bg, the lines intersect at point d. I say that the ratio of ad to de is as the ratio of bd to dz, the first part being twice the remaining part. The proof follows [see Figure 4.6]. From point a draw ai equidistant from line bg, and extend line bz until it meets ai at i {**p. 113**}. Thus there are two similar triangles, azi and gzb. Whence as line az is to zg, so is iz to zb, for ia to bg equals az to zg.[27] Therefore iz to zb equals ia to bg. Again, because triangles

---

[26] "Preparatoria in diuidendis trigonis per datum punctum infra triangulum" (112.34, *f.*, 69r.19).
[27] The Latin phrase is "*az ex zg*" (113.2, *f.* 69r.31).

adi and *bde* are similar, as *ia* is to *be* so is *ad* to *de* and *id* to *db*. But line *bg* equals line *ia*. Whence as line *ad* is to *de* and *id* to *db*, so *bg* is to *be*. Now *bg* is twice *be*. Whence *ad* is twice *de* and *id* is twice *db*. And because *iz* equals *zb*, if line *zd* is added to both, then line *bi* equals the two lines *bz* and *zd*. Therefore it has been shown that *id* is twice *db*. Whence the two lines *bz* and *zd* are twice line *bd*. If line *bd* is taken from both, line *bd* remains equal to twice *dz*. Whence *bd* is twice *dz*. For it has been shown that line *ad* is twice *de*. Therefore as *ad* is to *de*, so is *bd* to *dz*, which had to be shown.

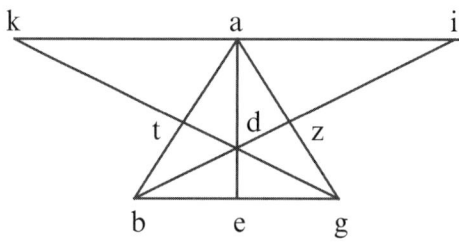

Figure 4.6

[7] If from angle *g* line *gt* is drawn through *d*, I say that side *ab* has been divided in two equal parts at point *t*. Extend line *gt* outside triangle *abg* until it meets point *k* on line *ik*. Then triangles *adk* and *edg* are similar. Whence as *ad* is to *de*, so is *ak* to *ge*. But *ad* is twice *de*. Whence line *ak* is twice line *ge* that is half of line *bg* but equal to *be* and *ge*. Since together they are twice each other and alone equal to each other, line *ak* equals line *bg*. Again, because triangles *atk* and *btg* are similar, as *ak* is to *bg* so is *at* to *tb*. For *ak* equals line *bg* and *at* equals lines *tb*. Therefore line *ab* has been divided in two equals parts by line *gt*, as had to be shown.

[8] Now from angles *abg* and *agb* in triangle *abg*, two lines *bz* and *gt* [are drawn that] intersect each other at point *d* and bisect the sides opposite the angles. It follows that the ratio of the parts is *bd* to *dz* as *gd* is to *dt*. I say it again that if a line is drawn from the remaining angle through the middle of the opposite side, it will pass through the point of intersection of the other two lines falling on the midpoints of their bases. For example: draw triangle *abg* again, and locate midpoints *t*, *e*, and *z* on the sides [see Figure 4.7]. Draw lines *ae* and *bz* intersecting at point *d*. I say that if a line is drawn from point *g* to point *t*, it will pass through point *d*. Because if it does not, then the line from *g* to *t* will pass through line *bz* between points *b* and *d* or between points *d* and *z*. First let it pass through line *bd* at some other point *f*. Because the two lines *bz* and *gt* from angles *abg* and *bga* meet the midpoints of the opposite lines *ab* and *ag*, they intersect at point *f*. Thus *bf* is twice *fz*. But *ad* is twice *dz*. Therefore as *bf* is to *fz*, so is *bd* to *dz*. Alternately therefore as the shorter *bf* is to the longer *bd*, so is the longer *bd* to the shorter *dz*. But that is inconvenient {p. 114} because the line drawn from *g* to *t* through *bz* will not pass between points *b* and *d*. Likewise it will be shown that it cannot pass between points *d* and *z*, for

it has been said to pass through point *d*. From this it is obvious that there is no other point within the triangle except one through which the lines from the angles divide the triangle in two equal parts. If we wish to have the line pass through that point and divide the triangle in two equal parts, it must be one of the lines drawn from an angle to a midpoint of an opposite side that divides the triangle in equal parts. For example: in the above-mentioned triangle *abg* point *d* is given through which lines *ae*, *bz*, and *gt* were drawn from their angles, thereby dividing the triangle in equal parts. Any one of these lines does divide the triangle *abg* in two equal parts, For triangles *aeb* and *aeg* have equal bases, *be* and *eg*, and share the same altitude, the perpendicular drawn from *a* to *bg*. Whence triangle *aeb* equals triangle *aeg*. For the same reason triangles *bzg* and *bza* are equal to each other, and also triangles *gat* and *gtb*. Therefore triangle *abg* has been cut in two halves by one of these lines, *ae*, *bz*, or *gt*, passing through the given point *d*. Similarly, if the given point falls on line *ae* between *a* and *d* or between *d* and *e*, triangle *abg* is divided in two equal parts by line *ae* going through the given point within[28] the triangle. The same is to be understood if the given point were on the extension of either of the two lines *bz* or *gt*.

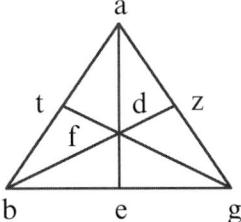

Figure 4.7

[9] Suppose the given point did not fall on the lines coming from the angles to the midpoints of the opposite sides. Then if we made the lines from the angles to the sides of the triangle to pass through another point, then the sections of the sides would not be equal to one another. From this it would follow that the two parts longer than the halves of the sides would be attached to one angle of the triangle and the shorter parts to another angle, and one long and one short section about the remaining angle [see Figure 4.8]. For example: in triangle *dez* let lines *da*, *eb*, and *zg* divide their respective sides equally at points *a*, *b*, and *g*, and let them intersect at point *i*. Now if another point is given within triangle *dez* that is not on lines *da*, *eb*, and *zg*, then it must be within one of the six triangles constituting the entire triangle *dez*, namely triangles *dib*, *dig*, *gie*,[29] *aie*, *aiz*, and

---

[28] Here (114.17, f. 70r.15) and above and below, the word *infra* appears. But the sense is clearly *within*; hence, *intra* seems to be the word.

[29] Text has *die*; context requires *gie* (114.29).

zib. First, let the point *f* be within triangle *dib* because we will draw a line from *d* to *f* so that it meets line *ez* between points *a* and *z*. And this will happen if point *f* is within triangle *daz* and the point will be *k*. Whence *zk* is shorter than half of line *ez* and *ek* is longer. Likewise we draw a line from angle *z* through point *f* to meet line *de* between points *d* and *g*, with point *f* within triangle *gzd*. We call this point *c*, so that *ec* is more than half of side *de* and *cd* is less than half. Therefore, there are two long parts, *ke* and *ec*, about angle *e*. Again, if we draw a line from angle *dez* through point *f* and falling on side *zd* between points *d* and *b*, and because point *f* is within triangle *deb*, the line drawn from *e* through *f* falls on point *h* [on line *dz*]. Whence *zh* is longer than the half of side *zd* and *hd* is shorter. Whence about angle *edz* there are two sections *cd* and *dh* that are shorter than half the sides, and angle *dze* remains contained by unequal sections, one longer the other shorter, than the halves of its sides. The longer side is *hz* and the shorter part is *zk*. Similarly if the given {**p. 115**} point *f* lies within any of the other five remaining triangles, you will find that the same thing will happen.

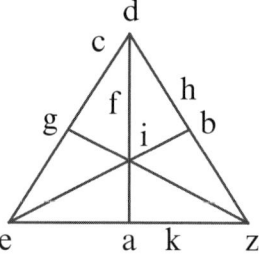

Figure 4.8

[10] **3** [*To divide a given triangle in two equal parts by a line passing through a point in the interior of the triangle.*] Suppose a point is given within a triangle that does not lie on one of the lines coming from an angle and meeting the midpoint of the opposite side. And you wish to divide the triangle in two equal parts by the line going through that point.[30] Concentrate on what has been said about finding an angle contained by unequal sections, because you must do what was said there. For example: in triangle *abg* let point *d* be given not lying on any line coming from any angle to the midpoint of an opposite side, and we wish to divide triangle *abg* into two halves by a line going through point *d* [see Figure 4.9]. First look at the parts of the lines coming from the angles that go through point *d* to the opposite sides. And notice angle *g* that is contained by unequal sides. Draw line *de* equidistant from one of those sides to the other side, and apply to it a plane equal to the half the area of *bg* by *ag*. Call this plane *de* by *gz*. Half the product of *bg* and *ag* divided by *de* produces *gz*. Then apply to line *gz* the parallelogram

---

[30] (115.3–5, f. 70v.11–13). Archibald 40.

deficient by a square figure equal to the plane *ge* by *gz*. This is what divides *gz* into two parts of which one multiplied by the other equals the product of *ge* and *gz*. This cannot be done in any other way unless the square on half of line *gz* is greater than the area of *ge* by *gz*. Or if it is equal, then as *zi* is to *ig*, so is *ge* to *gz*. Draw line *id* and let it continue on to point *t*. I say that triangle *abg* has been divided into two equal parts by line *it* passing through point *d*. The proof: because the product of *zg* and *eg* equals the product of *zi* and *ig*, *zg* is to *zi* as *ig* is to *eg*. Whence as *zg* is to another part of itself namely *ig*, so is *ig* to another part of itself, namely *ie*. But as *gi* is to *ie*, so is *gt* to *de*. Therefore as *zg* is to *ig*, so is *gt* to *de*. Whence the product of *ig* and *gt* equals the product of *zg* and *de*. But the product of *zg* and *de* is half the product of *ag* and *gb*. Therefore the product of *ig* and *gt* is half the product of *ag* and *bg*. Whence triangle *itg* is half of triangle *abg*, as we said.

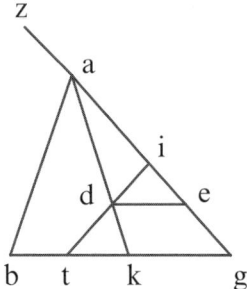

Figure 4.9

[11] We can show this with numbers. Let side *ab* be 13, *ag* 15, and *bg* 14. Draw cathete *ac* that measures 12. Then segment *bc* is 5 and *cg* is 9. Let point *d* lie between cathete *ac* and side *ag*. Construct to line *cg* cathete *dk* equal to 3. Extend line *ed* to point *l*,[31] and let the measure of *dl* be an eighth and a sixteenth part of one. Because line *dl* is equidistant from line *ck* and *dk* from *lc*, then *lc* equals *dk* and *dl* equals *kc*. Therefore *lc* is 3 and *kc* is $\frac{3}{16}$. Nine remains for *al*. Because in triangle *acg* line *le* was drawn parallel to line *cg*, then *al* is to *ac* as *le* is to *cg*. Whence *le* is $\frac{3}{4}$ 6. If $\frac{3}{16}$ is subtracted from $\frac{3}{4}$ 6, what remains is $\frac{9}{16}$ 6. If we divide this number into 105 the product of *ag* and half of *bg*, the quotient is 16 for the measure of line *gz*. Likewise, because *al* is to *ac* as *ae* is to *ag*, therefore *ae* is $\frac{1}{4}$ 11 that leaves for *eg* $\frac{3}{4}$ 3, a fourth part of line *ga*. If we multiply this by *gz*, we get 60. Therefore line *gz* must be divided in two parts of which one multiplied by the other makes 60. And for this to happen, subtract 50 from the square on half of line *gz*. Four remains whose root is 2 that must be added to half of line {**p. 116**} *gz*. Thus we have 10 for

---

[31] From [10] we know that point *e* is on *ag* and point *l* is on *ac*.

194   Fibonacci's *De Practica Geometrie*

line gi and 6 remains for line iz. Then if we divide 105 by gi, we will have the predicted $\frac{1}{2}$ 10 for line gt. I will show in another way that line gt is $\frac{1}{2}$ 10. Draw cathete imn from point i to line gb. And line in will be equidistant from line ac. Whence as gi is to ga so is 10 to 15. So in is to ac. Whence in is 8. Because the quadrilateral mk is a parallelogram, line mn equals line dk. Whence mn is 3 and im remains as 5.

[12.1] Again, because line in is equidistant from line ac, gi is to ga as gn is to gc. Whence gn is 6, and 3 remains for line nc. Since line ml is equals to nc, it too equals 3. If line dl or $\frac{3}{16}$ is subtract from this, then what remains for line md is $\frac{13}{16}$ 2. And because triangles imd and dkt are right triangles and similar to each other, then as im is to md (that is as 5 is to $\frac{13}{16}$ 2) so is dk or 3 to line kt. Whence if we multiply $\frac{13}{16}$ 2 by 3 and divide the product by 5, we get $\frac{11}{16}$ 1 for line kt. If from this we remove kc or $\frac{3}{16}$, what remains is $\frac{1}{2}$ 1 for line ct. If to this we add line cg or 9, we have $\frac{1}{2}$ 10 for line gt, as we said.

[12.2] Again I will show this in numbers [see Figure 4.10]. If in the same triangle we draw lines ao and bp through point d, segment gp is shorter than half of ga, and go is longer than half of gb. Because line de has been drawn equidistant from line og in triangle aog, then as ae is to ag, so is de to og. Therefore, as 3 is to 4, so is de or $\frac{9}{16}$ 6 to og. Whence if we multiply 4 by $\frac{9}{16}$ 6 and divide the product by 3, we get $\frac{3}{4}$ 8 for line go. Again, because line de has been drawn equidistant from line bg in triangle pbg, then as bg is to de so is gp to pe. Whence as the excess of bg over de, so is gc to ep. Therefore as $\frac{7}{16}$ 7 is to $\frac{9}{16}$ 6 (or 119 to 105 or 17 to 15), so is $\frac{3}{4}$ 3 to ep. Whence if we multiply 15 by $\frac{3}{4}$ 3 and divide the product by 17, we obtain $\frac{11}{68}$ 3 for line ep. Whence the whole line gp is less than $\frac{1}{2}$ 7 that is half of the entire line ga.

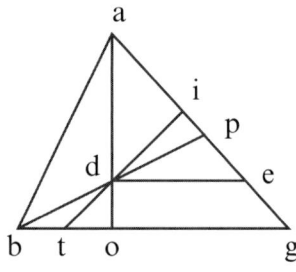

Figure 4.10

[12.3] Be it noted that you can always divide a triangle into two equal parts by a line passing through a point below some angle, the sides of which are unequal. Sometimes you can divide a triangle in two equal parts by a line through a given point below an angle, the sides of which are the longer segments. But you can never divide a triangle in two equal parts if the sides of the angle are the shorter segments. Therefore triangle abg has been divided in two equal parts by line it going through a given point t, and one of the parts is quadrilateral $abti$ and the other is triangle itg, as had to be proved.

[13] **4** *To divide a triangle in two equal parts by a line drawn from a given point outside the triangle.*[32] Given triangle *abg* and point *d* given outside the triangle, draw line *ad* cutting side *bg* at point *e*. Now if line *be* equals line *eg*, you have done what you proposed, because triangles *abe* and *aeg* sit on equal bases and have the same altitude. Whence triangle *abe* equals triangle *aeg*. But if line *be* is not equal to line *eg*, then one of them is longer; so let it be *be*. From point *d* draw line *zd* equidistant from line *be* **{p. 117}**. Extend line *ab* to *z*. Because line *be* is longer than half of side *bg*, the rectangular plane *ab* by *be* is more than half the rectangular plane *ab* by *bg*. By much more is the plane *ab* by *zd* larger than half the plane *ab* by *bg* since *zd* is longer than *be* [see Figure 4.11]. Let the plane *ib* by *zd* equal half the plane *ab* by *bg*. Because plane *ab* by *be* is larger than plane *ib* by *zd*, the ratio of *zd* to *be* will be smaller than the ratio of *ba* to *bi*. But the ratio of *zd* to *be* equals the ratio of *za* to *ab*. By disjunction the ratio of *zb* to *ba* is less than the ratio of *ai* to *ib*. Whence plane *zb* by *bi* is less than plane *ba* by *ai*. Adjoin[33] to line *bi* the parallelogram that exceeds by a square figure equal to plane *zb* by *bi*. That is, let line *bi* be adjoined to a certain line that multiplied by itself and *bi* is equal to the product of *zb* and *bi*. The parallelogram is the plane *ti* that you join to line *tkd*. Because plane *zb* by *bi* equals plane *bt* by *ti*, they are proportional according to the ratios of *zb* to *bt* and *ti* to *ab*. Therefore by conjunction as *zt* is to *bt* so is *bt* to *bi*. But as *zt* is to *bt*, so is *zd* to *bk*. Therefore as *zd* is to *bk*, so is *bt* to *bi*. Whence the plane *kb* by *bt* equals the plane *zd* by *bi*. But the plane *zd* by *bi* equals half the plane of *ab* by *bg*. Whence triangle *tbk* is half of triangle *abg*. Therefore triangle *abg* has been divided in two, cut by line *dkt* coming from point *d*. The two halves are triangle *tbk* and quadrilateral *tkga*, as we had to prove.

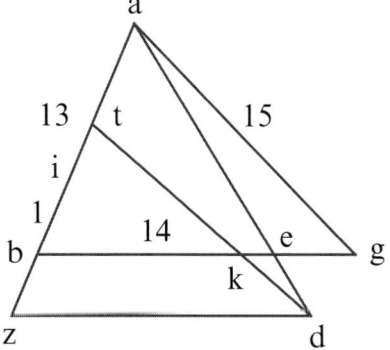

Figure 4.11

---

[32] "Demonstratio quomodo diuiditur trigonum in duo equa per lineam egredientem a puncto extra ipsam" (116.35–36, f. 71v.17–18). Archibald 47.

[33] That is, apply.

[14] Now I can do this with numbers: let side ab be 13, ag 15, bg 14, dz $\frac{2}{5}$ 10, and bz $\frac{3}{7}$ 1. Divide half of plane ab by bg or 91 by zd or $\frac{2}{5}$ 10 to obtain $\frac{3}{4}$ 8 for line bi. Then multiply zb by bi (that is, $\frac{3}{7}$ 1 by $\frac{3}{4}$ 8) to obtain $\frac{1}{2}$ 12. To this add $\frac{1\,1}{8\,8}$ 19, the square on half the line bi or multiplying $\frac{3}{8}$ 4 by itself, to obtain $\frac{41}{64}$ 31. The root of this number is $\frac{5}{8}$ 5 for the measure of line lt. If I add to this $\frac{3}{8}$ 4 or lb, then I have 10 for line tb. And because zt is to bt as zd is to bk, I multiply the product of zd and bt or $\frac{2}{5}$ 10 by 10 to obtain 104. Dividing this by $\frac{3}{7}$ 11 or line zt, I obtain $\frac{1}{10}$ 9 for line bk. Or, I can divide half of the plane ab by bg or 91 by 10 or bt to obtain the same $\frac{1}{10}$ 9 for line bk. Thus triangle btk will be half of triangle abg, as is necessary.

[15] **5** *To divide a triangle in two equal parts from a point on the extension of a side of a triangle*[34] [see Figure 4.12]. Suppose line ab were extended outside triangle abg to a given point z. If a line were drawn from the same point z dividing triangle abg in two equal parts, I can draw a line from point z that is equidistant from line bg. I extend line ag to point e and make the plane ze by gi equal to half the plane ag by gb. I apply to line gi a parallelogram and a square equal to the tetragon eg by gi. Let this be gt by ti.[35] And I join tlz. I say that triangle abg has been divided into equal parts by line tz coming from point z, one part being triangle tgl and the other quadrilateral $tabl$. For the proof: because the plane gt by ti equals the plane eg by gi, as eg is to gt so is ti to ig. Conjointly therefore as et is to tg so is tg to ig. But as et is to tg, so is ez to gl. *Per equale* therefore as ze is to lg, so is line gt to line gi. Whence the plane of lg by gt is equal to the plane {**p. 118**} ze by ig. But plane ze by ig is half of plane ag by bg. Whence plane tg by lg is also half of plane ag by bg. Therefore plane ag by bg is twice plane tg by lg. Therefore triangle abg is twice triangle tgl. Whence triangle tgl is half of triangle abg, as is necessary.

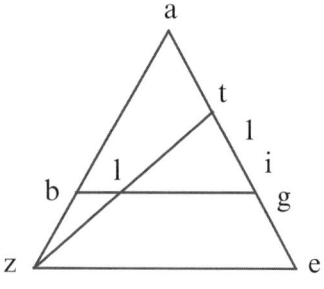

Figure 4.12

---

[34] "Et si unum ex lateribus trigoni extra trigonum protractum fuerit, et in ipso protractu punctus ille ceciderit, a quo recta egredi oporteat, que diuidat triangulum in duo equa." (117.30–32).

[35] This step produces a quadratic equation the solution of which yields the first numerical example. For details, see the next step.

[16] This can be demonstrated by numbers: again, let side ab be 13, side ag 15, base bg 14, and bz a fourth part of ab. Whence eg will be $\frac{3}{4}$ 3 or a fourth of ag. Whence ze adds a fourth part to bg. And thus ze will be $\frac{1}{2}$ 17 rods. If we divide this by half the product of ab and bg or 105, we obtain 6 for line gi. If to this we apply the plane gt by ti that is equal to the product of eg and gi, the plane gt by ti will indeed be $\frac{1}{2}$ 22. In order to know the measure of line ti, divide line ig in two equal parts at point h. Thus ih will be 3. Because line ig has been divided in two equal pars at point h, and line it has been added to it in a straight line, the product of it and gt with the square on line ih or 9 equals the square on line th. Whence th is the root of $\frac{1}{2}$ 31. If we add 3 or hg to this, then the whole line tg is 3 and the root of $\frac{1}{2}$ 31. Whence tg is a binomium, because, as was demonstrated, the product of lg and gt is equal to the product of ze and ig, and that is 105. One hundred five must be divided by the binomium gt, the parts of that are 3 and the root of $\frac{1}{2}$ 31. I will indicate how this division is done properly. Because line tg is a binomium, line ti is the recise of the binomium and equal to the root of $\frac{1}{2}$ 31 less 3.[36] In other words, hi that is 3 is removed from th the root of $\frac{1}{2}$ 31. Now the product of the binomium tg by the recise ti produces a rational number, namely $\frac{1}{2}$ 22.[37] If we divide $\frac{1}{2}$ 22 by the binomium gt, we obtain the recise ti. Therefore as much as $\frac{1}{2}$ 22 is in 105, by so much does the recise ti equal the quotient of 105 and the binomium tg. So, the division of 105 by $\frac{1}{2}$ 22 or 210 by 45 yields $\frac{2}{3}$ 4. Multiplying this by the recise ti produces the measure of line gl. The multiplication is done this way: multiplying $\frac{2}{3}$ 4 by line th[38] or the root of $\frac{1}{2}$ 31 yields four and two thirds roots of $\frac{1}{2}$ 31 or one root of 686. This comes from multiplying the square of $\frac{2}{3}$ 4 by $\frac{1}{2}$ 31. Now from the root of this product we must subtract the product of $\frac{2}{3}$ 4 by line ih or 14. And thus for line gl we have the root of 686 less 14 that is very close to a bit less than $\frac{1}{5}$ 12. And the binomium is very nearly $\frac{11}{68}$ 8. Multiplying this by $\frac{1}{5}$ 12 (that is, tg by gl) comes to about 105, as required. This is the procedure when we wish to divide some irrationalized number by a binomium. And note that if 105 need be divided by the recise it, do it the same way. From this division comes the binomium whose parts are the root of 686 and 14. Make every effort to do the same in similar cases.

[17] **6** *To divide a triangle in two equal parts by a line drawn from a point located outside the triangle but between two sides of the triangle, the sides having been extended from the same angle.*[39] In triangle abg extend sides ab

---

[36] Awkward as is this expression, it shows precisely that the recise is a quantity from which another quantity has been cut off or subtracted. In this case, the root of $\frac{1}{2}$ 31 is diminished by 3.

[37] Fibonacci understood conceptually, but could not express symbolically, that $(a - b)(a + b) = a^2 - b^2$.

[38] Text has ti; manuscript has th (118.25, f. 72v.14).

[39] "Sed protahantur duo latera trigoni extra trigonum continentia unum ex angulis trigoni; et punctus datus cadat infra lineas continentes angulum exteriorem, qui est equalis angulo interiori; at uolumus ab ipso puncto rectam protrahere diuidentem triangulum in duo equa" (118.34–37, f. 73r.1–5).

198   Fibonacci's *De Practica Geometrie*

and *bg* to points *d* and *e*. Within angle *ebd* let point *i* be given. From this draw a line dividing triangle *abg* in two equal parts by first joining points *i* and *b* and then extending line *ib* to point *z* on line *ag*. Therefore, if *az* is equal to *zg*, then triangle *abg* has been divided in two equal parts by line *iz*. But if *az* and *zg* are unequal, let the longer part be *az* [see Figure 4.13]. Extend *za* outside triangle *abg* to point *t*. And from point *i* draw line *it* equidistant from line {p. 119} *ba*. Because *za* is longer than half of side *ag*, plane *ab* by *az* is larger than half of plane *ba* by *ag*. Whence add [plane] *lt* by *ak* that is equal to half of plane *ba* by *ag*. Let plane *al* by *ak* equal plane *ta* by *ak*, and draw line *li*. Show this by what was said above,[40] that triangle *abg* was divided in two equal parts by line *il*, and that one of the parts is triangle *lac* and the other is quadrilateral *lcbg*.

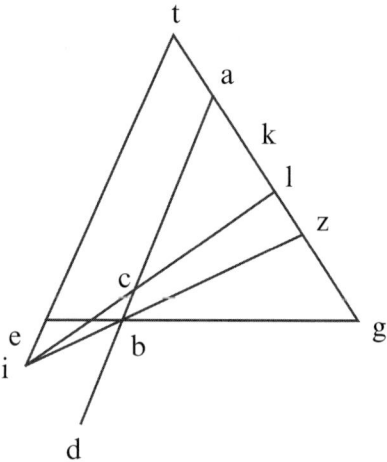

Figure 4.13

[18] **7** *To divide a triangle in two equal parts by a line parallel to a side.*[41] How this can be done will be shown in triangle *abg* by dividing it in two equal parts by a line equidistant to base *bg* [see Figure 4.14]. Extend line *ba* outside the triangle to point *d*, and let *ad* be equal to half of side *ba*. Let *ae* be the mean proportional for lines *ba* and *ad*,[42] that is, *ba* is to *ae* as *ae* is to *ad*. I say that triangle *abg* has been divided in two equal parts by line *ez*, triangle *aez* and quadrilateral *ebgz*. The proof follows. Triangles *aez* and *abg* are similar to each other because angles *aez* and *aze* are equal to the angles *abg* and *agb*, namely exterior angles to interior angles, and they have

---

[40] See [15] for a similar proposition.
[41] "Et si per equidistantem uni laterum trigonum aliquod diuidere uis" (119.7, f. 73r.18). Archibald 22.
[42] This is the construction that locates point *e* on line *abg*.

angle *bag* in common. Now the corresponding sides of the similar triangle are likewise similar, here in the ratio of two to one. Whence as *ba* is to *ad* so triangle *abg* is to triangle *aez*. Now *ba* is twice *ad*. Whence triangle *abg* is twice triangle *aez*. Therefore triangle *aez* is half of triangle *abg*. Or in another way: because when three lines are in continued proportion, as the first to the third, so the space that corresponds to the first is similar to what corresponds to the second, and similarly described.[43] Whence, as was said, as the first *ba* is to the third *ad*, so is that which corresponds to *ab* is twice that which corresponds to *ae*.[44] Therefore triangle *abg* is twice triangle *aez*, as I said. In another way. Because *ba* is to *ae* as *ae* is to *ad*, the product of *ba* and *ad* equals the square on side *ae*. But the square on *ab* is twice the square on *ae*. And because *bg* is equidistant from *ez*, and *ba* and *ae* are in the same proportion as *ga* and *az*, whence the square on *ba* is to the square on *ae* as the square on *ga* is to the square on *az*. Therefore the square on *ba* is twice the square on *ae*. Whence the square on *ga* is twice the square on *az*. Whence the product of *ba* and *ag* is twice the product of *ea* and *az*. And so triangle *abg* is twice triangle *aez*.

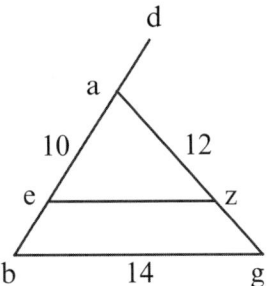

Figure 4.14

[19] If you wish to do this by numbers: let side *ab* be 10, side *ag* 12, and side *bg* 14. the product of *ab* by half of itself or 10 by 5 is 50 whose root is *ae*. Again, I multiply *ga* by half of itself and get 72 whose root is line *az*. Draw the line *ez*, and triangle *aez* is half of triangle *abg* because they share one angle. Let the product of *ea* and *az* be half that of *ba* and *ag*. Now the product of *ea* and *az*, the root of 50 and the root of 72, equals the root of 3600 whose root is 60, and that is half of 120, the product of *ba* and *ag*. And if you want to know the measure of line *ez*, multiply *bg* by its half to get 98 for the square on side *ez*. Do likewise in similar situations.

---

[43] "sicut prima ad tertiam, ita spatium, quod a prima ei, quod est a secunda simile, et similiter descriptum" (119.21–22); see *Elements*, VI.19 *porism*.

[44] The last clause is repeated by "id est, quod ab *ab* ei quod ab *ae*" (119.23–24).

On the Division of Triangles in Three Parts[45]

[20] **8** *To divide a triangle in three parts from one side*[46] **{p. 120}**. Given triangle *abg*, then divide side *bg* in three parts, such as *bd*, *de*, and *eg*, and draw lines *ad* and *ae* [see Figure 4.15]. I say that triangle *abg* has been divided in three equal parts that are triangles *abd*, *ade*, and *aeg*, because they sit on equal bases and share the same altitude. Therefore, they are necessarily equal to one another. And if I want a third part of triangle *bgd* from a given point *z*, I draw line *bz*.[47] And if *gz* is a third part of *gd*, then indeed triangle *bgz* is a third part of triangle *bgd*. But if *gz* is not a third part of line *gd*, then it will be more or less than a third. First let it be less [see Figure 4.16]. Whence I take a third part of *gd* that is *ga*, and I draw line *ai* equidistant from *zb*. Then I draw line *iz* that separates quadrilateral *ibgz* from triangle *bdg*. I will show that this is a third of the triangle *bdg*. For example: because line *ga* is a third part of the line *gd*, then triangle *bga* will be a third part of triangle *bdg*. And because triangles *bza* and *bzi* are between the equidistant lines *bz* and *ai* and on the same base *bz*, they are equal to each other. Let triangle *bgz* be added to both triangles. Then quadrilateral *ibgz* will be equal to triangle *bga*. Therefore quadrilateral *ibgz* is a third part of triangle *bgd*. What is left, namely triangle *izb*, can be divided into two equal parts by whichever of the above methods you want.

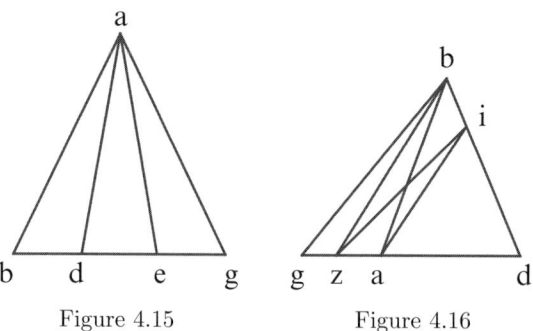

Figure 4.15   Figure 4.16

[21] Again, let triangle *abg* be divided in three equal parts that are triangles *abd*, *ade*, and *aeg*. Hence the sections of the base, *bd*, *de*, and *eg* equal one another. In the above figure, we had point *z* between *b* and *d*. It would be the same if we were to place it between *e* and *g*, and do the same thing. Now we put point *z* between *d* and *e* on side *bg* from which we want to draw a line cutting off a third part of triangle *abg* [see Figure 4.17]. First draw line *az*. Then through point *d* or *e* draw line *ei* equidistant from line *az*. Finally, draw *zi*. I say that line *zi* cuts a third part off of triangle *abg*; call it triangle *izg*. The proof follows.

---

[45] "Incipit de diuisione trigonorum in tres partes" (119.41, f. 73v.18).

[46] "Si trigonvm *abg* in tres equas partes diuidere uis per unum ex lateribus suis..." (119.43, f. 73v.20–21).

[47] Point *z* is not in Figure 4.15 (120 at the top, f. 73v below).

Because triangles *azi* and *aze* lie between equidistant lines and they have the same base, they are equal. If triangle *ade* is added to each, then trapezoid *adzi* equals triangle *ade*. But triangle *ade* is a third part of triangle *abg*. Hence quadrilateral *adzi* is likewise a third part of triangle *abg*. And triangle *abd* is another third part of triangle *abg*. What remains for the third part is triangle *izg* that was cut off of triangle *abg* by line *iz*. And this is what had to be done.

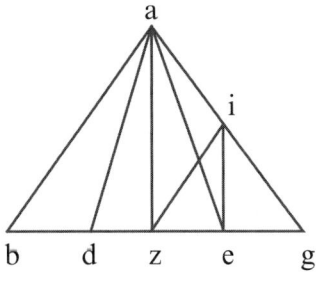

Figure 4.17

[22] **9** [*To divide a triangle in three equal parts so that a side belongs to each partner.*] Again, let there be a triangular field *abg* to be divided among three partners, each of whom wishes for a side of triangle *abg* to be in his part.[48] So, divide side *bg* in two equal parts at point *d*, draw line *ad*, and make *dc*[49] a third part of line *ad*. Draw lines *bc* and *cg*. I say that triangle *abg* has been divided in three equal parts, each on one side of a triangle, namely triangles *abc*, *acg*, and *bcg*. The proof follows [see Figure 4.18]. Because line *dc* is a third of line *da*, line *ac* is twice that of *cd*. Whence triangle *abc* is twice triangle *bcd*. For the same reason therefore triangle *acg* is twice triangle *gcd*. Again, because line *bd* equals line *dg*, triangles *cbd* and *cdg* are equal. Whence triangle *cbg* is twice each of the triangles *cbd* and *cdg*. And since it is twice each, the one equals the other. Whence triangles *acb*, *acg*, and *bcg* equal one another. Therefore triangle *abg* has been divided as required {**p. 121**}.

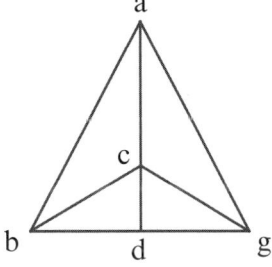

Figure 4.18

---

[48] "Adiaceat iterum trigonum *abg* quod oporteat inter tres consortes diuidere, quorum unusquisque uelit in sua portione habere unum ex lateribus triangulj *abg*" (120.16, f. 74r.18).

[49] Text has *de*; manuscript has *dc* (120.33, f. 79v.21).

202   Fibonacci's *De Practica Geometrie*

[23] **10** *To cut a given part from a triangle by a line drawn through a given point within the triangle.*[50] In triangle *abg* let an arbitrary[51] point *d* be given through which I wish to draw a line cutting off a given part, say a third, of the triangle. So I draw line *ae* through point *d*, and I consider whether the section, either *be* or *eg*, is a third part of side *bg*. Because if *be* is a third of line *bg*, then triangle *abe* is a third part of triangle *abg*. But suppose neither section is a third part of side *bg*. Then consider what I said about some line coming from the other angles and going to the other sides.[52] If I shall not have found some section of the sides to be a third of its side, then I shall draw through point *d* line *dz* equidistant from line *ge*. I shall make plane *dz* by *zi* equal to a third part of the plane *ag* by *bg*. I apply to line *gi* an equiangular plane less a square equal to *gz* by *zi*. Let it be *it* by *tg*. I shall construct line *td* and extend it to *k* [see Figure 4.19]. I say that line *tk* going through point *d* cuts off triangle *tkg*, a third part of triangle *abg*. The proof follows. Because the product of *gi* and *gz* equals the product of *it* and *tg*, *gi* is to *it* as *tg* is to *gz*. Alternately therefore, *ig* is to *gt* as *gt* is to *zt*. But *gt* is to *zt* as *kg* is to *dz*. Therefore as *kg* is to *dz*, so is *gi* to *tg*. Whence the product of *tg* and *gk* equals the product of *dz* and *gi*.[53] But the product of *dz* and *gi* is a third part of the product of *ag* and *gb*. Hence plane *ti* by *gk* is a third part of plane *ag* by *gb*. Whence triangle *tkg* is a third part of triangle *abg*, as was said.

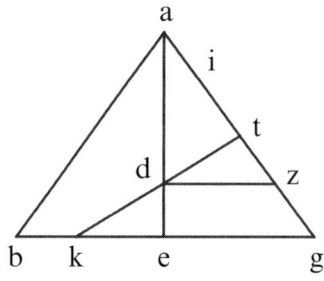

Figure 4.19

[24] **11** *To cut a given part from a triangle by a line drawn through a given point outside the triangle.*[54] Let point *d* be given outside triangle *abg*. It is

---

[50] "Demonstratio qualiter a trigono abscidatur pars data per lineam protractam per punctum datum infra triangulum" (121.1–2, f. 74r.31–33). Archibald 41.

[51] "fortuitu" (121.3).

[52] Possible reference to [12] and [13].

[53] Text has *zi*; manuscript has *gi* (121.19, f. 74v.15).

[54] "Demonstratio qualiter a trigono abscidatur pars data per lineam protractam per punctum datum extra triangulum" (121.22–23, f. 74v.17–19). Archibald 48.

necessary that a line going through point d cut a given part, say a third, from triangle abg. First I shall draw line ad cutting side bg at point c. Now if line bc or cg is a third part of line bg, then a third part of triangle abg has been cut off by line ad going through point d[55] [see Figure 4.20]. But if it does not, then extend lines ab and ag to points e and z outside triangle abg. Through point d I shall draw line ez equidistant from side bg. I make plane de by gi equal to a third part of plane ag by gb. I shall apply to line gi a parallelogram exceeded in area by a square equal to an area of eg by gi. Let the area be ik by kg. Through point k I draw line kmd. I say that triangle kmg is a third of triangle abg. The proof follows. Because plane eg by gi equals the plane kg by ki, then eg is to gk as ki is to ig. By conjunction therefore as ek is to gk, so is gk to ig.[56] But as ek is to kg, so is de to gm. Therefore as ed is to gm, so is gk to gi. Consequently plane gk by gm equals plane de by gi. But plane de by gi is a third part of plane ag by gb. Whence plane gk by gm is a third part of plane ag by gb. And because plane kg by gm is to plane ag by gb, so is triangle kgm to triangle agb. Therefore triangle kgm is a third part of triangle abg, as I said.

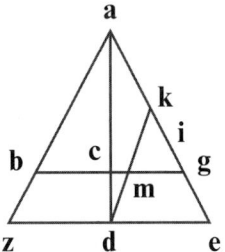

Figure 4.20

[25] In a similar way we can cut any required part from any triangle by a line drawn from a given point on the extension of a side outside the triangle to a side or through a point lying between the extension of two sides of the triangle {p. 122}. Again and in a similar way, if there are two given points d and e within triangle abg and we wish to divide the triangle in three equal parts by lines going through those points, then first we cut triangle zig from triangle abg by line zi going through point d [see Figure 4.21]. This contains two thirds of triangle abg, leaving quadrilateral $zabi$ for the remaining third of triangle abg. Next we divide triangle izg in two halves by line tk going through point e. One part will be quadrilateral $tzik$

---

[55] See [20].
[56] Text and manuscript have ki; context requires ig (121.35, f. 74v.).

and the other triangle *tkg*. Proceed in the same way if the given point is outside the triangle.[57]

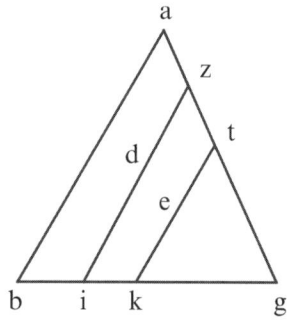

Figure 4.21

[26] **12** *To divide a triangle by lines equidistant from the base.*[58] Given triangle *abg* with base *bg*, extend side *ba* by a third of itself to point *d*. Then extend it farther straight on to point *e*, so that *de* equals *ad* and *ae* is two thirds of line *ba* [see Figure 4.22]. Make line *az* the mean proportional between *ba* and *ad*, and *ai* the mean proportional between *ba* and *ae*. Through points *z* and *i* draw lines *zt* and *ik* equidistant from base *bg*. I say that triangle *abg* has been divided in three equal parts, of which one is triangle *azt*, the second quadrilateral *zikt*, and the third quadrilateral *ibgk*. The proof follows. Because as *ba* is to *az*, so *az* is to *ad*. As *ba* is to *ad*, so triangle *abg* is to triangle *azt*, because the two triangles are similar.[59] Because line *ba* is thrice line *ad*, it follows that triangle *abg* is three times triangle *azt*. Therefore triangle *azt* is a third part of triangle *abg*. Again, because *ba* is to *ia*, so is *ai* to *ae*. Hence the triangle on *ea* is similar to the triangle on *ai*. Likewise the triangles *aik* and *abg* on sides *ai* and *ab* are similar. Whence, as *ea* is to *ab* (that is, as 2 is to 3), so is triangle *aik* to triangle *abg*. Therefore triangle *aik* is two thirds of triangle *abg*. If you take from triangle *aik* triangle *azt* that is a third part of triangle *abg*, what will remain is quadrilateral *zikt*, a third part of triangle *abg*. Also remaining is the similar quadrilateral *ibgk*, another third part. Therefore triangle *abg* has been divided in three equal parts, as required. Thus we have shown how to divide all kinds of triangles into three[60] or more parts.

---

[57] This is a very general rule requiring mastery of nearly all the preceding techniques.
[58] "Et si triangulum *abg* per equidistantes rectas basi *bg* diuidere uolumus..." (122.8, f. 75r.12–13). Archibald 23.
[59] *Elements* VI.19 *porism*.
[60] Text and manuscript have *quatuor*; context requires *tres* (122.27, f. 75r.32).

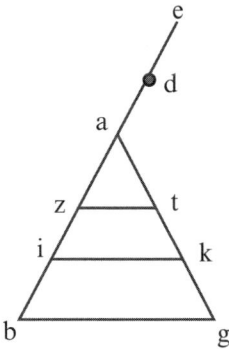

Figure 4.22

4.2 Division of Quadrilaterals[61]

[27] There are three kinds of quadrilaterals: parallelograms, trapezoids, and generic[62] quadrilaterals. For parallelograms with opposites sides and angles respectively equal, there are four kinds: the square (all side and angles equal), the rectangle (with each pair of opposite sides and all angles right angles), the rhombus (all sides equal and only opposite angles equal), and the rhomboid (only two opposite sides are equal, and the opposite angles are equal). Because there is one way of dividing these four species of parallelograms, I will use the same notation for each, so that what I say about one type is understood for the others. The letters *abcd* are used for the square, rectangle, rhombus, and rhomboid, and their diameters divide them in equal parts.

[27.1] **13** *To divide a parallelogram in two equal parts by its diameters.*[63] Given parallelogram *abcd*: draw diameters *ac* and *bd* as seen in the first figure; each cuts the figure in two equal parts [Figure 4.23]. For example: {**p. 123**} if we draw diameters *ac* and *bd*,[64] we obtain triangle *abc* equal to triangle *acd* because side *ad* equals side *bc*, side *ab* equals side *dc*, and the base *ac* is common to both triangles.

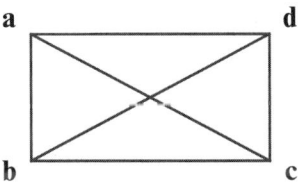

Figure 4.23

---

[61] "Incipit de diuisione quadrilaterorum" (122.28, *f.* 75r.33–34).
[62] "diuersilateral" (122.30).
[63] "... erunt utique in duas equales partes ab unoquoque dyametrorum ipsarum diuisa" (122.43, *f.* 75v.13–15).
[64] The text has only one diameter drawn; the context requires two.

206 Fibonacci's *De Practica Geometrie*

[27.2] **14** [*To divide a parallelogram in two equal parts from a midpoint on a side.*] In the second figure [Figure 4.24] suppose you wish to divide some side, say *bc*, in two equal parts. From midpoint *e* draw line *ef* equidistant from the other two sides *ba* and *cd*,. Thus you will have divided the quadrilateral *ac* in two equal parallelograms, *ae* and *fc*. For they are on equal bases *be* and *ec*, and the sides *ad* and *bc* are equidistant from each other.

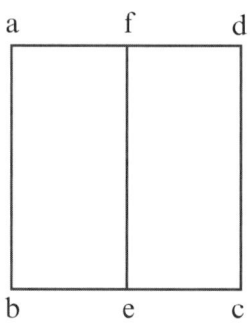

Figure 4.24

[27.3] **15** [*To divide a parallelogram in two parts by a line parallel to two sides and bisecting the other two sides.*] Similarly, if we draw line *gk* equidistant from lines *ab* and *ac* to cut lines *ad* and *bc* in half,[65] then the parallelogram has been divided in two equal parallelograms *gd* and *bk*, as shown in the third figure [Figure 4.25]. They are equal to each other because they have equal bases *cg* and *gb*, and sides *cd* and *ab* are equidistant from one another.

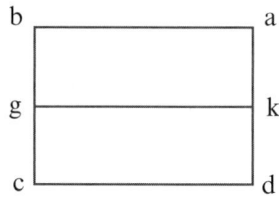

Figure 4.25

[27.4] **16** [*To divide a parallelogram in two parts from a given point on a side.*][66] In the fourth figure [Figure 4.26], let some point *h* be given between points *b* and *e*. Assign a point *i* opposite it between points *f* and *d* so that *fi* equals *eh*. Draw line *hi*. Then I say that each of the four quadrilaterals has been cut in two equal parts by line *hi*. The proof follows. Because *fe* and *ih* are between

---

[65] The text has position of the pairs of lines exchanged; the context requires the correction as seen in the figure and from what follows (123.15, f. 75v.25–27).

[66] (123.14, f. 75v.29–30). Archibald 27.

the equidistant lines *ad* and *bc*, angle *ifk* equals angle *keh* and angle *fik* equals angle *khe*. The vertical angles at point *k* are also equal, and the base *fi* equals the base *eh*. Therefore triangle *fki* equals triangle *khe*. Add to both triangles the common pentagon *kfabh*. Thus quadrilateral *iabh* equals quadrilateral *abef*, each of which is half of the whole quadrilateral *ac*. Or in another way, because the parallelograms *ae* and *fc*[67] lie between the two equal and opposite equidistant lines *ad* and *bc*, side *af* equals side *be* and *fd* equals *ec*. But *fd* is half of *ad*, and so *fd* equals *af*. Therefore *af* equals line *ec* and *fi* equals line *he*. Therefore *ai* equals *ch*, lines *di* and *bh* are equal, and *hi* is a common side. Therefore quadrilateral *iabh*, as we said, equals quadrilateral *ihcd*. Likewise if the given point were to fall between *e* and *c*, however much *e* is distant from *c* by so much do you measure *fa*, beginning from point *f*. And so in this way you can divide all parallelograms by a point given upon any of the sides.

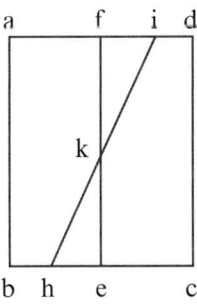

Figure 4.26

[28] **17** *To divide a parallelogram in two equal parts by a line drawn from a given point within the parallelogram.*[68] Given parallelogram *ag* within which is point *c*: through this point we want to draw a line that cuts the figure in two equal parts. In the middle of diameter *bd* select point *f*, join points *c* and *f*, and extend the line in both directions to points *e* and *z*. I say that the parallelogram *ag* has been divided in two equal parts of which one is quadrilateral *eabz* and the other quadrilateral *edgz*. The proof follows [see Figure 4.27]. Because two lines *ez* and *db* intersect between the equidistant lines *ad* and *bg*, angle *fde* equals angle *fbz*, angle *fed* equals angle *fzb*, and the [vertical] angles about point *f* are equal. Therefore triangles *fed* and *fbz* are equiangular. And because line *fb* equals line *fd*, lines *de* and *dz* are equal, and triangle *dfe* equals triangle *bfz*. After adding quadrilateral *dfzg* to both triangles, quadrilateral *ezgd* equals triangle *bdg*. But triangle *bgd* is half of parallelogram *ag* {**p. 124**}. Whence quadrilateral *ezgd* is half of parallelogram *ag*, as I said.

---

[67] The text and manuscript have "et *ic*"; the context does not allow it, because *ic* is not a parallelogram (123.23).

[68] "Et si paralilogramum aliquod in duo equa per lineam protractam a puncto dato infra ipsum secare desideras ..." (123.32–33, f. 76r.15–16).

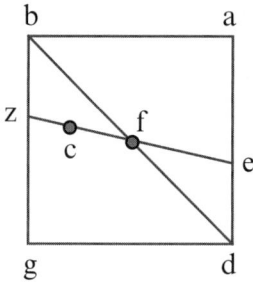

Figure 4.27

[28.1] **18** *To divide a parallelogram in two equal parts by a line drawn from a given point on one of its sides.*[69] For example: let the given point were *e*. In order for line *db* to be divided in two equal parts at point *f*, we draw line *ez* through this point to divide parallelogram *ag* in two equal parts, as we proved [see Figure 4.28].

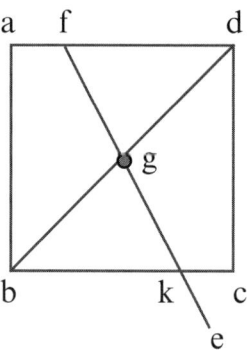

Figure 4.28

[28.2] **19** *To divide a parallelogram in two equal parts by a line coming through a given point outside the parallelogram.*[70] Select point *e* outside parallelogram *ac*. Then again draw diameter *bd* and divide it at point *g*.[71] Draw line *eg* and extend it to point *f* [Figure 4.28]. Thus parallelogram *ac* has been divided in to equal parts by a line coming from a given point *e*, which is proved in the foregoing figure. One half is quadrilateral *fabk*, the other is quadrilateral *fkcd*, as is obvious in the same figure.

---

[69] "hoc modo diuiduntur omnia paralilogramina erecta [linea] protracta a puncto dato super unum ex lateribus ipsius" (124.1–5, f. 76r. 29–30).

[70] "Et si paralilogramum aliquod in duo equa secare uis per lineam egredientem a puncto dato extra ipsum" (124.5–7, f. 76r.33–35). Archibald 31.

[71] Fibonacci implies that point *g* is a midpoint.

4.2 Division of Parallelograms in Many Parts[72]

[29] **20** *To divide a parallelogram into three equal parts on two of its given sides.*[73] Refer to the above list of the four kinds of parallelograms. Given parallelogram *abcd* with sides *ad* and *bc*. Divide side *ad* in three equal parts: *ae*, *ef*, and *fd*. Through points *e* and *f* draw lines *ef* and *fg* equidistant from sides *ad* and *bc*. I say that whichever of the above four parallelograms you choose, lines *eg* and *fh* will divided it in three equal parts. The proof follows. Given lines *ad* and *bc* are equidistant from one another [see Figure 4.29]. Draw equidistant lines *ab*, *eg*, *fh*, and *de*, all four being equal to each other. Whence quadrilaterals *ag*, *eh*, and *fc* are parallelograms with equal bases *bg*, *gh*, and *hc*. Because each is a third part of side *bc*, each of the parallelograms is a third part of the whole parallelogram *ac*, as seen in the first figure.

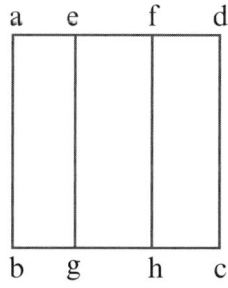

Figure 4.29

[30.1] **21** *To mark off a third of a parallelogram from a given point on a side.*[74] Given parallelogram *abcd* and point *e*: first divide side *ad* at points *e* and *f* in the order stated above, so that each section *ag*, *eh*, and *fe* is a third part of the whole line *ad*. Then by drawing line *eg*, a third of parallelogram *ac* will be cut off, namely parallelogram *ag*. Similarly, if the given point were *f*, then parallelogram *fc* would be a third of parallelogram *ac*. If the given point were neither *e* nor *f*, then let it lie be between *a* and *e* or between *e* and *f* or between *f* and *d*. First, let the given point *i* lie between *a* and *e*. I wish to remove a third of parallelogram *ac* by a line drawn from point *i* [see Figure 4.30]. First, I locate point *f* a third of the way along line *ad*. From point *f* I draw *fh*[75] equidistant from sides *ab* and *dc*. Thus parallelogram *fh* is a third of parallelogram *ac*. Whence parallelogram *ah* is twice the size of parallelogram *fc*. Consequently it is necessary to divide parallelogram *ha* in two equal parts by drawing a line

---

[72] "Incipit de diuisione paralilogramorum in plures partes" (124.13, *f.* 75v.5).
[73] "Et uolumus aliquod ipsarum [paralilogramorum] in tres equales partes diuidere super duo data latera eius" (124.14–15, *f.* 76v.6–7).
[74] "Et si super aliquem punctum datum super uno laterum rectam protrahere uis, que abscidat a paralilogramo dato tertiam partem ..." (124.24–25, *f.* 76v.17–18). Archibald 28.
[75] Text and manuscript have *fc*; context requires *fk* in both places (124.35, *f.* 76v.29).

from point *i*. So, the same distance that *i* is from *a*, so far must I make *k* from *h*. Then I connect points *i* and *k* to form quadrilateral *iabk* that is half of parallelogram *ah*. Therefore line *ik* has cut off a third of parallelogram *ac*. Further, it is divided in equal parts that are quadrilaterals *iabk* and *ikhf* and parallelogram *fc*, as seen in the second figure [see Figure 4.31]. If the given point *i* lies between points *f* and *d*, the same procedure can be used on another part to cut a third from parallelogram *ac*, namely parallelogram *ag* {**p. 125**}. Then I divide parallelogram *ec* in two equal parts by line *id* drawn from point *i*. It cuts off a third part of parallelogram *ac*, namely quadrilateral *ilcd*, as can be seen in the third figure. And if point *i* is between points *e* and *f*, I would cut parallelogram *fc* from parallelogram *ac*, and I divide what is left in two equal parts by line *im* drawn from point *i* [see Figure 4.32]. Let line *im* cut parallelogram *iabm* from parallelogram *ac*, the former a third of the latter, and the remaining two parts are quadrilateral *imhf* and parallelogram *fc*, as was demonstrated in the fourth figure, namely parallelogram *abcd*.

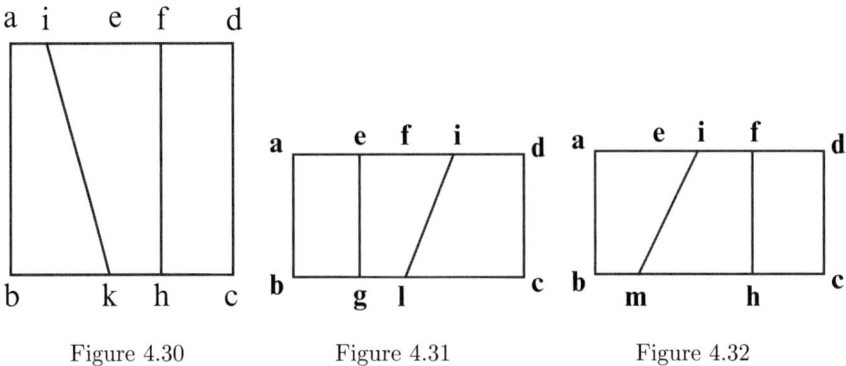

Figure 4.30   Figure 4.31   Figure 4.32

[30.2] **22** *To divide a parallelogram in four or more equal parts.*[76] For example: if you wish to divide some parallelogram in four equal parts, first divide it in two equal parts either by a diameter or by a right line equidistant from the two sides. Then divide each of the parts into two equal parts, and thus the parallelogram is divided in four parts.

[31] **23** *To divide a parallelogram in unequal parts.*[77] Let there be parallelogram *abgd* that the partners wish to divide for themselves in a half, a third, and a sixth. I divide the first parallelogram *ag* in two equals, *az* and *eg* [see Figure 4.33]. Then I cut from one of them a third part, parallelogram *ig*; that is, *id* cuts a third part of line *ed*. I say that parallelogram *ag* has been divided in three equal parts of which the half is parallelogram *az* and its sixth is parallelogram *ig*. And the rest,

---

[76] "Si uis paralilogramum aliquod diuidere in quatuor partes equales" (125.9, f. 77r.11–12).

[77] "De diuisione paralilogramorum in partes inequales" (125.13, f. 77r.15–16).

namely parallelogram *et*, is a third part of the whole *ag*. Because *id* is a third of *ed*, *id* is a sixth part of the whole *ad*. Whence I say that parallelogram *ig* is a sixth part of parallelogram *ag*. We can indeed take any part of the foregoing by a line extended from a given point, either *ad extra* or *ad intra* or even upon one of the sides of the parallelogram by the procedures set forth here.

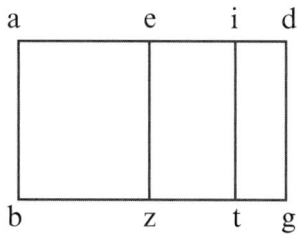

Figure 4.33

4.3 Division of Four Trapezoids with Only Two Equidistant Sides[78]

[32] There are four kinds of trapezoids: *semicaput abscisa, eque caput abscisa, diuerse caput abscisa, caput abscisa declinans*.[79] Because one method divides all four of these, I shall use the same letters and without any notes. Whatever I shall say about one of them must be understood as applicable to the others. Assign therefore to trapezoid *abgd* (the representative of the four species of trapezoids) the equidistant sides *ad* and *bg*.

[33] **24** *To divide a trapezoid [in two equal parts] by a line parallel to the bases.*[80] Given trapezoid *abgd* with equidistant bases *ad* and *bg*.[81] Let line *ad* be shorter than line *bg*. If we extend lines *ba* and *gd* from their end points *a* and *d*, they will meet at point *e*. Let the square on line *ze* be half the squares on lines *eb* and *ae*.[82] Through point *z* draw line *zi* equidistant from the base *bg*. I say that the trapezoid *abgd* has been divided in two equal parts by line *zi* equidistant from line *bg* [see Figure 4.34]. The proof follows. Because the squares on lines *eb* and *ae* are twice the square on line *ez*, triangles *ebg* and *ead* are twice triangle *ezi* since they are all similar. If we remove triangle *ebg* from triangle *ezi* (which equals itself), quadrilateral *zbyi* and triangle *ead*

---

[78] "Incipit de diuisione quatuor figurarum caput abscisarum, que duo tantum latera habent equidistantia" (125.25–26, f. 77r.26–28).

[79] These names do not translate well.

[80] "Volumus quamlibet ipsarum [figurarum] diuidere per rectam equidistantem basibus earum" (125.33, f. 77r.34). Archibald 25.

[81] The text and manuscript omit *et ad* (125.34).

[82] The purpose of this step is to determine the location of point *z*.

remain equal to triangle *ezi*. Remove triangle *ead* from each side and quadrilateral *zg* remains, the equal of quadrilateral *ai*. Therefore quadrilateral *abgd* has been divided in two equal parts by line *zi*, as required.

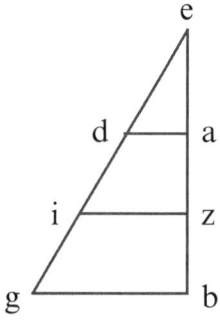

Figure 4.34

[34] Let us show this with numbers {**p. 126**}. Let side *ab* be 12, *bg* 12, *ad* 3, and *gd* 15 rods. Because side *da* in triangle *ebg* has been drawn equidistant from base *bg*, *ad* is to *bg* (that is, 3 is to 12) as *ea* is to *eb*. Whence as 3 is to 9, so is *ea* to *ab*. Therefore *ea* is 4 or a third part of *ab*. Therefore *eb* is 16. Consequently, after adding together the squares on *eb* and *ea* (or 256 and 16), we have 272 whose half is 136 the square on line *ez*. And because *ez* is to *eb* as *zi* is to *bg*, so the square on line *ez* is to the square on line *eb*, that is 136 to 256. Or in lowest terms 17 is to 32, the square on line *zi* to the square on line *bg*. Whence if we multiply 17 by 144 and divide by 32, we get $\frac{1}{2}$ 76 for the square on line *zi*. Because triangle *ebg* is a right triangle,[83] triangles *ezi* and *ead* will also be right triangles. So if we multiply half of *ez* by *zi*, we will have the area of triangle *ezi* that is 51, the root of 2601. The product of the fourth part of 136 and $\frac{1}{2}$ 76 the square on line *zi* is also 2601, the square of 51. If we remove triangle *ead* or 6 from this area (51), what remains is 45 the area of quadrilateral *azid*. Now 45 is half the area of trapezoid[84] *abdg*. If we multiply half of *ab* by the sum of *ad* and *bg*, that is, 6 by 15, or double 45, we have 90. This proves that the quadrilateral *ai* is half of the whole quadrilateral *abgd*. Or in another way: by removing *ez* from *eb* the root of 136 from 16 for line *zb*, what remains is 16 less the root of 136. Its half is 8 less the root of 34. If we multiply the half by the sum of *zi* and *be* or 12 and the root of $\frac{1}{2}$ 76, we also get 45 for the area of quadrilateral *zg*. The multiplication of a binomium by a recise proceeds in this way. As you can see in the margin [Figure 4.35], arrange first the terms of the numbers. Then multiply whole number by whole number (12 by 8) to get the whole number product (96). Add to this the product of 8 by the root of $\frac{1}{2}$ 76 to yield 96 and one root of the product

---

[83] Fibonacci merely suggested this in [33].
[84] "dimidium totius quadrilaterj semicaputascisci" (126.16).

of 64 by $\frac{1}{2}$ 76 or 4896. From this subtract the product of 12 by the root of 34 (or one root of 4896) to leave just 96. From this subtract 51, the product of the root of 34 by the root of $\frac{1}{2}$ 76, to leave 45 as remainder, as I said, for the area of quadrilateral *zg*.

integra	radix addita
12	1/2 76

	radix diminuta	
34		8

Figure 4.35

[35] **25** *To divide a trapezoid in two equal parts from a point on an equidistant side.*[85] Given trapezoid *abgd*, divide the equidistant lines *ad* and *bg* in two equal parts at points *t* and *k*. Then join points *t* and *k*, as seen in the second figure [Figure 4.36]. I say that quadrilateral *abgd* has been divided in two equal parts by line *tk*. The proof follows. Draw lines *tb* and *tg*, so that triangles *tkb* and *tkg* equal one another because they have the same height and equal bases. Again, because triangles *bat* and *gtd* have the same altitude, as *at* is to *td* so is triangle *bat* to triangle *gtd*. Because line *at* equals line *td*, triangle *bat* equals triangle *gtd*. It has therefore been proved that triangle *tkb* equals triangle *tkg*. So quadrilateral *ak* equals quadrilateral *tg*. Therefore quadrilateral *abgd* has been divided into two equal parts by line *tk*. It is clear from this, as we said, that for every quadrilateral having two equidistant sides, if some line cuts its equidistant sides proportionally, it will cut the figure in the same proportion. For, as we said, as *at* {**p. 127**} [is] to *td* so *bk* is to *kg*. Whence as *at* is to *td*, so is quadrilateral *ak* to quadrilateral *tg*.

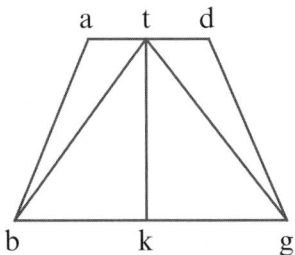

Figure 4.36

[36] **26** *To divide a trapezoid in two equal parts by a line from a given point on the shorter of the equidistant sides.*[86] Let the trapezoid *abgd* already be

---

[85] "Et si super latus *ad* aliquod ex supradictis caput abscisis in duo equa diuidere uis ..." (126.31–32, *f.* 78r.11).

[86] "Et si quadrilaterum duorum laterum equidistantium in duo equa diuidere uis per lineam protractam a dato puncto super minus latus equidistantium ..." (127.2–3, *f.* 78r.27–28).

divided in two equal parts by line *tk*. Let point *a* be given on line *ad*. On line *kg* I mark off line *kl* equal to line *ta* and line *bl* equal to half the sum of lines *ab* and *bg*. I join points *a* and *l*. I say that trapezoid[87] *abgd* has been divided in two equal parts by line *al* [see Figure 4.37]. The proof follows. Because line *at* is equidistant from line *kl* and line *al* intersects both, angle *tam* and *mlk* are equal. For the same reason therefore angle *atm* equals angle *mkl*, and the vertical angles at *m* are mutually equal. Further the bases *at* and *kl* are equal. Whence triangle *amt* equals triangle *lmk*. If we add quadrilateral *mabk* to both angles, then triangle *abl* equals quadrilateral *abkt* that is half of the whole *ag* as is seen in the third figure. In the same way if the given point *d* is on side *ad* and I make line *kn* equal to line *td*, then *gn* equals half of the sum of the sides *ad* and *bg*. I draw line *dn* that divides quadrilateral *abgd* into equal parts, as is seen in the fourth figure [Figure 4.38]. This was proved by what was said for the preceding figure. It also follows that if the given point is on line *bg* and between points *n* and *l*, the location of the dividing line will be on side *ad*.

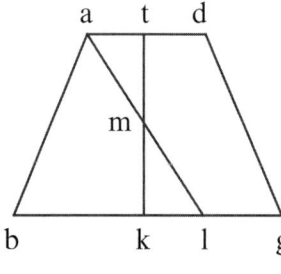
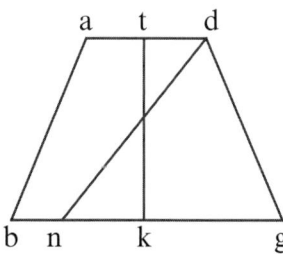

Figure 4.37   Figure 4.38

[37] But if the given point were between points *b* and *n* or between points *l* and *g*, we will have to show how the line drawn from it divided the whole quadrilateral in two equal parts. Consider again trapezoid *abgd* with two sides equidistant and line *dn* dividing the trapezoid in two equal parts. Let the given point be *b* on line *bg*. I join points *d* and *b* and make line *nc* equidistant from it. Finally I join points *b* and *c* [Figure 4.39]. I say that the trapezoid *abgd* has been divided in two equal parts by line *bc* drawn from the given point *b*. The proof follows. Triangles *ncb* and *ncd* are equal because they lie between the same equidistant lines *bd* and *nc* and are on the same base *nc*. If triangle *cng* is added to both, then triangle *cgb* equals triangle *dgn*. But triangle *dgn* is half of the trapezoid *abgd*. Whence triangle *cgs* is one half of trapezoid *abdg*. You must know that if the given point were between points *b* and *n* on line *bn*, then the line drawn from that given point will cut line *gd* between points *c* and *d*. The section can be found if the given point between

---

[87] In this section and for most of the time Fibonacci uses the term *quadrilaterum* where he is writing of a trapezoid. I have almost always used the term *trapezoid*.

b and *n* is joined with point *d*, and an equidistant line drawn from point *n*. Proceed in a similar way if the given point were on line *bg* and between points *l* and *g*. For example: consider again trapezoid *abgd* that had been divided in two equal parts by line *al*, and let the given point be *g* [see Figure 4.40]. In the order mentioned above I draw line *ga* and the equidistant line *lf*. I draw line *gf* that coming from given point *g* divides trapezoid *abdg* in two equal parts. This is proved easily by the method mentioned above. It has already been shown how a trapezoid can be divided by a line drawn from any given point on equidistant lines.

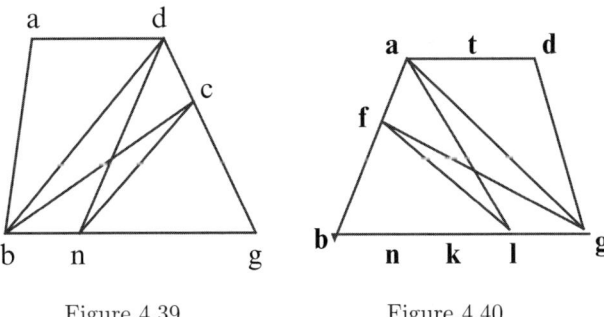

Figure 4.39   Figure 4.40

[38] **27** *To divide a trapezoid by a line drawn from a given point on the other side.*[88] Again consider trapezoid *abgd* that has been divided in two equal parts by line *zi* equidistant from the base *bg*. Also draw line *gf* as found in the above figure. Let the given point be on line *ab* wherever you choose. Let it be between *b* and *z* {**p. 128**} or between *z* and *f* or between *f* and *a*, or one of them. So if the given point were *b*, trapezoid *abgd* would be divided in two equal parts by line *bc*. Similarly, if the given point were *z*, line *zi* would divide trapezoid *abdg*, as was said. And if *f* were the given point, line *fg* would indeed cut trapezoid *abdg* in two parts, as was shown. And if it were *a*, then line *al* would cut it in two equal parts according to what has gone before [see Figure 4.41]. But if the given point were *h* lying between points *b* and *z*, I would draw line *hi*. And then through point *z* I would draw line *zc* equidistant from line *ih*, and finally draw line *hc*. I would say that trapezoid *abgd* has been cut in two equal parts by line *hc* that was drawn from point *h*. The proof follows. Because triangles *hic* and *hiz* lie between equidistant lines *hi* and *zc* and they share the same base *hi*, they are equal to each other. So if we add to each the same quadrilateral *igbh*, then quadrilateral *cgbh* is equal to quadrilateral *izbg* that is half of the whole trapezoid *abgd*, as was shown [see Figure 4.42]. And if a line is drawn from point *p* lying between points *z* and *f* [to point *q*], let points *p* and *i* be joined to form a line equidistant from line *zq*. Draw line *pq* to divide trapezoid *abdg* in two equal parts. This does not require a proof

---

[88] "Nunc vero ostendamus quomodo diuidantur a linea egrediente a dato puncto super reliqua latera" (127.39–40, *f.* 78v.34–35).

216 Fibonacci's *De Practica Geometrie*

because what has already been said should make it completely obvious [see Figure 4.43]. And if *r* be the given point between *a* and *f*, then draw *rg* equidistant from line *fs*. Next, draw line *rs* to divide trapezoid *abdg* in two equal parts. For example: triangles *fsg* and *sfr* equal one another. If triangle *sbf* is added to both, then triangle *rbs* equals triangle *fbg*. But triangle *fbg* is half of trapezoid *abdg*. Therefore triangle *rbs* is half of the same trapezoid. If the point is given on side *dg*, then the problem is solved in a similar way.

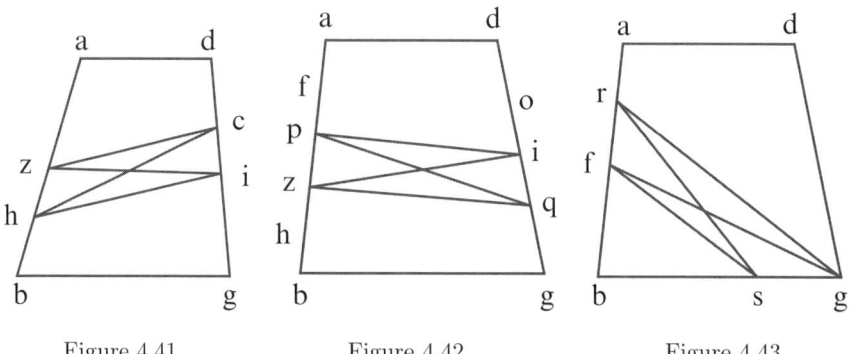

Figure 4.41　　　　　Figure 4.42　　　　　Figure 4.43

[39] **28** *To divide a trapezoid in two equal parts from its angles*[89] [see Figure 4.44]. Let sides *ad* and *bg* be the equidistant sides of trapezoid *abdg* with *ad* the smaller side. Draw diameters *ag* and *bd* intersecting each other at point *m*. Because sides *ad* and *bg* are equidistant, triangles *amd* and *bmg* are similar. Wherefore, as *bg* is to *ad*, so are *bm* to *md* and *gm* to *ma*. Because *bg* is longer than *ad*, therefore *bm* is longer than *md* and *gm* longer than *ma*. Therefore let diameters *ag* and *db* be divided in two equal parts at points *e* and *z*. Draw line *ec* through point *e* equidistant from diameter *bd*. Finally draw line *bc*. I say that line *bc* drawn from angle *b* divides quadrilateral *abcd* in two equal parts. The proof follows. Let lines *be* and *ed* be drawn. Because point *e* is in the middle of diameter *ag*, triangles *ade* and *deg* equal one another. Also equal are triangles *abe* and *bge*. Wherefore quadrilateral *edab* is half of the whole quadrilateral *abdg*. And because triangles *bdc* and *bde* are between the parallels *bd* and *ec* and share the same base *bd*, they equal each other. After adding triangle *abd* to both triangles, quadrilateral *abcd* equals quadrilateral *abed*. But quadrilateral *abed* is half of quadrilateral *abdg*. Whence quadrilateral *abcd*[90] is half of quadrilateral *abgd*. Similarly, if we draw line *zf* equidistant from diameter *ag* and draw line *gf*, then quadrilateral *abgd* has been divided in two equal parts by line *gf* drawn from angle *g*. Then, so that the division may proceed from the other angles, take the lines *bl* and *gn* that

---

[89] "Demonstrabo rursus alio modo qualiter quadrilatera duorum laterum equidistantium diuidi debeant ab angulis ipsius" (128.23–24, F. 79r.21–22).

[90] Text has *abgd*; manuscript has *abcd* (128.40, f. 79v.10).

are equal to half the sides *ad* and *bg*. Or, draw lines *ce* and *fz* that necessarily fall on points *l* and *n*. Because line *ne* noted above was drawn equidistant from the diameter *bd* {**p. 129**} and line *lf* drawn equidistant from diameter *ag*, I draw lines *al* and *dn*. What was proposed will come, as was demonstrated above. In summary then, I have shown you how to divide a trapezoid in two equal parts from a point on one of its sides.

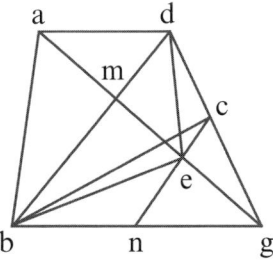

Figure 4.44

[40] **29** *To divide a trapezoid in two equal parts from a point outside the figure.*[91] Consider again trapezoid *abgd* divided in two equal parts from its angles by lines *al*, *dn*, *gf*, and *bc*. Let these lines be extended indefinitely in both directions and be represented by points *e*, *z*, *i*, *k*, *h*, *o*, and *p* on them. We will show that the division of the trapezoid cannot occur unless the point outside the figure is situated on one of these four lines or between two adjacent lines. Wherefore if the point is given on one of them, there will be a line from that point that divides trapezoid *abdg* in two equal parts. For instance, line *eal* coming from the given point *e* divides the trapezoid *abdg* in two equal parts. Keep the same thing in mind for the other points. Suppose the given point *q* falls on line *ad* between lines *ea* and *zd*, and I want a line coming from point *q* to divide the trapezoid *abdg* in two equal parts. Because *ad* and *bg* are equidistant, I can draw line *qm* and extend it to point *v*. I say that trapezoid *abdg* has been divided in two equal parts by line *rs* coming from given point *q* [see Figure 4.45]. The proof follows. Because lines *ad* and *nl* are equidistant from one another and line *rs* intersects them, angle *ars* equals angle *rsl*, angle *sla* equals angle *lar*, and lines *am* and *ml* are equal.[92] Now it has been shown that lines *ad* and *nl* equal each other.[93] Since straight lines intersect one another between equal and equidistant lines, *per equalia* it is shown in geometry that they intersect one another. Whereby lines *nm* and *md* are equal, as are lines *am* and *ml*, as I said. Whence triangles *arm* and *msl* equal each other, are equiangular

---

[91] "Demonstratum est ergo quomodo quadrilatera duorum laterum equidstantium diuidantur a puncto dato super quodlibet latus ipsius. Nunc dicamus qualiter diuidi debeant a dato puncto extra figuram" (129.2–4, 79v.18–20). Archibald 33.

[92] *Elements*, I.29.

[93] [36].

and equilateral. If we add to each the quadrilateral *absm*, then quadrilateral *absr* equals triangle *abl*, namely half of trapezoid *abdg*. Therefore trapezoid *abdg* has been cut in two equal parts by line *rs* drawn from point *q*. In a similar way, if the given point were under line *nl* yet between lines *nh* and *lk*, the same procedure must be followed, namely to join that point with *m* and draw {p. 130} the line to *ad*.

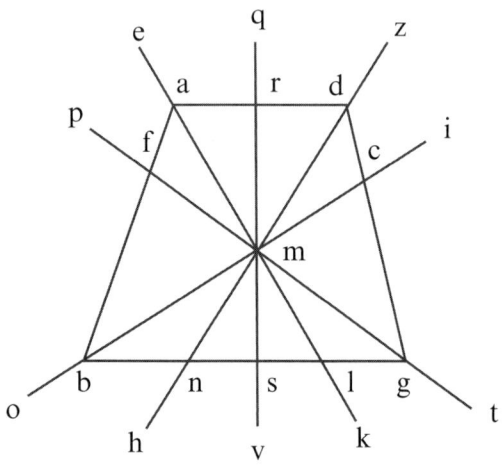

Figure 4.45

[41] For example: if the point were *v*, let trapezoid *abdg* be divided in two equal parts by line *rs* coming from the given point *v*. If a given point *x* were between the lines *nh* and *ob*, as seen in this other figure, then there would be two line-segments[94] [Figure 4.46]. One would extend to diameter *db* to terminate at point *d*; the other to extend from *b*[95] to *c* to terminate at point *c*.[96] Then these line-segments would be equidistant from each other.[97] Let the first line-segment be moved from *b* toward *n* intersecting line *bn*; and the other moved from *c* toward *d* cutting line *cd*, all the while in the movements the line-segments remain equidistant. This continues until the sections of the lines *bn* and *cd* are in a straight line with the given point *x*, and thus they make the line-segments drawn from point *x* go through those sections. And we have reached what we sought.

---

[94] "fila" (130.4). "Strings" does not seem appropriate; at least no such appear in the figure. I translate it as "line-segments" to separate them from the other lines.
[95] Text and manuscript have n_2; context requires *b* (130.5, f. 805.16).
[96] Text and manuscript have *n*; context (parallelism) requires *c* (130.5, f. 805.16).
[97] Fibonacci is writing about the "fila". Regardless, the lines cannot be parallel.

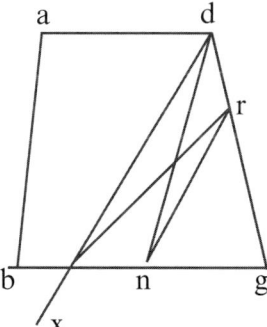

Figure 4.46

[42] For example: let the first line-segment be on line *dy*, the second on line *rn*, and let them be equidistant from one another. Points *r* and *y* are in a straight line with point *x*. Draw line *xyr*. I say that line *ry* cuts the trapezoid *abdg* in two equal parts. The proof follows. Because line-segments *nr* and *dy* are equidistant, and triangles *nrd*[98] and *nry* on base *nr* lie between them, the triangles are equal.[99] Now if triangle *rgn* is added to them, then triangle *rgy* equals triangle *dgn*. But triangle *dgn* is half of trapezoid *abgd*. Therefore triangle *rgy* is half of the same trapezoid. You would follow the same procedure if the given point *q* were between lines *dz* and *ci* and in a straight line connecting points *r* and *y* between the equidistant line-segments *dy* and *nr*. Whence if a line went straight from the point through points *r* and *y*, the line would divide the trapezoid in two equal parts, as was shown. The same procedure must be followed if the point falls between lines *gl* and *lk*, or between lines *fp* and *ae*. That is, to fix on point *a* a line-segment from *a* to *g*, and to fix on point *l* another from *l* to *f*, and to move the line-segments cutting lines *af* and *gl* until the sections are in a straight line with the given point. But suppose the given point were outside line *bf* and between lines *fp* and *ob*. How to do this is shown in another figure in which lines *oi* and *pt* are extended from angles *b* and *g*, dividing trapezoid *abdg*, and the given point *z* lies between lines *fp* and *ob* [Figure 4.47]. Now we wish to draw a line dividing trapezoid *abdg* in two equal parts. Extend the first line-segment from *c* to *f*, and let it terminate at *f* **{p. 131}**. Draw another line-segment to line *gb* and fix it at point *g*. Then move both equidistantly intersecting lines *bf* and *cg* until the sections are in a straight line with the given point *z*. For example: let points *x*,[100] *q*, and *z* be in a straight line, and lines *gq* and *fx* be equidistant. Connect points *x*, *q*, and *z*. I say that trapezoid *abdg* is cut in two equal parts by line *qx* that was drawn from the given point *z*. The proof follows. Because lines *gq* and *xf* are equidistant, triangles *qgx* and *qgf* are equal. If triangle *qbg* is added to each, then

---

[98] Text has *ncd*; context requires *nrd* (130.15).
[99] But triangle *ncd* does not have the same base!
[100] Point *x* is not in the figure in the manuscript (f. 80v).

220   Fibonacci's *De Practica Geometrie*

quadrilateral *qbgx* equals triangle *fbg* that is half of quadrilateral *abdg*. We would proceed in the same way if the given point were outside of line *cg* and between lines *ci* and *gt*. Therefore it has been demonstrated how to divide a trapezoid by a line coming from any point outside the figure.

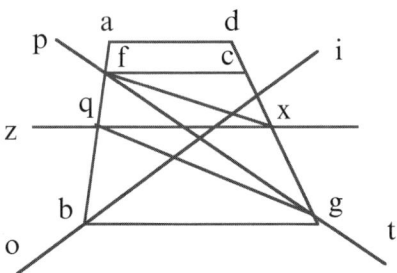

Figure 4.47

[43] **30** *To divide a trapezoid by a line proceeding from a given point inside the figure.*[101] Given trapezoid *abdg* with sides *ad* and *bg* equidistant from each other, with *ad* the shorter side. Let the trapezoid be divided in two equal parts by any of the lines *al*, *dn*, *gf*, or *bc*. First, let the given point *e* be within triangle *amd*. Draw the straight line *em* and extend it to points *h* and *i*. I say that the trapezoid *adgb* has been divided in two equal parts by line *hi* [see Figure 4.48]. The proof follows. Triangles *hma* and *mil* are equal. If quadrilateral *abim* is added to both, then quadrilateral *abih* equals triangle *abl* that is half of trapezoid *abgd*. Similarly, if the given point were to fall within triangle *nml*, then the same procedure must be followed; namely, we connect that point with point *m*, divide the line in two parts, and we have what we wanted. And if the given point were upon one of these lines *al*, *dn*, *gf*, or *bc*, then the line going through that point would divide trapezoid *abgd* in two equal parts. But if the given point were within quadrilateral *abnm* or *dmlg*, we would have to work with the line-segments. First, I would draw a line-segment between points *d* and *b* that I would fix on point *d*. Then beginning at point *n* I would draw another line-segment from *n* to *c*. Next I would move the two line-segments, keeping them equidistant and cutting lines *bn* and *cd* until they are in a straight line with the given point, if this is possible. If it is not possible, then I would draw line-segments between points *a* and *g* and between *f* and *l*, fixing them on points *a* and *l*. Then keeping them equidistant from one another, I would lead them along, cutting lines *fa* and *gl* until the sections are in a straight line with the given point. And if this is not possible, then I would draw line-segments *cg* and *fe*, fixing them at points *g* and *f*. Then keeping them equidistant from one another, I would lead the intersecting lines *bf* and *cg* until the sections lie in a straight

---

[101] "De diuisione eiusdem generis, qua quadrilaterorum per rectam transeuntem per punctum datum infra ipsum" (131.13–14, f. 80v.20–22).

line with the given point. And that without doubt is possible. Then by drawing the lines through the sections, you will have what was proposed, and that is proved by what we said above. Thus enough is said about dividing trapezoids in two equals parts. Now we shall show how they are to be divided in unequal parts according to a given ratio.

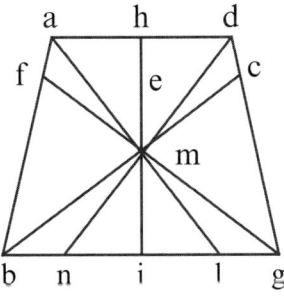

Figure 4.48

[44] **31** *To divide a trapezoid in two parts in a given ratio by a line equidistant from its base.*[102] Given trapezoid *abgd* and the given ratio *ez* to *zi*. Let lines *ba* and *gd* be extended from points *a* and *d* until they meet in *t* [see Figure 4.49]. Then let the ratio of the squares on lines *tl* and *at* {p. 132} be as the ratio of lines *zi* to *ez*. Let the ratio of the square on line *ht* to the squares[103] on lines *bt* and *tl* be as the ratio of lines *ez* to *ei*. Through points *l* and *h*, draw lines *lm* and *hk* equidistant from lines *bg*[104] and *ad*. I say that line *hk* has divided quadrilateral *ag* according to the ratio of the number *ez* to the number *zi*. The proof follows. Because triangles *tlm* and *tad* are similar, as the square on line *tl* is to the square on line *ta*, so is triangle *tlm* to triangle *tad*. For as the number *zi* is to the number *ez*, so the square on line *tl* is to the square on line *ta*. *Per equale* therefore, as *zi* is to *ez*, so triangle *tlm* is to triangle *tad*. Conjointly, therefore, as *ei* is to *ez*, so triangles *tlm* and *ead* are to triangle *tad*. Alternately, therefore, as *ez* is to *ei*, so triangle *tad* is to triangles *tad* and *tlm*. For as *ez* is to *ei*, so the square on line *ht* is to the squares on lines *bt* and *tl*. But as the square on line *ht* is to the squares on lines *bt* and *tl*, so triangle *thk* is to triangles *tbg* and *tlm*. *Per equale* therefore, as *ez* is to *ei*, so triangle *thk* is to triangles *tbg* and *tlm*. But quadrilateral *ak* and triangle *tad* are equal to triangle *thk*. Similarly, quadrilateral *ag* and triangles *tad* and *tlm* are equal to triangles *tbg* and *tlm*. Therefore as *ez* is to *ei*, so quadrilateral *ak* with triangle *tad* is to the conjunction of quadrilateral *ag* and triangles *tad*[105]

---

[102] "Adiaceat rursus quadrilaterum *abdg* quod diuidi oporteat per rectam equidstantem basi sue in data proportione *ez* ad *zi*" (131.41–42, *f.* 81r.15–17). Archibald 53.

[103] Text has *quadratum*; context requires *quadrata* (132.1).

[104] Text and manuscript have *hg*; context requires *bg* (132.3, *f.* 81r.21).

[105] Text and manuscript have *ead*; context requires *tad* (132.17, *f.* 81v.1).

and *tlm*. But as *ez* is to *ei*, so triangle *tad* is to triangles *tad* and *tlm*. Whence the ratio of quadrilateral *ak* to quadrilateral *ag* is as *ez* to *ei*. By separation therefore *ez* is to *ei* as quadrilateral *ak* is to quadrilateral *hg*, as required.

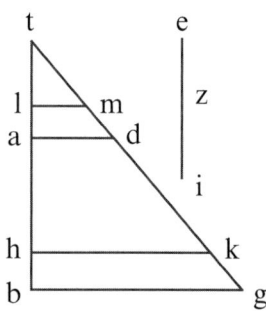

Figure 4.49

[45] This can be shown in numbers. Let side *ad* be 6, *ab* 12, *bg* 15, and *dg* 15, with a right angle at vertex *b*. After multiplying *ba* and *gd* by *t*,[106] *at* is 8 and *td* is 10. Whence *tb* is 20. The ratio of *ez* to zi[107] is as the ratio of 4 to 3. Therefore the square on line *tb* is 400 and the square on line *tl* is 48. Because *iz* is $\frac{3}{4}$ of *ze*, three fourths of 64 or the square on line *ta* is 48. Adding those two together yields 448. If we take from this four sevenths or the ratio of *ez* to *ei*, we obtain 256 for the square on line *th*. Therefore *th* is 16. And because line *th* is twice *ta*, therefore *hk* is likewise twice *ad*. Therefore *hk* is 12. So if we join *kh* with *ad* and we multiply the sum by 4 or half the cathete, we get 72 for the area of quadrilateral *ak*. Likewise having multiplied half of *bh* or 2 with the sum of *bg* and *hk* or 27, we obtain 54 for the area of quadrilateral *hg*. For the ratio of 72 to 54 is the same as 4 to 3 that is the given ratio of *ez* to *zi*.

[46] This can be done more easily if we let the ratio of *ms* to *ls* be as the ratio of the square on line *tb* to the square on line *ta* [see Figure 4.50]. Then having divided line *ml* at *n*, we make the given ratio be *ln* to *nm*. We let the ratio of the square on line *th* to the square on line *tb* be as the ratio of *ns* to *sm*. Having drawn line *hk* equidistant from base *bg*, we prove again that quadrilateral *ak* is to quadrilateral *hg* as line *ln* is to line *nm*, because as the square on line *tb* is to the square on line *ta*, so triangle *tbg* is to triangle *tad*. And the ratio of *ms* to *ls* is as the square on line *tb* to the square on line *ta*. Whence as *ms* is to *ls*, so triangle *tbg* is to triangle *tag*. Again, because *ms* is to *sn*, so the square on line *tb* is to the square on line *th*. For as the square on line *bt* is to the square on line *th*, so triangle *tbg* is to triangle *thk*. Therefore as *ms* is to *ns*, so triangle *tbg* is to triangle *thk*. Conversely as *sm* is to *nm* so triangle

---

[106] "Quare productis *ba* et *gd* in *t*, erit *at* 8 et *td* 10" (132.21–22); Paris Lat. 10258 (209.18) shows the line "corrected" to read as seen in this note.

[107] Text has *ei*; context requires *zi* (132.22).

tbg is to quadrilateral *hg*. But as *ms* was to *ls*, so triangle *tbg* is {**p. 133**} to triangle *tad*. Hence, by alternation: the number *ms* is to triangle *tbg* as the number *ls* is to triangle *tad*, and as the number *ns* is to triangle *tkh*. Therefore the ratio of the numbers *ls* and *ns* to the ratio of triangles *tad* and *thk* is the same as the ratio of the number *ms* to the plane *tbg* that equals the ratio of the number *nm* to quadrilateral *bg*. Whence after disjunction, the ratio of *sl* to *ln* is as the ratio of triangle *tad* to quadrilateral *ak*. For as *sn* is to *nm*, so quadrilateral *ak* is to quadrilateral *hg*. Therefore, the given quadrilateral *ag* has been divided by line *hk* equidistant from the base *bg* into the given ratio of the numbers *ln* and *nm*, as required.

Figure 4.50

[47] This can be shown in numbers. As we said before, let the square on line *tb* be 400 and on line *ta* 64, such that their ratio is 25 to 4. Further, if I subtract 4 from 25 or *ls* from *ms*, the remainder is 21 or *ml*. Having divided these in the given ratio of 4 to 3 at point *n*, we have 12 for *ln* and 9 for *nm*. Whence, making the ratio of the square on line *th* to the square on line *tb* as the ratio of *sn* to *sm* or 16 to 25, we find again that the square on line *th* is 256. Whence *th* is 16, as we found above. Or in another way: taking the square on line *ta* from the square on line *tb* 336 remains. Four sevenths of this is 192. If we add the square on line *ta*, we get again 256 for the square on line *th*.

[48] **32** *To divide a trapezoid in a given ratio by a line drawn through two equidistant sides.*[108] You are going to divide the sides in the same ratio and extend the line through the points of division. For example: we want to cut the equidistant sides *ad* and *bg* of trapezoid *abdg* in the given ratio *ez* to *zi*. Place points *t* and *k* on sides *ad* and *by* so that *ez* is to *zi* as *at* is to *td* and *bk* to *kg*. Join the points *t* and *k* [see Figure 4.51]. I say that ratio of quadrilateral *ak* to quadrilateral *tg* has been cut in the ratio *ez* to *zi*. The proof follows. Draw lines *tb* and *tg* from point *t* to points *b* and *g*. Thus triangles *tbk* and *tkg* share the same altitude. Whence as *bk* is to *kg*, so triangle *tbk* is to triangle *tkg*. Again, since triangles *bat* and *gtd* share the same altitude because they lie

---

[108] "Si per rectam protractam super duo latera equidistantia quadrilaterum caput abscisum in data aliqua proportione diuidere uis, ..." (133.17–19, f. 82r.13–14).

within the same equidistant lines, as *at* is to *td* (or as *bk* is to *kg*) so triangle *bat* is to triangle *gtd*. Therefore as *bk* is to *kg* (or as *ez* is to *zi*) so quadrilateral *ak* is to quadrilateral *tg*. Therefore line *tk* drawn through the equidistant sides has divided trapezoid *abdg* in the required ratio, as we had to prove.

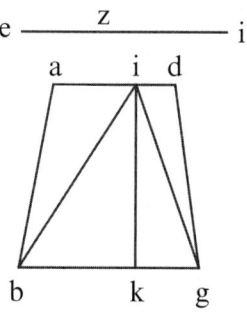

Figure 4.51

[49] **33** *To divide a trapezoid in a given ratio by lines drawn from two adjacent angles.*[109] Given trapezoid *abdg*, let the adjacent angles be *a* and *d*. Select points *n* and *l* on line *bg*. Set the distance *kl* equal to *ta* and the distance *gn* equal to *bl*, and draw lines *dn* and *al*, as shown in the other figure [Figure 4.52]. From what has been said we prove that triangle *abl* is equal to quadrilateral *ak*.[110] Whence *ez* is to *zi* as triangle *alb* is to quadrilateral *alyd*. Again, because the triangles *alb* and *dgn* have the same base and lie between equidistant lines *ad* and *bg*, they are equal. Whereby triangle *dgn* equals quadrilateral *abkt*. So what remains, namely quadrilateral *nbad*, equals quadrilateral *tkgd*. Consequently, as quadrilateral *abkt* is to quadrilateral *tkgd* in the ratio of *ez* to *zi*, so triangle *dgn* is to quadrilateral *nbad*. Therefore trapezoid *abdg* has been divided in the required ratio of *ez* to *zi* from the angles *a* and *d*, as required. And if you want to divide the trapezoid in the ratio of *ez* to *zi* by a line from another angle such as *b* or *g*, then you do it twice beginning with angle *b* {**p. 134**}. Identify point *n* [on line *bg*] to which the required divisor will be drawn from angle *d*. Make the area *bg* by *gc* equal to the area *dg* by *gn*. Draw line *bc*. Or, after drawing the diameter from *b* to *d*, draw an equidistant line from *n* to *c*, and then draw *bc*, as we said [see Figure 4.53]. Let us do the same thing from angle *g*. Point *l* on side *bg* will be the aforementioned point of division from angle *a*. I make plane *gb* by *bf* equal to plane *ab* by *bl*. Or, after drawing the diameter from *a* to *g*, I draw an equidistant line from *l* to *f* and join *g* and *f* to divide trapezoid *abdg* in the required ratio of *ez* to *zi*. Therefore triangle *fbg* is to quadrilateral *afgd*, and triangle *bgc* is to quadrilateral *abcd* as *ez* is to *zi*. This has been proved by

---

[109] "Si diuisionem in eadem proportione ab angulis *a* et *b* habere uolumus, ..." (133.30–31), the given ratio is from the previous rule.

[110] See [36].

what has already been said. Also we will divide the trapezoid from any point on whichever side and even from any point within or outside the figure.[111]

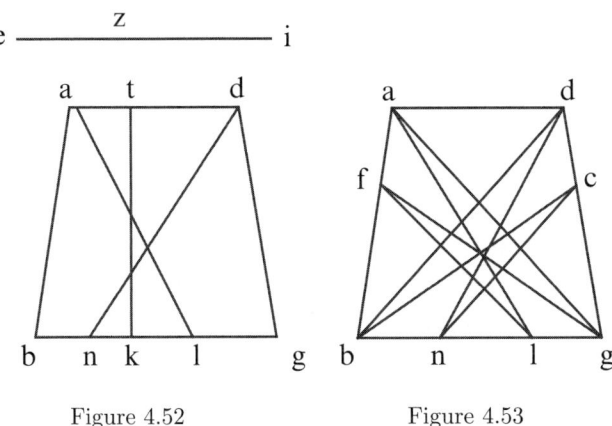

Figure 4.52   Figure 4.53

Division of Trapezoids into Many Parts[112]

[50] **34** *To divide a trapezoid in three equal parts by lines equidistant from its base.*[113] Given trapezoid *abdg* with equidistant sides *ad* and *bg* [see Figure 4.54]. From points *a* and *d* I extend the other sides *ab* and *dg* until they meet at point *e*. Also I draw line *zti* nearby so that the ratio of *zi* to *it* is as the ratio of the square on line *eb* to the square on line *ae*. Then I divide *tz* in three equal parts: *tk*, *kl*, and *lz*. And I make the ratio of the square on line *em* to the square on line *eb* be as *ki* is to *zi*. Again, I make the ratio of the square on line *en* to the square on line *eb* as the ratio of *li* to *iz*. Finally, through points *m* and *n* I draw lines *mo* and *np* equidistant from the base *bg*. I say that the trapezoid *ag* has been divided in three equal parts, namely quadrilaterals *ao*, *mp*, and *ng*. The proof follows. As we said, as the square on line *be* is to the square on line *ae*, so triangle *ebg* is to triangle *ead*. Whence as *zi* is to *ti*, so triangle *ebg* is to triangle *ead*. Likewise, because *zi* is to *ik*, so the square on line *be* is to the square on line *me*. But the ratio of the square on line *be* to the square on line *me* is as the ratio of triangle *ebg* to triangle *emo*. Therefore as *zi* is to *ik*, so triangle *ebg* is to triangle *emo*. For as *zi* is to *il*, so triangle *ebg* is to triangle *enp*. After disjunction: as *it* is to *tk*, so triangle *ead* is to quadrilateral *ao*. Also, as *tk* is to *kl*, so quadrilateral *ao* is to quadrilateral *mp*. Now *tk* equals *kl*. Whence quadrilateral *ao* equals quadrilateral *mp*. Again, as *kl* is to *lz*, so quadrilateral *mp* is to quadrilateral *ng*. Now *kl* equals *lz*, so

---

[111] I see this sentence as a transitional sentence. Archibald seems to agree; see his p. 45, lines 3–12. The Arabic–Euclid *On Divisions*, however, offers it as Proposition 12; see Hogendijk (1993), 148.
[112] "Incipit de diuisione quadrilaterorum in plures partes" (134.13, f. 82v.18).
[113] "Si uero aliquod caput ascisum in tres partes equales a rectis equidistantibus sue basi diuidere uis, ..." (134.14–15, f. 82v.19–20). Archibald 26.

quadrilateral *pm* equals quadrilateral *ng*. Therefore the three quadrilaterals *ao*, *mp*, and *ng* are equal to one another, as we said.

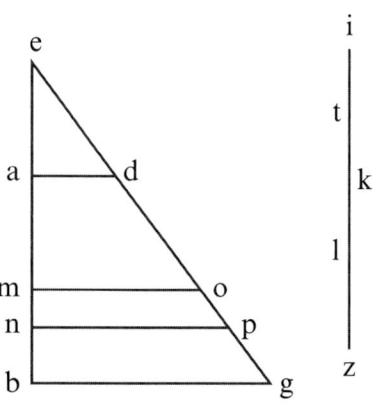

Figure 4.54

[51] This can be shown in numbers. Let *ad* be 6, *bg* 15, *ab* 12, and *dg* 15, with a right angle at point *b*. Whence *eb* measures 20. The ratio of its square to the square on line *ea* is as 25 to 4.[114] Let *zi* be 25 and *ti* 4. This leaves 21 for *zt*. Having divided 21 in three equal parts, each of the parts *tk*, *kl*, and *lz* is 7. Whence *ik* is 11. So, multiply 11 by 400 and divide the product by 25. Or take a twenty-fifth of the square on *eb* to get 16. Multiply by 11 to get 176 for the square on line *em*. Again, multiply *il* or 18 by 16 that is $\frac{1}{25}$ of the square on line *eb* to reach 288 for the square on line *en*. Therefore *am* is the root of 176 less 8, since *ea* is 8 and line *mn* is the root of 288 less the root of 176. Now line *nb* is 20 less the root of 288. And because as the square {p. 135} on line *ea* is to the square on line *ad* or 16 to 9, so is the square on line *em* to the square on line *mo*. If we multiply 176 by 9 and divide by 16, or if we take $\frac{1}{16}$ of 176 that is 11 and multiply by 9, we get 99 for the square on line *mo*. Again, if we multiply $\frac{1}{16}$ of 288 the square on line *en* or 18 by 9, we get 162 for the square on line *np*. Having found this, if you investigate the area of quadrilateral *ao*, *mp*, and *ng* according to what we taught here, you will find that the quadrilaterals equal one another. Another way: add twice the square on line *ea* to the square on line *eb*, and you will find that a third of the sum is the square on line *em*. Similarly, if we add twice the square on line *be* to the square on line *ea*, we get a third of the square on line *em*. Or, if we subtract the square on line *ea* from the square on line *eb* or 64 from 400, we get 366. If a third of this or 112 is added to 64, it yields 176 for the square on line *em*. Adding 176 and 112 produces 288 for the square on line *en*. The proof of this follows

---

[114] Fibonacci assumed that his reader was finding required values by himself or herself.

from what I said above where we taught how to divided trapezoids according to a ratio.

[52] Suppose side *ab* were a single road[115] along the property. Each partner, as is customary, wanted to have a third of the road. Then divide side *ab* in three parts at points *i* and *l* and draw the lines *io* and *lp* [see Figure 4.55]. Through point *m* draw line *mh* equidistant from line *io*, and draw line *nq* through point *n* equidistant from line *lp*. Draw lines *ih* and *lq*. And thus lines *ih* and *nq* have divided trapezoid *abdg* in three equal parts: *aihd*, *ilqh*, and *lbgq*. The proof follows. Because quadrilateral *amod* is a third of trapezoid *abdg*, if triangle *oim* were removed from it and triangle *oih* added to it, then quadrilateral *aihd* would be equal to quadrilateral *amod*. You can prove that the other divisions are equal in the same way.

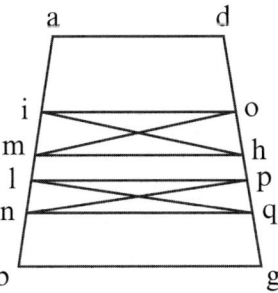

Figure 4.55

[53] Suppose we had divided sides *ad* and *bg* in three equal parts and had drawn straight lines, we would have divided the quadrilateral in three equal sections. For example: given the trapezoid *abdg* with sides *ad* and *bd* divided in three equal parts at points *e*, *z*, *i*, and *t*. Draw lines *ei* and *zt* [see Figure 4.56]. I say that trapezoid *abdg* has been cut in three equal sections, namely quadrilaterals *ai*, *et*, and *zg*. Their respective bases, top and bottom, are equal and they have the same altitude. Or, because *ae* is to *ed* as *bi* is to *ig*, therefore they are in the same ratio as quadrilateral *ai* to quadrilateral *eg*. If we join them, *bi* will be to *bg* as quadrilateral *ai* is to quadrilateral *ag*. Now *bi* is a third of *bg*. So also is quadrilateral *ai* a third of quadrilateral *ag*. Again, because *ac* is to *cz* as *bi* is to *it*, so does *ae* equal *ez* and *bi* equal *it*.[116] Whence quadrilateral *ai* equals quadrilateral *et*.[117] Therefore quadrilateral *ai* is a third of trapezoid *ag*. Also, quadrilateral *et* is a third of trapezoid *ag*. And the remaining part, *zg*, is also a third.

---

[115] "Et si latus *ab* fuerit secus uiam, ..." (135.16).
[116] "... est equalis *ae* ex *ez* et *bi* ex *it*." *ex* seems to be a synonym for *equalis* (135.35).
[117] "Quare et quadrilaterum *ai* ex quadrilatero *et* est equale." (135.35–36). Is "ex" in the other manuscripts? The context makes the two quadrilaterals equal.

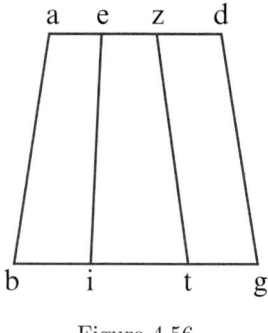

Figure 4.56

[54] **35** *To cut a third part of a trapezoid from an angle.*[118] Given trapezoid *abgd*, we bisect lines *ei* and *zt* at points *k* and *m* [see Figure 4.57]. Then we draw lines *ak* and *dm* and extend them to points *l* and *n*. Each of the triangles *abl* and *dgn* is a third of trapezoid *abgd*. The proof follows. Because line *ek* equals line *ki*, triangles *ake* and *ilk* are mutually equal and similar, since they have equidistant bases *ae* and *il*. Whence if {**p. 136**} quadrilateral *abik* is added to both, then triangle *abl* will be equal to quadrilateral *eb*. But quadrilateral *eb* is a third of quadrilateral *ag*. Whence triangle *abl* is a third part of quadrilateral *ag*. You can show in a similar way that triangle *dgn* is equal to quadrilateral *zg* that is a third of quadrilateral *ag*. Removing triangle *abl* from trapezoid *abgd*, trapezoid *algd* remains. If this is divided by any of the aforementioned techniques, then trapezoid *abgd* has been divided in three equal sections. Similarly, if we removed triangle *dgn* from quadrilateral *abgd*, quadrilateral *abnd* remains to be divided in two equal parts.

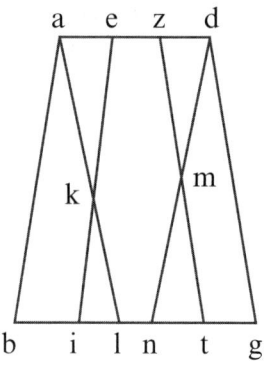

Figure 4.57

---

[118] "et si tertiam partem quadrilateri *ag* ab aliquod angulorum *a* et *d* secare uolumus, ..." (135.38–39, *f.* 83v.2–25).

[55] If we want to cut a third from quadrilateral ag by a line drawn from point b, then first I draw line dn cutting a third from quadrilateral ag. The third is triangle dgn.[119] Then I draw diameter bd, a line nc[120] equidistant from it, and finally line bc [see Figure 4.58]. I say that line bc cuts from quadrilateral ag a third that is triangle cgb. Because, if triangle cgn is added to triangle cbn that equals triangle cnd,[121] then triangle cgb equals triangle dgn. Therefore triangle cbg is a third of quadrilateral ag. In another way: because triangles abd and gbd share the same altitude—the perpendicular between equidistant sides ad and bg, they are therefore to one another as their bases. Whence as bg is to ad, so triangle bgd is to triangle abd. After composition, therefore, bg is to $bgad$, as triangle bgd is to quadrilateral ag. Add line ge that equals line ad. Therefore as bg is to be, so triangle dbg is to quadrilateral ag. Removing line bz a third of the whole line be, we have the ratio of line gc to line gd as bz is to bg. Then draw bc. I say that triangle bcg is a third of quadrilateral ag. For example: triangles bgc and bgd are to one another as their bases. Therefore, gc is to gd (that is, bz to bg) as triangle bgc is to triangle bgd. For as bg is to be so triangle bgd is to quadrilateral ag. Therefore as bz is to be, so triangle bgc is to quadrilateral ag. Now bz is a third of be. Whence triangle bgc is a third part of quadrilateral ag, as I said.

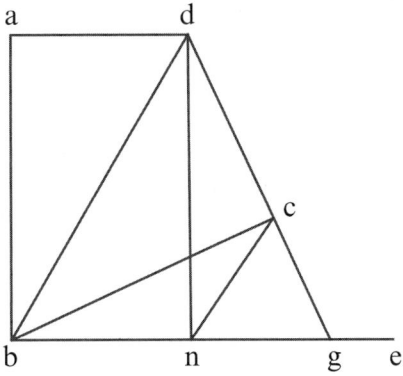

Figure 4.58

[56] We can show this with numbers. Let ad be 6, bg 24, and each of the sides ab an d dg 15. After adding 24 and 6, we have 30. Therefore as 24 is to 30 (that is, as bg is to be), so is triangle bdg to quadrilateral ag. And as 10 is to 30, so is a third part of quadrilateral $abgd$ to quadrilateral ag. Whence the ratio of 10 to 24 or of bz to bg is $\frac{5}{12}$. Taking this much from line dg, we have $\frac{1}{4}$ 6 for line gc. Whence if we draw line bc, we have the ratio of triangle bgc to triangle bgd as bg[122] is to gd. But gc is $\frac{5}{12}$ of gd. Whence

---

[119] Text and manuscript have agn; context requires dgn (136.10, f. 84r.7).
[120] Text and manuscript have ne; context requires nc (136.10, f. 84r.8).
[121] Fibonacci assumed that the reader has proved this equality.
[122] Text and manuscript have bc; context requires bg (136.36, f. 84.34).

triangle bgc is $\frac{5}{12}$ of triangle bgd which in turn is $\frac{24}{30}$ or $\frac{4}{5}$ of quadrilateral ag. Therefore triangle bgc is $\frac{5}{12}$ of trapezoid ag.[123] Therefore triangle bgc is a third of trapezoid ag since $\frac{5}{12}$ of $\frac{4}{5}$ of anything is a third of it. In like fashion if we draw diameter ag, the ratio of triangle abg to quadrilateral db is as bg to the sum of the lines bg and ad. If you take from this ratio the ratio that a third of the whole has to its ratio, you find that you have to take $\frac{5}{12}$ of line ba that is $\frac{5}{12}$ of line bd.

[57] With the foregoing explained, we will now show how to divide in two equal parts what remain of triangles cgb and fbg, namely {p. 137} quadrilaterals $abcd$ and $afgd$, when we discuss quadrilaterals with unequal sides.[124] Afterward, we discuss dividing the quadrilaterals into three equal parts from third parts of the sides. Then, division from any angle, as mentioned above. Finally, cutting a third or any part from similar quadrilaterals by a line from a given point; the point may be within or without the figure or even on one of the sides.

[58] **36** *To divide a quadrilateral proportionally in three different parts.*[125] Again, given trapezoid $abgd$ and the ratio of ez to it [see Figure 4.59]. We wish to divide the figure in three parts by lines equidistant from the base, such that the first line is to the second as ez is to zi, the second to the third as zi is to it. First, extend lines ba and gd to point k, and let the ratio of the square on line bk be to the square on line ak as tl is to el. Then have the ratio of the square on line bk to the square on line km be as tl is to lz. After this set the square on line bk to the square on line kn be as tl is to il. Finally, through points m and n draw lines mo and np equidistant from the base bg [see Figure 4.60]. You will find from what has been said above[126] that the ratio of quadrilateral ao to quadrilateral mp is as ez to zi and that the ratio of quadrilateral mp to quadrilateral ng is as zi to it.

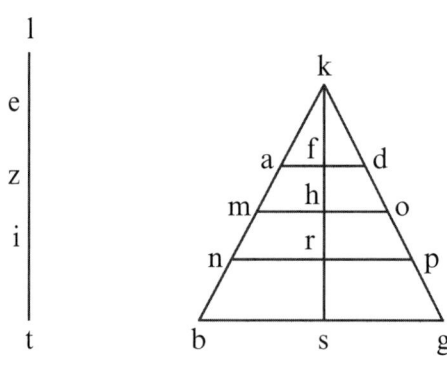

Figure 4.59 Figure 4.60

---
[123] "... *bgc est ex quatuor punctis ex toto quadrilatero ag*" (136.36).
[124] "... *de diuisione quadrilaterorum diuersorum laterum*" (137.1) excludes trapezoids.
[125] "*Vnde qualiter proportionaliter in tres diuersas partes diuidi debeant, demonstremus*" (137.6–7, f. 84v.15–16). Archibald 54.
[126] See [50].

[59] We can show this with numbers. Given trapezoid *abgd* as shown in the figure above, subtract *ad* from *bg* or 6 from 24 to leave 18. Thus 18 is to 6 as *ba* is to *ak*. Whence *ak* is 5. But *bk* is 20. Its square to the square on line *ak* is as 16 to 1. Whence I put 16 for line *tl*, 1 for *el*, which leaves 15 for *et* which is thus divided in the given ratio of *ez* to *it*.[127] Whence we make *ez* equal 4, *zi* 5, and *it* 6. And so as 4 is to 5, so *ez* is to *zi*; and as 5 is to 6, so *zi* is to *it*. And these are the given ratios. Further, because the square on line *bk* is to the square on line *mk*, so *tl* is to *lz*. For the square on line *bk* is to the square on line *ak*, as *tl* is to *el*. Alternately therefore, as *le* is to *lz*, so is the square on line *ak* to the square on line *mk*. For *zl* is five times *el*. Whence the square on line *mk* is five times the square on line *ak*. Therefore the square on line *mk* is 125. Similarly, as *il* is to *el* or 10 to 1, so the square on line *nk* is to the square on line *ak*. Therefore the square on line *nk* is 250. Similarly, the square on line *np* is ten times the square on line *ad*. Whence the square on line *np* is 360. Likewise because the square on line *mk* is five times the square *ak*, so the square *mo* is five times the square *ad*. Therefore the square on line *mo* is 180. You will find further that the cathete *ks* is 16. When the square on line *bs* is taken from the square on line *bk*, 256 remains for the square on line *ks*. And because *ka* is a fourth of *bk*, then *fk* is a fourth of *ks*. Whence *kf* is 4 and its square is 16. Five times this is the square on cathete *kh* or 80. Twice this or 160 is the square on line *kr*. Whence if we multiply half of *kh* by *mo* we get for the area of triangle *kmo* the root of 3600 or 60. Subtracting 12 or triangle *kad* from this leaves 48 for square *ao*. Again, if we multiply a fourth of the square on *kr* or 40 by 360 or a fourth of line *np*, we get 14400. Its root is 120 for the area of triangle *knp*. If from this we take 60 or triangle *kmo*, 60 remains for the area of quadrilateral *mp* {**p. 138**}. The 72 remaining from quadrilateral *ag* is for quadrilateral *ng*. The ratio of 48 to 60 is 4 to 5, that is as *ez* to *zi*. And the ratio of 60 to 72 is as 5 to 6 that is as *zi* to *it*, as is fitting. To summarize dividing by equidistant lines drawn from three given points on line *ba*: if we divide line *ad* in whatever the given ratio may be, in that ratio we divide line *bg*. By drawing lines between those points, we divide quadrilateral *abgd* in the required ratios. A similar summary fits proportional divisions from the vertices of a quadrilateral.

4.4 Division of Quadrilaterals with All Sides Unequal[128]

[60] **37** *To divide a quadrilateral with all sides unequal in length in two equal parts from a given angle.*[129] Given quadrilateral *abcd* that I wish to divide in two equal parts from a given angle *a* [see Figure 4.61]. First I draw diameter

---

[127] See Figure 4.59.
[128] "Incipit de diuisione quadrilaterorum diuersilaterum" (138.9, *f*. 85r.34).
[129] "Primum quidem demonstrare uolo quo modo diuidatur quadrilaterum diuersilaterum in duo equa ab angulo data." (138.10–11, *f*. 85r.33–35). Archibald 35.

bd opposite[130] angle bad. Then I intercept line bd by diameter ac at point e. Lines be and ed are either equal or they are not. Assume first that they are equal. Then because be and ed are equal, triangle abe equals triangle ade. Also triangle ebc equals triangle ecd. Whence the entire triangle abc equals the entire triangle acd. Quadrilateral $abcd$ has therefore been divided in two equal parts by diameter ac drawn from angle a, as required. On the other hand, assume that lines be and ed are not equal. Let line bz equal line zd. So I draw line zi equidistant from diameter ac, as seen in the second figure [Figure 4.62]. And I join ai. I say again that quadrilateral $abcd$ has been divided in two equal parts by line ai drawn from the given angle a. The equal parts are triangle abi and quadrilateral $aicd$. The proof follows. Draw lines az and cz. Thus triangles azd and dzc equal triangles abz and bzc. Whence quadrilateral $azcd$ is half of quadrilateral $abcd$. And because triangles aci and acz share the same altitude and lie within the equidistant lines ac and zi, they equal each other. If triangle acd is added to both, then quadrilateral $aicd$ equals quadrilateral $azcd$. But quadrilateral $azcd$ is half of quadrilateral $abcd$. Therefore quadrilateral $aicd$ is half of quadrilateral $abcd$, as required.

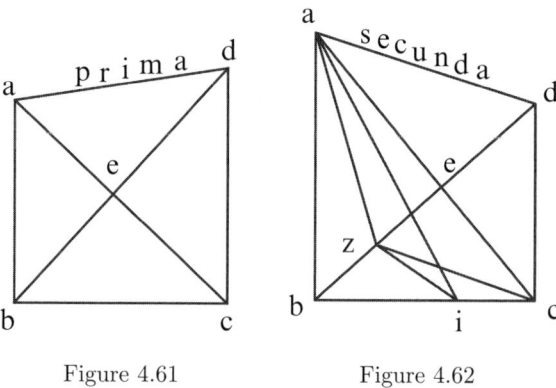

Figure 4.61 Figure 4.62

[61.1] **38** *To divide a quadrilateral in two equal parts by a line drawn from a point on one of the sides.*[131] Given quadrilateral ac that you wish to divide from a given point e on side ad. First, divide quadrilateral ac in two equal parts by line dt drawn from angle d.[132] Draw line et. Now line et is either equidistant from line dc or it is not. First, suppose that lines et and dc are equidistant, as appears in figure three [Figure 4.63]. Draw line ec. I say that quadrilateral ac has been divided in two equal parts by line ec drawn from the given point e. The proof follows. Because lines et and cd are equidistant, triangles ecd and

---

[130] "... bd subtendentem angulum bad" (138.12).
[131] "Rvrsvs si a puncto dato super unum ex lateribus quadrilaterum aliquod in duo equa diuidere uis, ..." (138.28–29, f. 85v.20–22). Archibald 37.
[132] By [60].

tcd are equal. But triangle *tcd* is half of quadrilateral *ac*. Whence triangle *ecd* is half of quadrilateral *ac*. Therefore quadrilateral *ac* has been divided into two equal parts by a line drawn from the given point *e*, as required. But, suppose that lines *et* and *dc* are not equidistant. Then I draw line *dz* equidistant from line *et*, as appears in figure four [Figure 4.64]. Quadrilateral *abcd* has been divided in two equal parts by line *ez*. The proof follows. Let triangles *tze*[133] and *dzt* be equal, because they share the same base and lie within the same equidistant lines. Now if triangle *dcz* is added to both {**p. 139**}, quadrilateral *ezcd* equals triangle *dct* that is half of the whole quadrilateral *ac*.

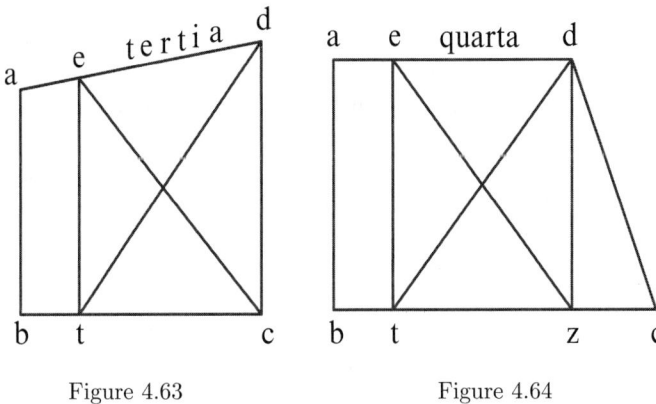

Figure 4.63   Figure 4.64

[61.2] It must be noted that if the diameter *bd* divides quadrilateral *ac* in two equal parts, what we said above about two figures pertains here. But if the division falls on side *ab* by line *di* that divided quadrilateral *abcd* in two parts, then another procedure is required[134] [see Figure 4.65]. I will divide quadrilateral *ac* by line *bk* drawn from angle *b*. And then if *k* is a given point on side *ad* from where dividing line *kb* must originate, then it divides quadrilateral *ac* in two equal parts. But if *k* is not the given point, then let the given point lie between points *k* and *d* or between points *k* and *a*. First then, let the given point *e* lie between points *k* and *d*. Draw line *be*. Through point *k* extend line *kl* equidistant from line *eb*. Draw line *el*, and it divides quadrilateral *ac* in two equal parts. The proof follows from what has been said before. Again, let the given point *e* lie between points *k* and *a*. Likewise, draw line *eb*. Through point *k* draw the equidistant line *km*. Draw line *em*, and it divides quadrilateral *ac* in two equal parts, quadrilaterals *am* and *ec*, as shown in the sixth figure [Figure 4.66]. Again, the proof follows from what has been said before.

---

[133] Text and manuscript have *dze*; context requires *tze* (138.42, f. 86r.1).
[134] See [60].

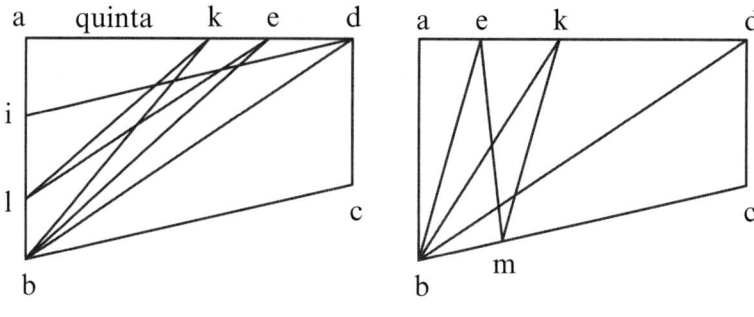

Figure 4.65 Figure 4.66

[62] **39** *To divide a quadrilateral with unequal sides from a point given on a side.*[135] Using the method above, let point *e* be given first in the middle of side *ab* of quadrilateral *abcd*. From point *d* I draw line *dz* equidistant from *ab*, dividing it in two equal parts at point *i*. Then I draw lines *ei*, *ci*, and *ec*, and extend line *it* equidistant from line *ec*. Next I draw line *et*. And thus quadrilateral *ac* is divided in two equal parts by line *et*. The proof follows [see Figure 4.67]. Quadrilateral *az* has two equidistant sides *ab* and *dz* that line *ei* cuts in two equal parts. Whence quadrilateral *ez* is half of quadrilateral *az*. Similarly, because the bases *zi* and *id* are equal, the triangles *czi* and *cid* are equal. Whence triangle *ciz* is half of triangle *cdz*. Because quadrilateral *ez* was half of quadrilateral *az*, therefore quadrilateral *ebci* is half of quadrilateral *ac*. Since triangles *cci* and *cct* are equal, if triangle *ebc* is added to both of these, then quadrilateral *ebct* equals quadrilateral *ebci*. But quadrilateral *ebci* is half of quadrilateral *ac*.[136] Finally, therefore, quadrilateral *bt* is half of quadrilateral *ac*, as was necessary. Having explained all of these things, it is clear that we can divide any quadrilateral in two equal parts from any given point on any of its sides. And if a line were drawn from any given point inside or outside the figure to divide a quadrilateral in two equal parts, we can manipulate these line-segments ingeniously.

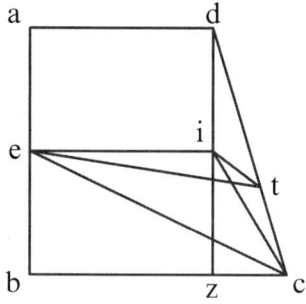

Figure 4.67

---

[135] "ADSignabo (*sic*) quidem per modum superius Demonstratum quomodo diuidantur diuersilateral quadratera a puncto dato super unum ex lateribus ipsius" (139.15–17, f. 86r. 16–17).

[136] Text has *ec*; context requires *ac* (139.30). Then follows a sentence with *bt* correctly substituted for *ebci* but with the erroneous *ec* in place of *ac* (139.31). I omitted this sentence as redundant.

[63] **40** *To divide a quadrilateral of unequal sides into required parts from a given angle.*[137] Again, given quadrilateral *abcd* from which I wish to cut a third part from angle *d*. Draw diagonal *ac* subtended by angle *d* [see Figure 4.68]. Then draw diameter *db* cutting *ac* at point *e*. Now line *ce* is either a third part of diameter *ac* or it is not. Suppose it is. Then diameter *bd* has separated triangle *bcd*, a third part, from quadrilateral *ac*. The proof follows. Because triangles *dce* and *dea*[138] share the same altitude, they are to one another as their bases[139] {p. 140}. For the same reason triangles *bce* and *abe*[140] are in the same ratio. Whence as *ce* is to *ac* so triangle *bcd* is to quadrilateral *ac*. For *ce* is a third of *ac*. Whence triangle *bcd* is a third of quadrilateral *ac*. Therefore line *bd* coming from the given angle *d* cuts off one third of quadrilateral *ac*, as required. On the other hand, if *ce* is not a third of *ac*, then I assign *cg* as a third of *ac* [see Figure 4.69]. Next from point *g* I draw line *gf* equidistant from diameter *bd*. By drawing line *fd*, I cut quadrilateral *fbcd* from quadrilateral *ac*. I will prove that it is a third of the whole quadrilateral *ac*. Because *cg* is a third of *ca*, triangles *bgc* and *cgd* that constitute quadrilateral *bcdg* are a third of triangles *abc* and *acd* or quadrilateral *abcd*. Hence by what has been proved before, this is equal to quadrilateral *fbcd*.

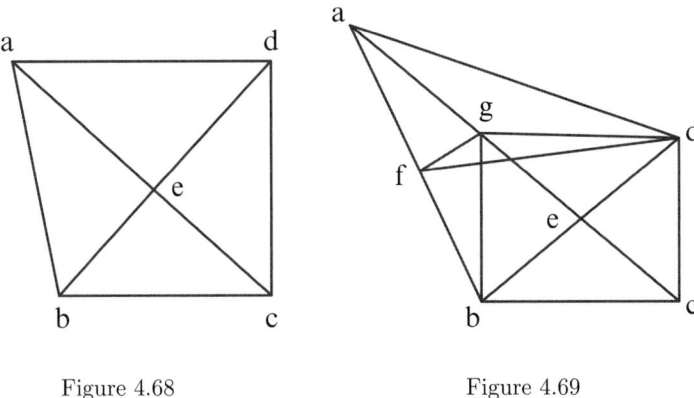

Figure 4.68   Figure 4.69

[64] **41** *To cut from a quadrilateral a given part by a line drawn from a given point on a side.*[141] Let the quadrilateral *abcd* be given with point *e* given on side *ad* and the required part a third of the quadrilateral [see Figure 4.70]. First I cut triangle *dcz* from quadrilateral *ac*, a third of quadrilateral *abcd*.

---

[137] "Demonstratio qualiter abscidatur a quadrilatero diuersilatero quantacumque pars data ab angulo dato" (139.36–37, *f.* 86v.8–9). Archibald 36; for page 140, read page 139.
[138] Text has *dca*; context requires *dea* (139.43).
[139] *Elements*, VI.1.
[140] Text has *abc*; context requires *abe* (140.1).
[141] "Et si uoluero a quadrilatero dato partem datam a puncto dato super unum ex lateribus abscidere" (140.11–12, *f.* 86v.26–27). Archibald 38.

236 Fibonacci's *De Practica Geometrie*

Then I draw line *ez*. Now *ez* is either equidistant from *dc* or it is not. Suppose first that it is. Then by drawing line *ec*, triangle *ecd* equals triangle *zcd*. Whence triangle *ecd* is a third of quadrilateral *ac*. But if *ez* is not equidistant from *dc*, then I draw line *id* equidistant from *ez*, and I draw *ei* [see Figure 4.71]. Then quadrilateral *eicd* equals triangle *zcd* that is a third of quadrilateral *ac*, as required. This is proved by triangles *zdi* and *edi* that equal each other because they share the same base and lie between the equidistant lines *di* and *ez*. If we add triangle *dci* to both triangles, then quadrilaterals *eicd* equals triangle *dcz*, as was said. Therefore quadrilateral *abie* remains as two third of quadrilateral *ac*. If this is divided in half, then quadrilateral *ac* has been divided in three equal parts. Again, let quadrilateral *abcd* have its side *ab* cut in three equal parts: *ae*, *ef*, and *fb* [see Figure 4.72]. Now I want to divide quadrilateral *ef* in three equal parts from points *e* and *f*. First I cut a third of quadrilateral *ac* by a line coming from point *f*, which I do by the second technique; namely, I make line *gd* equidistant from line *ab* and line *gh* a third of line *gd*. Then I draw lines *fh*, *ch*, and *fc*. And I construct line *hi* equidistant from line *fc*. Then by drawing line *fi*, quadrilateral *fbci* is a third of quadrilateral *ac*. Next I divide quadrilateral *afid* in two equal parts by line *ek*. And thus quadrilateral *abcd* has been divided in three equal parts: *ak*, *ei*, and *fc*. Knowing all of this, you can change the terms of the division at any point, as long as you do not forget the forgoing demonstration.

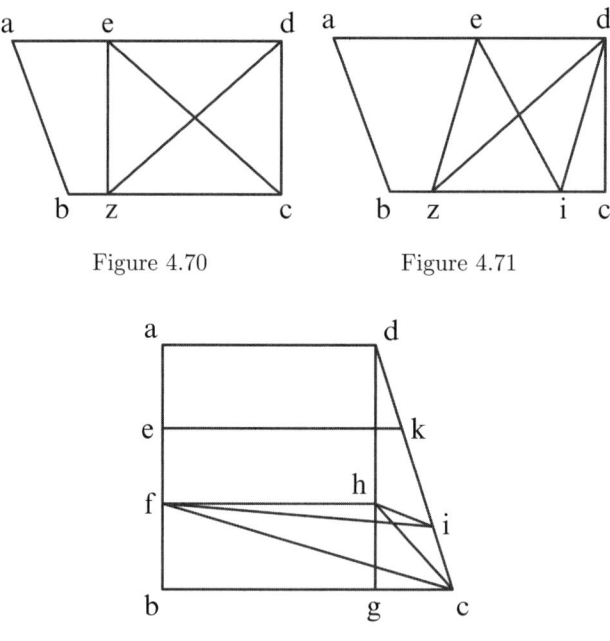

Figure 4.70   Figure 4.71

Figure 4.72

4 Dividing Fields Among Partners   237

[65] **42** *To cut a quadrilateral in two parts according to a given ratio, from either a given point or an angle.*[142] This is how you do it. Again, let quadrilateral *abcd* be given with the given ratio of *ez* to *zi*. I want to divide the quadrilateral in the ratio of *ez* to *zi* by a line drawn from angle *d* [see Figure 4.73]. First I draw the diameter *ac*. I also draw diameter *db* that goes first through point *t*. Then I make the ratio *ct* to *at* be as *ez* to *zi*. I say that line *bd* cuts quadrilateral *ac* in the ratio of *ez* to *zi*. For example: triangles *dct* and *dta* are to each other as *ct* to *ta*, their bases, for the triangles *cbt* and *tba* are in the same ratio {**p. 141**}. Whence as *ct* is to *ta*, so triangle *dcb* is to triangle *abd*. But *ct* is to *ta* as *ez* is to *zi*. Whence as *ez* is to *zi* so triangle *bcd* is to triangle *bda*. Therefore quadrilateral *ac* has been divided in two parts from the given angle according to the given proportion, as was required. On the other hand suppose diameter *bd* does not got through point *t*. Then it must pass between either points *c* and *t* or between *t* and *a*. First, let it pass between points *c* and *t*. Draw lines *bt* and *td*[143] [see Figure 4.74]. Then quadrilateral *tbcd* will be to quadrilateral *tbad* as *ct* is to *ta*, that is as *ez* is to *zi*. I draw line *tk* equidistant from the diameter *bd*. Drawing line *dk* makes quadrilateral *kbcd* equal to quadrilateral *tbcd*. This is proved by what has been said. Whence as *ct* is to *ta* (*ez* to *zi*), so quadrilateral *kbcd* is to triangle *dak*. Thus what was proposed is done. And if diameter *bd* goes between points *a* and *t*, then draw line *tl* from point *t* equidistant from diameter *bd* [see Figure 4.75]. Draw line *dl*. And again as *ct* is to *ta* (*ez* to *zi*), so triangle *dcl* is to quadrilateral *abld*. The foregoing proves this. If you wish to do this, divide line *ab* in the ratio of *bl* to *la* as *cz* is to *zi*. Then draw line *dm* equidistant from line *ab*. Let *mn* be to *nd* as *bl* is to *la* [see Figure 4.76]. Draw lines *ln*, *nc*, and *cl*. Make *no* equidistant from *lc*. Drawing *ol* makes the ratio of quadrilateral *ob* to quadrilateral *oa* as *bl* is to *la* (*cz* to *zi*). The foregoing proves this. If from point *l* the line equidistant line from *ab* does not fall within quadrilateral *ac*, then draw line *cp* equidistant from line *ab*. And make the ratio of *cq* to *qp* be as *bl* to *la* [see Figure 4.77]. Then draw the lines *lq*, *dq*, and *ld*. Through point *q* draw line *qr* equidistant from line *dl*. Having drawn line *lr*, quadrilateral *lbcr* will be to quadrilateral *alrd* as *bl* is to *la*, that is *ez* to *zi*. The foregoing proves this.

---

[142] "Et si quadratum aliquod in duas partes secundum datam portionem, a dato puncto uel angulo diuidere uis" (140.36–37, f. 87r.22–23). Archibald 55, 56.
[143] The previous sentence suggests a different letter for the new point. Fibonacci, however, called it *t*. Figure 4.74 offers some clarity.

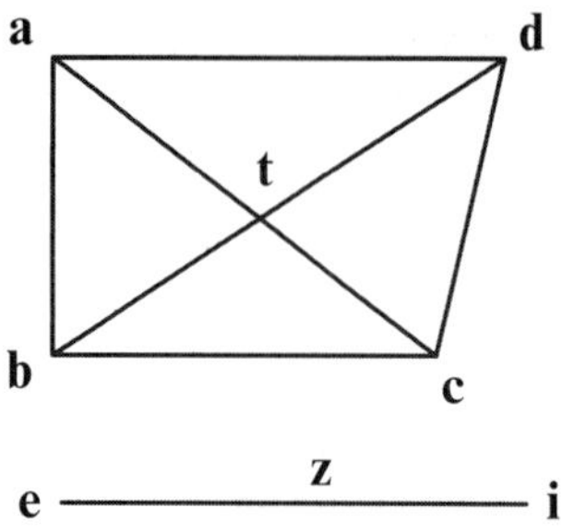

238 Fibonacci's *De Practica Geometrie*

Figure 4.73

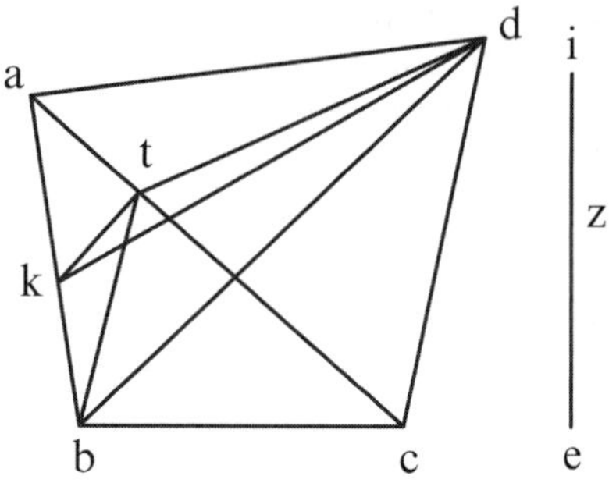

Figure 4.74

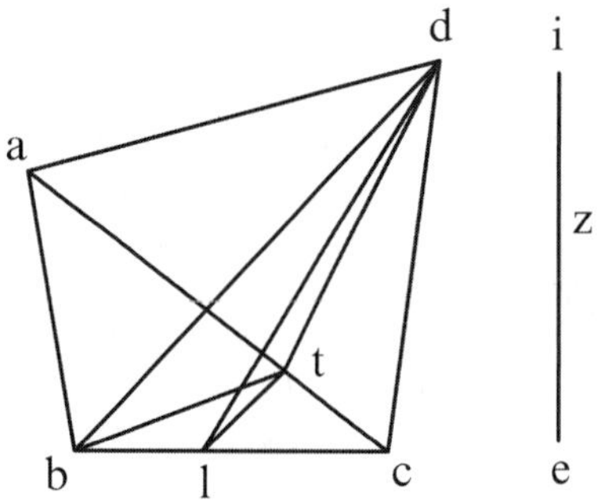

Figure 4.75

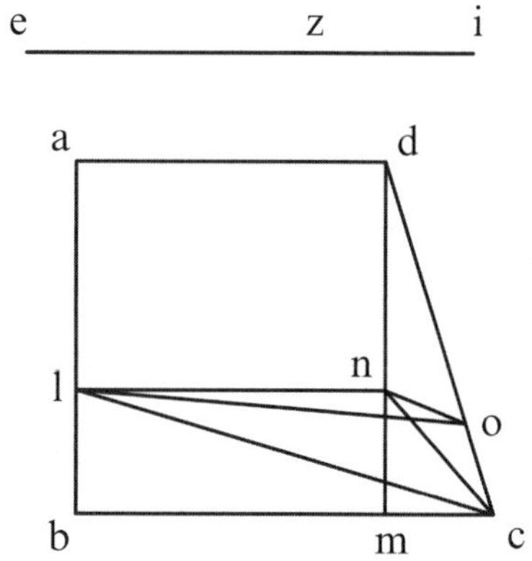

Figure 4.76

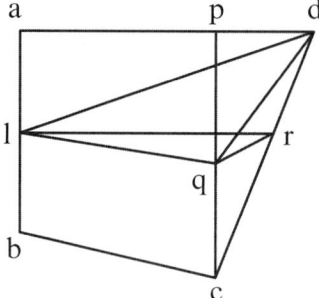

Figure 4.77

[66] **43** *To divide a quadrilateral in several parts according to given ratios.*[144] We wish to show how to do this. Let quadrilateral *abcd* be given with line *dg* drawn equidistant from side *ab*. Let the given ratio be *ez* to *it*. I will divide side *ab* at points *k* and *l* in these ratios: as *ez* is to *zi* so *bk* is to *kl*, and as *zi* is to *it* so *kl* is to *la*. Now I want to divide line *dg* at points *m* and *n* in these ratios. I draw lines *km*, *ln*, *cm*, *cn*, *ck*, and *cl*. And through point *m* I draw line *mo* equidistant from line *kc*. Through point *n* I draw line *np* equidistant from line *cl*. Having drawn lines *ko* and *lp*, quadrilateral *ac* is divided in the given ratios. The proof follows [see Figure 4.78]. Because *bk* is to *kl*, so *gm* is to *mn*. And as *kl* is to *la*, so *mn* is to *nd*. Then *bk* will be to *ka* as *gm* is to *md*. Whence as *gm* {**p. 142**} to *md*, so quadrilateral *gk* is to quadrilateral *ma*. For as *gm* is to *md*, so triangle *gcm* is to triangle *cmd*. Whence after being joined there will be *gm* to *md* as quadrilateral *kbcm* is to pentagon *akmcd*. But quadrilateral *kbcm* equals quadrilateral *bo*. Whence as quadrilateral *kbcm* is to pentagon *akmcd*, so quadrilateral *bo* is to quadrilateral *ao*. But as *gm* is to *md*, so *ez* is to *zt*. It can be shown in a similar way that quadrilateral *lo* is to quadrilateral *ld* as *mn* is to *nd*, that is, as *zi* is to *it*. Therefore as *ez* is to *zi* so quadrilateral *bo* is to quadrilateral *kp*. And as *zi* is to *it*, so quadrilateral *kp* is to quadrilateral *ld*. Therefore quadrilateral *ac* will be divided in the requested ratios, as required. We wish to offer a specific common way of dividing fields so that everything necessary will be done perfectly.

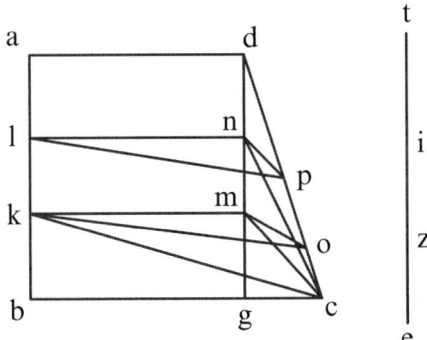

Figure 4.78

[67] **44** *To divide a quadrilateral in two equal parts from half of a side.*[145] Mark points *e* and *f* as the midpoints of lines *ab* and *gd* and draw line *ef*. This divides quadrilateral *ag* in two equal parts if sides *ab* and *gd* are

---

[144] "Vervm si quadrilaterum aliquod in plures partes et in proportiones datas diuidere uis" (141.24–25, *f.* 8v.23–24). Archibald 57, but not acknowledged there.

[145] "Vt si quadrilaterum *abgd* super dimidium lateris *ab* diuidere uolumus in duo equa" (142.11–12, *f.* 88r.113–14).

equidistant from each other. But if they are not, then the quadrilaterals *eg* and *af* are unequal. I will measure them by investigating their areas. Let quadrilateral *af* be less than quadrilateral *ge* by *h* rods. I will look for the number that measures the length of cathete *ei* that goes from point *e* to side *gd*. I will divide *h* by *ei* to get the quantity *kf*. I will draw line *ek*. I say that quadrilateral *ag* has been divided in two equal parts by line *ek* [see Figure 4.79]. The proof follows. Make *lk* equal to *kf* and join points *e* and *l*. Then the area of triangle *efl* will equal the area measured by cathete *ei* and line *fk*. Therefore the area of triangle *efl* equals the product of *ei* and *kf*. But the product of *ei* and *kf* is the number *h* by which quadrilateral *ebgf* exceeds quadrilateral *af*. Whence quadrilateral *ebgf* exceeds quadrilateral *af* by the size of triangle *elf*. Whence if triangle *elf* is removed from quadrilateral *bf*, quadrilateral *bl* remains equal to quadrilateral *af*. For the triangles *ekf* and *ekl* are equal to each other since they have equal bases and share the same altitude. And because "equals added to equals produce equals," quadrilaterals *kb* and *ak* are equal, as we said.

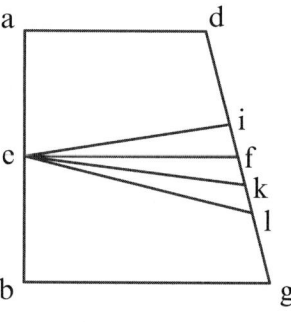

Figure 4.79

[68] In the same way we can divide a quadrilateral in two equal quadrilaterals that are similar to quadrilateral *abcd*. The latter has the form of an arrow head whose diameters can in no way intersect one another. First I draw the diameter *db*. Then, I find the areas of triangles *abd* and *cbd*. Now if they are equal, then I have divided quadrilateral *abcd* in two equal parts by line *bd*. But if they are not equal, then one area is larger than the other [see Figure 4.80]. Let triangle *cbd* be larger by some number *e*. From point *d* I draw cathete *dz* to side *cb*, and make the plane *dz* by *bi* equal to the number *e*.[146] After drawing line *di*, quadrilateral *abcd* has been divided in two equal parts, namely triangle *cdi* and quadrilateral *abid*. This was proved by what we said for the preceding figure.[147]

---

[146] Is this how point *i* is located on *cb*? Otherwise, the information is incomplete.

[147] The problem is not solved because it requires two quadrilaterals equal in area, and the solution produces a triangle dubiously equal to a quadrilateral; see the figure.

4 Dividing Fields Among Partners   241

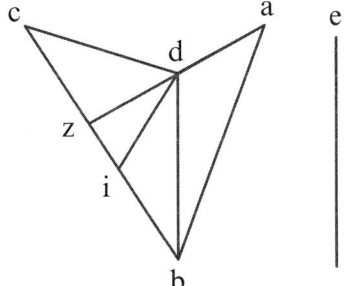

Figure 4.80

[69] But if we wish to do all of this geometrically, then draw line *ac* as seen in the adjacent figure. Find its midpoint *e* and draw line *be*. Now line *be* either goes through point *d* or it does not. First, let it go through [see Figure 4.81]. I say that quadrilateral *abcd* has been divided in two equal parts by line *db*. The proof follows. Triangles *abe* and *ebc* are equal; also triangle *aed* equals triangle *ced*. Whence if we remove equals from equals, what remains {**p. 143**} is equal. Therefore after removing triangle *aed* from triangle *aeb* and triangle *ced* from triangle *ceb*, what remain are triangles *abd* and *cdb* equal to each other, as I said. On the other hand, if line *eb* does not go through point *d*, then it cuts either side *ad* or side *cd*. So let it cut side *cd* at point *z*. Draw line *ed*. Let *dz* be to *zi* as *bz* is to *ze*. Draw line *bi*[148] [see Figure 4.82]. I say that quadrilateral *abcd* has been divided in two equal parts by line *bi*. The proof follows. Because triangles *ezd* and *biz* have equal vertical angles at *z*, the sides opposite the angles are mutually proportional. Whence the triangles are similar.[149] Thus *bz* is to *ze* so *dz* is to *zi*. Let triangle *cez* be added to both triangles. Then the two triangles *cez* and *biz* are similar[150] to triangle *ecd*. But triangle *ecd* equals triangle *aed*. Therefore triangles *cez* and *izb* are equal to triangle *ead*.[151] Now triangles *eab* and *ebc* equal each other. Whence if equals are taken from equals what remains is equal. Let therefore triangle *ead* be taken from triangle *eab* and triangles *cez* and *izb* from triangle *ebc*. What remain are quadrilateral *abzd* and triangle *edz* equal to triangle *cib*. But triangle *edz* equals triangle *biz*.[152] Whence quadrilateral *abid* equals triangle *cbi*, as I said.

---

[148] There are no instructions for locating point *i*.
[149] Text has *equalia*; context requires *similia* (143.8).
[150] Text has *equalia*; context requires *similia* (143.9).
[151] The conclusion is false because the required triangles are similar and not equal.
[152] Again, the conclusion is false, for the same reason.

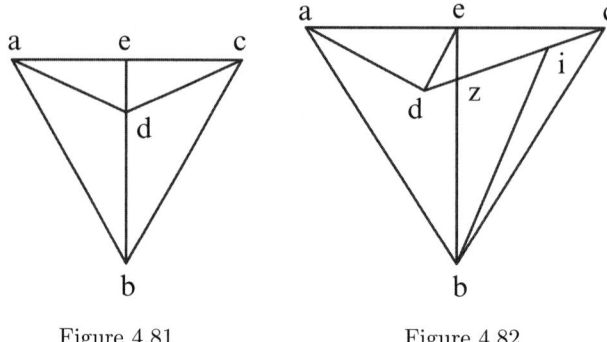

Figure 4.81   Figure 4.82

4.5 Division of Polygons[153]

[70] **45** *To divide a regular pentagon in two equal parts by a line drawn from a given angle to the middle of the opposite side.*[154] For example: given regular pentagon *adcde* and given angle *a*, draw line *af* to the middle of side *cd*. I say that pentagon *adcde* has been divided in two equal parts by line *af* [see Figure 4.83]. The proof follows. Draw lines *ac* and *ad*, and triangles *abc* and *aed* will be equal to each other and equiangular. Now triangles *afc* and *afd* are also equal. So quadrilateral *abcf* equals quadrilateral aedf. Therefore pentagon *adcde* has been cut in two by line *af* from the given angle *a*, as required. From this it is obvious that the line drawn from an angle in a regular pentagon to the middle of the side opposite the angle has cut it in two equal parts.

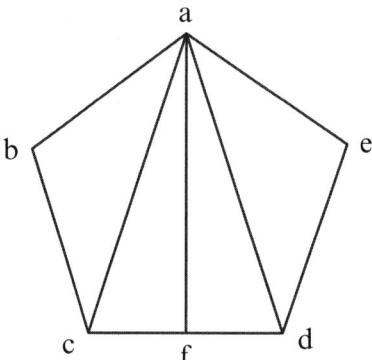

Figure 4.83

---

[153] "Incipit de diuisione figurarum plurium laterum" (143.17, *f.* 89r.1).

[154] "Si pentagonvum equilaterum et equiangulum ab aliquo angulorum dato in duo equa secare desideras, lineam ab ipso angulo per dimidium lateris sibi oppositi protrahe" (143.18–19, *f.* 89r.2–3).

[71] **46** *To divide a regular pentagon in two equal parts from a given point on a side.*[155] Consider regular pentagon *adcde* with point *g* in the middle of side *ab*. By drawing line *gd* from point *g* to angle *d*, then pentagon *adcde* has been divided in two equal parts. And if the given point were *h* between points *a* and *g* as in the second figure, then divide the pentagon in two by line *af* [see Figure 4.84]. Draw line *ai* equidistant from line *hf*. After drawing line *hi*, pentagon *adcde* has been divided in two equal parts. The proof follows. Because triangles *hfa* and *hfi* lie between two equidistant lines and share the same base *hf*, they equal each other. If quadrilateral *hbcf* is added to each, quadrilateral *hbci* will be equal to quadrilateral *abcf*. But quadrilateral *abcf* is half of pentagon *adcde*. Whence quadrilateral *hbci* is half of pentagon *adcde*.

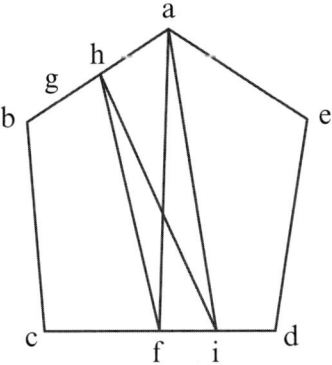

Figure 4.84

[72] **47** *To divide an irregular pentagon in two equal parts.*[156] First I draw line *eg* and then select *z* as its midpoint. From here I divide quadrilateral *abge* in two equal parts by line *zi*. I draw line *zd* that divides triangle *egd* in two equal parts. Now line *zd* can either be extended straight on to point *i* or it cannot be so extended {**p. 144**}. First let it be drawn straight on from one line to the other, as seen in the third figure. Then it is obvious that line *id* divides pentagon *abgde* in two equal parts. Now if you wish to divide the same pentagon from a given point *t* on side *ab*, then draw line *td* equidistant from line *ik*. Next, draw line *tk* to divide the pentagon in two equal parts. The proof is the same as above.[157] But if line *zd* is not in a straight line with *zi*, then draw line *id* equidistant from line *zl*. Next, draw line *il* to divide pentagon *abgde*

---

[155] "Et si a puncto *g* dato super latus *ab* eundem penthagonum in duo equa diuidere uis, ...(143.28–29, f. 89r.27–28).

[156] "Sed non sit penthagonum datum equilaterum et equiangulum, ut penthagonum *abgde*; et uolo ipsum diuidere in duo equa" (143.39–41, f. 89r.27–28).

[157] See [71].

244   Fibonacci's *De Practica Geometrie*

in two equal parts [see Figure 4.85]. The proof follows. Lines *iz* and *zd* divide pentagon *abgde* in two equal parts. Triangles *idz* and *idl* equal each other. If you add quadrilateral *ibgd* to them, then pentagon *ibgdl* equals pentagon *ibgdz* that is half of pentagon *abgde*, as required. Consequently we have a method for dividing a pentagon in two equal parts from any side or from any point given on a side, and even into as many parts as desired according to what is given. For instance, if we want to divide it in three parts by line *eg* and we have divided quadrilateral *abge* and triangle *egd* in equal parts by the points of section, then we proceed in the order already demonstrated and we have what we wished.[158]

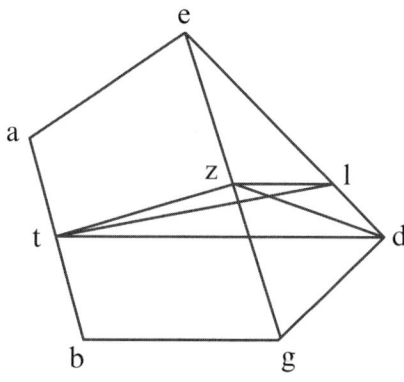

Figure 4.85

[73] If the pentagon has the form of a miter like pentagon *bcdef*, draw line *fa* at right angles to side *cd*. Investigate the areas of quadrilaterals *bcaf* and *fade*. If they are equal, then line *af* has divided pentagon *bcdef* in two equal parts. But if the areas are not equal, let the area of quadrilateral *ae* be larger by a given amount *g* [see Figure 4.86]. Let the area defined by *fa* and *az* be equal to the number *g*. Since the area defined by *fa* and *az* equals *g*, the length of *az* is known as is point *z* [see Figure 4.87]. Draw line *fz*, and the area of triangle *faz* equals half of the number *g*. Whence line *fz* divides pentagon *bcdef* in two equal quadrilaterals, *bczf* and *fzde*. Now this was done by the common method. To do it geometrically, draw line *be* and divide it in two equal parts at point *i*, as can be seen in the other figure [Figure 4.88]. Divide quadrilateral *bcde* in two equal parts by line *ik* that goes through point *f*. Thus line *fk* divides pentagon *bcdef* in two equal parts. Because if we take triangle *bif* from quadrilateral *bcki* and take triangle *ief* from quadrilateral *ikde*, what remain are quadrilaterals *fc* and *fd* equal to one another.

---

[158] This is an example of the previous statement, a generalization, and should not be considered a new rule. Similarly with [74] below.

4 Dividing Fields Among Partners   245

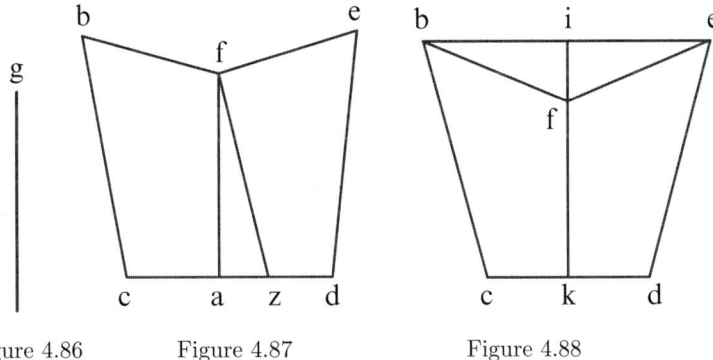

Figure 4.86   Figure 4.87   Figure 4.88

[74] If line *ik* cuts side *ef* at point *l*, as can be seen in the last figure, let *ml* be to *lf* as *lk* is to *il*. Draw line *km* to divide pentagon *bcdef* in two equal parts, of which one is quadrilateral *mkde* and the other pentagon *bckmf* [see Figure 4.89]. The proof follows.[159] Draw line *if*. Triangles *ifl* and *kml* are equal to one another, since the ratio of *ml* to *fl* is as the ratio of *lk* to *il*.[160] Triangle *ife* equals triangles *ile* and *klm*. Because triangle *ief* equals triangle *bif*, triangle *bif* equals triangles *iel* and *klm*. If we take triangle *bfi* from quadrilateral *bcki* and take triangles *ile* and *klm* from quadrilateral *ikde*, what remain are quadrilateral *bckl*[161] and triangle *ifl* equal to quadrilateral *kdem*. Because triangle *klm* equals triangle *ilf*,[162] pentagon *bckmf* equals quadrilateral *mkde*, as I said. We make the division from point *f* if we draw line *fk*, make line *ei* equidistant from line *mn*,[163] and draw line *fn*[164] **{p. 145}**.

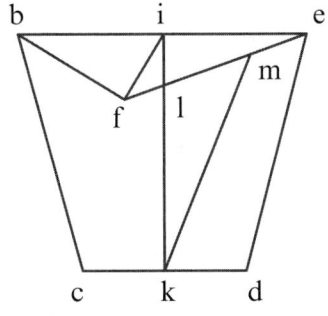

Figure 4.89

[75.1] **48** *To divide a pentagon in three equal parts.*[165] Use line *op* to take a third part from quadrilateral *bcnf*. Similarly with line *qd*, we take a third part

---
[159] The proof is fallacious because triangles *ifl* and *kml* are similar, not equal.
[160] This is patently false; the triangles are only similar. The remainder of the proof fails.
[161] Triangle *bfi* from quadrilateral *bcki* leaves a pentagon; see the figure.
[162] This is false; triangles are only similar.
[163] Impossible! Yet the reading is correct (144.43, f. 90r.12).
[164] This statement is irrelevant to the problem.
[165] "quod etiam penthagonum si in tres equas partes diuidere uolumus" (145.1–2, f. 90r.14–15).

of quadrilateral *fnde*. Thus pentagon *bcdef* has been divided in three equal parts: quadrilateral *bcpo*, pentagon *opdqf*, and triangle *qde*.[166]

[75.2] **49** [*To divide a hexagon in two equal parts.*] Note further that if you want to divide a hexagon, cut it in two quadrilaterals by an inside line with three sides on one half and the other three sides on the other half. Then on half of that line place a point from where you will divide the quadrilateral in two equal parts. For instance, in hexagon *adcdef* draw line *ad* to divide the hexagon in two quadrilaterals: *abcd* and *aedf*. Then on half of line *ad* find point *g* from which we can divide both quadrilaterals in two equal parts. Thereafter we will have repeated what was said above: we can divide the hexagon in two equal parts by one line. Also we can change the parts from any given point on any side. Also, if we follow this procedure for other polygons, we can easily divide them.

4.6 Division of Circles and Their Parts[167]

[76] **50** *To divide a circle in two equal parts by a line drawn through the center from a given point on the circumference or from a given point outside the circle.*[168] Given circle *abgd* with center *z* and point *e* outside the circle [see Figure 4.90]. Draw line *ez* and extend it to point *g*. Therefore line *ag* is the diameter of circle *abgd*, for the diameter of a circle connects its center with two points on the circumference. Line *ag* was drawn through the center *z*. Therefore *ag* is a diameter. Whence each portion, *abg* and *adg*, is a semicircle. Therefore circle *abgd* has been divided by a line drawn from a given point *e* that lies outside the circle. If the point lies within the circle such as the given point *i*, then draw line *iz* and extend it both ways to points *b* and *d*. Thus line *bd* is a diameter of the circle, cutting circle *abgd* in two parts, as before.

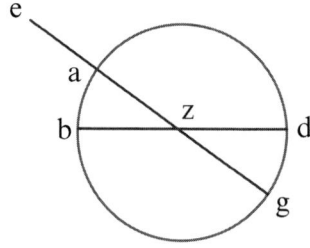

Figure 4.90

---

[166] Unfortunately, there was no helpful figure to assist understanding the procedures that begin with separating out a third of a quadrilateral.

[167] "Incipit de diuisione circulorum et eorum partium" (145.17, *F.* 90r.32).

[168] "Si CIRCVLVM aliquem in duo equa secare uolumus, per lineam protractam a puncto dato super periferiam eius, aut extra circulum, ipsum punctum cum centro copula et ipsam rectam usque ad periferiam ducere studeas" (145.18–20, *f.* 90r.33–35).

[77] **51** *To divide a circle in three equal parts.*[169] In a circle with center *d* draw an equilateral triangle *abg*. Then draw the lines *da*, *db*, and *dg*. They divide the circle in three equal sectors: *dab*, *dbg*, and *dga* [see Figure 4.91]. For example: The three lines *ab*, *bg*, and *ga* are equal and contain the three equal arcs *ab*, *bg*, and *ga*, because equal chords in equal circles subtend equal arcs. The lines *da*, *db*, and *dg* each drawn from the center are equal, and the angles at *d* the center are equal. Whence the sections *e*, *z*, and *i* equal one another. Therefore circle *abg* has been divided in three equal sections, as required. If you want to divide more parts from the center, divide the circumference is so many equal parts, as many as you wish. Then draw lines from the points of section, and you will have what you wanted.

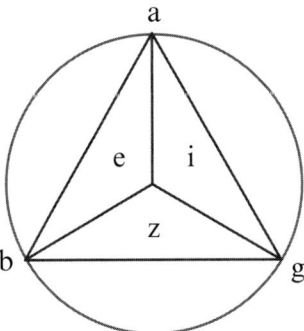

Figure 4.91

[78.1] **52** *To divide a circle in three equal parts by lines equidistant from one another.*[170] Without a great deal of work I know of no way of doing this. But I can show how to do it approximately. First, in circle *abg* draw line *ag* that is the side of an equilateral triangle inscribed in the circle. Consequently, arc *ag* is a third of the whole circumference of the circle {**p. 146**} *abg*. Then draw line *db* from the center *d* and equidistant from line *ag*. Next, draw lines *da*, *dg*, *ab*, and *bg*, resulting in triangles *dag* and *bag* being equal. Add to them the part of the circle between line *ag* and arc *ag*. Thus the figure between lines *ba* and *bg* and arc *ag* equals sector *dag* that is a third of circle *abg*[171] [see Figure 4.92]. Now if the portion of the circle contained by line *bg* and arc *bg* is added, then the portion of the circle between line *ab* and arc *agb* will be more than a third of circle *abg* by the size of the area between line *bg* [and arc *bg*]. This is how to find the size of this area. Divide it in half through line *gb*. Select point *e* just beyond the midpoint of arc *bg* to create arc *be* that is a little more than half of arc *bg* [see Figure 4.93]. Draw line *ae* so that the

---

[169] "Et si circulum aliquem in tres partes diuidere uolumus, ..." (145.29, f. 90v.11–12).
[170] "Si uero circulum per lineas equidistantes in tria diuidere uolumus" (145.39, f. 90v.18).
[171] This sentence is repeated immediately at (146.5–6, f. 90v.24–25).

portion *age* approaches a third part of circle *abg*. Next draw cathete *dz* from the center *d* to line *ae* and extend it to point *i* so that line *di* equals line *dz*. Through point *i* draw line *tk* equidistant from line *ae*. Then portion *tkl* equals portion *abe*. What remains contained by lines *ae* and *tk* and arcs *ke* and *at* will be equal to another third part.

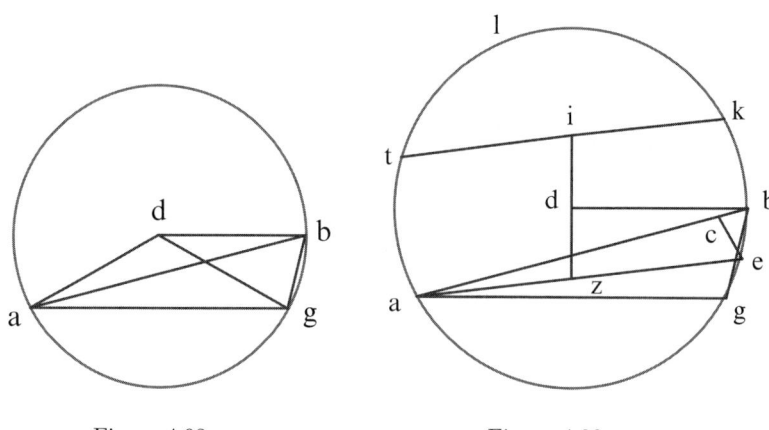

Figure 4.92   Figure 4.93

[78.2] The proof follows. From point *e* draw cathete *ec* to line *ab*. Let what results from the division be twice the area of portion, as measured by the product of line *bg* by line *ab*. After drawing line *eb*, triangle *abe* will equal a portion of the circle *bg*. After removing triangle *abe* from portion *agb* of the circle, the remaining portion *aeg* is a third part of circle *abg*. Because lines *tk* and *ae* are equidistant and equally distant from the center, they are equal to each other. So *tk* and *ae* are equal, the segments in the same circle being equal as stated in Euclid III.[172] Therefore portion *tkl* equals portion *aeg*. Whence section *tkl* is a third part. What remains between the equidistant lines *tk* and *ae* and arcs *ek* and *at* is the remaining third part, as required. Thus and as well as I could and as carefully, I investigated the ratio of line *dz* to the semidiameter of the circle. I found the ratio to be nearly 9 to 34. So if we wish to divide a circle in three equal parts by equidistant lines, we must find the diameter of the circle. With each half of the diameter divided in the stated ratio and the circle divided in two parts through its center, we shall drawn lines through the parts of the radius at right angles. Then indeed we shall have divided the circle in three equal parts.

[79] **53** *To divide a circle in four equal parts.*[173] Given circle *abgde* draw the diameters *az* and *ge* at right angles through *c*, the center of the circle.

---

[172] *Elements*, III.24.
[173] "si circulum *abgde* in quatuor equas partes diuidere uolumus" (146.32, *f.* 91r.29).

Then divide the semidiameter cz at point k in the ratio of 611 to 1512 [see Figure 4.94]. Next make line cl equal to line ck. Then draw lines bh and df through points l and k, to cut diameter az at right angles. Consequently each of the semicircles will have been divided quite closely in two equal parts.

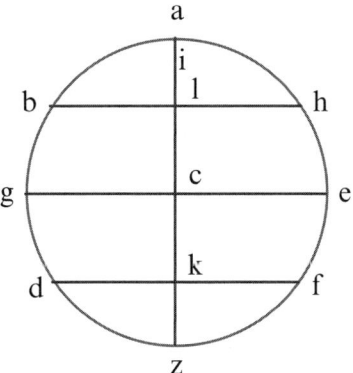

Figure 4.94

[80] **54** To cut from a circle a given part such as a third between two equidistant lines.[174] Given circle abg with center d and two equidistant lines, we wish to remove a third of it. Inscribe in the circle line ab as the side of an equilateral triangle [see Figure 4.95]. Through the center d draw line dg equidistant from ab. Draw line bg. Divide arc ab in two equal parts at point e. Draw line ez equidistant from line bg. I say that the figure contained by lines bg and ez and arcs eb and gz is a third of circle abg. The proof follows. Draw lines da, db, and ag. They form equal triangles {**p. 147**} gab and dab. By adding part abe to both triangles, the figure bounded by lines ga and gb and arc aeb equals sector $daeb$ that is a third part of circle abg. Therefore the figure bounded by lines ga and gb and arc aeb is a third part of circle abg. And because lines bg and ez are equidistant, arcs eb and gz are equal. But arc eb equals arc ae. Therefore arc ae equals arc gz. After adding arc bg to both, arc $aebg$ equals arc $ebgz$. Whence part $ezgb$ of the circle equals part $aqbe$ of the circle. By removing from both the part bounded by line bg and arc gb, the remaining figure bounded by lines gb and ez and arcs be and gz equals a third part of the circle, namely the figure bounded by lines ga and gb and arc aeb, as required.

---

[174] "Et si inter duas equidistantes a dato circulo abg cuius centrum sit d datam partem, que sit tertia, auferre uolumus..." (146.37–38, f. 91r.35–91v.2). Archibald 50; see also his page 68.

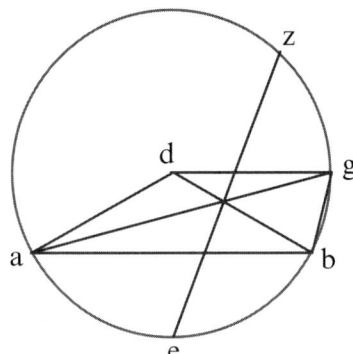

Figure 4.95

[81] **55** *To remove from a given circle some given part by a concentric circle.*[175] Given circle *abcd* with center *e*: I wish to remove a third of the circle by another circle equidistant from it. I draw diameter *ag* and make the square on line *ae* three times the square on line *ez*. At center *e* and with radius *ez*, I draw circle *zit*. I say that a third part of circle *abcd* has been removed by circle *zit*. The proof follows. Because line *ea* is half of diameter *ag*, the square on diameter *ag* is four times the square on line *ae*. For the same reason therefore the square on line *zt* is four times the square on line *ez*. Whence as the square on line *ea* is to the square on line *ez*, so the square on diameter *ag* is to the square on diameter *tz*. But as the square on diameter *ag* is to the square on diameter *tz*, so is circle *abcd* to circle *zit*, for "circles are to one another as the squares on their diameters," as shown in Euclid, XII.[176] Whence as the square on line *ea* is to the square on line *ez*, so circle *abcd* is to circle *zit*. Now the square on line *ea* is three times the square on line *ez*. Whence circle *abcd* is three times circle *zit*. And both circles are concentric because they have the same center. Therefore a third part of circle *abcd* has been removed by circle *zit*, as required.

Division of Parts of Circles[177]

[82] **56** *To divide a semicircle in two equal parts at a given point.*[178] Given semicircle *abg* with line *ag* divided in two equal parts at point *d* [see Figure 4.96]. At point *d* draw line *db* at right angles to *ag*. I say that semicircle

---

[175] "Ostendam rursus quomodo auferatur a circulo dato qualiscunque pars data per circulum equidistantem ei." (147.10–11, *f.* 91v.20–22).
[176] *Elements*, XII.2.
[177] "Incipit de diuisione partum circulorum" (147.28, *f.* 92v.5–6).
[178] "Si SEMICIRCVLVM *abg* in duo equa secare uis, divide rectam *ag* in duo equa super punctum *d*..." (147.29, *f.* 92r.6–7).

abg has been divided in two equal parts by line db. The proof follows. After drawing lines ba and bg, triangles bdg and bda are equal to each other, because sides ad and dg are equal. Now line bd is shared by both sides together with the angles at d that are right angles. Whence line bg equals line ba. Remove the equal lines from the sections. Whence section bg equals section ba, and triangles bdg and bda are equal. Whence section $bdai$ equals section $bdge$. Therefore the semicircle $abgd$ has been divided in two equal parts, as required.

[83] **57** *To divide a semicircle in two equal parts by a line equidistant from the diameter and a point on the radius.*[179] Divide radius az at point z so that the ratio of dz to zb is as 611 is to 1512 [see Figure 4.97]. Then line ei drawn through point z equidistant from the base divides the semicircle in two equal parts, as was shown above in the division of a circle in four parts.[180]

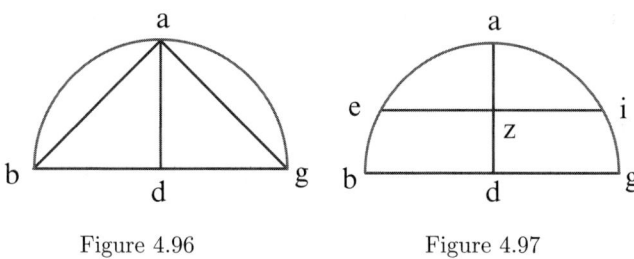

Figure 4.96   Figure 4.97

[84] **58** *To divide a part of a circle, whether larger or smaller than a semicircle, in two equal parts.*[181] To do this, draw the arrow through the midpoint of the chord. For example. Given a part of a circle larger than semicircle abg: draw arrow da through the midpoint of the chord **{p. 148}** and part abd of the circle will equal part adg. This is proved by what was said about the semicircle[182] [see Figure 4.98]. And if we draw line dc at right angles to line chord bg at point d, then part bcg of the circle that is smaller than a semicircle will be divided in two equal parts.

---

[179] "Et si ipsum cum recta equidistante basi ag, diuidere uis..." (147.3–7, *f.* 92r.16–17).

[180] See [79].

[181] "Et si portio circuli, siue sit minor siue major semicirculo, in duo equa diuidere uis..." (147.41, *f.* 92r.21–22).

[182] See [82].

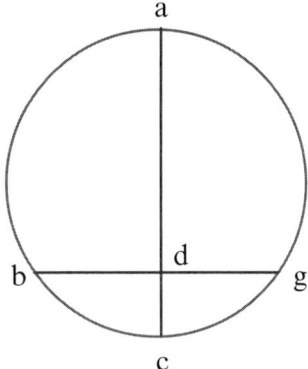

Figure 4.98

[85] **59** *To divide a semicircle in three equal parts.*[183] Divide line *bc* in two parts at point *d*. Then divide arc *bac* in three equal parts at points *a* and *g*. Draw lines *da* and *dg*. These divide semicircle *abc* in three equal parts [see Figure 4.99]. The proof follows. Because part *abc* is a semicircle and point *d* the center of the circle, half the circle is semicircle *abc*. Whence lines *db*, *da*, *dg*, and *dc* are all equal. And because arcs *ba*, *ag*, and *gc* are also equal, each sector is contained by two of those lines and one arc. Because all the sectors are contained by equal sides and arcs, they are necessarily equal. Therefore the semicircle has been divided in three equal sectors: *dab*, *dag*, and *dgc*, as required.

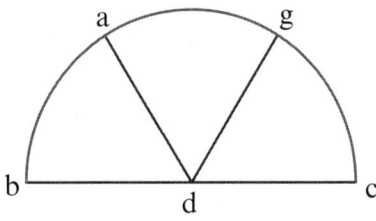

Figure 4.99

[86] **60** *To divide a sector in two equal parts.*[184] Given sector[185] *abdg* whose two sides *ab* and *ag* are straight lines, the remaining side *bdg* being an arc of a circle. First draw *bg* and divide it in two equal parts at point *e*. From here draw line *ae*. Then on line *bg* and from point *e* given on it, draw line *ed* at right angles.

---

[183] "Rvrsus si semicirculum *abc* in tria equa diuidere uis,..." (148.4, f. 92r.28).
[184] "Et si figuram aliquam contenctam sub duabus rectis et arc periferie in duo equa diuidere uis" (148.13–14, f. 92v.4–5). Archibald 49.
[185] Text and manuscript have *trigonum*; context requires *sector* (148.14 & .29, f. 92v.5).

Lines *ae* and *ed* are either in a straight line or they are not. Assume that they are [see Figure 4.100]. Then figure *abdg* has been divided in two equal parts by line *ad* because line *ae* has divided right triangle *abg* in two equal parts. Thus line *ed* divides part *bdg* of the circle in two equal parts. On the other hand lines *ae* and *ed* may not be in a straight line, as can be seen in this other figure[186] [Figure 4.101]. Then draw line *ad*. Next, through point *e* draw line *ei* equidistant from line *ad*. Draw *di*. [I say] that [*di*][187] divides the whole figure *abdg* in two equal parts. The proof follows. Because lines *ad* and *ei* are equidistant and are bases for triangles *dai* and *dea*, the triangles are equal. Add to them the figure they have in common described by lines *ab* and *ad* and arc *db*. Thus the figure bounded by lines *di*, *ia*, and *ab* and arc *db* equals the figure bound by lines *de*, *ea*, and *ab* and arc *bd*.[188] This is half of sector *abdg*, as required. Thus ends the fourth chapter on the division of fields among partners.[189]

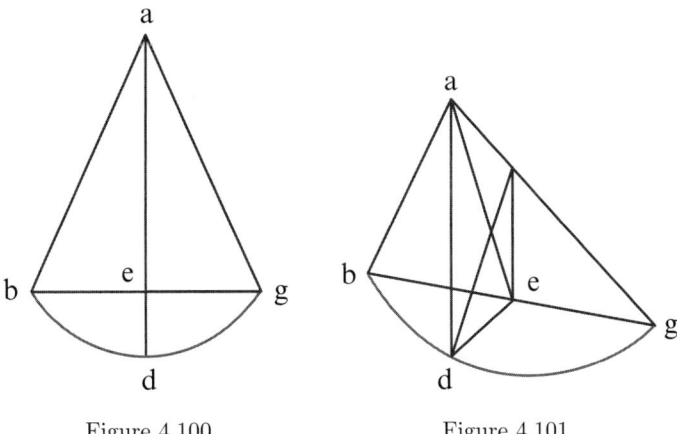

Figure 4.100 Figure 4.101

---

[186] Text and manuscript have *formula*; context requires *figura* (148.22, f. 92v.14).

[187] The two bracketed insertions are necessary to make sense of the statement and to match the proof in the Arabic version; otherwise, the proof would be assuming what it seeks to prove. See Hogendijk (1993), 158.

[188] Fibonacci left it to the reader to complete the proof, that *ai* has bisected the figure.

[189] "Explicit distinctio quarta de diuisione camporum inter consortes" (148.30, f., 92v.32–33).

5
Finding Cube Roots

COMMENTARY

The method for finding cube roots, as Fibonacci stated in [2], parallels the method for finding square roots. He uses the cubic identity that we would recognize as

$$z = (a+b)^3 = a^3 + 3a^2b + 3ab^2 + b^3, \text{ for } z < 1000.$$

Users of both techniques are required by Fibonacci to know the respective powers of the numbers from 1 to 10. He exemplified the cubic identity in [2] and [3]. Additionally, he exemplified a method for finding the difference between two cubes, without knowing them. The difference is necessary for finding b.

Where z is not a perfect cube, subtract the closest cube, $a^3 < z$, from z to leave a remainder r. Now a is the integral part of $\sqrt[3]{z}$. The task is to find b, the fractional part of the root. The next step is to multiply a by $a+1$, then by 3, and finally to add 1 to the product. Call the sum s. The fraction r/s is usefully close to b. For instance, in [7] to find the cube root of 900:

(1) $900 - 729 = 171 \qquad z - a^3 = r$; 729 is a^3, the largest cube less than 900; $a = 9$
(2) $3([9][10]) + 1 = 271 \quad 3(a[a+1]) + 1 = s$.
(3) $\dfrac{r}{s} = \dfrac{171}{271} \sim \dfrac{2}{3} = b.$
(4) Therefore $\sqrt[3]{900} = 9.6667$.

How Fibonacci adjusted the resulting root, $a + b = 9.667$, to get closer to the cube root of 900 awaits the reader in [7]. The process is repeated on r, and its cube root is added to $(a + b)$. The simple approximation procedure can be symbolized thus,

$$\sqrt[3]{N} \approx a + \frac{r}{3a(3a+1)+1}.$$

This procedure works well for finding cube roots of numbers less than 1000. For numbers greater than 1000, Fibonacci adjusted the method, much as he did for finding square roots of numbers greater than 100. For cube roots of numbers greater than 1000, b is a whole number, but Fibonacci was not very helpful in discovering its value. He only remarked, "You can find this digit only from much experience." The examples are worth reading.

A caution while reading the text: Fibonacci worked out in laborious, even tedious, detail each part of each step enumerated above. In some places it is difficult to follow the steps themselves because they require proficiency in the method of medieval division on the part of the reader. By letting the subtraction process light the way much of the time, the reader will probably find the steps that rise from each level to the next. As Fibonacci implied, "Finding cubic roots takes practice."[1]

The industrious reader who worked his or her way through all the examples might well wonder, "Cui bono?" The answer lies in the next section, to find two numbers in continued proportion between unity and a given number. The problem itself is the well-known Delic problem: how to construct a cubic altar twice the volume and similar in shape to the present cubic altar. Hippocrates of Chios (ca. 440 BC) found the answer: find two numbers in continued proportion between 1 and 2, one representing the volume of the current altar and two the volume of the proposed altar. His answer, however, did not state how to find the two numbers. There are many ways to do this.

Fibonacci offered three methods, all well known from classical times. The first is the method proposed by Archytas of Tarento (428–347 B.C.).[2] He showed that a perpendicular dropped from a point on the surface of a half cylinder to its base led to the desired cube root; see paragraph [12] and Figures 8.a and 8.b. The point was found by two moving figures. The first is a semicircle perpendicular to the diameter of the base of a circular half cylinder and rotating about one endpoint of the diameter, thereby cutting through the surface of the half cylinder. The other moving figure is a right triangle tangent to the base of the half cylinder with one side on and equal to the diameter of the base. With the base of the triangle fixed on the base or diameter of the half cylinder, the triangle is raised up, the hypotenuse cutting through the surface of the half cylinder. Where the two cuts intersect is the point from which the perpendicular is dropped.

The second method described by Fibonacci is that of Philo of Byzantium (ca. 250 B.C.) described in [13]. It too requires a movement, but only of a single line. The procedure is far simpler than that of either Archytas or of Plato (which follows). Figure 5.9 shows the completed Philomic procedure. It was constructed as follows. Beginning with a 1 by 2 rectangle, a diagonal is drawn to become the diameter of a circle. Then two adjacent sides of the rectangle are extended indefinitely to form two sides of a right triangle enclosing the rectangle. Finding the hypotenuse requires two steps: (1) place a ruler so that it intersects the two sides and touches point b, and

---
[1] "Quam figuram inuenire non poteris nisi ex usitato arbitrio" (150.34–35).
[2] See Knorr (1989), *passim*, for the clearest discussion of the three methods that I have found, particularly of the so-called Platonic method. His explanation was most instructive.

(2) move the ruler about point b until the length of the distance zb equals the distance eh.

The third method is attributed to Plato but uncomfortably so by many historians because he was not in favor of any movement in geometry.[3] The method is easier to understand after seeing the end result, because it shows the four lines in continued proportion: $dg : dm = dm : dz = dz : de$. The sequence is valid because the crucial angles at z, d, and m are right angles. How did he accomplish this?

Consider the two constructions shown in Figures 5.10 and 5.11. The first sets up the problem, to find two mean proportionals between line de and line dg. Line ez' seems to have been drawn at random. Through point ea line ep is drawn perpendicular to ez', distinct and separate from line eg and fixed to ez'. Line $p'm'$ is drawn through point g parallel to line ez' and not fixed to point g but capable of moving through it.

Now for the movement: the method requires three lines moving in unison, two lines together (ez' and its perpendicular), a third separately ($p'm'$). Consider Figure 5.11: as line ez' slides down dz', it moves through the pivot at point e carrying its perpendicular with itself. At the same time, point m' is sliding along dk pulling line gm' with it through pivot point g. The motion continues until the perpendicular to ez' becomes perpendicular to $p'm'$. At that instant, z' becomes simply z, m' becomes m, and thus the right angles have been reached. The conclusion follows.

What surprised me is that Fibonacci apparently did not modify what he had learned from the Banū Musā. If he had drawn the line $z'm'$, he might have had a simpler method of setting up what is seen in Figure 5.12. For by letting line ez' slide down line dz' as before, the angle at z' would be opening up. Then when the line ez' became parallel to ep', the angle would have reached 90°, line ez' would have reached point z, and the downward motion would stop. Thus he could have had what appears in Figure 5.12, the required result produced by his own method. The chapter concludes with straightforward instructions on multiplication and division, addition and subtraction of cube roots.

Fibonacci made few claims to originality in *De practica geometrie*. I suspect that he saw himself, rather, as a compiler of the best mathematics from Arabic and Grecian sources that would fit the aims of the treatise. In this chapter (in [21]), however, he did tell us that he found a definition of the line that is useful for subtracting radical lengths. Given straight line abg with $ab > bg$, then

$$(ag)^2 + (ag)(bg)^2 = 2(bg)(ag)^2 + (ag)(ab)^2.$$

He immediately applied the definition that we might want to call a property.

---

[3] Knorr suggests, "... such an error could be readily explained through Plato's association with Eudoxus, often spoken of as a disciple of Plato;" op. cit., 79.

SOURCES

The first part of Chapter 5, which focuses on finding cube roots and computing with them, is for the most part an excerpt from Chapter 14 of *Liber abaci*.[4] There as here you find the interesting statement, "Thinking about this definition a little more, I found this method of finding cube roots which I shall explain below."[5] Inasmuch as he wrote that he found the method, I would assume that he meant just that. Furthermore, this is certainly the thinking of Bortolotti who discussed the issue of originality, concluding with Eneström that the method was Fibonacci's creation.[6]

The source of the material on the Delic problem, however, is quite clear. On the one hand, Eutocius (ca. 560) had written a commentary on Archimedes' *Sphere and Cylinder* that had been translated into Arabic twice, first by Ishāq ibn Hunayn, then in a corrected edition by Nasīr al-Din al-Tūsī.[7] Because it contains several methods of duplicating the cube, it was available as a source for the second part of Chapter 5. On the other hand, Fibonacci certainly used sections of *Verba Filiorum* for propositions in Chapter 3 and would do so in Chapter 6, therefore he might at least have compared proofs by Archytas and Plato (ascribed to Menelaus by the Banū Musā) found in propositions XVI and XVII with other resources at hand. Finally, and not to be ignored, is *Istikmāl* (*Perfection*) by Yūsuf al-Mu'taman ibn Hūd who ruled Saragossa from 1081 until his death in 1085. This brilliant mathematician compiled in Arabic a source book of mathematics that includes *On the Sphere and Cylinder* of Archimedes.[8] This much is certain: a comparison of texts demonstrates that Archytas' method was not copied from the *Verba Filiorum*.[9] Apart from two lengthy omissions perhaps from between identical words and several possible typos, the Platonic method is taken word for word from Gerard of Cremona's translation called *Verba Filiorum*.[10] The word-for-word source for Philo's method might have been a translation by Gerard, as Clagett suspects, but not part of the Banū Musā opus.[11]

---

[4] *Liber abaci*, 378–387; Sigler (2002), 520–527. For a detailed analysis of Fibonacci's method, see Lüneburg (1993), 272–278,

[5] "Et cum super hanc diffinitionem diutius cogitarem, inueni hunc modum repperiendi radices, secundum quod inferius explicabo." *Liber abaci* 378.37–39; Sigler (2002), 520.27–29; *De practica geometrie* 149.11–14.

[6] Bortolotti (1929–1930), 40–41.

[7] Lorch (1995), I.95.

[8] Hogendijk (1996), 17.

[9] Clagett (1964), 334–340.

[10] Ibid., 340–44.

[11] Ibid., 658–661, 664.

TEXT[12]

[1] Since at the beginning of this treatise I had promised to discuss how to find cube roots, a topic to which I gave special attention in *Liber abaci*,[13] I rewrote the material for a regular chapter here so that I could more easily promote the purpose of this treatise. A number is a cube if it arises from the multiplication of three equal numbers or the square of a number and its root. For example: 8 and 27. 8 is the product of 2 times 2 times 2 or of four[14] times its root (2). 27 arises from three threes or from nine multiplied by its root (3). For the cube root of eight is 2 and the cube root of 27 is 3. So it is with the rest of the cubes and their roots. The other numbers that are not cubes can not have cube roots. Whence their cube roots are called surds.[15]

[2] Now it is possible to approximate the cube root of any number, as I shall show {p. 149}. First however I wish to show here how to find these roots. Given a line divided in two parts, the cubes of the parts added to three times the square of one section times the other[16] equals the cube of the whole line [see Figure 5.1]. For instance, divide line *ab* wherever you choose at point *g*. I say that the cubes of parts *ag* and *gb* with three times the square of part *ag* times *gb* and three times the square of *gb* times *ab* equals the cube of the whole line *ab*. This is obvious in numbers. Let *ab* equal 5, *ag* 3, and *gb* 2. The cubes of the parts are 27 and 8, and their sum is 35. Now three times the square of 3 times 2 is 54, and three times the square of 2 times 3 is 36. Thus we have 125 which is the cube of 5 or line *ab*. Now 5 is the cube root of 125, and its square is 25 which times 5 is 125.

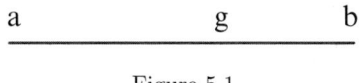

Figure 5.1

[3] Thinking about this definition a little more, I found this method of finding cube roots which I shall explain below. But first I want to show according to this definition how to find the cube of any number. For example, if you want

---

[12] "Incipit quinta de radicibus cubicis inueniendis" (148.31, f. 92v.33–34).
[13] Apart from a few insignificant changes in wording (e.g., *cuiusque* for *cuius*), the omission of a lengthy passage quite unrelated to finding cube roots on multiplying together three digit numbers (LA 379.18–380.6, Sigler 531.14–522.8), the addition of an equally unrelated lengthy passage on finding a series of numbers in continued proportion (DPG 153.16–155.31), Chapter 5 of *De practica geometrie* is the same, word for word, as the passage on finding cube roots and computing with them, in *Liber abaci* 378–387, Sigler (2002), 520–530.
[14] "ex multiplicatione quaternij" (148.38).
[15] The word "surd" is from the Latin "surdus" that means "deaf." It was often used to describe irrational numbers. Because rational numbers were expressed as a ratio and "ratio" in Latin also means "word," if the comparison of two numbers cannot be expressed, then it is "deaf."
[16] As the general example that follows indicates, the $3ab^2$ term is missing here (149.4, f., 93r.6).

260   Fibonacci's *De Practica Geometrie*

to cube 12, sum the cubes of 10 and 2, parts of 12, to 1008. Add to this three times the square of 10 times 2 or 600, and three times the square of 2 times 10 or 120. The final sum is 1728 which is the cube of 12. You can follow this procedure for cubing any number.

[4] Now I want to return to the method for finding the cube root of any number you choose. First you need to know which cube numbers are in the first place.[17] Now the cube of one is 1, of two is 8, of three is 27, of four is 64, of five is 125, of six is 216, of seven is 343, of eight is 512, of nine is 729, and of ten is 1000. Learn these by heart.[18] Then remember that the cube root of numbers of one and two and three digits is one digit, of four, five, and six digits is two digits, and of seven, eight, and nine digits is three digits. This order may be continued by adding one, two and three digits to the number which increases the digits in the root by one only.

[5] Once this is understood, I shall teach you how to find the difference or amount between two adjacent cube numbers. Multiply the root of one by the root of the other. Triple the product and add 1, (which is the cube of 1). The sum is the amount by which the root of the larger cube exceeds the root of the smaller. For example: I want to find out how much was added to the cube of 2 to get the cube of 3. Triple the product of 2 by 3 to get 18. Add 1 for 19, which is the desired amount. For, if you add 19 to 8 the cube of 2, you will have 27 the cube of 3.

[6] Now that this has been explained, let us approximate the cube root of 47. Subtract from 47 the closest cube, which is 27 with cube root 3. This leaves 20. Therefore the cube root of 47 is 3 with 20 left over. Now let 3 be represented by line ab [see Figure 5.2]. Next find the ratio of 20 and the amount by which the cube of 4 exceeds the cube of 3. You can find this amount by tripling the product of 3 by 4 and adding 1. Or you can subtract 27 from 64 to get 37; 20 is a little more than half of 37. So add a half or bg to line ab. Now to find the cube of the number ag. Cube the two parts ab and bg to obtain $\frac{1}{8}$ 27. To this add three times the square of ab multiplied by bg, and three times the square of bg multiplied by ab. That is, three times $\frac{1}{2}$ 13 increased by $\frac{1}{4}$ 2 to get $\frac{7}{8}$ 42. This sum is $\frac{1}{8}$ 4 less than 47. Therefore the cube root of 47 is $\frac{1}{2}$ 3 {**p. 150**} with remainder $\frac{1}{8}$ 4. If you make a ratio of this number to the number that comes from the product of thrice ag by 4 which is the root of the next cube, then the number is 42 of which $\frac{1}{8}$ 4 is nearly a tenth part. Whence add $\frac{1}{10}$ or gd to the number bg. Now subtract from $\frac{1}{8}$ 4 the cube of bg or $\frac{1}{1000}$, the product of thrice the square of ag and gd or $\frac{1\ 6}{4\ 40}$ 3, and the product of thrice the square of gd and ga or $\frac{1\ 10}{2\ 100}$. What remains is $\frac{344}{1000}$ or $\frac{43}{125}$. Therefore the cube root of 47 is $\frac{1\ 1}{10\ 2}$ 3 or $\frac{3}{5}$ 3 with a remainder of a bit more than $\frac{1}{3}$ of a whole unit.[19] If you take the ratio of this third to the product of thrice ad and 4, you get even closer to the root of 47.

---

[17] The word *place* is used exclusively to represent the place of a digit in a number. Because Fibonacci counted from right to left, the digit 4 in 234 is in the first place.
[18] "quibus per ordinem cordetenus cognitis" (149.22).
[19] Note the two forms of fractions in this section.

a	b	g	d

Figure 5.2

[7] Likewise: if you wish to find the root of 900, you already know that the number belongs to the third class and the root is a single digit. So subtract the cube of 9 or 729 from 900 to leave 171. Then find the difference between the cube of nine and the cube of 10, and it is 271. The ratio of 171 and 271 is a bit less than $\frac{2}{3}$. Subtract the cube of $\frac{2}{3}$ or $\frac{8}{27}$ from 171, leaving $\frac{19}{27}$ 170. Next multiply thrice the square of 9 or 243 by $\frac{2}{3}$; that is, take $\frac{2}{3}$ of 243 to get 162. Subtract this from $\frac{19}{27}$ 170 to obtain $\frac{19}{27}$ 8. Then square $\frac{2}{3}$ to reach $\frac{4}{9}$ that you triple to obtain $\frac{1}{3}$ 1. Multiply this by 9 to get 12. It cannot be subtracted from $\frac{19}{27}$ 8. So subtract $\frac{19}{27}$ 8 from 12 to leave $\frac{8}{27}$ 3 to be subtracted.[20] Therefore the root of 900 is $\frac{2}{3}$ 9, which errs by $\frac{8}{27}$ 3, because the cube of $\frac{2}{3}$ 9 is $\frac{8}{27}$ 903. You can check this by tripling $\frac{2}{3}$ 9 to obtain 29, cubing it, and dividing by 27 the cube of 3.[21] If you want to refine the cube root of 900, then multiply $\frac{2}{3}$ 9 by 10 and triple the product, or multiply thrice $\frac{2}{3}$ 9 by 10 to get 290. Now divide $\frac{8}{27}$ 3 by 290. Subtract the quotient from $\frac{2}{3}$ 9, and you will have what you wanted.

[8] Again: if you wish to find the root of 2345, you know that the integral part of the root has two digits. Whence the last digit must be put under the second place. I will show what this digit has to be [see Figure 5.3]. After setting aside the first, second, and third digits of 2345, the number 2 remains. The largest whole number root in 2 is 1, leaving a remainder of 1. Put the root 1 under the 4 and the remainder 1 above the 2. Join the 1 with 345 to make 1345. Thus for the root of 2345 you have 10, the 1 being in the second place. There remains 1345. The digit that you are going to place before the 1 will be under the 5. Whatever this digit is, it will be multiplied by thrice the square of the digit under the 4. Also the same digit will be multiplied by thrice the square of the digit you will position, and then it will be cubed. Subtract all of this from 1345. What remains is no more than thrice the product of the whole root thus far found (10) and the next digit in the sequence of numbers. You can find this digit only from much experience. It is 3 that you place under the 5. Triple the square of 3 and place it under the third digit, because when the second digit is squared, it becomes the third digit. Now multiply the 3 under the 5 by the 3 under the 3 to get 9. Subtract this from the 13 (the 1 under the 2 and the 3 that follows) to leave 4 that you place above the 3[22] in the third place. Triple the square of the 3 under the 5 to get 27 that you put in the second and first place. Because when the first digit is squared, it becomes the first digit or ends it. Now multiply 27 by the 1 under the 4 and subtract the product from the union of the 4 over the 3 and the following **{p. 151}** four (44) and 17 remains above the 44. Join the 17 to the 5 in the first place to get 175. Subtract the cube in the

---

[20] This is the error that appears in the next sentence.
[21] An interesting algorithm for cubing mixed numbers: (150.18–19).
[22] Text and manuscript have 5; context requires 3 (150.39, f. 94r.32).

third position under 5, namely 27, and 148 remains. This does not exceed thrice the product of the root that was found, namely 13 by the number following it, namely 14. Therefore the cube root of 2345 is 13 with 148 left over. Therefore, multiply 14 by thrice 13 and add 1 to get 547. Now 148 is a little more than a fourth of 547. Hence add $\frac{1}{4}$ to the root thus found, and there is $\frac{1}{4}$ 13. Subtract the cube of $\frac{1}{4}$ or $\frac{1}{64}$ from 148, leaving $\frac{63}{64}$ 147. Take thrice the square of 13 or 507 and multiply by $\frac{1}{4}$ to obtain $\frac{3}{4}$ 126. Subtract this from $\frac{63}{64}$ 147 to leave $\frac{15}{64}$ 21. Then take thrice the square of $\frac{1}{4}$ or $\frac{3}{16}$ which multiplied by 13 yields $\frac{7}{16}$ 2. Subtract this from $\frac{15}{64}$ 21 to leave $\frac{51}{64}$ 18. And thus for the root of 2345 you have $\frac{1}{4}$ 13 with $\frac{51}{64}$ 18 left over. Whence multiply thrice $\frac{1}{4}$ 13 by 14, the first following number, to obtain $\frac{1}{2}$ 556. Divide this into $\frac{51}{64}$ 18 to get about $\frac{1}{30}$ which you add to $\frac{1}{4}$ 13 to get the desired root: $\frac{17}{60}$ 13. Strive to do this in similar cases.

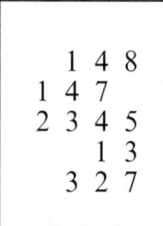

Figure 5.3

[9] If you wish to know the cube root of 56789, you already know that the root will have two digits. So ignoring the first three digits for the moment, subtract the cube of the root closest and below 56, which is 3. 29 remains [see Figure 5.4]. Put the 3 in the second place and the 29 over the 56. Triple the square of 3 to 27, which you put under the fourth and third places, because when a second digit is squared, it goes in the third place or ends in itself. Find a digit to precede the 3, so that when it is multiplied by 27 you can subtract the product from 203[23] (29 times 7). A number is left that joined with 8 the next digit you can subtract from the product of thrice the square of the digit so placed by 3 located as assigned. Thence the number remains which when joined with 9 in the first place, you can then subtract the cube of the number that will be placed. What remains is not more than thrice the product of the whole root that was found by the number that follows. Keep this in mind for every digit that must be positioned. Multiply the digit 8 placed at the first place by 27; that is first by 2 and then by 7. For 8 by 2 is 16, which subtracted from 29 leaves 13. 8 by 7 gives 56, which subtracted from 137 leaves 81. Afterwards square 8 to get 64 ending in the first place. Tripling 64 get 192 likewise ends in the first place. Whence you put them under the third, second, and first places. Multiply the last digit of 192 by the three under the second place, which puts 3 in the fourth place. Whence, subtract [3 from] the same 8, the last digit of 81, and 5 remains

---

[23] Text and manuscript have 297 that is not (29)(7) (151.22, f. 94v.25).

5 Finding Cube Roots   263

beyond the 8. Multiply 9 (the following figure in 192) by the same 3 to get 27 ending in the third place. Because when the second place multiplies the second place, it produces the third place. So, subtract 27 from 51 in the third place. 24 remains in the same place. Joining this to 8 produces 248. From this subtract [9] the product of the first place of 193 by the aforementioned 3, and 242 remains. Joining this to 9 produces 2429. From this subtract 512 the cube of 8 to leave 1917. Or in another way, square 8 to get 64. Place it below the second and first places {p. 152}. Multiplying 6 by 8 and subtracting the product from 242 leave 194 from the 242. Joining this to the following 9 makes 1949. From this subtract 32 (the first digits of the product of 64 by 8) and 1917 remains, which is exceeded by thrice the product of 38 by 39. Therefore the cube root of 56789 is 38 with a remainder of 1917. This adds $\frac{3}{7}$ to the 38 and a little more.

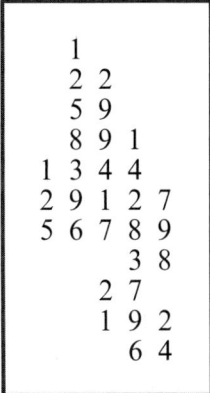

Figure 5.4

[10] Again: if you wish to find the root of 456789, set to one side the first three digits. The root of 456 (the remaining three digits) is 7 that you put in the second place. Put the remainder, 113, over 456. Put 147, thrice the square of 7 in the first place, so that it fits under the third place. Now by the method already described, try to find the digit that must be placed under the first place before the 7. You will find this to be 7; so put it under the first place [see Figure 5.5]. Now multiply it by the 1 of 147 to get 7 which you take from the 11 over the 45, leaving 4 above the fifth[24] place. Join this to the 3 that follows for 43. Subtract from this 28, the product of 7 and the 4 in 147, to leave 15 from the 43. Join 15 to 7 to get 157. Subtract from this the product of the same 7 in the first place and the 7 in 47. What remains is 108 of the 157. Then triple the square of the 7 in the first place to get 147 ending in the first place. You order the differences by the 7 in the second place according to what we taught about division.[25] Now 1 times 7 is 7. Subtracting this from

---

[24] Text has *quartum*; Figure 5.5 shows the 4 over the *fifth* place (152.12, f. 95r. 28).
[25] *Liber abaci*, 28; Sigler (2002), 54.

10 leaves 3 over the 0. 4 times 7 is 28. Subtracting this from 38 leaves 10 over the 38. 7 times 7 is 49. Subtracting this from 108 leaves 59 over the third and second places. Joining this to the 9 in the first place makes 599. Take from this 343, the cube of 7, and 256 remains. And thus you have found that the root is 77 with a remainder of 256.

```
              1
            3 2
          1 0 5
        4 5 0 5
      1 1 3 8 9 6
      4 5 6 7 8 9
                7 7
            1 4 7
              1 4 7
                4 9
```

Figure 5.5

[11.1] To continue: if you wish to find the root of 9876543, separate out the first group of three digits leaving 9876. After setting this aside, its cube root of 21 is found in the usual manner, leaving a remainder of 615. Position 21 under the third and second places because the root of a seven digit number has three digits. Put the remainder, 615, above 876 as shown here [see Figure 5.6]. Triple the square of 21 to get 1323, and place it so that it ends in the third place where also ends the square of the 1 that was put in the second place. With those numbers positioned, let the last digit be under the sixth place. Having found 4 by the previous instruction, put it before the 21. Then multiply 4 by each of the digits of 1323 in order. Begin by taking the 6 above the sixth place, because a digit in the first place times a digit in the sixth place puts the product in the sixth place. Therefore multiply 4 by 1, subtract the product from 6 leaving 2 above the 6. Then multiply 4 by 3 and subtract the product from 21 leaving 9 above the 1. Then multiply 4 by 2 [to get 8] to subtract from 95 leaving 87 above the fifth and fourth places. Next multiply 4 by 3 and subtract from 875 leaving 863 above the fifth, fourth, and third places. Then multiply the square of 4 by 3 to get 48 that you put in the second and third places. Multiply the 4 of 48 by the 2 of 21 located in the root of 8 that is subtracted from 86, the number ending in the fourth place. Because when a digit in the second place is multiplied by a digit in the third place, its product goes in the fourth place. 78 remains from the 86 above the fifth and fourth places. Multiply the same 4 by the 1 of 21 and subtract the product from 783, the number ending in the third place. Because when a digit in the second place {p. 153} multiplies a digit in the second place, it makes a digit in the third place. So 779 remains above the fifth, fourth, and third places. Then multiply the 8 that remains from the 48 by the same 21. That is, multiply 8 by 2 to get 16 that you subtract from the number in the third place. Because when you multiply a digit in the first place by a digit in the third place, you create a digit in the third place. So 763 remains over the fifth, fourth, and third places. [Secondly] multiply 8 by 1 to get 8 which you subtract from the number ending in the second place, namely from 7634. Join this with the 3 in the first place to make 76263. From this subtract 64, the cube of 4, and 76199 remains over the root that was found, namely 214.

Figure 5.6

[11.2] In the same way: if you wish to find the root of a number of eight or nine digits, set aside the first digits, and strive to find the root of the remaining digits by the method demonstrated here. Then after joining what remains of them with the three digits set aside, you can proceed along the way we have described, and you will find what you sought. If God wills, you will be able to find the roots of numbers with ten or more digits orderly and in the same way.

5.2 Finding Two Numbers in Continuous Proportion

[12] In order to find two numbers in continuous proportion between unity and some number, the first to be found will be the cube root of the last number, as has been shown in Geometry. Hence, if we want to find the cube root of some number geometrically, we do well to use the method of the Ancients.[26] They taught us how to put two quantities between two given

---

[26] The Banū Mūsā identified Menelaus as one of the Ancients; see Clagett (1964), 336.

quantities, so that all four are in the same ratio. Although this is not an easy thing to do, regardless we shall show the best way to do it[27] [see Figure 5.7]. Given two quantities, c and f. We wish to put two quantities between c and f in continuous proportion. Let f be greater than c. Construct circle $abgd$ with diameter ag equal to the quantity f [see Figure 5.8a]. Draw line gd equal to quantity c in circle $abgd$. Erect a semicolumn orthogonally to the semicircle adg so that its base sits on arc adg. Whence the diameter of the base of the column lies on diameter ag and its endpoints are a and g. Thus you have erected a plane surface of a semicolumn orthogonally upon arc ag. Within it I shall draw semicircle aeg whose plane is erected orthogonally upon the plane of circle $abgd$. From point a I draw line az making a right angle with line ga, meeting the extension of line gd at point z. I will rotate semicircle aeg from point a, and revolve triangle zag upon line ag which will be immobile during the movement. At the same time I shall move semicircle aeg about point g that remains immobile [Figure 5. 8b]. I shall do this until arc $a'e'g$ meets line gz' in the surface of the column at point h. Line gta' is the

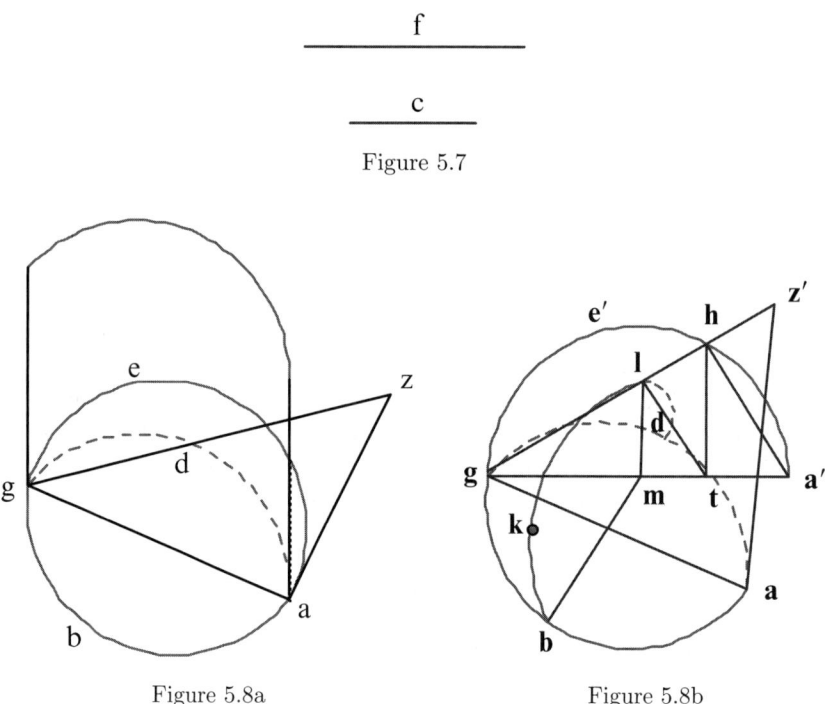

Figure 5.7

Figure 5.8a   Figure 5.8b

---

[27] What follows is Archytas' method for duplicating the cube; an English translation of the Greek text is in Thomas (1957) 285–289. As the text appears the letters identifying parts of figures that move are confusing, because the text does not differentiate initial and final positions of the figures. Hence, I have added primes to letters in the text for the final positions of the moving semicircle and triangle for setting up the ratios crucial to the derivation of the two mean proportionals. Furthermore, where the manuscript and transcription offer a single incomprehensible figure, I have prepared two figures (5.8a and 5.8b) to indicate the initial and final positions of the moving parts. It helps to imagine that triangle gaz' is transparent.

5 Finding Cube Roots

diameter of semicircle $a'e'g$. I will draw line ht to fall orthogonally along the surface of the column to arc atd, and I will draw line $a'h$. Then I understand that arc dkb has been marked by point d when triangle $z'a'g$[28] revolves, and point d falls on point b in its revolution. I shall draw line gh and mark the point where line gh meets arc dkb as l. I shall draw line lm to fall orthogonally upon circle $abgd$. I draw line lt. And because line lm in semicircle bkd has been drawn orthogonally to diameter bd, there is a surface bounded by bm and md (equal to lm). Now surface bm by md equals surface gm by mt, since lines db and gt intersect in circle {p. 154} $abgd$. Whence angle glt is a right angle. Now angle gth is a right angle because line ht is on the surface of the column. Because arc $ge'ha'$ is a semicircle, angle gha' is a right angle. Similarly, angle gml is a right angle. Therefore triangles gha', gth, glt, and gml are orthogonal, and they have one angle in common, namely that which lies under lgm. Therefore the triangles are similar to one another and have proportional sides about the common angle. Whence as $a'g$ is to gh, so is line hg to gt and tg to gl and lg to gm. The three lines gl, gt, and gh are indeed placed in continuous proportion between lines gm and ga'. Also, the two lines gt and gh are in continuous proportion between lines gl and ga'. Now line gl equals line gd. But gd equals line c. And ga' equal to ga equals line f. Therefore between quantities c and f two quantities gt and gh have been placed in continuous proportion. Whence if the quantity c is one, then the quantity gt is the cube root of the number f, as required.

[13] I will show you another way of finding two lines such that all the lines are in one continuous proportion.[29] Given two lines ab and bg at right angles to one another, I draw line ag. Then about line ag of triangle abg I draw circle abg. Next from point a I draw perpendicular ad to line ab and perpendicular dg from point d to line bg. Then I extend both lines straight on to points z and e, respectively. Next I make a transition rule for moving a line through point b but not apart from it, cutting the two immobile lines dz and de until one part of the line falling between points z and b equals the part that falls between points e and h.[30] Therefore line zb equals line eh. Therefore the product of zh and zb equals the product of be and he [see Figure 5.9]. But the product of eb and he equals the product of de and ge, and the product of hz and zb equals the product of dz and za. Therefore the product of dz and za is as the product of de and eg.

---

[28] Text and manuscript have *zaig*; context omits *i* (153.39, *f.* 96r.34).

[29] This is the method of Philo of Byzantium; see Drachman (1974), 587 col. b.

[30] This is the sense of what Fibonacci wrote: "deinde faciam transire regulam que moueatur super punctum *b*, et non separatur ab eo, et sic abscidens duas lineas *dz* et *de*, et non cesset moueri donec sit illud quod cadit ex eis inter *z b* equale ei quod cadit inter *e h*" (154.18–21).

Therefore the ratio of *de* to *dz* is as the ratio of *az* to *ge*. But the ratio of *ed* to *dz* is as the ratio of *ba* to *az*. And the ratio of *ba* to *az* is as the ratio of *az* to *ge*, and even as the ratio of *ge* to *bg*. And that is what we wanted to prove. This proposition was proved by the penultimate theorem in the third book of Euclid,[31] and by similar triangles, in this way: if line *gb* is set equal to some number and *ab* is set equal to one, then *az* will be the cube root of *gb*. For it has been proved that as *ba* is to *az* so is *az* to *ge* and as *ge* is to *gb*. Whence, between *gb* and *ba* fall two lines in continuous proportion, namely lines *ge* and *az*, Since line *az* follows one in the ratio, it is therefore the cube root.

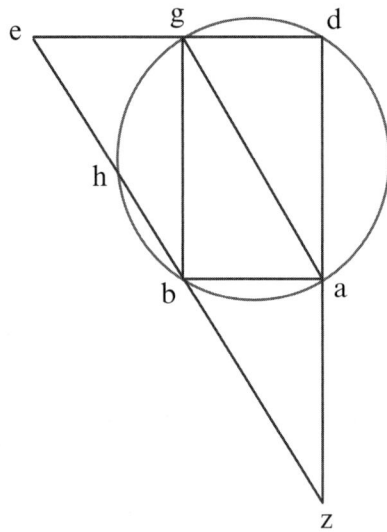

Figure 5.9

[14] Another way to find two quantities in continuous proportion between two other quantities *a* and *b* [see Figure 5.10]. Let line[32] *gd* equal quantity *a*, and draw line *de* at right angles to *gd*. And let line *de* equal quantity *b* and draw line *ge*. Extend the two lines *gd* and *de* straight on indefinitely. [At point *g* draw line *gm′* perpendicular to line *ge*.][33] At point *e* draw a line perpendicular to line *ge* and extend it until it meets the extension of line *gd*.[34] Call the point of intersection *z′*. Line *gz′* is equidistant from line *gm′*. Extend line *gm′* in the opposite direction to some point *p′*, such that the measure of line *m′p′*

---

[31] *Elements* III.36.
[32] Text and manuscript have *quantitatem*; context requires *lineam* (154.34, f. 97r.10).
[33] This sentence is not in the text; the construction is required for what follows.
[34] Hereafter I inserted′ to indicate initial position of certain points, because Fibonacci did not distinguish between initial and final positions in his explanation.

5 Finding Cube Roots 269

equals the measure of line ez'. Now move point z' down along line dz', all the while maintaining line ez as a straight line passing through point e fixed in place {**p. 155**}. Now mark point k on the extension of line ed. [As point z' moves down along line dz,] I imagine point m' of line $p'm'$ moving toward point k [along line dk] and inseparable from it. As in the movement of line ez' through point e, so line $p'm'$ moves through fixed point g. Further, I imagine the two lines ez' and pm' remaining equidistant as they move in unison [see Figure 5.11]. Additionally, I imagine that at e the end point of line $z'e$ a perpendicular has been erected orthogonally to line $z'e$, following it in its own movement. And I do not determine any endpoint to this line. This line continues to intersect line $m'p'$ by the motion of the two lines $z'e$ and $m'p'$. Since therefore the two lines $z'e$ and $m'p'$ are equidistant in their motion, their endpoints move along the two lines $z'd$ and $m'k$, as we stated. There is no doubt about the perpendicular to line $z'e$ moving with it and eventually intersecting line $m'p'$ at point p.

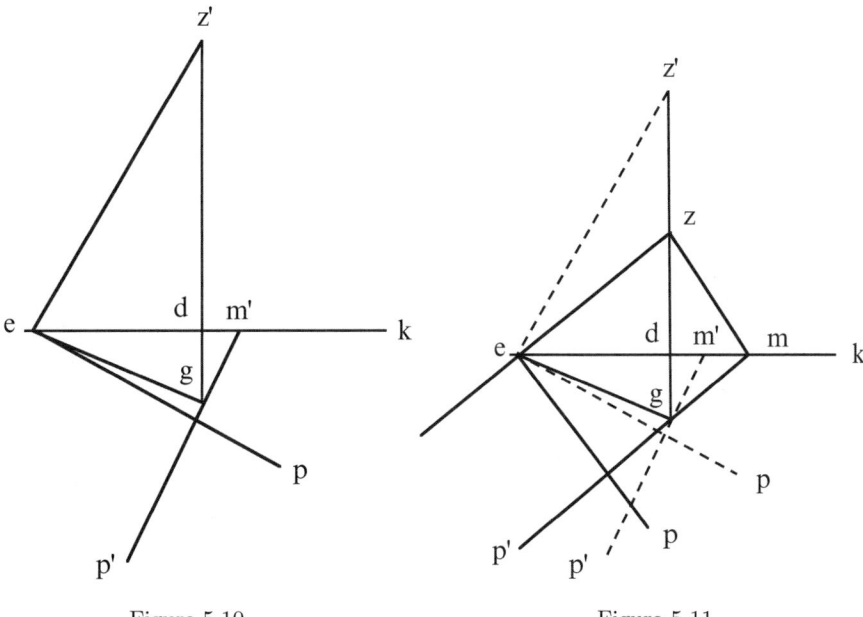

Figure 5.10   Figure 5.11

[15] When therefore the perpendicular line erected on point e arrives at point p, the two lines ze and mp also arrive in place, and we draw the two lines ep and zm. We know that line ep has been erected orthogonally to each of the two lines ze and mp, because at the start we erected it orthogonally on line $z'e$ that moved with it until it reached point p [see Figure 5.12]. I say that the two lines dm and dz are the two quantities which have fallen between the two quantities gd and de. And the ratio of gd to dm is as the ratio of dm

270   Fibonacci's *De Practica Geometrie*

to dz and dz to de. The proof follows. Because the two lines ze and mp are equidistant and equal, and the two angles zep and mpe are right angles, then line zm equals line ep. And each of the two angles ezm and mpe is a right angle. But line md is perpendicular to line zg and line zd is perpendicular to line me. Therefore the ratio of line gd to dm is as the ratio of dm to dz and dz to de. But line gd equals the quantity a, and line de equals the quantity b. Therefore the two lines dz[35] and dm fall between the two quantities a and b, and the same ratio is continued, as we wished to show.

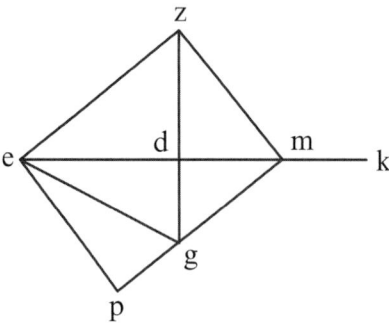

Figure 5.12

5.3 Computing with Cube Roots

Multiplication of Cube Roots by Themselves[36]

[16] If you wish to multiply the cube root of 40 by the cube root of 60, multiply 40 by 60 to get 2400. Then take its cube root to get what you want. And if you wish to multiply 5 by the cube root of 90, then cube 5 to get 125. Therefore if you wish to multiply the cube root of 125 by the cube root of 90, then multiply 125 by 90 and take the cube root of the product for what you want. Now, if you want to multiply two cube roots of 20 by three cube roots of 40, then rework them to one cube root of one number in this way: cube the 2 of the two roots of 20 to get 8 which you multiply by 20 to yield 160 the cube root of which equals two roots of 20. Likewise for the three roots of 40, cube 3 and multiply 27 by 40 reach 1080. Its cube root equals the three roots of 40. Finally, multiply 160 by 1080 and take the cube root to find what you want. Likewise, if you want to multiply the cube root of 20 by some number, such that the product equals a given number, say 10, then cube the 10 to get 1000. Divide this by 20 to yield 50 {**p. 156**} the cube root of which is what you want. And if you want to find two cube roots of numbers that are not cubes yet when multiplied together make a rational number, then cube

---

[35] Text and manuscript have de; corrected to dz in the margin of manuscript (155.29, f. 97v.18).
[36] "Incipit de multiplicatione radicum cubicarum inter se" (155.31, f. 97v.20–21).

5 Finding Cube Roots   271

the number as you wish. Find two numbers that multiplied together make a cubic number. Now the cube roots of these two numbers are sought. For example: cube 6 to get 216. Now find two numbers that multiplied together make 216, such as 9 and 24, and seek its cube root. Otherwise, take any two square numbers, such as 4 and 9, multiply each by the root of the other to get 12 and 18. The product of these numbers leads to the root of the cube number, as you wanted.

Division of Roots Among Themselves[37]

[17] If you want to divide the cube root of 100 by the cube root of 5, divide 100 by 5 to get 20 whose cube root is what you wanted. If you divide 5 by 20 you find $\frac{1}{20}$ whose cube root comes from the root of 5 divided by the root of 100. And if you want to divide 8 by the root of 32, take the cube of 8 or 512 and divide it by 32 to obtain 16 whose cube root is what you wanted. If you wish to divide the root of 80 by 2, then divide 80 by the cube of 2 to get 10. The cube root of this is what you wanted. Likewise if you wish to divide eight cube roots of 10 by three cube roots of 5, rework the large number of roots to just one root: for 8 roots of 10 equals the cube root of 5120, and for 3 root of 5 equals the cube root of 135.

Addition and Subtraction of Roots[38]

[18] The addition and subtraction of cube roots share characteristics with square roots, namely that some of them can be added or subtracted together and others cannot be. When cube roots have the same ratio among themselves, as one cube number to another, then the roots can be added and subtracted. Consequently if you wish to add cube roots that have the same ratio to one another as the cube numbers, then add the roots (of the numbers in the ratio) and cube the sum. Next multiply the cubed sum by the multiple that the cube roots are to the ratio of the cubes. For example: if you wish to add the cube root of 16 to the cube root of 54, and the ratio of their numbers is as the ratio of the cubes 8 and 27, then add the cube root of 8 to the cube root of 27 or 2 to 3 to get 5. Then cube this to get 125. Double this answer because 16 is twice 8 and 54 is twice 27 to reach 250. Whence, the cube root of this is the desired sum.

[19] If you wish to subtract these roots, namely 54 and 16, then subtract the root of 8 from the root of 27 to yield one. Multiply its cube or 1 by the aforesaid multiple, namely 2, and you have the root of 2 for the remainder of the desired subtraction. Likewise if you wish to add the cube root of 4 with the cube root of 32, the ratio of these numbers is 1 to 8, for each is four times the ratio of its cube. So, add the cube root of 1 to the cube root of 8 to get 3. Cube this

---

[37] "Incipit de diuisione radicarum inter se" (156.10, f. 98r.9).
[38] "Incipit de aggregatione et disgregatione radicarum" (156.20, f. 98r.20–21).

272   Fibonacci's *De Practica Geometrie*

to yield 27 which you quadruple to 108. The cube root of this number is the sum you wanted. And if you want to subtract the root of 4 from the root of 32, subtract the root of 1 from the root of 8 to get 1. Quadruple its cube to get 4 whose cube root remains from the desired subtraction.

[20] In another way: given line *ab* as the cube root of 32 and *bc* as the root of 4. I wish to know the sum *ac* {**p. 157**}. Because the line *ac* is divided at point *b*, there are the two cubes, *ab* and *bc*, with three times the product of *bc* and the square of *ab* and three times *ab* and the square on *bc*, all equal to the cube of line *ac* [see Figure 5.13]. Whence adding the cubes of the parts *ab* and *bc* or 32 and 4, gives 36. Now the square of *ab* (which is the cube root of 32) and the square of the cube root of 32 equal the cube root of 1024. Tripling this root or multiplying 1024 by 27 yields the cube root of 27,648. Multiply its root by *bc* or the cube root of 4 yields the cube root of 110,592 or 48. Or another way: the product of *bc* and the square of line *ab* always produces a cube number in similar circumstances. Whence, multiply 1024 by 4 to get 4096. Is cube root is 16 which multiplied by 3 is 48. Add this to 36 to reach 84. Likewise, the square of line *bc* namely the cube root of 16 multiplied by *ab* or the root of 32 produces the cube root of 512 which is 8. Multiply this by 3 to get 24 which added to 84 produces 108 for the cube of the whole line *ac*. Therefore *ac*, the sum of the root of 32 and the root of 4, is the cube root of 108, as we found in another way.

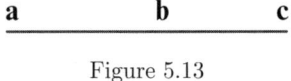

Figure 5.13

[21] There is another way to subtract the root of 4 from the root of 32. It requires an understanding of a particular definition of the line that I thought of. First, cube the whole line. Then add to it the product of the square of the [smaller] part and the whole line. (The product is called a solid number.) This sum equals the sum of twice the product of the smaller part and the square on the whole line (another solid number) and the product of the whole line and the square on the longer part (again a solid number) [see Figure 5.14]. For example: Let line *ab* be divided at *g* so that *ag* is 3 and *gb* is 2. Therefore the cube of line *ab* is 125. And the solid number that comes from the product of line *ab* and the square on line *gb*, is 20. Add this to 125 to get 145. Now 145 equals twice the solid number that comes from the product of line *bg* and the square on line *ab* together with the solid from the product of line *ab* and the square on line *ga*. The square on *ab* is 25. Multiplying this and *gb* gives 50. Double it for 100. Add 45, the product of lines *ga* and *ab*, to obtain 145, as required.

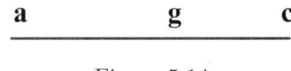

Figure 5.14

[22] With these definitions understood, to find the root of 32 consider the line *de*. Take part *ez* or the cube root of 4 from it, leaving *zd* as the unknown we want to find. Build on line *de* the equiangular, equilateral quadrilateral *cdef*. Mark point *b* on line *ef* so that *eb* is equal to line *ez*, and through point *b* draw line *ba* so that *ad* equals line *be*. Now, from point *z* draw line *zg* and let *gf* equal *ze* [see Figure 5.15]. With this done, cube line *de* to get 32. By squaring *ze* the cube root of 16 is found for the square *hzeb*. Erecting on its surface line *de*, that is, multiplying the plane surface by a line equal to *de* erected on it (namely the root of 16 by the root of 32). What results is the cube root of the product of 16 by 32. But 16 by 32 is the same as half of 16 by twice 32 or 8 by 64. But 8 by 64 is a cube number because 8 and 64 are cubes. Its root comes from the root of 8 by the root of 64 or 2 by 4. Thus we have 8 for the solid number, the product of the square on line *ez* times line *ed*. Adding this 8 to 32 the cube of line *ed* makes 40. If we subtract from this twice the solid number formed by the surface *ade* and the line *ed* (that is, two solids which come from the products of the aforementioned surface and the surface *zefg* times line *ed*) there will remain the solid {p. 158} which is the product of the unknown square *za* (or *cahg*) times the length of line *de*. For the solid number that is the product of surface *adzh* times *de* is the product of *be*, *ed*, and *ed*, which is the product of the square on *ed* by the length *ez*. Therefore multiply the root of 32 by itself to obtain the root of 1024. Multiply this by *be* or by *ez* which is the root of 4 to get the root of the number that results from multiplying twice 4 by half of 1024, that is 8 by 512. But the root that results from 8 by 512 is what you get from multiplying 2 by 8, namely from the root of 8 times the root of 512. Therefore the solid number that is the product of *de* by *ez* multiplied by *de* is 16. Subtract twice this from 40 to leave 8 for the solid number that arises from the square on line *zd* multiplied by the line *de*. Whence, if we divide 8 by the line *de* the cube root of 32, we get the cube root of 16 for the square on line *zd*, that is for the squared surface *ag*. Whence line *ha* or *zd* is the square root of the cube root of 16 or the cube root of 4. Therefore if we take the cube root of 4 from the cube root of 32, what remains is the cube root of 4, which we have now found by another method. And because line *zd* has been found equal to line *ze*, then the whole line *de* is twice line *ez*.

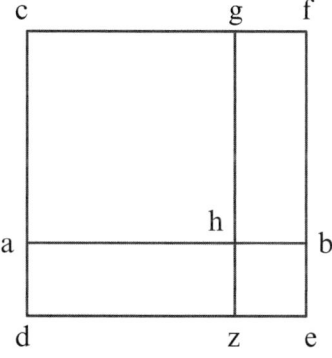

Figure 5.15

[23] From all of this it is clear that where some number is eight times another, then the cube root of the larger number is twice the cube root of the smaller. Whence the cube root of 32 is reduced to two roots of 4. When you wish to add a root of 32 to a root of 4, then you wish to add two roots of 4 to one root of four; their sum is three roots of 4. That is, the root of the number that came from multiplying 27 by 4. Similarly, if you want to subtract the root of 4 from the root of 32, then take one root of 4 from two roots of 4, to leave one root of 4 in the way we found. In order to express this better: add the root of 135 to the root of 1715, because their ratio is the same as the ratio of the cubes 27 and 343, for each of these is a fifth of its corresponding cube. Whence, take the roots (3 and 7) of the cubes. Say, for instance, that you want to add three roots of 5 to 7 roots of 5. Their sum is 10 roots of 5 or one root of 5000. If you wish to subtract the root of 135 from the root of 1715, subtract three roots of 5 from seven roots of 5, to leave four roots of five or one root of 320. Understand that there is a certain commonality[39] among these roots. The other roots for which there is no ratio cannot be added or subtracted. Whence if you want to add the cube root of 5 to the cube root of 3, the addition produces only the root of 5 and the root of 3. And if you wish to subtract them, what you have is the root of 5 less the root of 3. There is no nicer way to express this. And so I bring this chapter to an end {p. 158b}.

---

[39] "... inter se communicantibus" (158.30).

6
Finding Dimensions of Bodies

COMMENTARY

The second longest chapter, Chapter 6 may be described as Fibonacci's contribution on solid geometry. It offers a plethora of guidelines for finding the various dimensions of all sort of three-dimensional bodies, with alternate methods for many cases not to overlook proofs for many assertions. After a set of definitions that identifies foci of interest, Fibonacci announced the purpose and tripartite division of the chapter, to measure the dimensions of parallelepipeds, pyramids, and spheres. Thereafter he named all the Euclidean propositions from *Elements*, Books XI through XV [*sic*], that he might use. From there on he subjected his reader to methods for finding the surface areas and volumes of these figures: parallelepipeds, wedges, columns, pyramids, cones, spheres, and other solids. Noteworthy is that he used both arithmetic and algebraic tools to find dimensions. Fibonacci concluded the exercise with a discussion about finding the volumes of solids inscribed in one sphere, whereby knowing the volume of one inscribed solid the volume of another inscribed in the same or another sphere could be found. A few details about the three parts enhance their value.

Perhaps the first thing to notice about Part 1, *Measuring Parallelepipeds*, is that wedges, columns both polygonal and circular, and silos are included. In order to keep to the same computational method of area of base times height, Fibonacci positioned the wedge on a triangular side as the base. Furthermore, if the parallel quadrilateral bases of a solid are irregular, then he advised the reader to cut one base into so many triangles and proceed accordingly.

The next part detailing the measurement of pyramids both polygonal and conical,[1] begins with the standard method and then reminds the reader that the wedge is half of a parallelepiped and thrice the pyramid with triangular base. As in the previous part he investigated finding altitudes that are both within and without the pyramid, given only necessary lengths of edges. Toward the end of Part 2 is an investigation of sections of truncated pyramids and cones. In [42] Fibonacci finds a circle between and parallel to the bases of a truncated cone, and proves that the three circles are in continued proportion.

---

[1] Fibonacci used the same word *piramis* for both pyramid and cone. I used both words as called for.

In [43] he does the same for a truncated triangular prism, a third triangle in continued proportion between and parallel to the bases. In [44] he cuts a truncated triangular prism into three triangular prisms, and proves that they are in continued proportion. Figure 6.23 illustrates this last arrangement.

The third part begins with the measurement of spheres, the techniques being those well known for spheres, cones, and their parts. The figures in both texts are somewhat confusing because they are two-dimensional representations of solids. I found it more helpful to draw my own figures after the instructions in the text. At [51] Fibonacci once again skated close to infinite processes.[2] The problem is to measure the longer side of a right triangle formed by the extension of a radius of a semicircle, the shorter side being another radius with the hypotenuse joining the two endpoints. The solution is derived from the bases of three trapezoids constructed within the semicircle, one atop the other with a segment of the semicircle resting on the pile. Fibonacci proved that the external part of the extended radius equals the sum of the upper bases of the three trapezoids. The suggestion of an infinite process arises from the possibility of using an indefinite number of trapezoids to accomplish the same result.

In addition to offering the standard procedures for computing the volumes of the usual figures, Fibonacci developed the ratios of their volumes. In particular, he envisioned nested solids: cube, column, sphere, double cone, and octahedron, the first containing the next and so on. For any two he developed the ratio of their volumes, as shown in Table 6.1.

For the most part, the development of the ratios is straightforward except in the discussion of the sphere. Where at [70] Fibonacci laid the foundation for the ratio of the column to the sphere, unexplained letters slipped into the text, notably m and n. In order to understand them, their place and significance, I created Figure 6.55 displaying the letters. This suggests that Fibonacci wanted to emphasize the distinction between the axis of the column and the diameter of the sphere (hence, k and m are the same endpoint of the line-segment) yet to maintain their real identity (hence, i is the other endpoint). Another unusual feature: why did he recommend multiplying the

Table 6.1 Table of Ratios of Volumes

↓: →	Cube	Column	Sphere	Double Cone	Octahedron
Cube		14:11	21:11	42:11	6:1
Column	11:14		3:2	3:1	33:7
Sphere	11:21	2:3		2:1	22:7
Double cone	11:42	1:3	1:2		11:7
Octahedron	1:6	7:33	7:22	7:11	

---

[2] See Chapter 3 [246], but both proofs depend on a finite number of cases.

6 Finding Dimensions of Bodies   277

surface area of a sphere by a sixth of its diameter to find the volume? The answer lies in [54] where the surface area is found by squaring the diameter and then multiplying by $3\frac{1}{7}$.

The measurement of the side and chord of a regular pentagon is the focus of the near final section of Part 3. To accomplish this, Fibonacci turned to the early propositions in *Elements*, XIII, particularly Prop. 8 that measures a side by a part of the chord. That is, if the chord is cut in mean and extreme ratio, the longer part measures the pentagonal side. The method is simple: given the measure of the square on the chord, say 12, increase its size by a fourth. Now the side of the new square is $\sqrt{15}$. Then the longer length is the root of the new side less the root of the fourth, $\sqrt{15}-\sqrt{3}$. That is to say, $\sqrt{12}$ is to $\sqrt{15}-\sqrt{3}$ as $\sqrt{15}-\sqrt{3}$ is $\sqrt{15}$ to $(\sqrt{12}-\sqrt{3})$.

SOURCES

Again, one of the Arabic translations of Euclid's *Elements* played a prominent part in the preparation of this chapter, particularly a copy containing the extraneous Chapters XIV and XV. No place else in this treatise did Fibonacci quote at such length so many propositions from this classical work. Furthermore, paragraphs [45]–[47], [49]. and [51]–[54] reflect Propositions VIII to XV of *Verba Filiorum* by Bānū Musā. In paragraph [45] Fibonacci referred to the *Spherics* of both Menelaus and Theodosius, probably Book I of the latter.[3] Without ignoring the possibility that he consulted Abraham's *On Mensuration and Calculation*, I found nothing that suggested even moderate dependence on the Spaniard's treatise.

TEXT[4]

[1] There are many kinds of bodies we discuss here: solids, serated bodies, pyramids, columns, spheres together with their parts and other bodies identified by their bases and described about the spheres. Properly speaking a solid has length, width, and height resulting in six sides, such as bricks, cabinets, cisterns, and the like.

[2] A SERATED BODY [or prism] is half of a solid consisting of three parallel edges and two triangles, formed by a plane cutting through the solid [here, a prism] along two diameters resulting in six bases for two equal and separate parts called SERATES or WEDGES {p. 159}. A PYRAMID is a figure brought from a base of whatever form to one point. A COLUMN is a figure erected perpendicularly

---

[3] Because Fibonacci wrote "in libro Miles et Theodosii," it may be that both tracts were in the same codex (179.12).

[4] "Incipit distinctio VI^a in dimensione corporum" (158.35, If. 99v.30).

upon a circular base and ending in a circle equal to its base, or it is formed by rotating a rectangle around one fixed edge back to its starting position. It is also called a WELL ROUNDED FIGURE. The PYRAMID OF A COLUMN is formed by rotating a right triangle completely around a fixed side of the right angle.[5] If the fixed side equals the other side, then the figure is a right pyramid. If it is longer, then it is an acute pyramid. If it is shorter, then it is an obtuse pyramid. The base always is a circle. A SPHERE is a figure that is entirely round, commonly called a ball.[6] It is formed by rotating a semicircle completely about a fixed diameter, and its center is the center of the semicircle, from which all lines going from it to its surface are equal. A HEMISPHERE is half of a sphere, the base being a major circle rotated about its diameter, and endpoints called the poles of the sphere. A SPHERICAL PART is more or less than a hemisphere with a minor circle as its base. POLYHEDRA classified according to their faces[7] are of many kinds: those with 8 faces, 12 faces, and 20 faces. Euclid discusses the construction of these within a sphere in Book XIIII.[8] There is nearly an infinite number of other kinds of bodies whose dimensions can be found by techniques that we discuss here for the aforementioned bodies, methods that measure the bodies perfectly.

[3] We divide this chapter into three parts. In the first we will measure solids whose sides are equidistant and sit upon their bases so that the forms are raised on high from equal bases but with dissimilar sides. In the second part we will measure pyramids and their parts. In the third we will measure spheres, their parts, and polyhedra lying within spheres. Note that when we speak of a body having so many ells or palms or digits or ounces or some other unit of measurement, then understand that these are corporal or cubic measures that fit the three dimensions of width, length, and height. A corporal or cubic palm is for right angled figures and indicates one palm in width, length, and height. The same holds for the ell and other units of measure. So that when we wish to talk about them, they will be shown with certain proofs.

[4] Certain of these figures are in the eleventh and following books of Euclid and proved most appropriately, among which are these.[9] A straight line cannot be in a plane and on high (XI.1). When two lines intersect each other, they are in one plane, and every triangle is one plane (XI.2). The intersection of two planes has a straight line in common (XI.3). If a line is perpendicular to the intersection

---

[5] Obviously this is a cone, a word I used throughout the chapter (159.5).
[6] "...quam vulgo pallam dicamus" (159.9). "palla" is an Italian word meaning "ball, bullet, bowle, or round packe" Florio (1611), 352.
[7] Fibonacci wrote "Solida multarum basium," clearly polyhedra. Hence, where he used the Latin word for *base*, I employ *face*.
[8] Folkerts corrected this to Book XIII; see his (2006) IX, 12n49. On the other hand, Fibonacci may have had Book XIV in mind because of the propositions listed there; see [9] above.
[9] Fibonacci listed so many theorems that it seemed more useful to number them within the text rather than by footnotes. Thereby they can be compared with the propositions in Busard (1987) 331 ff., keeping in mind that the contents rather than the wordings are similar.

point of two straight lines, it is perpendicular to the whole plane (XI.4). Where a straight line stands at the intersection of three lines forming right angles, these three straight lines are in one plane (XI.5). If they are perpendicular to the same plane, they are equidistant from each other (XI.6). Between two equidistant straight lines another straight line can pass in their plane (XI.7). Any line perpendicular to any plane of equidistant lines is perpendicular to the same (XI.8). Lines equidistant to any other line but not all in the same plane are equidistant from each other (XI.9). If two lines containing an angle are equidistant from two other lines containing an angle {p. 160} but not in the same plane, then they are equal[10] (XI.10). It is impossible to erect two perpendicular lines from the same point in any plane (XI.13). If two lines not both in the same plane contain two planes and are equidistant from two other lines containing another plane, then the planes are equidistant (XI.15). If the planes intersect two equidistant planes, then their common sections are equidistant (XI.16). If equidistant planes intersect two lines, their common lines of intersection are proportional (XI.17). If several planes are drawn through a line perpendicular to some other plane, they are perpendicular to that plane (XI.18). If several planes intersecting one another are perpendicular to some other plane, their common line is perpendicular to that plane (XI.19). If a solid angle contains three planes, any two taken together are larger than the third (XI.20). All plane angles contained by a solid angle are less than four right angles (XI.21). If equal lines contain three plane angles of which two taken together are greater than the third, a triangle can be formed from their chords (XI.22).

[5] For all solids with equidistant faces, the faces opposite one another are equal (XI.24). If a plane with equidistant sides intersects a solid with equidistant sides, it necessarily intersects two opposite and equidistant sides of the other plane according to the ratio of their faces (almost XI.25). If a plane intersects a solid with equidistant sides through two diameters of opposite faces, it necessarily cuts the solid in half (XI. 28). Solids with equidistant sides sitting on the same base and having one and the same altitude are equal (XI.29). Solids with equidistant sides sitting on the same base but not on the same line yet having the same altitude are equal (XI.30). Solids with equidistant sides sitting on the same base with equal lines drawn from the base and having the same altitude are equal (XI.31). Solids with equidistant sides sitting on the same base but without lines drawn perpendicularly from the base yet with the same altitude are equal (similar to XI.25). The ratio of solids with equidistant sides sitting on the same base having the same altitude equals the ratio of their bases (XI. 32). For solids with equidistant sides: if they are equal to the lines erected perpendicularly from the bases, then the bases and altitudes are mutually proportional. For any solids with equidistant sides: if they are equal, then their bases and altitudes are

---

[10] "they" refers to the angles. The text and manuscript omit the verb "continebunt" (my "are") as read in Busard (1987), 316 (160.1, f. 100v.23).

mutually proportional. And if the bases and sides are mutually proportional, then the solids are equal. Between similar solids of equidistant faces: the ratio of the solids to one another is triple the ratio of their sides (XI.33). Those solids are similar that have equal angles, and the sides about the angles are proportional.

[6] If two lines are raised up from two equal plane angles to form two angles above with the angular lines below, they are equal to the two angles contained similarly placed here and there (XI.35). If two perpendicular lines are dropped on a plane containing angular lines from points however selected in lines raised up, then the points on which the lines fall contain the lines within those angles; further, these lines make equal angles with the lines that were raised up (XI.34-porism). Between solids with two equidistant sides of which one contains three proportional lines, the other has equal sides, their halves make three proportionals, their angles are equal, and they are equal. If the ratio of similar solids with equidistant sides are constructed from the same information, they are proportional. If similar solids with equidistant sides are similar, the lines used to construct them are proportional {**p. 161**}.

[7] If two planes are described by all the sides of two cubes with opposite sides divided through the middle cutting the cube itself, then their common section cuts the diameter of the cube in half (XI.38). If two wedges of equal height but one with a parallelogramic base and the other with a triangular base equal to the parallelogram, both wedges are necessarily equal. A triangular pyramid with equal bases is similar to a pair of wedges having the same height; the ratio of the bases is that of one wedge to the other. The ratio of one pyramid to another, both with the same height, is that of the base of one to the base of the other (XII.5). Every wedge can be divided into three equal pyramids with triangular bases (*nearly* XII.7). If their triangular bases are equal, then the bases and altitudes of all such pyramids are mutually proportional. If the bases and altitudes [of pyramids] are mutually proportional, they are considered[11] equal pyramids. Similar pyramids with triangular bases are in triplicate ratio as their corresponding sides (XII.8). Columns on one base or on equal bases have the same ratio at their axes (XII.11). Well turned columns of equal bases and well formed, dense pyramids[12] with the same base have thrice the ratio of their heights (XII.14). Every column and well formed pyramid with the same base and axis have thrice the ratio of the diameters of their bases (XII.12). If every pyramid and well turned column of one base and axis have the same altitude as the pyramid, then the ratio of the pyramid to the column equals the ratio of their bases. If a pyramid and well turned column on one base and with the same axis equal another pyramid and well turned column with one base and the same axis, then their altitudes are mutually proportional (XII.14). But if their

---

[11] Here and below: "...equales esse conuertitur" (161.12–23).
[12] Here (161.15) and in the next several propositions where Fibonacci joins pyramids to columns, Euclid has cones and columns.

bases and altitudes are mutually proportional, then they are considered to be equal (XII.15). The ratio of spheres is thrice the ratio of their diameters (XII.18).

[8] Given a line divided in mean and extreme proportion: if the longer part is extended to equal half the divided line, then the square on the half is five times the square on the half line (XIII.1). Given a line so divided in two parts: if the square on the whole line is five times the square on the shorter part, then the longer part has been prolonged to twice the shorter part. Given a line divided in mean and extreme ratio: the longer part is the first part of that line. Given a line divided in mean and extreme ratio: if half of the longer part is considered equal to the shorter part, the square on the line is five times [the square] on half the longer part (XIII.3). Given a line divided in mean and extreme ratio and extend by the length of the longer part, then the whole line is again cut in mean and extreme ratio with the original line being the longer part (XIII.5). Given a line divided in mean and extreme ratio: the square on the line together with the square on the shorter part is three times the square on the longer part (XIII.4). For every rational line divided in mean and extreme ratio, each part is an apotome[13] (XIII.6). If the three angles of an equilateral pentagon are equal, then all the angles are equal (XIII.7). For a triangle inscribed in a circle, the square on a side is three times the square on radius (XIII.12). If the side of a hexagon is added to the side of a decagon that has been inscribed in a circle and the increased line be divided in mean and extreme ratio, then the longer side is the side of the hexagon (XIII.9). The side of an equilateral pentagon inscribed in a circle can be placed over the side of a hexagon and the side of a decagon. If the chords of two angles of an equilateral pentagon inscribed in a circle intersect each other, they do so necessarily. At the point where a line has been divided in mean and extreme ratio, and the longer part equals the side of a pentagon {p. 162} inscribed in a circle with a rational diameter, then the longer part is irrational and is called the minor line (XIII.11). If a solid of 20 equilateral triangular faces is constructed within a sphere with a rational diameter, then its side is irrational and is called the minor line (*within* XIII.16). If a solid of 12 pentagonal faces is constructed within a sphere with a rational diameter, then its side is rational and is called apotome (XIII.17).

Theorems from Book Fourteen[14]

[9] If the side of a hexagon is divided in mean and extreme ratio, the longer part is the side of a decagon (XIV.1). If a perpendicular is drawn from the center of a circle to the side of the pentagon, it is half the side of the hexagon and half the side of the decagon (XIV.2). The square on the side of a pentagon together with the square on the chord from its angle is five times the square on the side of the hexagon (XIV.3). We assume these things demonstrated, because the same circle contains the solid pentagon of 12 pentagonal faces

---
[13] Fibonacci used the word "residuum" from the Italian "residuo" (161.35).
[14] "Incipiunt theoremata quartidecimi libri" (162.6, *f.* 102r.26).

and the solid of 20 triangular faces constructed in the same sphere (XIV.4). In an equilateral pentagon inscribed in a circle the surface bound by the side of the pentagon and the perpendicular from the center of the circle to the side of the pentagon is 30 times the surface of the solid with 12 pentagonal faces (XIV.5). It can be proved in a similar way, because if the circle contains an equilateral triangle and a perpendicular is drawn to its side from the center of the circle, the surface contained by the side of the triangle and the perpendicular is 30 times the surface of the solid of 20 triangular faces (XIV.6). With these things established we assume that as the ratio of the area of the dodecahedron is to the area of the icosahedron (both constructed n the same sphere), the side of the cube in the sphere is to the side of the icosahedron (XIV.7). Then we assume, because of the ratio of the side of the cube to the side of the icosahedron (both inscribed in the same sphere), in the line divided in mean and extreme ratio the square on the longer part is to the square on the shorter part. Because of this we conclude because of the ratio of the side of the cube to the side of the icosahedron, the dodecahedron is to the icosahedron (XIV.9). For whatever determines which lines will be divided in mean and extreme proportion, the same holds for every line so divided (XIV.10).

[10] After the proofs of all these propositions, we need to know the following operations described in the same books of Euclid. To erect a perpendicular on a given surface from a given point above to another given point above. To erect a perpendicular from a given point on a surface. Given a three plane angle of which any two are greater than the third, to construct a solid angle; whence the three angles are necessarily less than four right angles. On a designated point on a given line, to construct a solid angle equal to a designated linear angle. To construct a solid on a given line that is similar to a solid whose faces are equidistant. To construct a multiangle figure between two circles with one center and barely tangent to the inner circle. To construct a multifaceted solid between two spheres with a common center and tangent to the smaller sphere. To construct solids having four, six, eight, twelve, and twenty faces all together within a sphere. Finally, to construct certain of these five solids within their spheres, as is described in the fifteenth book[15] {p. 163}.

6.1 Measuring Parallelepipeds

[11] When you want to measure any body belonging to this first part, look carefully for the area of the base by a method that I taught above in the section on areas. Multiply it by the height of the body, and what is produced will be the volume[16] of the body. The altitude of the bodies belonging to this

---

[15] From Book XV and all solids being regular: (1) to construct a tetrahedron within a cube; (2) to construct an octahedron within a tetrahedron; (3) to construct an octahedron within a cube; (4) to construct a cube within an octahedron; (5) to construct a dodecahedron within an icosahedron.

[16] Fibonacci used the word "embadum" here for volume (163.3). The Latin word is used for both volume and area, distinguishable according to the context.

first part is that straight line perpendicular to the base and terminating in the plane of the other, equidistant base. This is easily shown. Consider first the cube *aeg* of 10 palms[17] in length, width, and height [see Figure 6.1]. Its six faces are squares with right angles and equal sides. Its base is the square *ac* equidistant from square *eg*. The area of the base is 100. So multiply it by line *ae* the altitude of the cube. And you have 1000 cubic palms for the volume of this cube. If you want to find line *hb*, the diameter of the cube, square the two edges *ba* and *ad*, add them together, and you have 200 for line *db*. Add to this 100, the square on line *dh* for 300 the square on the diameter *bh*. Therefore the diameter *bh* is the root of 300. Note that the line *bh* is also the diameter of the sphere in which the cube *aeg* is inscribed. Knowing only the diameter *bh* and you want to find the side of the cube and its area, then square the diameter, take a third of the product and you have the side of the cube that is a square. If you multiply the area of the square by its root, you have the volume of the cube.

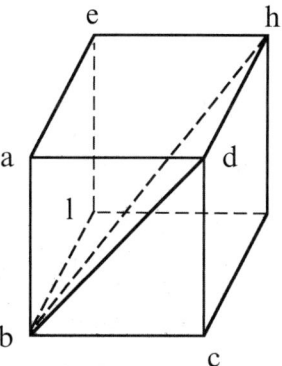

Figure 6.1

[12] If someone should say: I have joined the square on the diameter of the cube with the square on its side and got 400. Then I set the edge *ab* equal to thing that I square to get a *census*. To this I add three times the square on the diameter to obtain four *census* to equal 400 drachmas. Divide the 400 by 4 to get 100 for the square on *ab* the edge of the cube, and the edge is 10. Likewise I add the square on the diameter with its edge and obtain 310. Again, I set the edge *ab* equal to thing or root. Whence the diameter *bh* equals three *census* or thrice the square on line *ab*. Therefore you have three *census* and one root equal to 310 for the square on the diameter of the cube and its edge. Return this to one *census*. That is, take a third part of what you added to get a *census* and a third of the root equal to $\frac{1}{3}$ 103. Half of that third of the root is $\frac{1}{6}$ that squared is $\frac{1}{36}$. Add this to $\frac{1}{3}$ 103 to obtain $\frac{13}{36}$ 103. Its root is $\frac{1}{3}$ 10. Subtract the sixth for that which you multiplied and 10 remains for edge

---

[17] A unit of measurement equal to 12 inches, common throughout Italy (Zupko [1981] 183).

284   Fibonacci's *De Practica Geometrie*

ab.[18] Its square is 100. Triple this and you have the square of the diameter bh. In this way we have found several subtle and similar things about cubes.

[13] Again: there is a solid whose base is a square with an edge equal to 10 feet. The height is more or less than the edge of the base. Multiply the area of the base by is altitude that is 12 to obtain 1200 feet for the volume of the solid. You have found its diameter to be the root of 344, assuming that you added the square of its altitude to 200 that is twice the square on its edge, to get the 344. If it is proposed that the diameter of the solid with equidistant sides is the root of 344, its base is a square, and the altitude is two more than the edge of the base, what then is the edge of the base and the volume of the solid? [See Figure 6.2.] Consider solid aei with diameter tb. Make the edge ab the root to which you add the square on line ab with the square on line ad, to produce two *census* for the square on line db. Mark point c on line dt, and let ct be two. Therefore dc remains equal to line da. Whence dc {**p. 164**} is the root. And because line dt has been divided in two at point c, the two squares on lines dc and ct with twice the product of tc by cd equals the square on line td. For the square of dc is a *census*, the square of ct is 4, and twice tc by cd is 4 roots. Therefore the square on line td is a *census*, 4 roots, and 4 drachmas. Add to this two *census* or the square on line db, and there is three *census*, 4 roots, and 4 drachmas for the square on the diameter tb to equal 344. If 4 drachmas are taken from both parts, what remains is three *census* and 4 roots equal to 340. Whence if all of this is returned to one *census*, we have *census* and $\frac{1}{3}$ 1 roots that equals $\frac{1}{3}$ 113. Whence if we halve $\frac{1}{3}$ 1, we get $\frac{2}{3}$ that squared and increased by $\frac{1}{3}$ 113 yields $\frac{7}{9}$ 113. The root of this is $\frac{2}{3}$ 10. Taking $\frac{2}{3}$ away we are left with 10 for side ab. Whence the altitude dt is 12. If we multiply this by 100, the square ag, we get 1200 for the volume of solid aei.

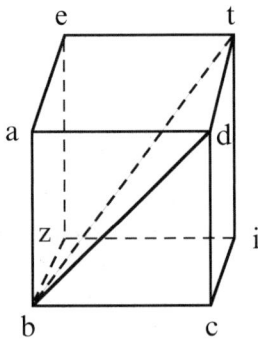

Figure 6.2

[14] Again, let the base ag be a square with height dt two more than side ab. I add the squares on lines ba, ad, and dt to the square on the diameter tb. The total is 688. Because the square on the diameter tb is the result of adding the squares on lines ba, ad, and dt, if we halve 688 we get 344 for the square on

---

[18] Fibonacci rounded the answer to 10.

the diameter *tb* that we worked with above. Now, add the lines *ab*, *bd*, and *dt* to the square on diameter *tb*, and the sum is 376. Again I set edge *ab* as the root. Whence the sum of *ab*, *bd*, and *dt* becomes three roots and two. Adding this to the square on diameter *tb*, namely to three *census* and 4, we have three *census*, seven roots, and 6 equal to 376. If 6 is subtracted from both groups, what remains is three *census* and seven roots equal to 370. Returning this to one *census* produces *census* and $\frac{1}{3}$ 2 roots equal to $\frac{1}{3}$ 123, and so on.

[15] For a similar solid we can pose various problems that will be solved in this way [see Figure 6.3]. Let there be a rectangular parallelepiped[19] *aeg* whose sides are raised perpendicularly from the base. Let its width *ba* be 10, length *ad* be 11, and height *dh* be 12, and let the quadrilateral *ac* be a square. Whence if its volume is the product of *ba* by *ad* by *dh*, then we have 1320 for the volume of the solid. By adding the squares on lines *ba*, *ad*, and *dh*, we obtain 365 for the square on the diameter. And if in a similar way we make the diameter equal to the root of 365, the length of the base exceeds its width by one, and the altitude *dh* exceeds the same width *ab* by 2. And since we want to find the base and altitude of the solid, we make *ab* the root. Whence the edge *ad* is the root increased by one, and *dh* the altitude is similarly the root increased by two. Whence the *census* results from the square on line *ab*, that is, *ab* times *ab*. The square of *ad* yields a *census* and two roots increased by one. And the square on *dh* gives a *census* and 4 roots and four. Thus we have three census and six roots increased by 5 for the square on the diameter *hb* that equals 365. If 5 is taken from both groups, then 360 remains equal to three *census* and six roots. Whence a *census* and two roots equal 120. Squaring half the root yields one that added to 120 is 121. Take one from its root gives 10 for *ba*. Whence *ad* is 11 and *dh* is 12.

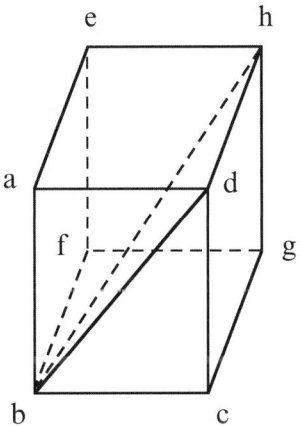

Figure 6.3

---

[19] "solidum equidistantium laterum" (164.25). The context makes it clear whether the rectangular parallelepiped is entirely rectangular or just two pairs of opposite bases are rectangular, the other pair being parallelograms. In any case, all pairs of opposite sides are equidistant.

[16] If something similar were said: I add the squares on lines *ab*, *ad*, and *dh* with the square on diameter *hb* so that the total is 730 {**p. 165**}. Half of this is 365 for the square on diameter *hb*. Then you work on this as above. Suppose it were said: I added the height *dh* with the square on the diameter *hb* and got 377. Further, *dh* is one more than *da*, and *da* is one more than *ab*. Likewise I call *ba* thing. Whence *ad* is thing increased by one and *dh* is thing increased by two. The squares on the three lines, as we showed above, are three *census* and six roots and 5 equal to the square on diameter *hb*. If the altitude *dh* is added to this and being a root increased by 2, there are three *census* and seven roots increased by seven that equal 377. Take 7 from both groups, and there remain three *census* and seven roots equal to 370. Whence a *census* and $\frac{1}{3}$ 2 roots equal 123 and $\frac{1}{3}$. Then proceed as above.

[17] Now if you want to know how it is that the volume of similar solids[20] results from the product of the area of the base by the altitude, consider this. Since one foot measures the altitude of a solid and the boundaries of the face equidistant from the base are similarly measured in feet, then there are as many cubic feet between the base and the face as there are in the area of the base. Similarly, if we double the number of feet in the height between the base and another similar equidistant face, then between that face and the base there will be as many cubic feet as there are in twice the number of feet in the base. Similarly, if we triple the altitude to a third similar face, there are as many cubic feet in this solid as there are in thrice the number of feet in the base. Therefore, because the height of any solid is a multiple of a solid whose altitude is one foot, so the number of feet in the volume is a multiple of the feet in the area of the base. Therefore when the area of the base is multiplied by the height of the solid, the result is the volume of the solid, as was stated. Although there are solids whose sides are not perpendicular to their bases, if their bases and altitudes are equal, then they are equal.[21] Indeed this is the same case where a base has been raised to create the volume of a solid. The same can be understood for any body in this first part.

[18] Again, consider the rectangular parallelepiped *dhf* with a rectangular base *abcd* from which are drawn lines *ae*, *bf*, *eg*, and *dh* not at right angles to the base, so that angles *abf* and *dcg* are acute angles. Whence angles *eab* and *hdc* are obtuse. Further the planes *af* and *dg* are both rhombi or rhomboids. The remaining two planes *fc* and *ed* may or may not be rectangular. But whatever they are, if we find the height of this solid and multiply it by the area of the base, we will definitely have the volume of the solid. Now I shall briefly indicate how to find the altitude of this body [see Figure 6.4]. From point *f* draw cathete *fi* to line *ab*. Now *fi* is either a cathete to plane *ac* or it is not. If it is, then line *if* is the height of the body. But if line *if* is not the

---

[20] Fibonacci is referring to a rectangular block of height 1.
[21] *Elements*, XI.31.

cathete, then draw through point i a line ko at right angles to line ab. On this line [ko] draw a cathete from point f, and there is the height of the body. Or in the common fashion: drop a plumb line from point f to the base ac. Thus you have the cathete from point f, whether it falls on the base ab or outside it. But if the body is so dense that the plumb line cannot fall at right angles within the body from point f to the base ac, then select some point h or e so that it will fall outside the body. But if we want to do this according to the art {**p. 166**} we have to use the instrument of sticks.[22] Erect a stick at point b perpendicular to the base ac; call it bm and its length is as noted.[23] At m fix another stick mn at right angles. Let the line of be the cathete for the solid dhf. The ratio of bm to of is as bn to bf. Therefore the length of line bf is known and, consequently, the length of cathete fo. We can show all of this in numbers. Let the edges ab be 18 and bc be 12 and each of the edges ae, bf, cg, and dh be 20. We wish to find the cathete fo. Let the measure of stick bm be 4 and of stick mn be 3. Whence bn is 5. I draw the line bo, and triangle fob is similar to triangle bmn. Whence as fb is to bn, so fo is to mb. The quadruple of fb of bn equals the quadruple of fo of mn. Therefore fo is 16. The product of ab and bc is 216. Multiplying this by 16 or fo yields 3456 for the volume of solid dhf.

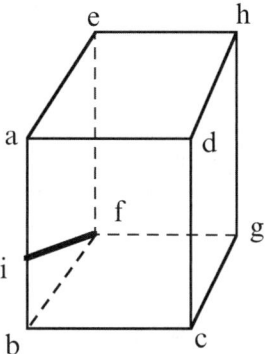

Figure 6.4

[19] If we want to measure a wedge,[24] which some people call "cunus", we multiply [the area of] its triangular base by its altitude, as is done with the solid of six bases. To make this clear: let there be a wedge $abgdez$ with triangular base abg. And let lines ad, bc, and gz equal one another and be at right angles to the triangle abg. Whence triangles dez and abg are equal and equidistant. Further,

---

[22] "uti arundinum instrumento" 166.1. Some places this is translated simply as "reeds"; see Victor (1979)312–313.
[23] This reference to a figure finds nothing comparable in the margin of Boncompagni (1862).
[24] "seratile" (166.13).

line *de* equals line *ab*, *dz* equals *ag*, and *ez* equals *bg*. Therefore the quadrilaterals are parallelograms: *ae*, *eg*, and *gd*. I will find the area of triangle *abg* and multiply it by *ad* the altitude of the wedge to find its volume. The proof follows. From point *a* draw line *ai* equal to and equidistant from line *bg*. Draw line *ig*. Raise the perpendicular line *ik* equal to lines *da*, *eb*, and *zg*. Draw lines *dk* and *kz*. Therefore you have *ike* a solid of six equidistant bases that is divided in two equal parts by quadrilateral *dagz*, for base *ib* is a parallelogram. Whence the diameter *ag* divides the base in two equal parts. Therefore triangle *abg* equals triangle *agi*. Similarly, triangles *dez* and *dzk* are equal. Whence wedge *abgdez* equals wedge *agidzk*. Therefore wedge *abgdez* is half of solid *ike*. The product of the area of quadrilateral *ib* by the altitude *ad* is twice the product of the base of triangle *abg* by its altitude *ad*. But the product of the base *bi* by *ad* yields the volume of solid *ike*. Whence from the product of the area of triangle *abg* by its altitude *ad* we have the volume of the wedge, as I said.

[20] Be it noted that triangles *abg* and *dez* are both either equilateral, isosceles,[25] or scalene. Hence, whatever the form of each, the area of triangle *abg* is to be found by the method demonstrated above in the tract on triangles. If we want to find this in numbers, let edges *ab* be 13 and *ag* 15, and the basal edge *bg* 14 [see Figure 6.5]. Whence also *de* equals 13, *ez* 14, and *dz* 15. Draw cathete *al* equal to 12 in triangle *abg*. Multiply this be half of *bg* to get 84 for the area of triangle *abg*. Multiplying this by 20 the height *ad* produces 1680 for the volume of wedge *abgdez*. And if the lines *da*, *eb*, and *zg* are not perpendicular to triangle *abg* but lean somewhat, try to find the altitude by a method mentioned for the preceding solid. And if [the base] of wedge *abgdez* lies in the plane of the base *ebgz*, multiply its area by half the cathete *al*, and you have what you sought. For example, because the quadrilateral *eg* is perpendicular, multiply *bg* by *be* (or 14 by 20) {p. 167} and you get 280 for the area of the base of quadrilateral *eg*. Multiply this by 6 or half of *al* as above to reach 1680 for the volume of wedge *abgdez*.

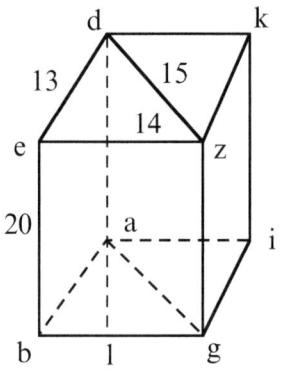

Figure 6.5

---

[25] "equicuria" (166.32).

[21] Now if a body like a silo[26] is made from a solid of six equidistant bases and a prism,[27] add the volume of the solid to the volume of the prism, and you have what you wanted. Or, add half the cathete of the triangle to the altitude of the solid of equidistant sides, and multiply what results by the base of the silo [see Figure 6.6]. For example: there is a silo with a rectangular quadrilateral *abcd* for its base.[28] At its angles are perpendicular lines *ae, bf, cg*, and *dh*, all equal to one another. Lines *le* and *ih* are drawn from points *e* and *h* to form angle *eih*. Similarly, lines *fk* and *kg* are drawn from points *f* and *g* to form angle *fkg*. Hence, pentagon *aeihd* is one base and pentagon *bfkgc* is the other base, both pentagons being equal to one another. Draw lines *ik, ef, fg, gh*, and *he*. Quadrilateral *eg* is equidistant from and equal to base *ac*; likewise quadrilaterals *af* and *gd*, and quadrilaterals *fc* and *ah*.[29] Therefore *dhf* is a solid of equidistant sides whose volume is the product of *ab* by *da* by *ae*. The volume of prism *efghik* is the product of plane *ef* by *eh*; that is to say, the product of *ba* by *ad* by half the cathete of triangle *fkg*. For example: let *ab* be 8, *ad* 6, *ae* 9, and one of the sides of triangles *eih* and *fkg* be 5. Whence cathete *kl* will be 4. Multiply *ba* by *ad* to get 48. The product of this with *aeg* and 11 or half of *kl* yields 528 for the volume of the whole silo.

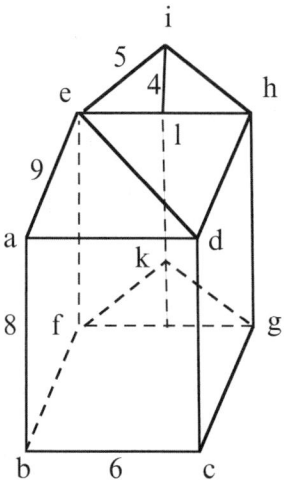

Figure 6.6

[22] Suppose the solid and prism were combined in another way so that one of the planes upon which the base was set is higher than the plane equidis-

---

[26] "ut sunt arce in quibus tenetur frumentum" (167.4).
[27] Fibonacci uses both *cunus* and *seratile* in this paragraph, but the instruction for construction indicate that a prism is needed, not a wedge.
[28] Note the ambiguous use of the word *base*.
[29] Text and manuscript have *ad*; context requires *ah* (167.14, f. 105v.35).

tant from it, and the other two planes are joined to them as in a house, or as shown in the following figure [Figure 6.7]. Then erect plane *abgd* perpendicular to plane *af*. Let each of the lines *ab* and *gd* be 20 ells. The height of plane *ag* or *bd* is 15 ells. Set plane *efch* equidistant from plane *ad*. Let each of the lines *ec* and *fh* be 8 ells, and each of the lines *ae* and *bf* be 24 ells. Therefore each of the lines *gc* and *dh* will be 25. This is found as follows. Through points *c* and *h* draw lines *ci* and *hl* equidistant from lines *ea* and *fb*. Each of these will be 24. If the square of each or 576 is added to the square on line *ig* or 49, the result is 625 for the square on line *gc*. Its root is 25 for the length of line *gc*. Whence if you want to know the volume of this body, join lines *ec* and *ag* as one, multiply their half that is $\frac{1}{2}$ 11 by 480 the area of base *af* to reach 5520, for the volume of the whole body.

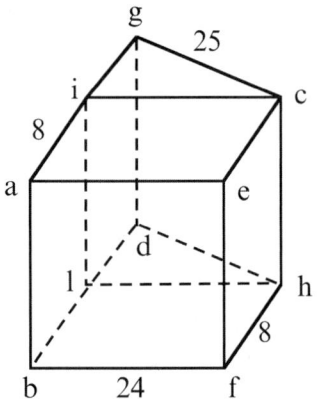

Figure 6.7

[23] Now suppose the base of a body in this first set has a quadrilateral base that is not rectangular and the edges are not equidistant but end in similar, equal, and equidistant planes. The volume can be found by multiplying the area of the base by the height of the body. This is easily seen if a plane cuts through the diameters of the base and the base equidistant from it, so that the whole body is cut in two prisms. Their volumes can be found, as has been shown, by multiplying each triangular base by the same height. In a similar way, if the base of a body in this first set is pentagonal or polygonal, it can {p. 168} be reduced to three or more prisms having triangular bases, because any rectilinear plane figure can be reduced to triangles numbering two less than the number of sides in the original figure. For instance, the pentagon can be reduced to three triangles, the hexagon to four, and so on. Whence if you multiply the area of all the triangles within the base by the height of the body, that is the area of the base by the altitude, you will indeed find the volume of the body. Because there is much that we can say apart from what we have already said about finding the

volume of multilateral figures [with bases] inscribed in circles, in order to make this work perfect, I want to add some more immediately.

[24] The area of an equilateral and equiangular hexagon inscribed in a circle equals the product of the diameter by three fourths of an angular chord. The chord is the side of an isosceles[30] triangle inscribed in the same circle. Its square is three times the square on the radius of the circle that is the side of the hexagon [see Figure 6.8]. To make this clear: let hexagon *abgdez* be in circle *abd*. Draw chord *ag* that is cut in two equal parts by the diameter *be* at point *k*. Therefore triangle *iab* is a sixth part of hexagon *abgdez*, and chord *ag* is the side of an isosceles triangle inscribed in circle *agd*. Because arc *abg* is a third of the circumference of the circle and *kg* is divided in two equal parts at point *l*, therefore *ak* is twice *kl*. The product of *ak* by *bi* is twice the area of triangle *abi*. Whence the product of *al* by *bi* is thrice the area of triangle *abi*. But the diameter *be* is twice *bi*. Whence the product of diameter *be* by *al* yields six times the area of triangle *abi*. Therefore six times of the area of triangle *abi* is the area of the whole hexagon. For *al* is a half and a fourth, that is $\frac{3}{4}$ of chord *ag*. Therefore the product of the diameter of the circle by three fourths of the chord of the hexagonal angle produces the area of the hexagon, as I said. The same thing can be found by multiplying three fourths the diameter by the whole chord of the hexagonal angle. By a similar discovery I found that the area of a square can be found by multiplying the diameter of the circumscribing circle by seven eights of the chord of the hexagonal angle. Further the area of an octagon equals the product of the diameter of the circumscribing circle by the whole octagonal chord that is a side of the square lying in the same circle. Thus by this same method of discovery the areas of the other polygonal figures circumscribed by circles can be found.

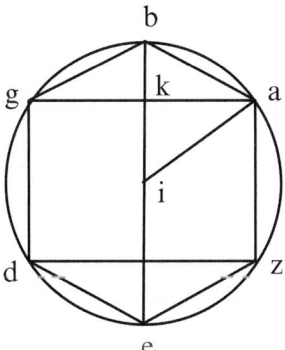

Figure 6.8

---

[30] Text has *equilateral*; context requires *isosceles* (168.10 and 12).

[25] Suppose you want to measure a column the base of which has a diameter of 7 feet and height or axis of 30 feet. Then multiply $\frac{1}{2}$ 38 the area of the circular base by the altitude to get 770 feet for the volume of the column. Note that for all bodies of whatever form that are similar to each other, the ratio of the volume of one to the volume of the other is as the cubic ratio of one similar edge to the similar edge of the other. For example: if the length of one solid is 2 feet and the length of another is 3 feet, then ratio of the smaller solid to the larger is the cubic ratio of two to three. For the cubic ratio of two to three is as a cube of two to the cube of three or 8 to 27. For as 2 is to 3, so 8 is to 12 and 12 is to 18, and 18 is to 27.[31] Thus the ratio of 8 to 27 is the cubic ratio of two to three. Whence if you want to know the volume of the smaller body, multiply it by 27 and divide the product by 8, and you will indeed have the volume of the smaller solid. And so on. Be it noted, that which was said about solids or columns, the same is understood about wells and pits, for these have length, width, and height {**p. 169**}.

6.2 Measuring Pyramids[32]

[26] If you wish to find the volume of a pyramid, multiply its base regardless of the form by a third of its altitude, and the product is the volume of the pyramid. This rule follows from what is demonstrated in Euclid,[33] namely that the solid is twice the wedge; the wedge is thrice its pyramid having a triangular base[34] [see Figure 6.9]. To make this clear: Given equilateral pyramid *abgd* with triangular base *abg*. Its height is that of the perpendicular falling from point *d* to the base *abg*. So we need to find where it fell in triangle *abg*. This is what you do. Draw circle *abg* with center *e* about triangle *abg*. I say that point *e* in the plane of triangle *abg* is where the perpendicular fell from point *d*. If it is not, then let it be point *z*. So draw lines *dz*, *za*, *zb*, and *zg*. Because *dz* is perpendicular to the plane *abg*, each of the angles *dza*, *dzb*, and *dzg* is a right angle. Therefore the sides *dz* and *za* meet at line *da*. Similarly, the squares on the lines *dz* and *zb* equal the square on line *db*. And in the same way, the square on *dg* equals the squares on lines *dz* and *zg*. But the sides *da*, *db*, and *dg* equal one another. Therefore the two squares on lines *dz* and *za* equal the two squares on lines *dz* and *zb*. If the square on *dz* is removed from both groups, then the squares on lines *za* and *zb* are equal. Whence line *za* equals line *zb*. In a similar way it can be shown that line *zg* equals both of the lines *za* and *zb*.[35] Therefore there are three lines equal to one another from *z* to the

---

[31] The significance of this sequence escapes me.
[32] "Incipit pars secunda de dimensione pyramidum" (169.1, f. 107r.11).
[33] *Elements*, XI.28.
[34] Ibid., XII.7.
[35] Text has zb_2; manuscript has *zg* (169.19, f. 107r.31).

periphery *abg*. So point *z* is the center of circle *abg*. But this is impossible because it was given that point *e* is the center of the circle. Consequently, if we multiply the area of triangle *abg* by a third of the cathete *de*, we will have the volume of pyramid *abg*.

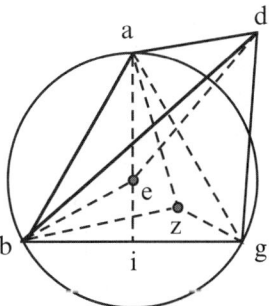

Figure 6.9

[27] A numerical example. Let each of the edges[36] of a pyramid be 12. Draw lines *ae*, *be*, and *ge*, all equal to one another. Extend lines *ae* and *be* to points *i* and *t*. Let point *i* be the midpoint on line *bg*, and angles *bai* and *gai* equal to one another [see Figure 6.10]. Because lines *ab* and *ag* equal one another, the cathete *ai* forms a right angle at *bg*. Because line *bt* meets line *ag* at its midpoint, it also intersects the cathete *ai* at a third from its endpoint, as I demonstrated above in Chapter 3 on triangles. Having removed the square on edge *bi* from the square on edge *ab*, that is, subtracting 36 from 144, what remains is 108 for the square on edge *ai*. Because *ei* is a third of *ai*, the square on edge *ei* is a ninth of the square on line *ai*. Therefore the square on edge *ei* is 12. If we add the square on line *ei*, the square on line *be* will be 48. Or, since *ae* or *be* is $\frac{2}{3}$ of *ai*, then the square on edge *be* is 48 or $\frac{4}{9}$ of the square on line *ai*. If this is subtracted from the square on line *db*, what remains is 96, the square on cathete *de*. Now the root of 3888 comes from the product of *ai* and *bi* for the area of triangle *abi*.[37] If this is multiplied by a third of *de* (that is a ninth of its square that is $\frac{2}{3}$ 10), we get 41472 for the square of the volume of pyramid *abgd*. Its root is $\frac{3}{5}$ 203 and a bit more, somewhat less than $\frac{1}{20}$ but more than $\frac{1}{21}$. From this it is obvious that in every equilateral pyramid the square on the altitude is $\frac{2}{3}$ of the square on one edge. Also the square on the lines coming from the angles to the base of the altitude of the pyramid is a third of the square on each edge.

---

[36] Fibonacci used *latus* throughout for both *edge* of a pyramid and *side* of a triangle. This distinction can become unsettling where in one sentence the edges of a pyramid are the referents and in the next sentence they are the sides of its face.

[37] Text has *abg*; context requires *abi* (169.37).

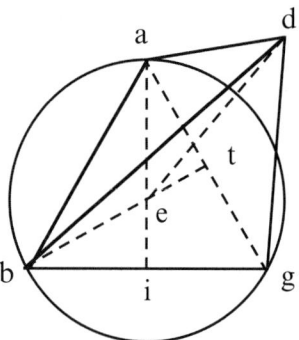

Figure 6.10

[28] We can investigate all of this as follows {**p. 170**}. Let bi be half of bg. Whence the square on bi is a fourth of the square on bg. Whence the square on ai is $\frac{3}{4}$ of the square on line ab. Then, line ae is $\frac{2}{3}$ of ai. Whence the square on ae is $\frac{4}{9}$ of the square on ai. Therefore the square on ae or be or ge is $\frac{4}{9}$ of $\frac{3}{4}$ of the square on side ab. But $\frac{4}{9}$ of $\frac{3}{4}$ of any thing is the same as $\frac{3}{4}$ of $\frac{4}{9}$ of the same thing, that is, $\frac{1}{3}$. Therefore the square on any of the sides ae, be, or ge is a third of the square on side ba equal to side da. Whence if the square on line ae is taken from the square on side da, what remains is $\frac{2}{3}$ of the square on side da for the square on the height de, as I said. Be it noted that for every pyramid with a triangular base having its equal edges raised up from the vertices to a summit point, the foot of the altitude is located at the center of the circumscribing circle that contains the base of the pyramid.

[29] Again, let pyramid $abcd$ have equal edges da, db, and dc. Its basal triangle abc can be any triangular figure. We want to find the numerical location of the foot of the altitude falling from point d to triangle abc so that it will be the center of the circle containing triangle abc [see Figure 6.11]. First draw cathete ae in triangle abc, dividing it in parts bounded by sides ba and ac. What results is the diameter of the circle containing triangle abc, as I demonstrated above in the tract on circle.[38] For example:[39] let side ab be 13, side ac 15, and side bc 14. Whence the cathete is 12. If we divide 195 the product ba and ac by 12, we get $\frac{1}{4}$ 16 for the length of the diameter circumscribing triangle abc. Half of this is $\frac{1}{8}$ 8, the difference of the distance from the foot of the altitude of the pyramid and any angle of the base. Let us set each edge da, db, and dc equal to 12. Whence if we subtract $\frac{1}{64}$ 66 (the square of $\frac{1}{8}$ 8) from 144 (the square of 12), what remains is 78 less $\frac{1}{64}$ for the square on cathete dg. If we multiply a ninth of this by the 7056, the square [of the area] of triangle abc, we will have the square [of the volume] of the whole pyramid. Be it noted that if the base of similar pyramids are acute angled,

---

[38] 3 [227].
[39] "Ad cuius rei similitudinem" (170.17–18).

then the location of the foot of its altitude is within triangle *abc*. If angle *bac* is a right angle, then *bc* is the diameter of the circle containing triangle *abc* and the foot of the altitude will be at the midpoint of *bc*. And, if angle *bac* is obtuse, then the foot will be outside the triangle on the *bc* side.

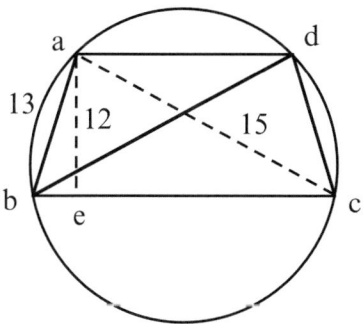

Figure 6.11

[30] Likewise: suppose I have a pyramid [*abgd*] with an equilateral or isosceles triangular base, the edges being *da*, *db*, and *dg* of which only two *db* and *dg* are equal. I want to show how to find the altitude of this pyramid [see Figure 6.12]. Because two sides *db* and *dg* of triangle *dbg* are equal, if a cathete *de* is drawn from point *d* onto the base line *bg*, it divides the base in two equal parts at point *e*. Similarly, because the two sides *ab* and *ag* of triangle *abg* are equal, if line *ea* is drawn from point *e*,[40] it too will be a cathete on base line *bg*. Now extend *ae* from both endpoints to the indefinite points *f* and *c*. Then draw the cathete from point *d* to line *fc* to be the altitude of pyramid *dabg*. Sometimes the foot of the altitude falls on line *ae*, other times it falls outside the triangle and between points *a* and *f* or between *e* and *c*, and occasionally on points *a* or *e*. I can show all of this with numbers. Let the edges *ab* and *ag* be 10 and edge *bg* 12. First by subtracting the square on the line *be* from the square on edge *db*, I find that the cathete *de* equals the root of 160. Similarly {p. 171}, cathete *ae* will be 8. Then I will find the cathete of triangle *dae* on base *ae* thus: I subtract 144 from 160, that is the square on *da* from the square on *de*. What remains is 16 that if divided by *ae* leaves 2. Adding this to *ae* gives 10 whose half or 5 is the major segment on the larger side *de*. So I put *h* at the endpoint of the segment so that *eh* is 5. What remains for *ah* is 3. Therefore if the square on line *ah* is taken from the square on line *da* or the square on line *eh* is taken from the square on line *de*, what remains is 135 the square on cathete *dh*. Note that if angle *dah* is a right angle, then *da* is the cathete descending from point *d*. If angle *dah* is greater than a right angle, then the cathete descending from point *d* falls outside the triangle and

---

[40] Text has *c*; manuscript has *e* (170.35, *f.* 108r.28).

296   Fibonacci's *De Practica Geometrie*

on line *af*. Similarly, if angle *deh* is a right angle, then *de* is the altitude of the pyramid. If angle *deh* is obtuse, then the cathete descending from point *d* falls outside the triangle between points *e* and *c*. And if each of the angles *dah* and *deh* is acute, then the perpendicular falls on line *ae* between the points *a* and *e*.

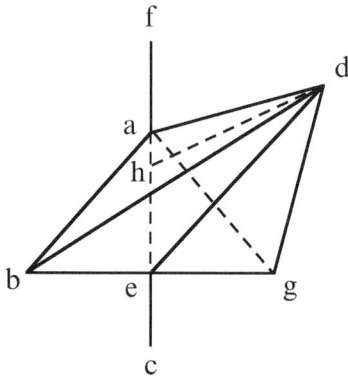

Figure 6.12

[31] Again, let there be a pyramid with summit *a* and base *bcd* a scalene triangle with edge *bc* 13, edge *cd* 14, and edge *db* 15 with cathete line *be*. Of the remaining edges coming from point *a* to the points *b*, *c*, and *d*, two edges are equal: *ac* and *ad* each of which is 10 feet. The remaining edge *ab* is 15. I want to find the cathete falling from point *a* upon the triangular base *bcd* [see Figure 6.13]. First I draw cathete *af* in triangle *acd*. Then in the plane of triangle *bcd* I draw line *fg* from point *f* making right angles *gfd* and *gfe*. After drawing line *ag*, I draw the cathete falling from point *a* to line *fg* in triangle *afg* that is on the plane of triangle *bcd*. To find this cathete numerically: since angle *gfd* is a right angle, line *fg* is equidistant from cathete *be*. Whence as *df* is to *de*, so *dg* to *db*, with *de* equal to 9 and *df* equal to 7 half of *ed*. Since triangle *acd* is isosceles, therefore as 7 is to 9, so the unknown *dg* is to the known *db* that is 15. So if we multiply 7 by 15 and divide by 9, we get $\frac{2}{3}$ 11 for line *dg*. Similarly, if we multiply *df* by *be* or 7 by 12 and divide by *de*, we obtain $\frac{1}{3}$ 9 for line *fg*. Then I draw cathete *ah* to line *bd* in triangle *abd*. The longer segment *bh* is $\frac{2}{3}$ 11 and the shorter segment *dh* is $\frac{1}{3}$ 3. Whence the square on cathete *ah* will be $\frac{8}{9}$ 88. If we add this to the square on line *hg* or $\frac{4}{9}$ 69, we obtain $\frac{1}{3}$ 158 for the square on line *ag*. Again, if the square on line *ad* is taken from the square on *df* or 49 from 100, what remains is 51 for the square on line *af*. Therefore line *af* is the root of 51, line *ag* is the root of $\frac{1}{3}$ 158, and line *fg* is $\frac{1}{3}$ 9. Afterwards we know the sides of triangle *afg*. We can know the cathete descending from *a* onto line *fg* by the method that we taught in the measurement of areas of triangles.

6 Finding Dimensions of Bodies   297

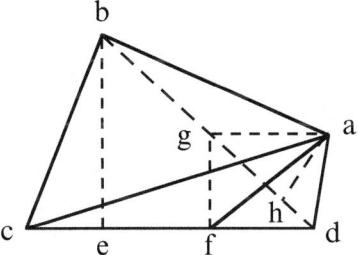

Figure 6.13

[32] Again, let there be pyramid *abgd* with base *abg* a scalene triangle. Unequal edges descend from point *d* to points *a*, *b*, and *g*. While one of the edges is perpendicular to triangle *abg*, the others incline to it [see Figure 6.14]. For example: let edge *ab* be 10, *ag* 9, *bg* 5, and *da* 15. Edge *db* is 13 and *gd* is 12. In this pyramid the straight line *dg* is the cathete because the squares on lines *bg* and *gd* equal the square on line *db* {**p. 172**}. Also, the squares on lines *dg* and *ga* equal the square on line *ad*. Therefore if a third of *dg* is multiplied by the area of triangle *abg*, we have the volume of pyramid *abdg*.

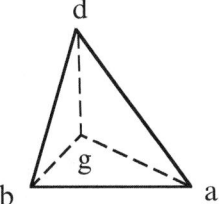

Figure 6.14

[33] Suppose the cathete of pyramid *abcd* is not any of the lines coming down from the summit, and its edges are *ab* equal to 13, *bc* to 14, and *ac* to 15. Further edge *bd* is 13, *dc* 15, and *da* 14. First draw cathete *ae* in triangle *abc*, draw cathete *df* in triangle *dbc*, and from point *f* draw line *fh* equidistant from cathete *ae* so that angle *hfc* is a right angle. Then in triangle *dac* draw cathete *dg* to line *ac*. From all of this line *dh* is known. Whence the sides of triangle *dhf* are known. Therefore the cathete falling on it from point *d* to line *fh* is known, and it will be the altitude of the pyramid [see Figure 6.15]. For example: cathete *ae* is 12, segment *be* is 5, and segment *ec* is 9. Likewise, cathete *df* in the plane of triangle *dbc* is 12, segment *fc* is 5. Because *fh* is equidistant from cathete *ae*, *cf* is to *ce* as *fh* is to *ae* and *ch* to *ca*. Whence *fh* is $\frac{2}{3}$ 6 and *ch* is $\frac{1}{3}$ 8. Then, in order to find cathete *dg*, I subtract 169 from 196, the square on *dg*[41] from the square on *da*, to get a remainder of 27. Dividing this by

---

[41] Text has *dc*; context requires *dg* (172.14).

ac yields $\frac{4}{5}$ 1 that I add to *ac* to get $\frac{4}{5}$ 16. Its half is $\frac{2}{5}$ 8 for the major segment *ag*, and the minor segment *gc* is $\frac{3}{5}$ 6. From these we have $\frac{1}{5}$ 11 for cathete *dg*. With its square added to the square on *gh*,[42] we have $\frac{4}{9}$ 128 for the square on line *dh*. Whence *dh* is $\frac{1}{3}$ 11. What remains is to measure the cathete falling from point *d* onto line *fh* in triangle *dfh*. The longer segment *fi* will be $\frac{1}{2}$ 4. Because the cathete *di* is the altitude of the pyramid and equals the root of $\frac{3}{4}$ 123: if we multiply this by 28 a third of the area of triangle *abc*, we will have the root of 97020 for the volume of pyramid *abcd*.

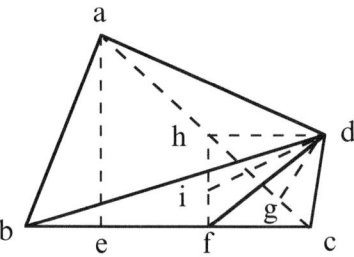

Figure 6.15

[34.1] Be it noted that if angle *dcb* in triangle *dbc* is obtuse, then cathete *df* falls outside of triangle *dbc*. In this case another formula is called for. We let edge *ac* be 13, *bc* 9, *ab* the root of 160, edge *da* 19, edge *db* 17, and edge *dc* 10 [see Figure 6.16]. Whence cathete *ae* will be 12, segment *be* 4, and *ec* 5. If we subtract the square on *dc* from the square on edge *db*, what remains is 189. Dividing this by 9 or *bc* we get 21. Adding this to the same 9 and dividing the sum in two equal parts produces 15 for the segment *bf*. Therefore point *f* is not on edge *bc*. Further *cf* is 6.[43] Removing its square from the square on edge *dc* leaves 64 for the square of cathete *df*. Whence *df* is 8.

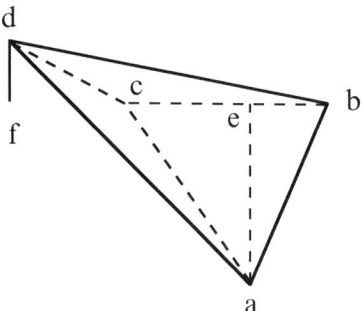

Figure 6.16

---

[42] Where was "the square on *gh*" computed? (172.18). Apparently, it is 3.
[43] Text has 36; context requires 6 (172.30).

[34.2] Then I draw line *fh* equidistant from cathete *a* and make it meet [the extension of] line *ba* at point *h*. Draw line *dh*. Right triangle *hbf* is similar to triangle *abe* [see Figure 6.17]. Thus as *be* is to *ea* so *bf* is to *fh*. Hence *fh* is 45. If we add its square or 2025 to 225 the square on the base line *bf*, we get 2250 for the square on line *bh*. Or, because *be* is to *bf* as *ba* is to *bh*: if we multiply the square on *bf* by the square on *ba*, that is, 225 by 160, and divide the product by 16 the square on *be*, we get once again 2250 for the square on line *bh*. Then in order to find the measure of line *dh*, look for the cathete of triangle *dba* falling from *d* to point *k* on side *ba*, in this way. Subtract the square on *db* from the square on *da*, 289 from 369, to get 72. Divide this by edge *ba* the root of 160. Then divide 5184 the square of 72 by 160 to get $\frac{2}{5}$ 32. Add this to 160 to reach $\frac{2}{5}$ 192. If from this we subtract 144 (twice the root of the product of $\frac{2}{5}$ 32 and 160) {**p. 173**}, what remains is $\frac{2}{5}$ 48 for *bk* the square of twice the shorter segment. So the square on *bk* is $\frac{1}{10}$ 12 a fourth of $\frac{2}{5}$ 48. Taking $\frac{1}{10}$ 12 from the square on line *db*, $\frac{9}{10}$ 276 remains for the square of cathete *dk*. Then in order to know the measure of line *kh*, add the square of *bk* with the square of *bh*, namely $\frac{1}{10}$ 12 to 2250 to get $\frac{1}{10}$ 2262. From this subtract 330, twice the root of the product of $\frac{1}{10}$ 12 and 2250. What remains is $\frac{1}{10}$ 1932 for the square on line *kh*. If to this we add $\frac{9}{10}$ 276 the square on *dk*, we obtain 2209 for the square on line *dh*. Therefore *dh* is 47. Then in order to find the cathete of triangle *dfh* that falls from point *d* to side *fh* or 45 and edge *df* is 8 and working with the procedure described above, we will find that the cathete falls outside of line *hf* by a measure of $\frac{1}{3}$ 1. Let this be the segment *fl*. Subtracting $\frac{7}{9}$ 1 from 64 the square on *fl* from the square on *df*, the remainder is $\frac{2}{9}$ 62 for the square on line *dl*, It is the cathete of pyramid *abcd*.

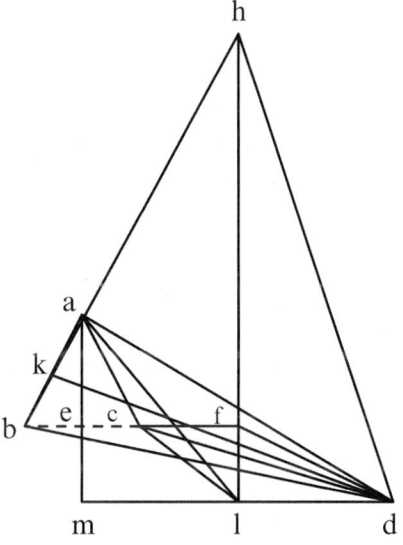

Figure 6.17

[35] You can find the same thing if you subtract the square on segment hl from the square on edge dh. Or if you subtract the square on line cl from the square on line dc. The square on line cl equals the two squares on lines cf and fl. Or in another way, extend line ae to point m so that em equals fl. Therefore the entire line am is $\frac{1}{3}$ 13. Draw lm that equals 11 or ef. Add the squares on lines am and ml, and you have the square on line al. If you subtract this from the square on line da or 361, what remains is still $\frac{2}{9}$ 62 for the square on cathete dl. From this it is obvious that the line dl is perpendicular to the plane fbh since it makes right angles with lines lh, la, and lc. Then if we multiply the cathete dl by 18, a third of the area of triangle abc, we will have 20160 for the volume of the whole pyramid $abcd$. Thus you should strive to proceed with all pyramids. For you can find exactly the altitude of all pyramids with the instruments mentioned above, the sticks or plumb line.

[36] If the base of a pyramid is a quadrilateral or some other polygon, multiply a third of the height by the area of the base. Because the base, regardless of its size, can be divided into triangles, and lines can be drawn from the vertices of the angles to the top of the altitude, the whole pyramid can be sectioned into as many triangular pyramids as there are triangles in the base, all having the same altitude [see Figure 6.18]. Hence, if a third of the altitude is multiplied by the area of each of the triangular bases in which the whole base had been divided, you have the volume of the whole pyramid. And if some pyramid is removed from another pyramid by a plane equidistant from its base, and you want to know the volume of what is left, namely the truncated pyramid, subtract the volume of the part cut off from the volume of the whole pyramid. What remains is the volume of the truncated pyramid. For example: given pyramid $dezi$ with triangular base ezi cut from pyramid $dabg$ with triangular base abg, and the bases are equidistant. We want to find the volume of the truncated pyramid $abgezi$. Subtract the volume of pyramid $dezi$ from the volume of pyramid $dabg$. What remains is the volume of the truncated pyramid $abgezi$. Or in another way: because triangle ezi is equidistant from triangle abg, and the sides of triangle ezi are cutting the triangular planes dab, dbg, and dga, the sides of these triangles are equidistant: side ez to side ab, side ei to side ag, and side zi to side bg. And because {p. 174} in triangles dab, dag, and dbg lines ez, ei, and zi are drawn equidistant from the base lines ab, ag, and bg, the exterior angles are equal to their corresponding angles, namely angle dez to angle dab. Also angles dze, dzi, diz, die, and dei are equal to angles dba, dbg, dga, and dag. Further, angles zei, eiz, and ize are equal to angles bag, agb, and gba, because plane ezi is equidistant from plane abg. So the solid angles at e, z, and i of pyramid $dezi$ are equal to the solid angles at a, b, and g. Whence it is obvious that pyramid $dezi$ is similar to pyramid $dabg$. Now, similar pyramids are to one another as the triple ratio of their corresponding edges. Wherefore the ratio of pyramid $dabg$ to pyramid $dezi$ is

thrice the ratio of edge bg to edge zi [see Figure 6.19]. For this reason we can set bg to zi as the quantities c to f. And as c is to f, so f is to x and x to y. Now the ratio of c to y is thrice the ratio of edge bg to edge zi. Whence as c is to y, so pyramid $dabg$ is to pyramid $dezi$. Removing the quantity y from the quantity c leaves the quantity h. Then as c is to y, so the quantity $h+y$ is to the quantity y. Therefore as the quantity h increased by y is to the quantity y, so pyramid $dabg$ is to the pyramid $dezi$. Therefore by separation[44] as h is to y, so the truncated pyramid $abgezi$ is to pyramid $dezi$. So if triangles abg and ezi together with the altitude of pyramid $dabg$ are known, then also known by what has been said thus far is the volume of truncated pyramid $abgezi$.

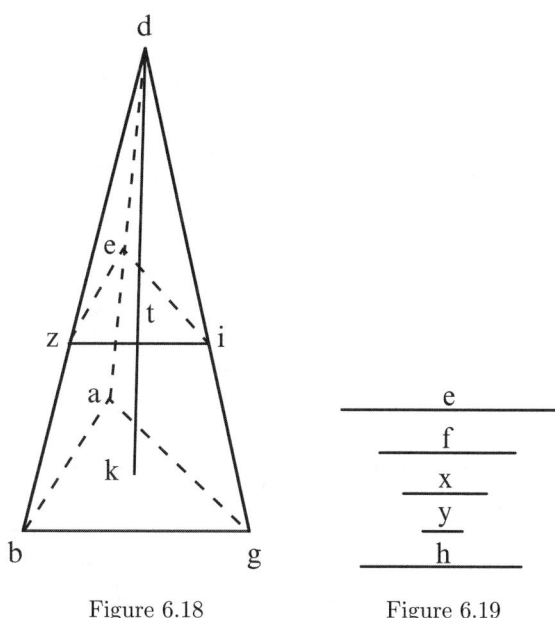

Figure 6.18   Figure 6.19

[37] Let us show this by numbers. For pyramid $dabg$ let edge ab be 13, ag 15, and bg 14 together with altitude dk 24. At the midpoints of edges da, db, and dg set the plane triangle czi in which side cz is $\frac{1}{2}$ 6, side ci is $\frac{1}{2}$ 7, and side zi is 7. Let the cathete dk be 24 and be cut by the plane ezi in two halves at point t. Further, set the quantity c equal to 8. Because edge bg is twice side zi, the quantity c will be twice the quantity f, f twice the quantity x, and x twice the quantity y. Whence f is 4, x is 2, and y is 1. If we remove an

---

[44] "disiunctim" (174.16).

amount equal to y from c, that is 1 from 8, what remains is 7 for the quantity h. Because we found that h is to y as truncated pyramid *abgezi* is to pyramid *dezi*, h is seven times y. Now the volume of pyramid *dezi* is 84 that we found by multiplying 4 by 21, that is, two thirds of the altitude dt by the area of triangle *ezi*. So if we multiply 84 by 7, we get 588 for the volume of the truncated pyramid *abgezi*. Or if we subtract 84 the volume of pyramid *dezi* from 672 the volume of the whole pyramid resulting from the product of 8 and 84 (two thirds of the altitude by the area of triangle *abg*), what remains is still 588 for pyramid *abgezi*.

[38] Consider again pyramid *dabg* with summit point d. Remove from it pyramid *dezi* with base *ezi* equidistant from base *abg*, and draw cathete *dtk* in it from point d. I say that the volume of truncated pyramid *abgezi* results from the product of two thirds of tk[45] its altitude and the product of the areas of the upper[46] base and the plane that is the mean proportion between the areas of the two bases. The proof follows [see Figure 6.20]. Construct rectangle *bm* on edge *bg* with area equal to the area of triangle *abg*. Draw line *ng* equal to line *zi*. Apply the area of rectangle *nopg* (equal to the area of triangle *ezi*) to line *ng* {**p. 175**}, and extend line *po* to *q*. First of all, I say that plane *np* is similar to plane *bm*. The proof follows. Because, as in the preceding figure, triangles *abg* and *ezi* were shown to be equiangular, they are similar. Now similar triangles are in duplicate ratio as their similar sides.[47] Sides *bg* and *zi* are similar sides. Therefore the ratio of triangle *abg* to triangle *ezi* is the duplicate ratio of the sides *bg* and *zi*. Consequently as *bg* is to *zi*, so *zi* is to *u*. For which reason, as *bg* is to *v*, so triangle *abg* is to triangle *ezi*. But rectangle *bm* equals triangle *abg* and plane *np* equals triangle *ezi*. Whence as *bg* is to *v*, so plane *bm* is to plane *np*. And line *ng* equals line *zi*. Whence as *bg* is to *ng* so *ng* is to *v*. So three lines are in continued proportion: the first is to the third, as the first figure is to the second similar and similarly described figure. Whence as *bg* is to *v*, so quadrilateral *bm* is to the quadrilateral described on line *gm* and similar to quadrilateral *bm*. On the other hand,[48] if quadrilateral *np* is not similar to quadrilateral *bm*, then another quadrilateral is described on line *ng* and at angle *ngm*. It is either larger or smaller than quadrilateral *np* to which quadrilateral *bm* described on line *bg* has that ratio that *bg* has to *v*. But the ratio of quadrilateral *bm* to quadrilateral *np* has been shown to be the ratio of *bg* to *v*. Therefore quadrilateral *bm* has the same ratio to two different quadrilaterals. And that is impossible. Therefore quadrilateral *np* is similar to quadrilateral *bm*, as I said.

---

[45] Text and manuscript have *ek*; context requires *tk* (174.39, f. 111r.15).
[46] "in summam arearum basis et capitis illius et superficiei" (174.39–40).
[47] *Elements*, VI.20.
[48] "Vnde" (175.14) does not provide the disjunction required here.

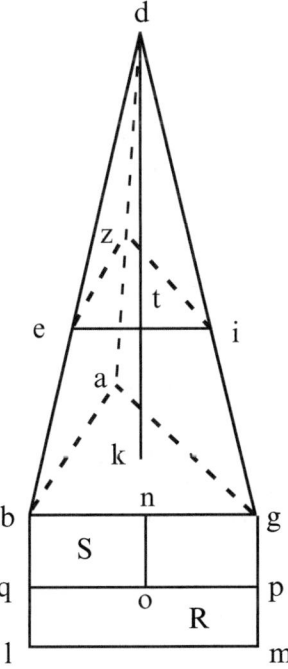

Figure 6.20

[39] Similar figures with equal angles have similar sides. Whence as bg is to gm, so ag is to gp. Alternately therefore as bg is to ng, so mg is to pg. But mg is to pg as plane bm is to plane bp. Because bg is to ng (this is as mg is to pm), so plane bp is to plane pn. Therefore as plane bm, plane bp, and plane pn are in continued proportion, plane bp is the mean proportional between planes bm and pn, that is, between the plane abg and triangle ezi. So it has been shown that the product of the third part of tk together with [the sum of] the areas of triangles abg and ezi and the area bp (that is, together with areas bm, bp, and pn) make the volume of the truncated pyramid $eziabg$.

[40] First, it was shown that the volume of the whole pyramid $dabg$ follows from the product of the third part of altitude dk by the area of triangle abg or rectangle bm. Consequently the product of dk by the area of bm produces thrice the area of pyramid $dabg$. But the product of dt by tk by the area of bm equals the product of dk by the area of bm. Therefore thrice the volume[49] of pyramid $dabg$ comes from the sum of the products of dt by bm and tk by bm. If we subtract the volume of pyramid $dezi$ from the threefold volume, that is, the product of dt by the area of triangle ezi or the area of np, what remains

---

[49] Text and manuscript have *aree*; context requires "volume" (175.35, *f.* 111v.21–22). Note that below Fibonacci uses the phrase *area cubica* for volume.

is the product of tk by the area of bm and the product of dt and the area R and the area S for thrice the volume of the truncated pyramid eg. Having proved all of this, join the equidistant lines bk and zt in the plane of triangle dkb. Whence, as bd is to zd, so kd is to dt. But as bd is to zd, so bg is to ng and mg is to pg. For by separation: bn is to ng or as mp is to pg, so kt is to dt. But as mp is to pg, so the plane qm is to the plane bp. So the product of tk by the plane bp {p. 176} is equal to the product of dt by the area R that is the plane qm. Again, because bn is to ng as tk is to dt, so bn is to ng as area S or area bo is to plane np. Whence the product of tk by the area np equals the product of dt by the area S. Therefore the product of tk joined with the areas bp and pn equals the product of dt by the area R plus S. Adding to both the product of tk by the area bm, the product of tk by the areas bm, bp, and pk equals the product of tk by bm and the product of de by the areas R and S. But the product of tk by the area bm and dt by the areas R and S[50] yields thrice the volume of the truncated pyramid eg. Therefore the product of tk by the areas bm, bp, and pm (triangles abg and ezi and area bp that is the mean proportional between the triangles) produces thrice the volume of the truncated pyramid eg. So by multiplying a third of tk by the areas of triangles abg and ezi and the area bp, we reach the volume of the proposed truncated pyramid. And that is what I wanted to demonstrate.

[41] By the numbers. Let edge bg be 12 palms, the cathete falling from point a to side bg in triangle abg 15, side zi 4, and altitude tk 12. Whence as 3 is to 1, so bg is to zi, for bg is thrice zi. And as bg is to zi, so zi is to v. Whence bg is nine times v. And because bg is to v so the area of triangle abg is to the area of triangle ezi. Therefore triangle abg is nine times triangle ezi. For the area of triangle abg is 90 square palms. This was found by multiplying the aforementioned cathete by half the base bg. Whence a ninth of 90 palms is 10 for the area of triangle ezi. But the plane falling between triangle abg and triangle ezi as mean proportional is 30, because 90 is to 30 as 30 is to 10. Adding these three 90, 30, and 10 together yields 130. Multiplying this by 4, a third of tk, yields 520 cubic palms for the volume[51] of the truncated pyramid eg. We reach this sum if we take pyramid $dezi$ from the whole pyramid $dabg$ as follows. Because bg is to zi (that is as bg is to ng), so dk is to tk. *Separando* therefore, bn is to ng as kt is to td. Whence if we multiply ng by kt or 4 by 12 and divide by ba or 8, we get 6 for cathete dt. Whence the whole line dk is 18. A third of it multiplied by 90 or the area of triangle abg produces 540 for the volume of the whole pyramid $dabg$. If we take from this the volume of pyramid $dezi$ or 20 equal to 2 times 10, the product of a third of cathete dt by the area of triangle ezi, what remains is the same 520 as above for the volume of the truncated pyramid eg.

---

[50] Text has miniscule rs (176.9).
[51] *area* again for volume.

[42] A similar demonstration shows that if the base of a truncated solid is quadrilateral, polygonal, or circular, everything said above is valid [see Figure 6.21]. To make this more complete, consider a truncated cone[52] *eg* with circular lower base *abcd* and circular upper base *efgh*. Its altitude is line *ik* with endpoints in the center of the circles. Draw diameters *bd* and *fh*. The circular bases are equidistant. So [to find the volume of the truncated cone],[53] multiply a third of *ik* by the sum of the bases *abcd* and *efgh* and the area of the plane that is the mean proportional between the two basal circles {p. 177}. For example. put the semicircle[54] on *ml*[55] between the two of them as the mean proportion *lm*[56] between the radius *bk* and the line *if*, the area of which contains the circle *mno* [see Figure 6.22]. First I say that the circle *mno* is the mean proportion between circle *abc* and circle *efg*. The proof follows. Because as line *bk* is to line *lm*, so line *lm* is to line *if*. Whence as *bk* is to *fi*, so the square on *bk* is to the square on line *lm*. But as the square on *bk* is to the square on *lm*, so the square on diameter *bd* is to the square on diameter *pm*. For as the square on diameter *bd* is to the square on diameter *pm*, so the area of circle *abcd* is to the area of circle *mnpo*. Therefore as the square on the semidiameter *bk* is to the square on semidiameter *ml*, so circle *abc* is to circle *mno*. Because *bk* is to *lm* as *lm* is to *if*, so will the square on *bk* be to the square on *lm*,[57] and so the square on *lm* will be to the square on *if*. Whence as the square on *ml* is to the square on *if*, so circle *abc* is to circle *omn*. For as the square on *lm* is to the square on *if*, so the circle *omn* is to the circle *efg*.[58] For it was shown that as the square on *ml*[59] is to the square on *if*, so the circle *abc* is to the circle *omn*. Therefore as the circle *abc* is to the circle *omn*, so circle *omn* is to circle *efg*. Therefore circle *omn* is the mean proportional between circle *abc* and circle *efg*, as was required. In order to have a sum from the products of these three circles, we add together the squares on the radii *bk*, *ml*, and *fi*. Multiply the sum by $\frac{1}{7}$ 3 and we have the sum of the areas of the three circles. If we multiply this by a third of the altitude *ik*, we have the volume of the truncated cone *ec*.

---

[52] Fibonacci used *pyramis* in this example, but he meant cone as the context makes clear.
[53] This parenthetical expression is not in the text (177).
[54] Text and manuscript have only *semi* (177.1, f. 110r.9).
[55] Text and manuscript have *bd*; context requires *ml* (177.1, f. 112r.9). From here throughout [42] there are many misuses of letters that represent points and segments.
[56] Add to (177.1) because (177.2) has *iflm* which is meaningless in the context. So I separated *lm* from *iflm* to create what is here.
[57] Text has *nqm*; context requires *lm* (177.11).
[58] Text has *ofg*; context requires *efg* (177.13, f. 112r.21).
[59] Text has *mb*; context requires *ml* (177.19, f. 112r.25).

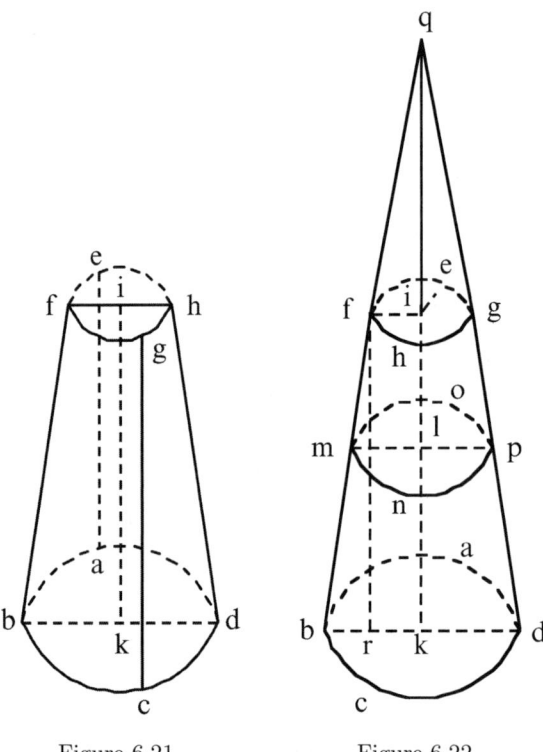

Figure 6.21   Figure 6.22

[43] By the numbers. Let the semidiameter *bk* be 4 and the semidiameter *if* be 1. Whence the semidiameter *lm* is 2, because 4 is to 2 as 2 is to 1. Add 16, 4, and 1, the squares on these semidiameter. Multiply the sum by $\frac{1}{7}$ 3 to get 66 and then by 5 a third of the altitude *ik*, to obtain 330 for the volume of the truncated cone.[60] If we want to know the volume of the whole cone *qabcd*, we understand that triangle *qgd*[61] cuts cone *qabcd* in two parts. Cathete *ik* is in its plane. Extending the cathete to *q*, line *kq* will be the cathete of triangle *qbd*. If in it we draw line *fr* equidistant from line *ik*, then line *fr* is equal to line *ik*. Since lines *fi*, *ml*,[62] and *bk* are equidistant, *rk* will be equal to line *fi*. And triangles *qif* and *frb* will be similar to one another. Whence if we remove *kr* equal to *if* from *kb*, what remains is *br* equal to 3. And because *br* is to *rf*, so *fi* is to *iq*. If we multiply *rf* by *if* and divide by *br*, we obtain 5 for the cathete *qi*. Whence *qk* is 20, the altitude of cone *qabcd*.

---

[60] Inasmuch as the context focuses on the truncated cone, the use of *totius pyramidis* at (177.25) is incorrect and out of place; *cone* is used in place of *pyramid* throughout [44], as the context requires.

[61] Here (177.26) and below, Fibonacci identifies triangles only by a side.

[62] Text and manuscript have *ke*; context requires *ml* (177.29, f. 113r.9).

[44.1] Again, another demonstration is possible that the product of the altitude of a truncated pyramid by the sum of the areas of its bases and the plane that is the mean proportion between the bases equals three times the volume of the truncated pyramid. Consider the truncated pyramid *abgdez* with triangular bases *abg* the larger and *dez* the smaller. Let line *it* be the mean proportional between *ab* and *de*. Construct on line *it*[63] triangle *kit* similar to each of the triangles *abg* and *dez*. Further, as *ab* the first term is to *de* the third term, so triangle *abg* is to triangle *kit* [see Figure 6.23]. Again, as *it* is to *de* (or, *ab* to *it*), so *de* is to line *L*.[64] Alternately, as *bg* is to *de*, so *it* is to *L*. But as *it* is to *L*, so triangle *kit* is to triangle *dez*. Therefore the ratio of triangle *abg* to triangle *kit* {p. 178} equals the ratio of triangle *kit* to triangle *dez*. Therefore the area of triangle *kit* is the mean proportional between the areas of triangles *abg* and *dez* [see Figure 6.24]. Erect on triangle *kit* cathete *km* equal to the cathete between triangles *dez*[65] and *abg*. Draw lines *mi* and *mt*. Let the altitude of pyramid *mkit* equal the altitude of the truncated pyramid *abgdez* [see Figure 6.25]. Then draw lines *ae*, *eg*, and *dg*, so that within the truncated pyramid there are three pyramids in continued proportion of which the larger is *abge*, the middle *aged*, and the smaller *dgze*.[66]

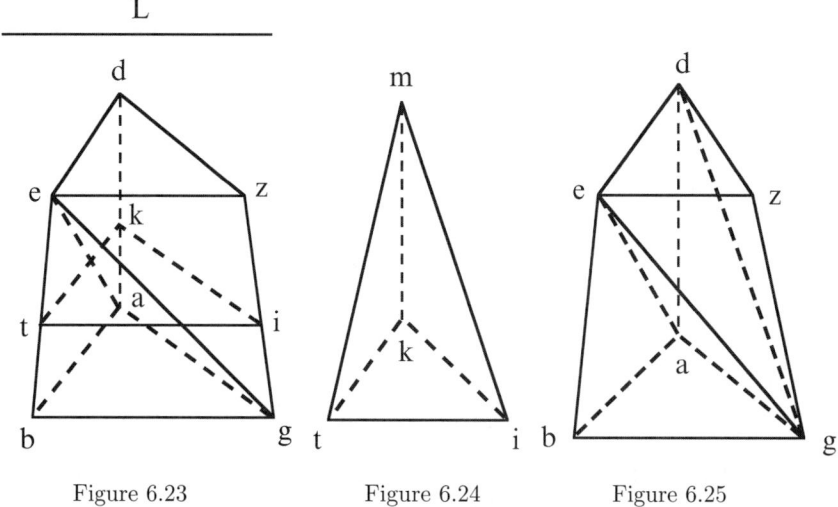

Figure 6.23   Figure 6.24   Figure 6.25

[44.2] The proof follows. Because lines *de* and *ab* are equidistant and parts of triangles *abe* and *ade*, there is one altitude for both triangles. Whence as *ab* is

---

[63] *Constituatur super rectam it* is from Paris 7223, f. 162r.29. and is arguably the correct reading.
[64] Text has *i* and *b* (next sentence); Paris 7223 has *L* (177.42, f. 162v.3 and 4.). The auxiliary line *L* appears with the figure in the margin of Boncompagni 177 and Urban 292 f. 113r.
[65] Text and manuscript have *daz*; context requires *dez* (178.3, f. 113r.29).
[66] The permutation of the four letters in this section of the manuscript is ignored, and the arrangement shown here is used throughout this translation.

308   Fibonacci's *De Practica Geometrie*

to *de*, so triangle *abe* is to triangle *ade*. Thus the larger pyramid *abge* is to the middle pyramid *aged*, for both pyramids have one altitude drawn from point *g* to their plane *dabe*. Therefore as *eba* is to *dea*,[67] so pyramid *abge* to pyramid *aged*. Again, because lines *ag* and *dz* are equidistant, so triangle *agd* to triangle *gdz*. But as triangle *agd* is to triangle *gdz*, so the middle pyramid *aged* to the smaller pyramid *dgze*. Since they all have the same altitude, they also have the same topmost vertex *e* and their bases are in the one plane containing quadrilateral *dagz*. Therefore, as *ag* is to *dz*, so pyramid *aged* to the smaller pyramid *dgze*. Now, as *ag* is to *dz*, so *ab* to *de*, since triangles *abg* and *dez* are equidistant. For it has been shown that as *ab* is to *de*, so the greater pyramid *abge* to the middle pyramid *aged*. And as *ab* is to *de* (that is, as *ag* is to *dz*), so the middle pyramid *aged* to the smaller pyramid *dgze*. Whence as pyramid *abge* is to pyramid *aged*, so pyramid *aged* to pyramid *dgze*. Whence the whole truncated pyramid *abgdez* has been divided into three pyramids in continued proportion, according to the ratio of line *ab* to line *de*. But as *ab* is to *de*, so was triangle *abg* to triangle *kit*. But as triangle *abg* is to triangle *kit*, so pyramid *abge* is to pyramid *mkit*. For both have equal altitudes. Therefore as *ab* is to *de*, so pyramid *abge* is to pyramid *mkit*. For it was shown that as *ab* is to *de*, so pyramid *abge* is to pyramid *aged*. Therefore pyramid *abge* is to the two pyramids *mkit* and *aged* that have the same ratio. Whence pyramids *mkit* and *aged* are equal. And so as pyramid *aged* is to pyramid *dgze* (that is, as *ab* is to *de*), pyramid *mkit* is to pyramid *dgze*. Therefore pyramids *abge*, *mkit*, and *dgze* are in continue proportion, have the same altitude, and together equal the whole pyramid *abgdez*. Whence, if their altitude is multiplied by each of the bases, triangles *abg*, *kit*, and *dez*, the product is thrice the entire truncated pyramid *abgdez*, as required.

6.3 Measuring Spheres[68]

[45] If there is a point within a sphere from which originate four equal straight lines that terminate on the surface of a sphere and their end points are not in the same plane, then the originating point is the center of the sphere. For example, let point *z* be within sphere *ab*, from which lines *zb*, *zg*, *zd*, and *ze* originate, are equal, and are not in the same plane, I say that point *z* is the center of sphere *ab* {**p. 179**}. The proof follows. Between point *b* and points *g*, *d*, and *e* draw lines *bg*, *bd*, and *be*; then draw lines *dg* and *de*. All these lines are within the sphere. Because every triangle is in one plane, as noted in the eleventh book of Euclid,[69] points *b*, *d*, and *g* are in one plane, and points *b*, *d*, and *e* are in another [see Figure 6.26]. Around these points are drawn circles

---

[67] Text and manuscript have *de*; context requires *dea* (178.12, f. 113v.13).

[68] "Explicatis his que ad areas cubicas pyramidum et earum partium pertinent. Nunc tractemus de his que spectant ad mensurationem et earum partium" (178.37–38, f. 113v.33–34). The context requires the use of the word "spheres".

[69] *Elements*, XI.2.

bdg and bde. Further, perpendicular lines zi and zt are drawn from point z to the planes of circles bgd and bde and are extended from both parts to points k, l, m, and n. And the lines ib, ig, and id are drawn through point i. Because zi stands orthogonally to the plane of circle bgd, angles zib, zig, and zid are right angles. And because lines zb, zg, and zd equal one another and they have line zi in common, therefore lines ib, ig, and id equal one another. Whence as stated in Euclid,[70] point i is the center of circle bgd. Because kl passes orthogonally through the center of circle bgd, line kl is the diameter of the sphere, and points k and l are the poles of circle bgd, as is proved in the book of Menelaus[71] and Theodosius.[72] Because the center of the sphere is on line kl, it is similarly shown that point t is the center of circle bde. Because of this, line mn is proved to be the diameter of sphere ab. Whence the center of the sphere is on line mn. And because point z is common to both diameters, kl and mn, it is obviously the center of the sphere, as I said.

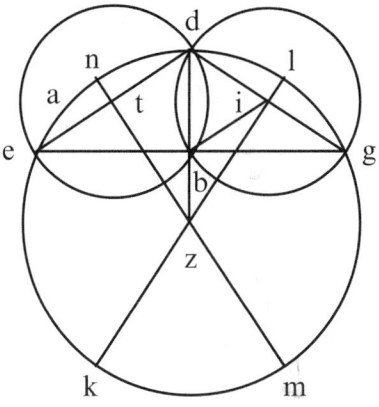

Figure 6.26

[46] If the line drawn from the vertex of a cone to the center of its base is a perpendicular, then all lines drawn from the same point to the circumference of the circular base are equal. The product of one of these lines by half the basal circle is the area of the surface of the cone, that is, the area between the vertex of the cone and the circumference of the base. For example, let a be the vertex of cone $abgd$ and circle bgd with center e its base. Further, let line ae be erected perpendicularly to the plane of the circle bgd. Finally, draw as many lines ab, ag, ad as you choose, all on the surface of cone $abgd$ and attached to the circle bgd. I say that these lines ab, ag, ad, and as many as you choose are all equal. The proof follows [see Figure 6.27]. Draw lines

---

[70] Ibid, III.9.
[71] *De Sphaericis* I.1.
[72] Heath (1921) II, 246–252.

310   Fibonacci's *De Practica Geometrie*

eb, *eg*, and *ed* from the center *e* that makes all of them equal to each other. Because *ae* is perpendicular to the plane of circle *bgd*, the angles *aeb*, *aeg*, and *aed* are right angles. Whence triangles *aeb*, *aeg*, and *aed* are orthogonal, have equal bases *eb*, *eb*, and *eg*, and have side *ae* in common. Whence the sides *ab*, *ag*, and *ad* subtending the right angles are equal. Because of this it is obvious that all lines drawn from *a* to the circumference of circle *bgd* are equal.

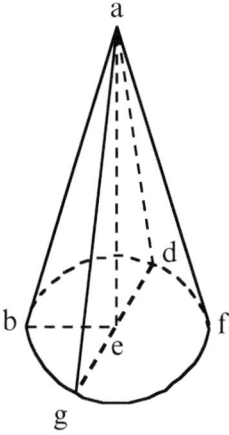

Figure 6.27

[47.1] Likewise I say that the product of *ab* and half the circumference *bgd* gives the area of the surface of cone *abgd* from its circular base to its vertex. If you do not think that is so, then let the product of line *ab* by a quantity equal to *iz* that is either longer or shorter than half the circumference *bgd* be the area of the surface of cone *abgd*. Let *iz* be twice the length of the circumference *bgd*. I construct on circle *bgd* a rectilinear figure with equal sides and angles. Let the sum of the sides less twice line *iz* be figure *tkl*. Now I draw lines *at*, *ak*, and *al*. I will show that line *ab* is perpendicular to line *bk* in this way. I draw line *et*. The squares on lines *eb* and *bt* equal the square on line *et*. Add the square of perpendicular *ae* to both {**p. 180**}. Then the squares on lines *ae* and *et* (that is, the square on line *at*) are equal to the squares on lines *ae*, *eb*, and *bt* (that is, the squares on lines *ab* and *bt*). Whence angle *abt* is a right angle. Therefore *ab* is perpendicular to line *tk*.[73]

[47.2] It is similarly shown that line *ag* is perpendicular to lines *ka* and *kl* and to line *ti*. And because lines *ab*, *ag*, and *ad* equal one another, the product of one of them, say *ab*, by half the sides of triangle *tkl* equals the surface area of cone *atkl*. This is greater than the surface area of cone *abgl*. Since it contains

---

[73] If the reading of this sentence (180.3–4, f. 114r.19) and its predecessor is correct, then in the next sentence there is a problem: "*ab*, *ag*, and *ad* are equal to one another." Point *a* is the vertex and points *b*, *g*, and *d* are on the circumference. Hence the lines are slant heights. Therefore we have a contradiction: *ab* is both a perpendicular and a slant height.

6 Finding Dimensions of Bodies 311

itself, namely what is between the circumference and point a,[74] half the sides of triangle tkl is less than quantity iz. Therefore the product of line ab that is the smaller line iz is greater than the surface area of the cone. And that is impossible. Therefore it is not possible that the product of line ab and the line that is longer than half the [circumference of the] circle bgd is the area of cone $abgd$. Again, let line iz be less than half the circumference of circle bgd. If it is possible, let the product of line ab and iz be the surface area of cone $abgd$. If that is so, then it follows that the product of ab by half the circumference of circle bgd produces the surface area of a cone larger than cone $abgd$, namely cone $acfh$ with vertex a and circular base fch. I will draw within circle fch a rectilinear figure cfh that is not tangent to circle bgd. I will draw cathete el from point e to line cf to divide line cf in two equal parts. I will draw lines ac, al, af, and ah. By what has already been said, it is shown that line al is perpendicular to line fc, and it must be equal to the perpendiculars falling from vertex a to lines cf and fh. So line al is longer than line ab. Because el is longer than eb, half the sides of figure cfh is greater than half the circumference of circle bgd, and half the circumference of bgd is greater than line iz. The product of al by half the sides of the rectilinear figure cfh produces the surface area of pyramid $acfh$ whose base is triangle cfh. The product of ab (that is shorter), al, and iz (that is shorter than half the sides of triangle cfh) produces the surface area of cone $acfh$ whose circular base is cfh. And this is inconvenient. The aforementioned cone $acfh$ with circular base contains pyramid $acfh$ with triangular base. For the product of line ab by the longer or shorter circumference of circle bgd does not produce the surface area of cone abg. Wherefore we may conclude that the product of ab by half the circumference of circle bgd yields the surface area of pyramid $abgd$ that lies between its vertex and circle bgd. And this is what I wanted to demonstrate. Whence if we set the perpendicular equal to 24 and the semidiameter eb equal to 7, then line ab is 25. If we multiply this by 22 or half the circumference of circle bgd, then we get 50 for the surface area of cone $abgd$.

[48] If a plane equidistant from the base cuts a cone, the section[75] of the plane and the cone is a circular line. Through its center passes the axis of the cone, a straight line from the vertex of the cone to the center of its circular base. The proof follows: let there be cone $abcd$ with vertex a, circular base $bcde$, and center f. Plane $hikl$ cuts the cone, equidistant from circle $bcde$. I say that the section of cone $bcde$ is the circular plane $hikl$. The proof follows. I mark two points b and c on circle bcd {p. 181}. Let arc bc be less than a semicircle. From points b and c draw diameters bd and ce. Draw lines ab, ac, ad, ae, and be. Then mark points h, i, k, and l on lines ab, ac, ad, and ae of the section $hikl$ [passing through the lines; see Figure 6.28]. Because triangles abd and

---

[74] The text repeats "Since it contains itself" which is not needed in translation (180.7).
[75] Here and below Fibonacci used the expression *communis sectio* (180.37). I do not translate the adjective.

ace intersect each other at points *a* and *f*, their common section is a straight line that falls perpendicularly from *a* to *f*. Therefore line *af* is the axis of cone *abcd* so that plane *hikl* cuts through point *m*. Again because triangles *abd*, *ace*, and *abe* intersect the two equidistant planes *bcde* and *hikl*, their common sections are equidistant. That is, line *hk* is equidistant from line *bd*, line *il* from *ec*, and line *hi* from *bc*. Because line *kh* is drawn equidistant from the base *bd* in triangle *abd*, so *ab* is to *ah* as *bd* is to *hk*. For as *ba* is to *ah*, so *ac* to *ai*. And since line *hi* has been drawn in triangle *abc* that is equidistant from the base *be*, as *ac* is to *ai* so *ce* to *il*. Therefore as *ab* is to *ah*, so *bd* to *dk*. Per equale therefore as *bd* is to *hk*, so *ce* to *il*. Alternately therefore, as *bd* is to *ce*, so *hk* to *il*. Now *bd* equals *ce*. Whence line *hk* equals line *il*. And because lines *hm*, *im*, *km*, and *lm* were drawn equidistant from the base lines *bf*, *cf*, *df*, and *ef* in triangles *afb*, *afc*, *afd*, and *afe*, as *bf* is to *fa*, so *km* to *ma*. As *cf* is to *fa*, so *im* to *ma*. Also, as *df* is to *fa*, so *km* to *ma*. And as *ef* is to *fa*, so *lm* to *ma*. But the ratio of lines *bf*, *cf*, *df*, and *ef* to *fa* is one ratio because these four lines all equal one another since they were drawn from the center of the circle. Whence the ratio of lines *hm*, *im*, *km*, and *lm* to *ma* is likewise one ratio. Because of this it is clear that they are equal to one another. They were drawn from point *m* to the common section of plane *hikl* and cone *abcd*. Whence it is obvious that the section *hikl* is a circle. Its center is *m* through which the axis *ax* passes. This is what I wanted to prove.

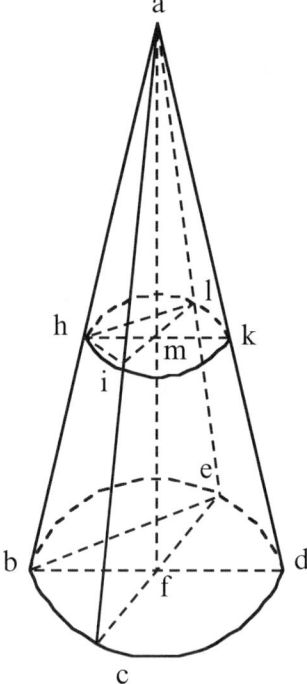

Figure 6.28

[49.1] In every part of a cone both of whose bases are circular and equidistant from each other, the line joining the two centers is perpendicular to each base. Thus if two diameters are drawn in the equidistant bases and their endpoints are joined, what lies between the two lines and the two diameters is continuous. Then the product of half the sum of circumferences of the two bases by the slant height[76] between them equals the area of the surface between the circles [see Figure 6.29]. For example: given the truncated cone *abgdz* with lower base circle *abg* and upper base circle *dez*, both circles being equidistant from one another. Draw line *it*, their line of centers, which is perpendicular to both circles. Draw the two diameters *bg* and *ez*. Then connect their endpoints to form lines *be* and *gz*. I say that the product of one of these lines, *be* or *gz*, and half the sum of the circumferences *abg* and *dez* gives the area of the surface between the circles *abg* and *dez*.

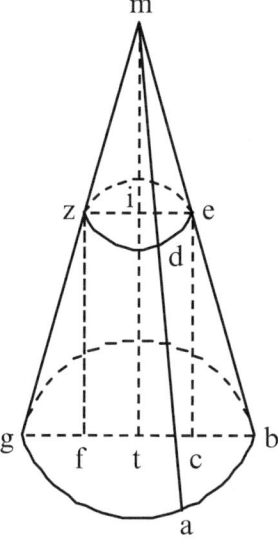

Figure 6.29

[49.2] The proof follows. Complete cone *mabk* by extending lines *be* and *gz* to point *m*. Thus lines *mb* and *mg* are equal to each other. Further lines *mc* and *mz* also equal each other. Whence lines *be* and *gz* are equal. And because every circle adds to its diameter three and a seventh {**p. 182**}: as circle *abg* is to its diameter *bg*, so circle *dez* is to its diameter *ez*. Because of this as arc *bag* or half of circle *abg* is to diameter *bg*, so arc *edz* or half of circle *dez* is to diameter *bg*. Alternately therefore as arc *bag* is to arc *edz*, so *bg* is to *ez*. But

---

[76] Fibonacci does not use the expression *slant height*. He simply says that *one of the lines* is an actor (181.31–32, f. 115v35).

as bg is to ez, so mb is to me, since in triangle mbg line ez is equidistant from the base line bg. By cutting off arc ag equal to arc edz from arc bag, the ratio of arc bag to arc ag is as line mb to line me, because by disjunction arc ba is to arc ag as line be is to line em. Therefore the product of me and arc ba is as the product of be and arc bag or arc edz. If the product of be and arc bag is added to both, then the products of be and arc bag and of be and arc edz will equal the product of be and arc bag and of em and arc ba. Adding again to both the product of em and ag or em and arc edz, you will have the product of all of mb and arc bag equal to the products of be and arcs bag and edz and of em and arc edz. But the product of mb and arc bag is the lateral area of that part of cone $mabg$ that lies between circle abg and point m. Therefore, the products of be and arcs bag and edz with the product of em and arc edz produces the same area. But the product of me and arc edz leads to the lateral area of cone $medz$ that lies between circle dez and point m. Whence the remaining lateral area of cone abc that lies between the circles abg and dez results from the product of be and arcs bag and edz. These arcs are half the sum of the circumferences of the two circles. And this is what I wanted to prove.

[50] By the numbers: set diameter bg equal to 14, diameter ez to two fifths of that or $\frac{3}{5}$ 5, and lines be equal to 15 and it to $\frac{2}{5}$ 14. Draw line mi. Through points e and z draw lines ec and zf equidistant from line it. Let lines ct and tf be equal as are lines ei and iz. Let the other lines bc and gf also be equal to each other. Therefore triangles ecb and zfg are similar and equal to each other. The angles at b and g equal one another. Therefore triangle mbg is isosceles with sides bm and mg equal. Because line ez is equidistant from line bg, triangle mez is similar to triangle mbg. Therefore triangle mez is isosceles with equal angles at e and z. Whence line mi is the cathete to line ez with point i at the midpoint of line ez. Since angle tib is a right angle, it is patent that line mit is straight. Therefore mt is the perpendicular within cone $mbag$ and passes through the center of circle edz. Because lines mb and mg are equal, if me and mz are taken from them, what remains is eb and zg, both equal to each other. Therefore zg is 15. And because ez is to bg as 2 is to 5, and because me is to mb as 2 is to 5, whence me is $\frac{2}{5}$ of eb or 10. The whole line mb is therefore 25. If the square of this is taken from 49 the square on tb, what remains is 576 for the square on cathete mt. Now arc bag is 22, the product of tb and $\frac{1}{7}$ 3, and arc edz is $\frac{4}{5}$ 8 or $\frac{2}{5}$ of arc bag. The sum of these two arcs is $\frac{4}{5}$ 30. Multiplying this by 15 or line eb produces 462 for the lateral area between circles abg and dez. The same answer 462 for the same lateral area between the bases edz and bag can be found by subtracting the lateral area of cone mez (found by multiplying 10 by $\frac{4}{5}$ 8 or line me by arc dz) from the lateral area of cone $mabg$ (found by multiplying 25 by 22 or line mb by arc bag). And this is what I wanted to show **{p. 183}**.

[51.1] Given a circle with its diameter extended outside the circle. Draw a perpendicular from the center to the circumference, thereby dividing the semicircle in two equal arcs. Then divide one of the quarter arcs into however so many equal parts. Next, draw a line from the top of the perpendicular through the closest point of division on the circumference to intersect the extension of the diameter. Then draw chords in the circle equidistant from the diameter and through every point of division on the quarter arc. Then the straight line lying between the intersection of the two extended lines and the center of the circle equals the sum of the radius and equidistant chords [see Figure 6.30]. For example: draw diameter *ab* in circle *abcd* with center *e*. Draw line *ec* perpendicular to the diameter *ab*, dividing arc *acb* in two equal parts. Let arc *cda* be divided in however so many equal parts, such as arcs *cd*, *df*, *fg*, and *ga*. Now draw chord *cd* and extend it outside the circle until it meets the extension of diameter *ab* at point *h*. Through points *d*, *f*, and *g* draw lines *di*, *fk*, and *gl* equidistant from diameter *ab*. I say that line *he* equals the [semi]diameter *ea* together with the sum of the chords *gl*, *fk*, and *di*.

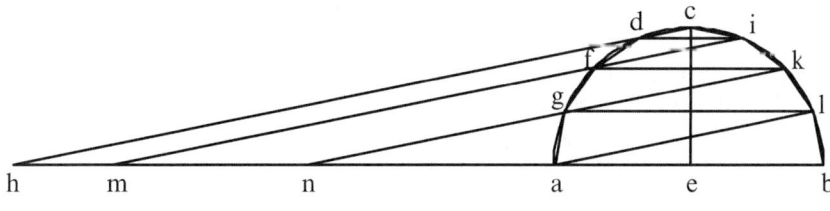

Figure 6.30

[51.2] The proof follows. Draw lines *al*, *gk*, and *fi*. Further, draw lines *if* and *kg* through the points *f* and *g* until they meet line *ah* at points *m* and *n*. Because line *al* meets the equidistant lines *ab* and *gl*, angles *bal* and *alg* are equal. The angles intercepting equal arcs are equal, whether drawn from the center or the periphery.[77] Hence arc *bl* equals arc *ag*. Likewise, because line *gk* falls on the equidistant lines *gl* and *fk*, angles *lgk* and *gkf* are equal. And therefore arcs *lk* and *gf* are equal. Similarly, arcs *ki* and *fd* are equal. And since arcs *ag*, *gf*, and *fd* equal one another, arcs *bl*, *lk*, and *ki* equal one another. Since arc *bl* equals arc *ag*, arc *lk* equals *gf*, and *ki* equals arc *fd*, therefore the whole arc *bi* equals the whole arc *ad*. The remaining arc *ic* equals arc *cd* as well as the arcs *ik*, *kl*, and *lb*. Whence angles *cdi*, *ifk*, *kgl*, and *lab*[78] equal one another. Again, because line *ch* falls on the equidistant lines *hb* and *di*, exterior angle *cdi* equals its opposite and interior angle *chb*. It is shown in a similar way that angle *imb* equals the exterior angle *ifk*, and that angle *knb* equals angle *kgl*. Because angle *lab* equals each of the afore-mentioned angles, therefore angles *lab*, *knb*, *imb*, and *chb* all equal one another. Therefore lines *la*, *kn*, *im*, and *ch*[79] are equidistant from one another, because line *bh* that falls across them makes exterior angles equal to the opposite interior angles. And because lines *la* and *gn* meet two equidistant

---

[77] Similar to *Elements*, III.27.
[78] Text has *gab*; context requires *lab* (183.27).
[79] Text has *ck*; context requires *ch* (183.34).

316   Fibonacci's *De Practica Geometrie*

lines *na* and *gl*, then *na* equals *gl*. Likewise, because the equidistant lines *kn* and *fm* meet the two equidistant lines *mn* and *kf*, then line *mn* equals line *kf*. Again, because lines *im* and *dh* meet the equidistant lines *mh* and *id*, then line *mh* equals line *id*. Whence line *ah* equals the sum of *gl*, *fk*, and *di*. If the semidiameter *ea* is added to both, then semidiameter *ea* with the lines *lg*, *kf*, and *id* equal line *eh*. And this is what I wanted to prove.

[52.1] Suppose in a hemisphere there is a pyramidlike body made of however so many columns sitting atop one another with a perfect cone on top.[80] Its apex meets the top of the pole perpendicular to the center of the base of the hemisphere[81] {p. 184}. The bottom of the lowest column is a major circle of the sphere and the base of the hemisphere. All the bases of the columns are equidistant from the major circle. The perpendicular pole goes through the centers of the circular bases. Straight lines joining the endpoints of the diameters of each of the parts are equal to one another. Then, the surface area of the whole composite of the columns and the perfect cone is less than twice the area of the base of the hemisphere that contains the body. It is also greater than twice the area of the major circle of the hemisphere contained by that body[82] [see Figure 6.31].

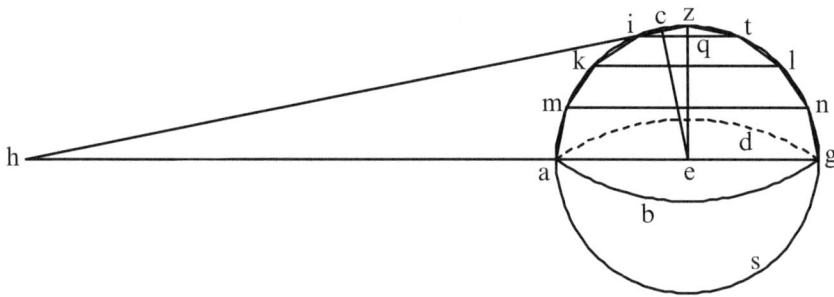

Figure 6.31

[52.2] The proof follows.[83] Let a major circle of a hemisphere be circle *abgd* with center *e* that is also the center of the sphere, since the center of the sphere is the center of all major circles on a sphere. Let the pole of circle

---

[80] "In medietate spere componatur corpus ex quotcumque portionibus pyramidum columnarum, et superior harum sit pyramis perfecta" (183.43–184.1). If you remove the diagonal lines in Figure 6.30 and rotate it 180 degrees, you would create the figure described in the Latin statement.

[81] Fibonacci piles truncated cones atop one another before the topmost perfect cone *izt*. The word "truncated cone" is not found in the Latin. Rather what appears is "corpus ex quotcumque portionibus pyramidum columnarum." But because a perfect cone sits atop "the pyramid," truncated cone is a more appropriate translation.

[82] This additional conclusion is addressed in [53] below.

[83] For Figure 6.31, the reader is asked to imagine circular bases about diameters *mn*, *kl*, and *it*.

$abgd$ be z. Further, let a major circle perpendicular to circle $abgd$ go through pole z. Whence the diameter ag of both circles is their common section. Let arc azg be a semicircle and equal to semicircle abg. Now the base of the hemisphere is circle $abgd$ and its pole is z. Construct within the hemisphere a body composed of three parts of columns and of one cone. The first part is a column with circular base $abgd$ and upper circular base with linear diameter mn. Let the lines am and gn be drawn from the ends of the diameters. The second part has a circular base with linear diameter mn and upper circle with linear diameter kl. Let the lines mk and nl be drawn from the ends of these diameters. The third part has a circular base with linear diameter kl. Its upper circle has the linear diameter it. The circular base of the cone with linear diameter it has at its apex the pole of the sphere. Straight lines falling from it to the endpoints of the basal diameter are zi and zt. Let lines am, mk, ki, and iz be equal to one another. Whence lines zt, tl, ln, and ng equal those lines because the upper and lower bases of the parts are mutually equidistant. All the lines are on the one plane that cuts the whole body in two halves at the diameters of the parts. The perpendicular ze comes down passing through the centers of the circles. Then, extend the diameter ga and chord zi to point h. Line eh will be equal to the sum of the radius ea and the diameters mn, kl, and it.[84] Draw hz. Cathete ec joins the midpoint of chord zi to the center e. And because the perpendicular ec is drawn in the orthogonal triangle hez on a side that subtends a right angle, the ratio of hz to ze is as ze to zc. Whence the product of hz and zc equals the square on line ez. But line ez equals the radius ea. Therefore the product of zh and zc equals the square on the radius ae. But zi is twice zc. Consequently the product of hz by zi is twice the square on the radius ae. If then you multiply twice this by $\frac{1}{7}3$, you get twice the area of circle $abgd$. Therefore, twice the area of circle $abgd$ is the result of multiplying zh by zi and then by $\frac{1}{7}3$. Because line hz is longer than line hi,[85] {**p. 185**} the product of zi and he multiplied by $\frac{1}{7}3$ yields somewhat less than twice the area of circle $abgd$. Because he equals the sum of the lines ae, mn, kl, and it, the product of zi and the sum of the lines ae, mn, kl, and it multiplied by $\frac{1}{7}3$ produce a bit less than twice the area of circle $abgd$.

[52.3] I will show that the area of the surface of the whole composite body comes from this product. First, the area of the surface of the first part of the cone with circular base $abgd$ and upper circular base with diameter mn arises from the product of radius ae and line om multiplied by $\frac{1}{7}3$ and then by am or iz.[86] Then, the surface area of the following part[87] with circular base,

---

[84] By [51].
[85] A marginal note in the text and manuscript would incorrectly substitute "hi" for "he" in this context (183, f. 118v).
[86] Text has $he.miz$; context requires hoc est iz (185.8).
[87] A marginal note in the text and manuscript adds "cuius basis est circulus $abgd$ et eius superior est circulus cuius diameter est mn" (185, f. 118r).

318   Fibonacci's *De Practica Geometrie*

diameter mn, circular upper base and diameter kl comes from the product of $\frac{1}{7}3$ and the semidiameters on and pl and all of that by the line mk or iz. Similarly, the surface area of the third part with circular base and diameter kl and upper circular base with diameter it comes from the product of $\frac{1}{7}3$ and the semidiameters pk and iq multiplied by ik or iz.[88] Again, the surface area of the perfect cone with circular base and diameter it with apex at z comes from the product of $\frac{1}{7}3$, iz, and tq. Therefore the surface area of the whole body composed of the aforementioned parts of the three columns and one cone is the result of multiplying the sum of lines ae, mn, lk, and it by line iz, and that product by $\frac{1}{7}3$. This proves that the surface area is somewhat less than twice the area of circle.

[53] Similarly, if this pyramidlike body contains a hemisphere,[89] it will be shown that its area is more than twice the area of the basal circle of the hemisphere with radius[90] equal to line ec. The arc of the semicircle is constructed orthogonally on the diameter and touches the midpoints of lines am, mk, ki, and iz. The pole of the basal circle of the hemisphere will be on line qz. Because of this we make line er equal to line ec. And we enclose circle rsv in the space er. Line rv is the diameter of the basal circle of the [inner] hemisphere. Because cathete ec is drawn to side hz[91] opposite angle hez in triangle hez, so line hc is to ce as line ec is to cz [see Figure 6.32]. Hence the product of hc and cz is as the square on ec or the square on er. And because iz is twice cz, the product of hc and iz is twice the square on the radius er. But hc is longer than he. Whence the product of he by iz is more than twice the square on er. Therefore the product of he, iz and $\frac{1}{7}3$ is more than the product of $\frac{1}{7}3$ and twice the square on the radius er. But the product of $\frac{1}{7}3$, he, and iz[92] yields the surface area of the body containing the hemisphere, as was shown above. And twice the area of the basal circle rsv of the [inner] hemisphere is twice the product of $\frac{1}{7}3$ and the square on er. Therefore the surface area of the body composed of parts of the cone containing the hemisphere is larger than twice the area of the circle of the hemisphere contained by that body. Thus it is less than twice the area of the circle of the hemisphere containing the body, as shown above. And this is what I wanted to prove.

---

[88] A marginal note in the text and manuscript puts "hoc est ducta in $\frac{1}{7}3$" in place of zi, which seems repetitious (185, f. 118v).

[89] The figure requires a cone izt supported by three disks or columns: $amng$, $mkln$, and $kitl$ as in the two previous figures. Because only lines from the cross-section of the cone appear in the proof, I have adjusted the figure to show clearly the three nested cross-sections.

[90] Text has *diameter;* context requires *radius* (185.20).

[91] Text has bz; context requires hz (185.25).

[92] A marginal note in the text and manuscript correct ic to iz (185.31, f. 118v).

6 Finding Dimensions of Bodies   319

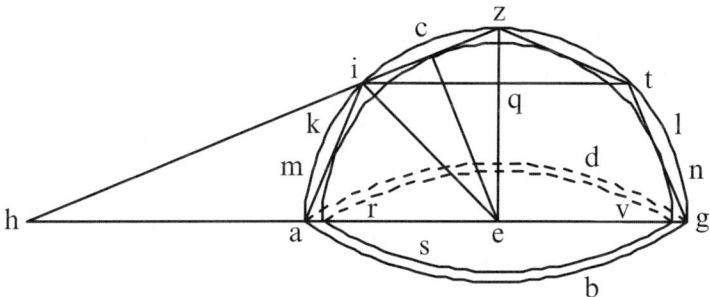

Figure 6.32

Volume and Surface of Spheres[93]

[54] Because all of this has been shown in a clear and open way, we say that the surface area of half of any sphere is twice the area of a major circle of a sphere whose diameter is the diameter of the sphere. Whence the area of the whole sphere is four times the area of that circle. Now to show this in numbers. If you wish to compute the surface area of a sphere with diameter equal to 7, then quadruple $\frac{1}{2}$ 38 the area of one of its major circle {**p. 186**}. Or, square the diameter to get 49 and multiply this by $\frac{1}{7}$ 3 to obtain 154 for the area of the surface of the sphere. And if you multiply half the square of the diameter by $\frac{1}{7}$ 3 or double the area of the aforementioned circle, you will have 77 for the area of the given hemisphere. And if you multiply the area of the whole sphere or 154 by $\frac{1}{6}$ 1 or a sixth of its diameter, or if you multiply a third of 154 by $\frac{1}{2}$ 3 or half the diameter you will obtain $\frac{2}{3}$ 179 for the volume.[94] The procedure proved by the scholars is to multiply a third of the surface area of the sphere by half its diameter to find the volume of the whole sphere.

[55] To prove this, consider the sphere *ab* whose radius is line *ag* with center at *g*. I say that the product of *ag* by a third of the surface area of sphere *ab* equals the volume of sphere *ab*. The proof follows. If the product of *ag* by a third of the surface area of sphere *ab* does not equal the volume of the sphere, then it is the volume of a sphere that is either larger or smaller than sphere *ab* [see Figure 6.33]. First, let sphere *de* be larger than sphere *ab*, both spheres having the same center. Therefore it is possible that sphere *de* is a corporal figure with many bases none of which touches the surface of sphere *ab*. Whence, each of the perpendicular lines going from the center to the surface of the sphere will be longer than line *ag*. These lines through the center meeting the surface of the sphere form pyramids with vertices at the center and

---

[93] "De embado et superficie rotunde spere" (185.38, *f.* 118v.29–30).
[94] "area magnitudinis" (186.6).

bases on the surface of the sphere. The volume of each one of these pyramids is computed by multiplying the perpendicular to the base by a third of the area of its base. Because of this, line *ag* the radius of sphere *ab* is shorter than any one of those perpendiculars. Further, the product of line *ag* by a third of each base will be smaller than the volume of the pyramid to which the base belongs. Therefore, the product of line *ag* by a third of the surface of that body is less than the volume of the body. But the surface area of that body is greater than the surface area of *ab*. Therefore the product of line *ag* by a third of the surface of that body is much less than the volume of the body. And yet it was proposed that the product of *ag* by as third of the surface area of sphere *ab* equals the volume of sphere *de*. Therefore it is necessary that sphere *de* be a much smaller body than it is. And that is impossible. Consequently the product of line *ab* by a third of the surface area of sphere *ae* is not larger than the sphere *ab*.

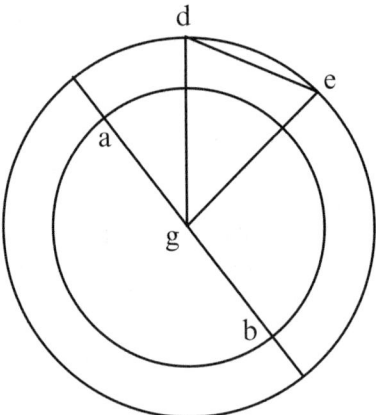

Figure 6.33

[56] I say again that the sphere *ae* is not smaller than sphere *ab*. But if it were possible, then let sphere *zh* also with center *g* be less than sphere *ab* [see Figure 6.34]. And again if possible, let there be a body of many bases in sphere *ab*, none of which touches the surface of sphere *zh*. Whence each of the perpendicular lines [*gz*] going from the center of sphere *ab* to the surface of that body will be shorter than half the diameter *ag* of sphere *ab*. Therefore the product of *ag* by a third of each surface area is greater than the volume of the pyramid whose base is that area and whose vertex is at the center *g*. The product, therefore, of line *ag* by a third of the surface area of sphere *ab* is larger than the volume of the body. But it was proposed that it would be equal to the volume of sphere *zh*. Therefore sphere *zh* is much larger than this body which lies between it. And this is impossible. Therefore the product of line *ag* that is half the diameter of sphere *ab* by a third of the surface area is larger than the body. Therefore it is equal to the body, as required. Since this has been established,

if we want to have half of some sphere, we multiply {**p. 187**} the surface area by a sixth of its diameter, or half its diameter by a third of its surface area. For example: the diameter of a sphere is given at 10. If we multiply this by its half, we get 50. Multiplying further by $\frac{1}{7}$ 3, we obtain $\frac{1}{7}$ 157 the surface area of the hemisphere. If we multiply this by a sixth of the diameter, namely $\frac{2}{3}$ 1, or if we multiply a third of $\frac{1}{7}$ 157 by five or half the diameter, we obtain $\frac{16}{37}$ 261 for the volume of the hemisphere.

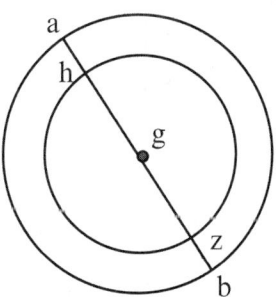

Figure 6.34

[57] If it is necessary for us to measure a part of a sphere, that may be more or less than a hemisphere, such as round fonts and other similar vases, we need to find the ratio between the polar distance *dg* from the center of the circular mouth of the object[95] and the radius of the sphere. Thereby we will know the proportion of the surface area of the sphere and its volume, and we will have what we want [see Figure 6.35]. For example: let *ab* be the diameter of the font, *g* its center, and point *d* on the pole of the circle. Whence line *dg* is perpendicular to the area of the circle with diameter *ab*. If we square half the diameter and divide by *gd*, we will have what remains from the whole diameter of the sphere above the line *gd*. For example: let the diameter *ab* be the root of 160. Whence *gb* the half of *ab* is the root of 40 whose square is 40. If we divide this by line *gd* or 4, we obtain 10 for what remains from the diameter above line *gd*, namely line *ge*.[96] Then the diameter of the sphere *de* will be 14. If we divide this in two equal parts, then the center of the major circle *aebd* of the sphere will be *z*. I shall make a ratio of line *gd* with the semidiameter *zd*, namely 4 with 7. Since the ratio of *gd* to *dz* is $\frac{4}{7}$, I will take $\frac{4}{7}$ of 308 the surface area of the hemisphere to obtain 176. This results from multiplying the area *dz* by *de* and then by $\frac{1}{7}$ 3 to obtain the surface area of that part of the sphere whose base is a circle of diameter *ab* with pole at *d*. The arc covering that portion of the great circle, which in turn falls on the sphere, is arc *abd*.

---

[95] "a centro circuli oris ipsius ad punctum poli eiusdem circuli" (187.9–10).
[96] *Elements*, VI.13.

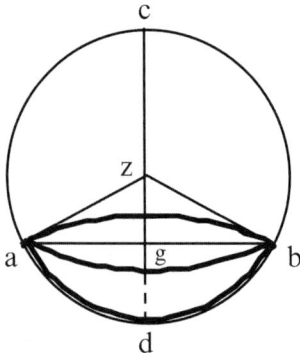

Figure 6.35

[58] The volume of the part can be found by multiplying a third of the surface area by 7 or the radius zd, and then subtracting [the volume of] the cone with apex at the center z. Its base is a circle of diameter ab, altitude zg is 3, and volume of the cone is $\frac{5}{7}$ 125. What remains for the area of the required part is 285 less $\frac{1}{21}$. If we want to know the area of the greater, remaining part of the sphere with the same base, diameter ab, and altitude eg called "an arrow" of length 10, we must proceed in the same way that we did in computing the smaller portion. That is, we divide the square on line gb that is 40 by the arrow ge, and we have 4 for line gd that is the arrow of the remaining part of the sphere. Whence ed the total diameter of the sphere is 14. If we multiply this by the arrow eg and then the product by $\frac{1}{7}$ 3, or by the surface area of the hemisphere that is 308, we obtain $\frac{1}{7}\frac{0}{9}$ for the ratio of arrow eg to radius ez. Thus we obtain 440 for the area of this larger part of the sphere. If we multiply this by a sixth of the diameter of the sphere or by a third of it and the radius, we will obtain $\frac{2}{3}$ 1026. If we add to this $\frac{5}{7}$ 125 the area of the cone mentioned above, we will get $\frac{2}{3}\frac{1}{7}$ 1152 for the volume of the larger part.

[59] In order to demonstrate these things geometrically, consider the hemisphere bzd with circular base $abgd$ and point z atop its pole {**p. 188**}. Designate the great circle[97] azg on hemisphere bzd, and let it be perpendicular to circle $abgd$. Their common section ag is a diameter to both circles. Let the perpendicular line ze come from point z atop the pole through the plane of circle[98] azg to the diameter ag [see Figure 6.36]. Let the point e be the center of both circles. Divide arc az in however so many equal parts, such as arcs ai, it, and tz. Through points i and t, draw chords ik and tl equidistant from the diameter ag. Extend chord zt and the diameter ga outside the circle to point m. Also extend outside of circle azg one of the two chords that are equidistant from the diameter ag, until it meets line mz. Whence, if we extend chord ki, it will meet line mz at

---

[97] A marginal note in the text and manuscript has "arcus semicirculi magni azg" (188, f. 120r).
[98] A marginal note in the text and manuscript has "semicirculi" (ibid.).

point n. Circumscribe circle $ihkx$ equidistant from circle $abgd$ with z its pole. Let its center be f and its diameter ik. This circle cuts part xzh from hemisphere dz. Within the portion lies the arrow or perpendicular line zf. I say that the ratio of part of the surface area of the portion xzh to the surface of hemisphere dzb is as zf to ze the radius of the sphere. The proof follows. From the foregoing we know that line me equals the sum of the radius ea and the chords ik and tl. In a similar way it can be shown that line nf equals the sum of the radius fi and chord itl.[99] Because line nf in triangle mez has been drawn equidistant from the base em, as zf is to ze so nf is to me. But as line nf is to me, so the surface zt in nf is to the surface zt in me. But as the surface zt in nf is to the surface zt in me, so the product of the surface zt in nf and $\frac{1}{7}3$ is to the product of $\frac{1}{7}3$ and the surface zt in em. But the product of me in tz and $\frac{1}{7}3$ produces the surface area of the body composed of the three parts of the cone within the hemisphere all of whose altitudes is line ze, and their greater base is circle $abgd$.

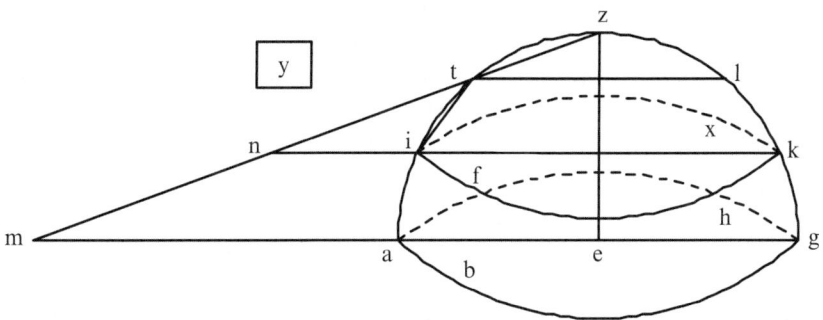

Figure 6.36

[60] In a similar way it can be shown that the product of nf, tz, and $\frac{1}{7}3$ produces the surface area of the body composed of two parts of the cone[100] lying within a part of sphere hzx. Therefore as zf is to ze, so the body composed of the two parts of the cone with altitude zf is to the surface of the body composed of the three parts of the cone with altitude ze. Then I make the ratio of the area of plane y to the surface of the hemisphere dzb be as the ratio of the surface area of the body with altitude zf to the surface of the body with altitude ze. Then alternately, as the surface of the body with altitude ze is to the surface of the hemisphere dzb, so the surface of the body of altitude zf is to the area of plane y. But the surface area of the plane with altitude ze is within the hemisphere; hence its surface area is less than the surface area of the hemisphere. And since the body itself contains a hemisphere, the surface area of the body is larger than the surface of that hemisphere. Whence, when the surface area

---

[99] Text has only l; context requires lk (188.17).
[100] A marginal note in the text and manuscript has "pyramidum columnae" (188, f. 120v).

of the body with altitude *ze* is less than the surface of the hemisphere, then the surface area of the body with altitude *zf* is less than the area of plane *y*. And where one is greater, so is the other greater. But since the surface of the body with altitude *ze* is less than the surface of the hemisphere *dzb*, then the surface area of the body with altitude *zf* is likewise less than the surface of the portion of sphere *hzx*. Again we have the relation of greater and greater. Because of this, the area of *y* and the area of that part of the sphere to the surface area of the body with altitude *zf* are one and the same portion. So, it is obvious that the surface of part of sphere *hzx* equals the area of the plane *y*, and that the ratio of the surface of the body of altitude *ze* to the surface of the hemisphere *dzb* is as the ratio of the surface of the body with altitude *zf* to the plane *y* {p. 189}. Hence, as the surface area of the body with altitude *ze* is to the surface area of the hemisphere *dzb*, so the surface of the body with altitude *zf* is to the surface of part of sphere *hzx*. When this is permuted, the ratio of the part of *hzx* to the surface of hemisphere *bzd* will be as the ratio of the surface of the body of altitude *zf* to the surface of the body of altitude *ze*. But the ratio of these two bodies is as the ratio of line *nf* to line *me*. The ratio of line *nf* to line *me* is as the arrow *zf* to the spherical radius *ze*. Therefore the ratio of the surface area of part of *hzc* to the surface area of hemisphere *dzb* is as *zf* the altitude of the part to the radius *ze*. And this is what I wanted to prove.

[61][101] If a cube is constructed in a sphere together with a pyramid of four equal triangular faces and equal edges, then the cube is thrice the pyramid. To prove this, consider [the figure with] diameter *ab* of a given sphere divided at point *c* so that *bc* is twice *ca* [see Figure 6.37]. Construct a semicircle [about points] *a*, *d*, and *b*. Draw line *cd* from point *c* on *ab* to form right angles. Draw lines *bd* and *ad*. Line *ad* will be an edge of the cube, and line *bd* an edge of the triangular pyramid in the sphere the diameter of which is *ab*. Because triangles *adb*, *dcb*, and *dac* are similar, the square on diameter *ab* is triple the square on line *ad*, the square on line *bd*, and the square on line *dc*, as Euclid showed.[102] Also, the square on line *db* is twice the square on line *ad*, and the ratio of the square on line *bc* to the square on line *ad* is 4 to 3. For example: because triangles *dab* and *dac* are similar and similar triangle have equal angles and the sides opposite those angles are proportional, then as *ba* is to *ad*, so *ad* to *ac*. When three lines are in continued proportion, as the first is to the third,[103] so the square on line *ab* is to the square on line *ad*. But *ba* is thrice *ac*. Whence the square on *ab* is thrice the square on line *ad*. Likewise, because triangles *adb* and *dcb* are similar, the square on line *ba* is to the square on line *ad* as the square on line *bd* is to the square on line *dc*. Since angles *dab* and *cdb* equal each other, it follows that

---

[101] Paragraphs [61] through [65] are a unit. During their development, Fibonacci utilizes Figures 6.37, 6.38, and 6.39.

[102] *Elements*, XIII.15.

[103] A marginal note in the text and manuscript has "ita quadratum quod fit a prima ad quadratum quod fit a secunda, ergo est sicut *ba* ad *ac*" (189, *f.* 121v).

the square on line *bd* is thrice the square on line *dc*. Likewise, because line *dc* is the mean proportion between lines *bc* and *ca*, so as *bc* the first term is to *ca* the third term, so the square on line *bc* is to the square on line *cd*. But as *bc* is to *cd*, so *bd* is to *da*. Whence as *bc* is to *ca*, so the square on line *bd* is to the square on line *da*. Now *bc* is twice *ca*. Whence the square on line *bd* is twice the square on line *da*, as was said above. Likewise, the ratio of the square on line *bc* to the square on line *ad* is composed of the ratio of the square on line *cb* to the square on line *ba* and of the ratio of the square on line *ba* to the square on line *ad*. Now *bc* is to *ba* as 2 is to 3. Whence the square on line *bc* is to the square on line *bz* as 4 is to 9. The ratio therefore of the square on *ba* to the square on *ad* is 9 to 3. Since the square on line *ba* is thrice the square on line *ad*, therefore the ratio of the square on line *bc* to the square on line *ad* is composed of the ratio of 4 to 9 and of the ratio of 9 to 3. Whence it is obvious the ratio of the square on *bc* to the square on *ad* is 4 to 3.

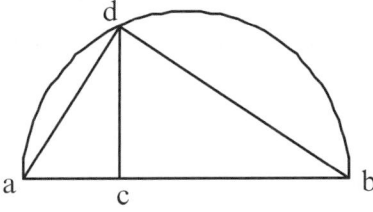

Figure 6.37

[62] Now with all of the foregoing understood, I make line *ef* equal to line *ad*, and I construct on line *ef* the cube *efghiklm* [see Figure 6.38]. Further, I place line *no* equal to line *cd*, and I circumscribe circle *opq* about the space *no*. I inscribe in circle *opq* equilateral triangle *opq* {**p. 190**}. I extend *on* to *r* to make *or* the diameter of circle *opq* [see Figure 6.39]. And I draw lines *pr* and *qr*, and triangles *opr* and *oqr* are equal and similar.[104] Whence angle *por* equals angle *qor*. Therefore as *po* is to *oq* so *ps* is to *sq*. Since *po* equals *oq*, therefore *ps* equals *sq*. Adding to both sides the line *os*, the two lines *ps* and *os* equal the two equal lines *qs* and *os*. The base *op* equals the base *oq*. Whence triangles *osp* and *osq* are equal to one another, and angle *osp* equals angle *osq*. The cathete *os* therefore falls on line *pq*. Because the square on the side of the equilateral triangle inscribed in the circle is triple the square on the semidiameter of its circle, the square on line *op*, therefore, is thrice the square on the semidiameter *on*. For, line *on* equals line *dc*. Therefore the equivalent[105] squares on lines *op* and *bd* have the same ratio. Whence *op* equals *db*. Then on the center I erect a cathete *nt* equal to line *bc*, and I draw lines *no*, *np*, and *nq*. Each of these equals line *db*, that is, line *op*. For the squares on lines *tn*

---

[104] Note the lack of word for congruency (190.2).
[105] Fibonacci used the word *adequalia* (190.10–11).

and *no*, or *tn* and *np*, or *tn* and *nq* are equal to the squares on lines *bc* and *cd*. Whence each of the squares on lines *to*, *tp*, and *tq* equals the square on line *bd*. Therefore the pyramid *topq* has been constructed from four equal equilateral triangles, of which each edge equals line *bd*, as required.

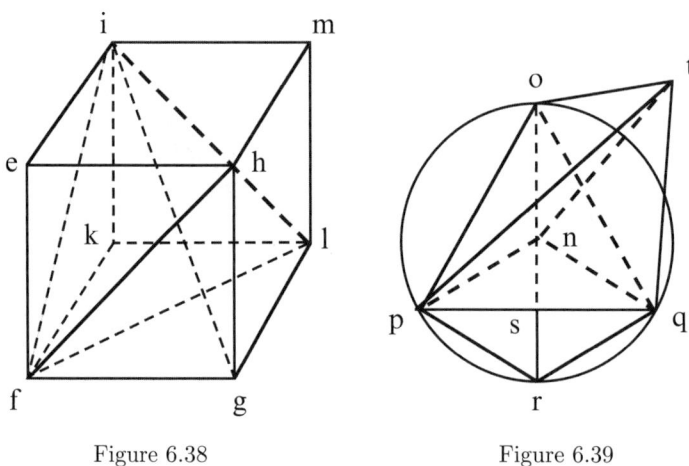

Figure 6.38   Figure 6.39

[63] I say, therefore, that the cube *eil* is three times the pyramid *topq*. The proof follows. Because line *ps* is half of line *op*, the square on line *op* is four times the square on line *ps*. But the square on line *op* is twice the square on edge *ef* of cube *eil*.[106] Since *op* equals *bd* and *ef* equals *da*, hence *op* is to *ef* as *ef* is to *ps*. Whence the right angled plane *op* by *ps* equals the square on line *ef*, that is, the square *eg*. Whence the ratio of square *eg* to the right angled plane *os* by *sp* is as the ratio of the surface area *op* by *ps* to the plane *os* by *sp*. But the ratio of the right angled plane *op* by *ps* to the right angled plane *os* by *sp* (that is, to the area of triangle *opq*) is as *op* to *os*. Therefore as *op* is to *os*, so square *eg* is to triangle *opq*. Whence the square of square *eg* to the square of triangle *opq* is as the square on line *op* to the square on line *os*. But the ratio of the square on line *op* to the square on line *os* is sesquitertial. Because if the square on line *ps* is subtracted from the square on line *op* (that is one part from four parts), what remains is three fourths of the square on line *op* for the square on line *os*. Therefore the ratio of the square on line *op* to the square on line *os* is 4 to 3. This ratio is called sesquitertial. Therefore the ratio of the square of square *eg* to the ratio of the square of triangle *opq* is as 4 to 3. But the ratio of 4 to 3 equals the ratio of the square on line *cb* to the square on line *ad*. Now the perpendicular *tn* equals *cb*, and the altitude *ei* of the cube equals line *ad*. *Per equale* therefore as 4 is to 3, so the square on

---

[106] Here and for the next four readings of *ef*, as the context requires, the text has *if* (190.21).

the perpendicular *tn* is to the square on the altitude *ei*. Therefore as the ratio of the square of square *eg* is to the square of triangle *opg*, so the square on the perpendicular *tn* is to the square on line *ei*. *Per equale* therefore as square *eg* is to triangle *opq*, so the perpendicular of pyramid *topq* is to the altitude of cube *hmk*, that is, *tn* to *ei*. Whence the product of triangle *opq* and perpendicular *tn* equals the product of square *eg* and altitude *ei*. But the product of square *eg* and altitude *ei* equals the area of cube *kmh*. And the product of triangle *opq* and perpendicular *tn* yields three times the area of pyramid *topq* {p. 191}. Therefore cube *eil* has been constructed in a sphere of diameter *ab* and it is thrice the pyramid constructed in the same sphere. This is what we wanted to prove.

[64] Another way: in cube *eil* draw lines *if*, *ih*, *fh*, *fl*, *lh*, and *il*. In cube *hmk* construct pyramid *lhif* of four plane triangles and equal sides of which each is the diameter of one of the six square sides of the cube. Whence the square on each of the sides is twice the square on the edge of the cube, as shown above. Whence the edges of pyramid *lhif* equal the edges of pyramid *topq*. So it is obvious that pyramid *lhif* equals pyramid *topq* [see Figure 6.40]. If pyramid *lhif* is removed from cube *hmk*, what remains of the cube are four pyramids, each equal and similar to one another, namely pyramids *ihef*, *lhgf*, *fikl*, and *himl*.[107] The base of any one of the first two pyramids is half the square *eg*, and their altitudes *ei* and *gl* are equal. The other two pyramids contain square *km* which is equal to square *eg*, and their altitude is the height of the cube. Since their altitudes are *mh* and *kf*, draw line *gi* in the cube. Because pyramids *ihef* and *lhgf* are equal, their sum is twice pyramid *ihel*. And because square *eg* is twice triangle *hif*, pyramid *ihefg* with square base *eg* is twice pyramid *ihef* with triangular base *hef*. And both have the same altitude, line *ei*. Whence pyramid *ihefg* equals the two pyramids *ihef* and *lhgf*. But pyramids *ihefg* and *topq* have bases and altitudes that are mutually proportional. For as square *eg* is to triangle *opq* so altitude *tn* to altitude *ie*. Further, the pyramids with mutually proportional bases and altitudes equal one another. Therefore pyramid *ihefg* equals pyramid *topq* [see Figure 6.41]. Now pyramids *ihef* and *lhgf* equal pyramid *ihefg*. Since they are equal to the same one, they are equal to each other. Therefore pyramids *ihef* and *lhgf* equal pyramid *topq*. In a similar way it can be demonstrated that the remaining two pyramids *fihl* and *himl* equal pyramid *topq*. Therefore it has been demonstrated that cube *kmh* contains the three pyramids equal to pyramid *topq*. Whence, it is clear that cube *kmh* is three times as large as pyramid *topq*. And this is what I wanted to show.

---

[107] The "other" manuscript lacks most of this sentence (191.25–27, f. 123r).

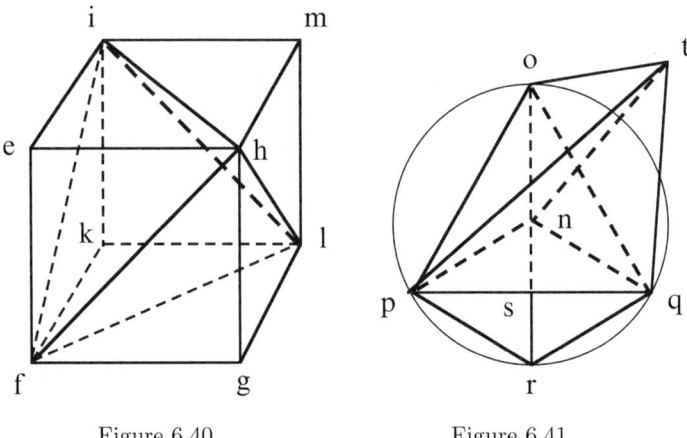

Figure 6.40   Figure 6.41

[65] To make this clearer: let the diameter of the sphere be 6, the edge *ef* of the cube be the root of 12, the edge *op* of the pyramid be the root of 24, and the square on the edge *ef* be the square *eg* or 12. Multiplying this last by width *ei* the root of 12 produces 12 roots of 12 that equals the root of 1728 for the volume[108] of the cube *hmk*. A ninth part of this square is 192 for the square of pyramid *topq*. For example: if we take a fourth of the square on the edge *op*, we have 6 for the square on line *ps*. If we take 6 from 24 the square on line *op*, what remains is 18 for the square on cathete *os*. If we multiply this by 6 the square on line *ps*, we get 108 for the square of triangle *opq*. And if we multiply the square of a third of the height or a ninth of the perpendicular *tn* that is the square 16, or if we multiply a ninth of 108 or 12 by 16, we get again 192 for the square of pyramid *topq*. But if we multiply by 9, we get 1728 as above for the square of cube *hmk* {**p. 192**}. This I wanted to demonstrate. For 42 is a bit more than the square root of 1728, and 14 a third of 42 is a bit more than the square root of 192. This shows that the cube is thrice the pyramid.

[66.1] *If a quadrilateral is circumscribed about a circle and its sides are tangent to the circle and meet at right angles, the quadrilateral that results will be a square. Further, the outside square will be twice the inside square.* So let quadrilateral *efgh* be around circle *abcd*, touching it at the points *a*, *b*, *c*, and *d*. Then draw lines *ab*, *bc*, *cd*, and *da*. I say that quadrilateral *abcd* is a square, and that the exterior quadrilateral *eg* is twice the square *abcd*. Let *i* be the center of circle *abcd*. Join lines *ia*, *ib*, *ic*, and *id* [see Figure 6.42]. Because line *eh* touches circle *abcd* at point *a*, from this point of contact the line *ia* was drawn. Line *ia* is therefore the cathete to line *eh*, as is understood in Geometry. Therefore both angles *iae* and *iah* are right angles. It is shown in a similar way that angles *ibe*, *ibf*, *icf*, *icg*, *idg*, and *idh* are right angles.

---

[108] The text has *pro quadrato* here and in the next sentence (191.35, f. 123r.15). The context however indicates that the volume of the cube is the end product of the computation.

Because line *ea* intercepts lines *be* and *ia*, the angles at *e* and those under *eai* are right angles. Further, lines *eb* and *ia* are equidistant. It is shown in a similar fashion that lines *ib* and *ae* are equidistant because the angles at *e* and under *ibe* are right angles. It was shown that lines *ia* and *ba* are equidistant. Whence line *ia* equals line *be*, and *ib* equals *ae*. Therefore quadrilateral *ebia* is equilateral. Further I say that it is right angled, for one of the angles under *ibe*, *bei*, and *eai* is a right angle. Whence the remaining angle under *bia* is a right angle. Therefore *ei* is a square. It is shown in a similar way that quadrilaterals *fi*, *gi*, and *hi* are squares. Whence, all the angles about the center *i* are right angles. Therefore it is clear that line *ai* is continuous with line *ic* and line *bi* with line *id*. Therefore lines *ac* and *bd* are diameters of circle *abcd* and they intersect each other at right angles. Whence each of the lines *ab*, *bc*, *cd*, and *da* are chords of a fourth of the circle. Therefore lines *ab*, *bc*, *cd*, and *da* equal one another. Consequently quadrilateral *abcd* is equilateral.

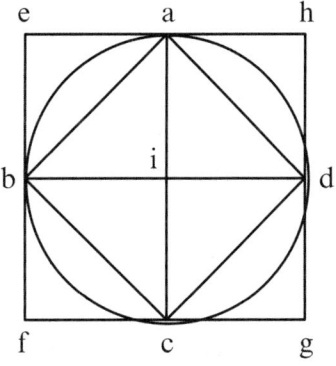

Figure 6.42

[66.2] I say again that it is right angled. Because triangle *aib* is a right triangle with two equal sides *ib* and *ia* both drawn from the center to the circumference, angles *iba* and *bai* equal each other, and angle *bia* is a right angle. So each of the angles under *iba* and *iab* is half of a right angle. It will be shown through triangles *bic*, *ida*, and *icd* that each of the angles under *ibc*, *icb*, *icd*, *idc*, *ida*, and *iad* is also half of a right angle. Whence angles *abc*, *bcd*, *cda*, and *dab* are right angles. Therefore quadrilateral *abcd* is right angled. It has been shown that it is equilateral. Therefore equilateral *abcd* is a square. I say it again that square *eb* circumscribed about the circle is twice square *abcd* inscribed in the circle. For each of the squares *ie*, *if*, *ig*, and *ih* is a parallelogram, and their diameters are *ab*, *bc*, *cd*, and *da* that cut them in two [equal parts]. So triangle *aib* equals triangle *iae*. In the same way triangle *bic* equals triangle *fbc*, triangle *icd* equals triangle *gcd*, and triangle *aid* equals triangle *adh*. Whence triangles *aeb*, *bfc*, *cgd*, and *ahd*, together with the four other triangles *aib*, *bic*, *cid*, and *dia* are all equal to one another and they make up the whole square *abcd*. So it is clear that square *eg* is twice square *abcd*.

[67] In another way: because segment *dab* is a semicircle lying under angle *dab*, this angle is therefore a right angle. And because the sides *ad* and *ab* are equal, each of the angles under *abd* and *adb* is half a right angle {**p. 193**}. For angle *iab* is half a right angle. And the angle under angle *aib* is a right angle. Whence triangles *abd* and *abi* are similar to one another and have an angle in common, namely angle *dba*. Therefore as *db* is to *ba* so *ba* is to *bi*. Consequently as the first *db* is to the third *bi*, so the square of the first *db* to the square of the second *ab*. For any one of the sides of the tetragons *ef* or *gh* equals diameter *db*. Therefore as *db* is to *bi*, so the tetragon on *ef* is to the tetragon on *ab*. Now *db* is twice *bi*. Consequently the tetragon on *ef* is twice the tetragon on *ab*. But the tetragon on *ef* is the tetragon *efgh*. And the tetragon on *ab* is tetragon *abcd*. Therefore the tetragon on *eg* is twice tetragon *abcd*. As required.

[68] Given that a column is inscribed in a cube of the same height and a sphere within the column, let a double cone be inscribed so that the common base is in the center of the column and the two vertices touch its basal centers.[109] Additionally, let the girth[110] be a circle cutting the column and the sphere in two halves, and let that circle be equidistant from the circles of the base and the top of the column. Moreover, in the double cone inscribe a solid octahedron the girth of which is a tetragon touching the surface of the aforementioned circle that cuts the column and sphere in halves. The four edges of the girth of the octahedron make right angles at the center of the double cone, with all sides of the octahedron being equal to one another.[111] I will therefore explain the ratios of all these bodies to one another. The ratio of the cube to the column is 14 to 11, the column to the sphere is 3 to 2, the sphere to the double cone is 2 to 1, and the double cone to the octahedron is 11 to 7. Assembling these, we have the ratio of the cube to the sphere is 21 to 11, the cube to the double cone is 42 to 11, the cube to the octahedron 6 to 1, the ratio of the column to the double cone is 3 to 1. The ratio of the column to the octahedron is 33 to 7, and the ratio of the sphere to the octahedron is 22 to 7.

[69] To demonstrate all of these ratios [see Figure 6.43]: inscribe circle *efgh* within square *abcd* with common center *i*. Construct line *ik* perpendicular to the plane of square *ac* and equal to any of its sides. Be aware that square *ac* is one of the six faces of the cube, that circle *efgh* is the base of the column, and that *ik* is the height of both bodies. Whence *ik* is the axis of the column, and point *k* is the center of the top of the column. Now in order to determine

---

[109] "pyramidis columne duorum capitum cuius due extremitates sint in centra circulorum basis et capitis columne." (193.12–13). Eventually there will be four solids within the cube.

[110] "uenter" (193.13) and in several places hereafter.

[111] This sentence was squeezed from "cuius [refers to the square cutting through the center of the cube] tetragoni angulis ad centra columne circulorum basis et capitis ascendentis quattuor recte utrique facientes super ipsum tetragonum octo trigona equilatera et equiangula, quarum unumquodque latus est equale uniuscuiusque lateris ipsius tetragoni" (193.15–19).

the volume of the cube, you must multiply the square of *ac* by *ik* its altitude, and in order to find the volume of the column, you must multiply the area of circle *efgh* also by *ik*. Therefore the ratio of the cube with square base *ac* and altitude *ik* to the column with circular base *efgh* and altitude *ik* is that of the tetragon to the circle. But the ratio of square *ac* to circle *efgh* is 14 to 11, as found above for the area of circle. Or by another way, draw the radius *if* that is half of side *ab*. Whence the square on line *if* is a fourth of the square on line *ab*. But the square on *ab* is tetragon *ac*. Therefore the tetragon on line *if* is a fourth of tetragon *ac*. Now if the square is multiplied by $\frac{1}{7}$ 3, you have the area of circle *efgh*. Therefore, the ratio {**p. 194**} of circle *efgh* to square *ac* is 11 to 14. Conversely therefore, the ratio of the square to the circle is 14 to 11, as I said. And the ratio of the cube to the column is also 14 to 11. As required.

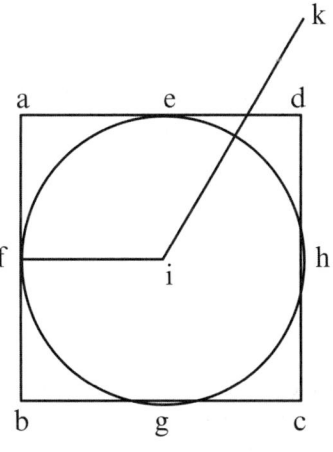

Figure 6.43

[70] Again, given *ik* as the diameter of a sphere inscribed within a column. Divide *ik* in two equal parts at point *l*. Let point *l* be the center of a major circle of the sphere and equal to and equidistant from circle *efgh*. A circle with center *l* cuts the column in two equal parts [see Figure 6.44]. Because the surface area of the sphere is four times the area of its circle that equals circle *efgh*,[112] the volume of the sphere is found by multiplying the surface area by a sixth of *im* the diameter of the sphere. Therefore if circle *efyh* is multiplied by four times the sixth of the diameter of the sphere, the quadruple being line *in*, the volume of the sphere is indeed found. But the product of circle *efgh* and the diameter of the sphere or of the altitude of the column yields the volume of the column. Therefore the ratio that line *ik* has to line *in* is the same as the ratio of the aforementioned column to the aforementioned sphere. And because *ik* is three halves of *im* and *in* is four times the same *im*, the ratio of

---

[112] Text has *efgd*; manuscript has *efgh* (194.8, f. 129v.32).

ik to *in* is 6 to 4 or minimally, 3 to 2. Whence, the ratio of the column to the sphere is 3 to 2, the sesquialter ratio. And because the ratio of the cube to the column is 14 to 11, the ratio of the same bodies is thrice 14 to thrice 11, that is 42 to 33. Because the ratio of the column to the sphere is 3 to 2, the ratio of the same bodies is eleven times 3 to eleven times 2 or 33 to 22. Because the ratio of the cube to the column is 42 to 33, the ratio of the column to the sphere is 33 to 22. And *per equale* 42 to 22 is the ratio of the cube to the sphere. As required.

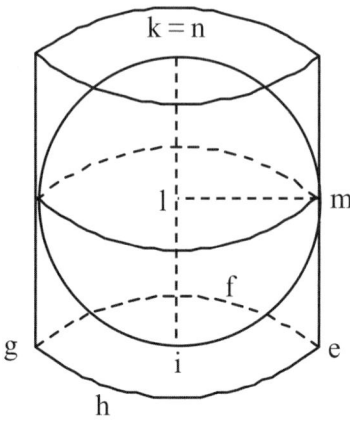

Figure 6.44

[71] Likewise, given a circle with center *l* and the two poles of the sphere at points *i* and *k*, create from a single cone a double cone, the altitude of one half being *lk* and of the other *li*, the diameter of the sphere being *ik*, as we said before. To find the volume of this body, multiply the circle with center at *l* by a third of *ik* its altitude. Now the product of the circle of center *l* by two thirds of diameter *ik* or *in* gives the volume of the sphere. Therefore the ratio of the sphere to the double cone is as the ratio of two thirds of the diameter of the sphere to a third of the same diameter. But the ratio of two thirds of anything to a third of the same thing is 2 to 1. Therefore the ratio of the sphere to the double cone is 2 to 1. And because the ratio of the column to the sphere is 3 to 2 and the ratio of the sphere to the double cone is 2 to 1, so the ratio of the column to the double cone is 3 to 1. As required.

[72] Again, construct a circle around center *l* equal to the aforementioned circle *efgh*, and inscribe in it a square equal to the aforementioned square *efgh*. Draw four lines, one from each vertex of that square to the points *i* and *k*, the poles of the sphere, thereby creating an octahedron within a sphere, a procedure described by Euclid in his thirteenth book.[113] The solid is divided in two

---

[113] *Elements*, XIII.14.

pyramid at the square inscribed in the circle of center l. Each of these has a square base, their sides are equilateral triangles, and the height of one is line kl, of the other line il {**p. 195**}. These altitudes equal one another, for each is half of line ik the diameter of the sphere or axis of the column or height of the cube. Thus you find the volume of the whole octahedron by multiplying the square inscribed in the circle of center l by a third of line ik. The volume of the double cone containing the solid octahedron is the product of the area of the circle containing the square and the same third of line ik. Therefore the ratio of the double cone to the octahedron is as the circle to the square inscribed in the circle, which is 11 to 7. Since the ratio of the square containing the circle to the circle itself is 14 to 11 and the outside square is twice the inside square, therefore the ratio of the double cone to the octahedron is 11 to 7. The ratio of the cube to the double cone is 42 to 11 and the ratio of the double cone to the octahedron is 11 to 7. Therefore the ratio of the cube to the octahedron is 42 to 7, a six fold ratio. Again, the ratio of the column to the double cone is threefold, 33 to 11. And the ratio of the double cone to the octahedron is 11 to 7. So, the ratio of the column to the octahedron is 33 to 7. Likewise, because the ratio of the sphere to the double cone is twofold, that is, 22 to 11, the ratio of the double cone to the octahedron is 11 to 7, and the ratio of the sphere to the octahedron is 22 to 7, the ratio that the circumference of any circle has to its diameter.[114]

[73] Having shown these things, we note that given the volume of any one of the five solids, we can use it to find the volume of any of the other solids. To demonstrate this: given a cube with edge ab equal to 6 so that its square is 36 which multiplied by ik or 6 its height is 216 the volume. In other words, we cube one edge. Now the ratio of the cube to the column is 14 to 11. If we take is the ratio of the column to the cube or $\frac{11}{14}$ of 216, we have $\frac{5}{7}$ 169 for the volume of the column. Likewise, if we use the ratios that the sphere, the double cone, and the octahedron have to the cube or 216, we find for the volume of the sphere $\frac{1}{7}$ 113, for the volume of the double cone $\frac{4}{7}$ 56, and for the octahedron 36. Likewise the volume of the column is $\frac{5}{7}$ 169. If we multiply this by 14 and divide by 11, we get 216 for the volume of the cube. If we take the ratio of the sphere to the column or $\frac{2}{3}$ of $\frac{5}{7}$ 169, we get $\frac{1}{7}$ 113 for the volume of the sphere. And if we take the ratio of the double cone to the column or a third of $\frac{5}{7}$ 169, we get $\frac{4}{7}$ 56 for the volume of the double cone. Likewise, if we take the ratio of the octahedron to the column or $\frac{7}{33}$ of $\frac{5}{7}$ 169, we get 36 for the volume of the octahedron. You can determine the volumes of the remaining solids in the same way.

[74.1] To find the volume of a dodecahedron inscribed in a sphere, you must know the diameter of the sphere. Let pentagon $abcde$ be one of the twelve

---

[114] Compare the foregoing with *Elements*, XIII.18.

equal pentagonal faces of a dodecahedron. Circumscribe a circle of center f about the pentagon, and draw its diameter ag and chord be that encloses a pentagonal angle. Let the center of the sphere be h. Draw line ah. Draw a semicircle of a major circle ali about the sphere. Extend line ah to point i. Let ai be the diameter of the sphere and of semicircle ali {**p. 196**}. Let its measure be six ells. Cut from line ai a third to be line ak. Draw line kl at right angles to line ai. Draw lines al and il. Then draw lines bh, ch, dh, and eh from the angles of pentagon $abcde$ to the center of the sphere [see Figure 6.45]. Thus you will have a pyramid, $habcde$, with pentagonal base $abcde$ and vertex at h, and it is a twelfth part of the dodecahedron inscribed in the sphere of diameter ai. Now if the lines are drawn to all the angles from h the center of the sphere, then the entire dodecahedron has been divided in twelve equal pyramids of which one is pyramid $habcde$ which we multiply by 12 if we want to find the volume [of the dodecahedron]. Then we have the volume of the whole solid.

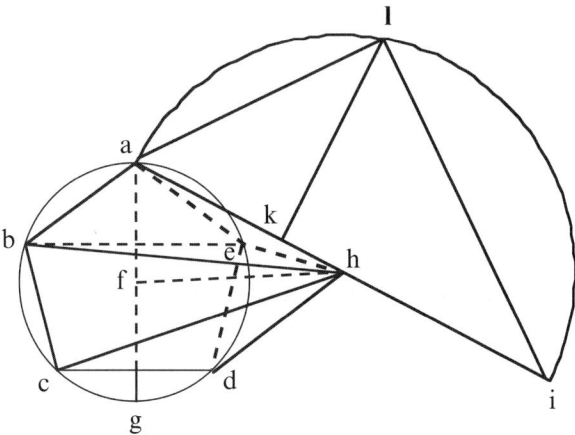

Figure 6.45

[74.2] Let us consider how we ought to proceed properly. First consider that the triangles ali and akl are similar to each other, because angles ali and akl are right angles[115] and they share the angle at point a. Further the angle at i equals angle alk. Therefore similar triangles have proportional sides around the common angle. Whence as ia is to al, so al is to ak. Since three lines are proportional and the first is to the third, then the square on the first is to the square on the second. Therefore as ia is to ak so the square on diameter ai is to the square on line al. Now ai is three times ak. Whence the tetragon on ai (that is 36) is three times the tetragon on al. Therefore the tetragon on al is 12, as stated in Euclid's thirteenth book.[116] al is the edge of the inscribed

---

[115] In Figure 6.45, points k and e are atop each other.
[116] *Elements*, XIII.12.a.

cube of diameter *ai*. Now the edge of the cube in the sphere is the chord of a pentagonal angle of one of the twelve pentagonal faces of the dodecagon. Therefore the square on chord *be*, that is also the chord of pentagonal angle *bae*, is 12. So that we may find one of the edges of pentagon *abcde*, we must divide chord *be* in mean and extreme proportion. As stated in the thirteenth book of Euclid, if the chord of a pentagonal angle is divided in mean and extreme proportion, the larger segment is the side of the pentagon.[117]

Method for Dividing a Line into Mean and Extreme Proportion

[75] The method for dividing chord *be* of length the root of 12 in mean and extreme proportion is to increase its square by a fourth, so that the new length is the root of 15 {**p. 197**}. Then remove from this half of line *be* to obtain the root of 3. Thereby you have the root of 15 less the root of 3 for the longer part.[118] Therefore the pentagonal side *ae* is the root of 15 less the root of 3. Now we read in the fourteenth book of Euclid that the squares on the chord and the side of a pentagon can be five times the side of a hexagon.[119] So in order to find the side of a hexagon, I add 12 (the square on chord *be*) to 18 less the root of 180 (the square on side *ae* as I shall demonstrate below)[120] to obtain 30 less the root of 180. Then, a fifth of this, 6 less the root of $\frac{1}{5}$ 7, is the square on *af* the side of a hexagon. Now the side of the hexagon equals the semidiameter of a circle. If its square or 9 is removed from the square on the semidiameter of sphere *ah*, what remains is 3 and the root of $\frac{1}{5}$ 7 for the square on perpendicular *hf*. Since triangle *ahf* is a right triangle, line *hf* is orthogonal to the plane of circle *abgd*. Now in the book of Menelaus and in the Almagest we read: when a plane cuts a sphere, the common section is a circle.[121] And when a line is drawn from the center of the sphere to the center of the cutting circle, that line is orthogonal to the plane of the circle.

[76.1] Using the letters *af* for the semidiameter and *hf* for the perpendicular, we now show how to measure pyramid *habcde* [see Figure 6.46]. First I divide *fg* in two equal parts at *o*. The measure of *ao* is a *dodrans*,[122] that is three fourths of diameter *ag*. By removing *bn*, a sixth of chord *be*, what remains is *en* a *destans*,[123] that is, five sixths of chord *be*. Because as stated in the fourteenth book,[124] the product of the dodrans *ao* and the destans *en* yields the

---

[117] *Elements*, XIII.8.
[118] *Elements*, II.11.
[119] *Elements*, XIV.3.
[120] In the margin is written "in the last chapter where is the paragraph 'I will demonstrate again by line *rae* that such is so'" (197.7, f. 127r).
[121] Folkerts ([2006] IX, 12) remarked that he could not find this theorem in either cited source but that it is in Theodosius' *Sphaerica* I.1.
[122] Nine twelfths of anything (197.19).
[123] Ten twelfths of anything (197.20).
[124] *Elements*, XIV.8.

pentagonal plane *abcde*. If the plane is multiplied by the perpendicular *hf*, the result is thrice pyramid *habcde*. And multiplying this threefold by 4[125] produces the volume of the whole dodecahedron. Its twelfth part is the volume of pyramid *habcde*. Because the ratio of dodrans *ao* to the radius *af* is 6 to 4 a sexquialter ratio, plane *af* by *fh* multiplied by *en* and then the product by 6 is as the plane *ao* by *en* multiplied by *hf* and then the product by 4. Likewise the ratio of *be* to *en* is 6 to 5. Consequently the plane *af* by *be* multiplied by *fh* and the product by 5 is as the plane *fa* by *en* multiplied by *hf* and the product by 6. But the result of plane *fa* by *en* multiplied by *hf* and the product by 6 gives the volume of the dodecahedron. Therefore the result of plane *af* by *be* multiplied by *fh* and the product by 5 is the volume of the same dodecahedron. For which reason I first multiplied the square on *af*, that is 6 less the root of $\frac{1}{5}$ 7, by 3 and the root of $\frac{1}{5}$ 7 to obtain $\frac{4}{5}$ 10 and three roots of $\frac{1}{5}$ 7.

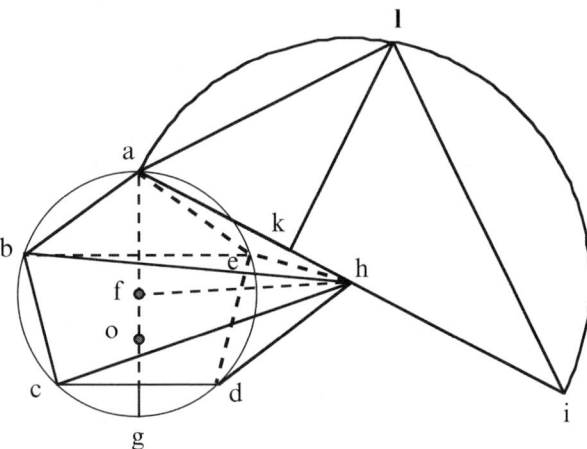

Figure 6.46

[76.2] For example: the product of 6 and 3 is 18. The product of the added root of $\frac{1}{5}$ 7 and the diminished root of $\frac{1}{5}$ 7 is the diminished quantity $\frac{1}{5}$ 7. Combining this with 18 produces $\frac{4}{5}$ 10. Further, the product of 3 by the diminished root $\frac{1}{5}$ 7 gives 3 diminished roots of $\frac{1}{5}$ 7. Now the product of 6 by the added root of $\frac{1}{5}$ 7 is 6 added roots of $\frac{1}{5}$ 7. Consequently, taking the three diminished roots from the six added roots {p. 198} leaves three added roots of $\frac{1}{5}$ 7. Thus we have $\frac{4}{5}$ 10 and three roots of $\frac{1}{5}$ 7 the latter being one root of $\frac{4}{5}$ 64. And then I multiply the square on *be* by the square of 5 to get 300. Multiply this by $\frac{4}{5}$ 10 and the root of $\frac{4}{5}$ 64. Now 300 multiplied by $\frac{4}{5}$ 10 is 3240, and 300 multiplied by the root of $\frac{4}{5}$ 64 is the root of the number from the square of 300, namely the root of 90000 times $\frac{4}{5}$ 64. The result is [the root of] 5832000. Therefore, the square of the volume of the dodecahedron is 3240 and the root of 5832000. This square is a first binomium because its larger term can be more than the smaller term accord-

---

[125] Text has *h*; manuscript has 4 (197.23, f. 127r.25).

ing to the quantity of the squared number. Whence its root is a binomium as defined in Book X of Euclid,[126] namely the root of 2700 and the root of 540. If we manipulated these roots carefully, we will have a little less than $\frac{1}{5}$ 75 for the volume of the dodecahedron. Or if we rather take the root of 5832000, we will find it to be a bit less than 2415. If we add 3240 to this, we will have a bit less than 5655. The root of this is a little less than $\frac{1}{5}$ 75. Be it noted that when we said that the root of 3240 and the root of 5382000 is the root of 2700 and the root of 540, then we have divided 320 into two parts, of which one part is 2700 and the other 540. The product of these two parts is a fourth of 5832000. Euclid had instructed us to operate in this way.

[77] I will demonstrate again how to find the square on line *ae*, the pentagonal side. Take again line *ae* and add line *ra* the root of three to it. Now the entire line *er* is the root of 15. Consequently, *ae* is the root of 15 less the root of 3. This is the *ae* whose square we wish to find. Because line *er* has been divided in two parts at some optional point *a*, the squares on lines *er* and *ar* are equal to twice the plane *ar* by *er* and the square on line *ae*. So, if we take twice the plane *ar* by *er* from the squares on lines *er* and *ar*, what remains is the square on line *ae*. Now *er* is the root of 15; so its square is 15. And *ar* is the root of 3; so its square is 3. Therefore the sum of the squares on lines *er* and *ar* is 18. Further, the product of the roots of *ar* and *er*, or of the root of 3 and the root of 15 is one root of 45. So twice the product of the roots of *ar* and *er* is two roots of 45. Subtracting this from 18 that you already found leaves 18 less two roots of 45.[127] The two roots, of course, as just one root of 180. Therefore the square on line *ae* is 18 less the root of 180, as we found above.

Finding the Volume of an Icosahedron Inscribed in a Sphere of Diameter Six[128]

[78.1] In order to find the volume of an icosahedron inscribed in a given sphere, consider the equiangular and equilateral triangle *abc* as one of the 20 faces of the solid [see Figure 6.47]. Circumscribe circle *abc* about the triangle so that its diameter *ad* cuts side *bc* at point *e*. Call the center of the circle *f* and make the center of the sphere *g*. Draw line *ag* and extend it to point *h* on the surface of the sphere. Describe a great spherical semicircle as *aih*. Draw lines *bg* and *cg*. Make *gabc* a pyramid with triangular base *abc* and vertex *g*. The volume of this pyramid is $\frac{1}{20}$ of the whole icosahedron. The reason is this {**p. 199**}: straight lines drawn from point *g* the center of the sphere to the vertices of all 20 triangles of the icosahedron divides it in 20 equal pyramids. One of these is pyramid *gabc*. As we said, if we multiply the volume of one pyramid by 20, we have the

---

[126] *Elements*, X. *def*.II.1.
[127] Note again that although $18 - 2\sqrt{45}$ appears as a binomial, it is really the number 4.58+.
[128] "De magnitudine solidi 20 basium triangularium constructi in eadem spera, cuius diameter sit 6" (198. 30–31, *f*. 1218r.15–16).

volume of the entire icosahedron. Further, as Euclid remarked,[129] the circle that contains one of the triangles of the icosahedron equals the circle that contains one of the pentagons of the dodecahedron, because both solids are constructed in the same sphere. This is the reason why, as we said above, that the semidiameter of circle *abc* is known. This square on *af* will be 6 less the root of $\frac{1}{5}$ 7, and the square on the perpendicular *gf* will be 3 and the root of $\frac{1}{5}$ 7. For *ag* the semidiameter of the sphere is 3 and its square is 9. Because the square on the side of the equilateral triangle inscribed in the circle is three times the square on the semidiameter of the same circle: if we triple the square on radius *af* that is 6 less the root of $\frac{1}{5}$ 7, we have 18 less three roots of $\frac{1}{5}$ 7 for one side of triangle *abc*. Then I draw lines *db* and *dc*, so that triangles *abd* and *acd* are orthogonal and equal. Because side *ab* equals side *ac*, arc *ab* equals arc *ac*, and the whole arc *abd* equals the whole arc *acd*, it follows that arc *bd* equals arc *cd*. Whence lines *bd* and *cd* are equal. Now triangles *abd* and *acd* are both equiangular and equilateral. Whence angle *bad* equals angle *cad*. Whence all the angles at vertex *e* equal one another. And if equal, they are also right angles. Because diameter *ad* in circle *abc* intersects a certain line *bc* at right angles, it also cuts the line in equal parts. Therefore line *be* equals line *ec*. Therefore *be* is half of *bc* and of *ab*. Whence the square on *ab* is four times the square on *be*. But the squares on line *ae* and *eb* equal the square on line *ab*. Therefore the squares on line *ae* and *eb* are four times the square on line *eb*. Whence, the square on line *ae* is thrice that square on line *eb*. Therefore and because cathete *ae* has been drawn from angle *b* in right triangle *abd* to the base *ad*, *ae* is to *eb* as *be* is to *ed*. So, as the square on side *ae* is to the square on side *eb*, so *ae* is to *ed*. But the ratio of the square on side *ae* to the square on side *ed* is threefold. Therefore *ae* is triple *ed*. Whence it is obvious that *ed* is a fourth of diameter *ad*. Now *af* is half of diameter *ad*. The remaining part *ef* is another fourth of *ae*. Line *ae* contains three fourths of diameter *ad*. Whence ratio of *ae* to *af* is sesquialter.

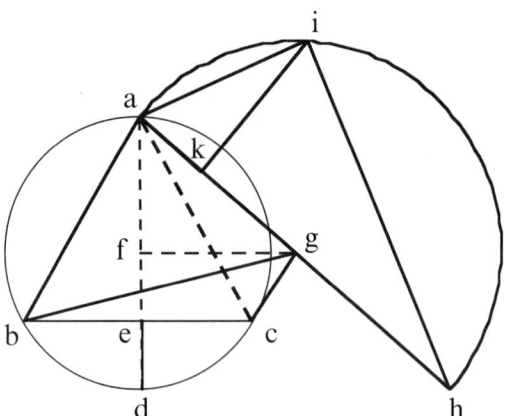

Figure 6.47

---

[129] *Elements*, XIV.4.

[78.2] It has been demonstrated that line ae is the cathete to line bc. Whence the product of ae and bc yields twice the area of triangle abc. If we multiply this by 10, we will have the area of the icosahedron. [Another way:] since the ratio of ae to af is 15 to 10, the area of all twenty triangles equals the product of af and bc multiplied by 15. Now if we multiply this entire product by a third of gf (the altitude common {p. 200} to the twenty triangular pyramids), we have the volume of the icosahedron. This is the same as multiplying af by bc and then by 5 and again by gf to reach the volume of the whole body. Thereupon if we multiply the square on af by the square on bc and all of that by the square of 5 or 25, then multiply by the square on gf, we will have from all of this the square of the volume of the solid. For by multiplying the square on line af by the square on line fg (that is, 6 less the root of $\frac{1}{5}$ 7 by 3 and the root of $\frac{1}{5}$ 7), we obtain $\frac{4}{5}$ 10 and the root of $\frac{4}{5}$ 64, as we found above. These must be multiplied by the square on the side of triangle bc (18 less 3 roots of $\frac{1}{5}$ 7 or one root of $\frac{4}{5}$ 64). The multiplication proceeds as follows: first multiply $\frac{4}{5}$ 10 by 18 or rational number by rational[130] number to obtain $\frac{2}{5}$ 194. Subtract from this the product of the root of $\frac{4}{5}$ 64 and the diminished root of $\frac{4}{5}$ 64 to reach $\frac{4}{5}$ 64, leaving $\frac{3}{5}$ 129. Likewise, multiply the root of $\frac{4}{5}$ 64 by 18 to obtain 18 roots of $\frac{4}{5}$ 64. Then multiply the diminished root of $\frac{4}{5}$ 64 by $\frac{4}{5}$ 10 to obtain $\frac{4}{5}$ 10 diminished roots of $\frac{4}{5}$ 64. After subtracting the $\frac{4}{5}$ 10 diminished roots from 18 roots, what remains is $\frac{1}{5}$ 7 roots of $\frac{4}{5}$ 64. Therefore by multiplying the square af by fg[131] by the square on bc, we obtain $\frac{3}{5}$ 129 and $\frac{1}{5}$ 7 roots of $\frac{4}{5}$ 64. Finally, multiplying this by the square of 5 we obtain 3240 and 180 roots of $\frac{4}{5}$ 64 for the square of the whole icosahedron.

[79] If we want to reduce those 180 roots to the root of just one number, we proceed this way: square 180 to get 32400 and change $\frac{4}{5}$ 64 to 324 fifths (that we got by multiplying 64 by 5); then, multiply this by a fifth of 32400 or 6480 to obtain 2099520. Therefore the square of the solid is 3240 and the root of 2099520. This is the fourth binomium the major term of which is 3240 and rational. The minor term is 2099520 (not a perfect square)[132] the root of which is a bit less than 1449. Adding this to 3240 yields a bit less than 4689. And the root of this is a bit less than $\frac{1}{2}$ 68 for the volume of the icosahedron.

[80] If we want to find the measure of the radius in another way, take kh as the quadruple of ak. Through point k, draw line ki at right angles to ah the diameter of the sphere. Draw lines ai and ih. According to what was said above, ha is to ai as ai is to ak. Whence as ha is to ak, so the square on diameter ah is to the square on line ai. For ah is the quintuple of ak. Consequently 36, the square on ah, is the quintuple of the square on ai. Therefore the square on ai is $\frac{1}{5}$ 7. And ai is the radius of the circle for which the inscribed pentagonal

---
[130] Text has "integra per integram" (200.11).
[131] Apparently af equals fg because they are radii of the same circle.
[132] "secundum quantitatem numeri non quadrati" (200.24–25).

side is also the side of triangle *abc* noted above, as stated in Euclid.[133] The radius of a circle is the side of a pentagon if we add a fourth of $\frac{1}{5}$ 7 to itself and obtain 9. Now the square of half the radius is the fourth part of the square of the radius. Then by subtracting half of the radius or $\frac{4}{5}$ 1 from 3 the root of 9, we have 3 less the root of $\frac{4}{5}$ 1 {**p. 201**}, and this is the side of a decagon. If we add $\frac{1}{5}$ 7 the square of the radius to $\frac{4}{5}$ 10 less the root of $\frac{4}{5}$ 64 the square on the side of the decagon, we obtain 18 less the root of $\frac{4}{5}$ 64 for the square of one side of triangle *abc*. Since this is thrice the square of the radius *af*, it must be divided by 3 to obtain 6 less the root of $\frac{1}{5}$ 7 for the square on radius *af*, as we found above. Because Euclid found that as the edge of the cube or the pentagonal chord is to the side of the equilateral triangle inscribed in the same circle,[134] so the ratio of the potent line on the longer part of the line divided in mean and extreme ratio is to the potent line on the shorter part of the same line. Thus, as the dodecahedron is to the icosahedron inscribed in the same sphere, it is possible to find the volume of one solid from the volume of the other.

[81] For example: we found above that the chord of a pentagonal angle is the root of 12, the side of the triangle is 18 less the root of $\frac{4}{5}$ 64, and the volume of the dodecahedron is $\frac{1}{5}$ 75. Therefore as the root of 12 is to the root of 18 less the root of $\frac{4}{5}$ 64, so $\frac{1}{5}$ 75 is to the volume of the icosahedron. Whence as 12 is to the root of 18 less the root of $\frac{4}{5}$ 64, so the square of $\frac{1}{5}$ 75 (that we found above to be 5655) is to the square of the volume of the icosahedron. Computing the root of $\frac{4}{5}$ 64 as closely as possible, we find it to be $\frac{1}{20}$ 8. After subtracting this from 18, what remains is 10 less $\frac{1}{20}$. Therefore as 12 is to 10 less $\frac{1}{20}$, so 5655 is to the square of the icosahedron. Therefore we multiply 5655 by 10 less $\frac{1}{20}$ to obtain 56267. Dividing this by 12 leaves a little less than 4689 for the square of the icosahedron, as we found above. If the square of the icosahedron were known, then we would multiply it by 12 and divided by 10 less $\frac{1}{20}$ to obtain the square of the dodecahedron.

[82] If we want to find the ratio of the potent segment divided in mean and extreme proportion to the potent segment or shorter part of the same divided line, take line *ab* and divide it in mean and extreme proportion at point *g*, the longer part being *gb*. Whence as *ab* is to *bg*, so *bg* is to *ga* [see Figure 6.48]. Now I draw line *gd* at right angles to line *ab* at point *g*, and I make it equal to line *ab*. Joining lines *da* and *db*, potent line *db* is connected to lines *dg* and *gb*. Line *ad*[135] will be potent[136] on lines *dg* and *ga*. Because *dg* equals *ab*, *db* is potent on lines *ab* and *gb*; this is on the longer part of the whole divided line.

---

[133] *Elements*, XIII.3.
[134] *Elements, XIV.9.
[135] Text has *db*; context requires *ad* (201.30).
[136] The word potent (*potens*) often refers to the square on a line, as in the first sentence of this paragraph. However here and in what follows, I have left it as potent to indicate that it would become the square later.

So line *ad* is the potent line on the same line *ab* and its shorter[137] part. Therefore as *db* is to *da*, so the volume of the dodecahedron is to the volume of the icosahedron. For this reason as the square on line *db* is to the square on line *da*, so the square of the dodecahedron is to the square of the icosahedron.

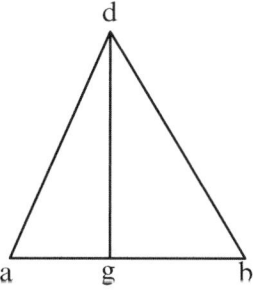

Figure 6.48

[83] We can express this in numbers. Let *ab* be 10. Then *bg* will be the root of five times the square of half of line *ab* less its half. Therefore *bg* is the root of 125 less 5, and *ag* is thrice half of *ab* less the root of five times the square of half the same line (that is, *ag* is 15 less the root of 125). We compute the square of *gb* to be 150 less 10 roots of 125, that is one root of 12500. Adding the square of line *dg* or 100 to its square, and since *dg* equals *ab*, we have 250 less the root of 12500 for the square on line *db*. Likewise adding the square on *gd* to 350 less 30 roots of 125 the square on line *ag* {**p. 202a**}, we obtain 450 less the root of 112500 for the square on line *da*. Because any of the squares on lines *db* and *da* is an abscise or recise, and so that we do not produce a large error from a small error,[138] we compute as carefully as possible the roots of 12500 and 112500. Then we subtract these from the larger terms that themselves are recises. Thus the ratio of the square on line *db* to the square on line *da* will be rational. For example: the root of 12500 is $\frac{4}{5}$ 111. If this is subtracted from 250, what remains is a little less than $\frac{1}{5}$ 138 for the square on *db*. Likewise the root of 112500 is a little more than $\frac{2}{5}$ 335. If this is subtracted from 450, what remains is a little less than $\frac{3}{5}$ 114 for the square on line *da*. Therefore the ratio of $\frac{1}{5}$ 138 to $\frac{3}{5}$ 114 equals the ratio of the squares of the volumes of the dodecahedron and the icosahedron. Further this ratio equals the aforementioned one of the squares on the pentagonal chords to the square on the side of the triangle inscribed in the same circle.

---

[137] In this sentence, text (201.33) has *db* and *maiorem*; context and Paris 7223 (*f*. 183v.24) have *ad* and *minorem* which are correct.

[138] "ut non habeamus ignotius per ignotum" (202.3).

[84] Knowing therefore the volumes of all the solids inscribed in one sphere, it is possible to find the volumes of similar solids inscribed in some other sphere. Since the ratio of similar solids equals the threefold ratio of similar edges or of the diameters of the spheres in which the solids are inscribed, the triple ratio of the diameter of one sphere to the diameter of another is as the ratio of the cube of the diameter of one sphere to the cube of the diameter of the other sphere. For example: if we want to find the volume of a dodecahedron inscribed in a sphere of diameter 7: as the cube of six is to the cube of 7 (or 216 to 343), so the known volume of $\frac{1}{5}$ 75 of one solid is to the unknown volume of another solid. Thus multiply $\frac{1}{5}$ 75 by 343 and divide by 216 to get a bit less than $\frac{5}{12}$ 119. In another way, if the square of 343 is multiplied by 3240 the square of the solid and the root of 5832000, and the product divided by the square of 216, you have the square of the volume of the inscribed dodecahedron of diameter 7. Use the same procedure for the icosahedron and other solids.

7
Measuring Heights, Depths, and Longitude of Planets

COMMENTARY

The title of Chapter 7 promises more than it delivers, "Incipit septima distinctio de inuentione altitudinum rerum eleuatarum et profunditatum atque longitudinum planitierum." Regardless of promises, the chapter offers only trivial exercises in measuring heights of trees, to judge their suitability as masts for ships. Not a word about finding depths, say of wells. It is curious that some of the Italian translations do offer examples of measuring depths of wells. Could this be evidence of a defect in the exemplary text followed by the copyist of, say, Urbino 292?

The focus of the chapter is on measuring heights, using similar triangles, as is done in most middle schools today, although in two different forms. The first is a quadrant, also called by him "an oroscope," for which he presented detailed instructions for both construction and use. The second is "a wooden triangle." It appears in [4] as a 3-4-5 right triangle for measuring heights, again by similar triangles. Despite the practicability of these instruments, the incompleteness of the chapter is bothersome: there is nothing about finding depths or longitude of planets. All the other chapters are quite complete; a few are even overflowing with information. Why does this chapter seem to have been clipped of some contents? The Latin manuscripts offer no suggestion. The Italian manuscripts, on the other hand, do offer problems seeking depths of wells. Yet none of the material in the corresponding chapter of the Riccardiana manuscript appears here. Furthermore, the figures do not always match the text, and the text was in need of severe correction, especially in [7]. New figures were drawn. Note the use of the word *equiangular* (corresponding angles are equal) that is so different from modern use (all angles are equal). Finally, there is a single example where the Pythagorean theorem is used.

The last word of the title suggests astronomy, the measurement of the location of planets and the distances between them. For this, one must have tools for measuring chords and arcs, occasionally very small chords and arcs. An understanding of sines and versed sines would be helpful for an astrologer. Unaccountably, Fibonacci mentioned both these relationships in Chapter 3, in an offhand sort of way, simply defining them and noting their utility for astrology.[1] Neither really belongs in Chapter 3; both are pertinent to Chapter 7. Furthermore, Chapter 3 is quite complete, be it ever so long, with a useful discussion of chords, arcs, and sectors of circles without needing a table of chords and arcs.[2] Such a table appears in Chapter 3; but it would be very necessary in Chapter 7 for astronomical purposes.

Inasmuch as I think it inexplicable as well as unacceptable that Fibonacci would foist an incomplete chapter on his good friend Dominic from whom he probably expected a few favors, I hypothesize that when Fibonacci wrote Chapter 7, he placed therein the information about sines and versed sines together with the well-developed and exemplified *Table of Chords and Arcs*, as I show after [9]. Later, some practitioner of the geometric craft edited a copy of *De practica geometrie* according to his bent, excising the definitions and table from Chapter 7 and placing them (paragraphs [211] through [220] where they are found in many manuscripts) in Chapter 3. Finally, the *Table* interrupts the flow of context that would measure chords, arcs, and areas of sectors by lodging them between arcs and areas, offering no assistance whatsoever for computing areas. My hypothesis, of course, is moot until at least one representative of another family of manuscripts is found, preferably (according to my expectation) without the second appendix.[3]

Returning to the subject of measuring chords: Fibonacci was obviously familiar with Ptolemy's theorem for finding chords from a general quadrilateral and its diagonals inscribed in a circle.[4] His references to the *Almagest* imply his knowledge of Ptolemy's *Table of Chords*, a set of numbers somewhat forbidding to practitioners of plane geometry. I suggest that Fibonacci wanted to introduce his readers to the use of a table of chords, somewhat simpler than Ptolemy's, just as yesteryear's elementary students were introduced to the use of a simplified table of trigonometric functions. Hence, he constructed his own *Table of Chord Lengths*.

How did he do this? What is obvious is that the table was based on a circle of diameter 42 rods and circumference divided into 132 equal parts developed from the proportion D : C = 7 : 22. For practical purposes, the table is restricted to the semicircle. Because each chord is shared by two arcs, a minor arc and its conjugate major arc, the table needs only 66 rows. Although Fibonacci refers to several theorems and properties of arcs and chords, he offers no explicit statement about the construction of the table. The closest to a rule is this remark, "the ratio of the arc of a seventh part of a circle to its chord is almost (*sicut*) 22 to 21."[5] This is nearly correct for chords of arcs

---

[1] Ch. 3, [206]; (94.18–20).

[2] (95.32–100.10).

[3] The first appendix is a problem that produces an indeterminate equation, completely irrelevant to the scope of the treatise.

[4] *Almagest*, I.10.3.

[5] (94.6–7).

from 1 to 22 parts of the semicircle. Afterwards the ratio becomes more and more than that of 22 to 21. Furthermore, a numerical comparison of pairs of ratios from numbers in the table demonstrates that there is an irregular increase in the lengths of the chords as the arcs increase regularly by one. It is entirely possible that he actually constructed a huge semicircle of diameter 42, divided it into 66 equal parts, constructed the required chords, and measured them to create the table.

The table has two uses: one for finding arc lengths given the chord, the other for computing the measure of chords given the arc. For the first there are three steps, given the length of a chord (crd) and the diameter (d) of the circle:

1. Multiply crd by 42; then divide by d.
2. Find the quotient in the third column from the left in the table.
3. Multiply by d the number in the first column from the left that is on the same line as the answer in step 2, then divide by 42.

The answer to step 3 is the measure of the arc over the given chord. Finding the chord given the arc reverses the process.

SOURCE

The initial material seems to have been taken or adapted from *Geometria incerti auctoris* (GIA) by Gerbert[6] (?). Paragraphs [1] and [2], which discuss similar triangles and the use of an isosceles right triangle, easily reflect GIA, 317 n. 16. The technique of the two arrows in [3] is found in GIA, 334 n. 26. The use of a measuring stick also in [3] is described in GIA, 325 n. 12. And finally, instructions similar in [5] for making a quadrant can be read in GIA, 330 n. 19. Because the chapter focuses on indirect measurement, readers may appreciate the excursion into common instruments for such activities as tangential.[7]

The *Table of Chord Lengths* is apparently original as Fibonacci claims, "...sequentes tabulas composui...."[8] Fibonacci certainly had access to Ptolemy's *Almagest* and possibly Menelaus' *Spherics*. On the other hand, in view of the fact that he referred to both "in alio libro...," it is possible that Arabic translations of the Greek texts were in a single volume.

---

[6] Bubnov (1899).
[7] Simi (1996).
[8] Benno van Dalen, who has researched ancient tables more than any other, wrote me, "...it now seem almost impossible to me that this very table has an Islamic source" (22.01.07). He also noted the typographical error in line 65 67 that should read 41 5 16 14.

TEXT[9]

[1] {p. 202} If you wish to measure a height, fix a staff perpendicular to the ground. Step away from the staff and the object you wish to measure. Stoop down to ground level from where you can see the top of the object across the top of the staff,[10] and mark the place from where you looked. Of course, if your line of sight from the ground through the top of the staff does not meet the top of the object, then move forward or backward until it does. Then the ratio of the ground distance from where you looked to the base of the object is to its height as the ratio of the distance to the base of the staff is to its height [see Figure 7.1]. For example: let the altitude be ab from the plane in which lies the line bc. Angle abc is a right angle. Erect staff de perpendicular to bc. Let point c be where you made the sighting ac through e the top of the staff. Thus triangle abc consists of the height ab and the ground line bc; line ac is your line of sight. Triangle edc consists of the staff de and the distances cd and de.[11] Consequently, triangles {p. 203} abc and edc are similar to each other because they are equiangular. Both angles abc and edc are right angles, and they share the angle at c. The other angle at a equals angle ced. Therefore triangles abc and dec are equiangular and similar. Similar triangles have proportional sides about the equal angles: cd is to de as cb is to ba. Whence, and considering ce the line of sight to the staff de, if line cb is known, then equally known is the height ab. Because if cd equals de, then cb equals ba. As the longer is to the longer, so the shorter is to the shorter.[12]

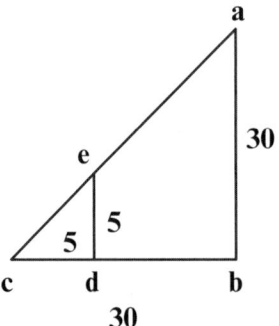

Figure 7.1

---

[9] "Incipit septima distinctio de inuentione altitudinum rerum eleuatarum et profunditatum atque longitudinum planitierum" (202.27–28, f. 130v.28–29).
[10] "pones oculum in terra prospiciens per summitatem aste" (202.31).
[11] Text has ce; context requires de (202.43).
[12] "si maior, maior; et si minor, minor" (203.8).

[2] This can be shown with numbers. Let ed be 5 palms and equal to the distance cd. Let the distance cb equal 30 ells. Because of this the height ab is similarly 30 ells, since cd equals de, as is obvious in the figure above. Again, let cd be longer than the staff ed. Because of this, the distance cb is longer than the height ab, as is obvious in this figure [Figure 7.2].[13] Hence we let cd equal 12 palms, the staff ed 8 palms, and the distance cb 60 ells. Whence, as cd is to de, so 12 is to 8. Or, in smaller numbers, as 3 is to 2, so cb is to ba. Whence, if we multiply 60 by 2 and divide by 3, we get 40 ells for the altitude ab. Again, let cd be shorter than de. Whence the distance cb will be less than the height ab, as in this figure [Figure 7.3]. We let cd the ground distance to the staff be 9, the staff ed 12, and the distance cb 45. Whence as much as dc is longer than ed, by so much is the distance bc longer than the height ab. Now the ratio of ed to dc is sesquitertial. Whence the height ab is 15 or a third more [than cb]. And thus the height ab is 60. Or, if cb is 45, then multiply it by 12 (or ed) and divide the product by 9 (or cd) to get 60 for the height ab. Or, if you multiply cb by a third of ed and then divide by a third of cd, you will also get 60 for ab. Or, if you multiply a third of cb by a third of ed, you will ge 60 for the height ab.

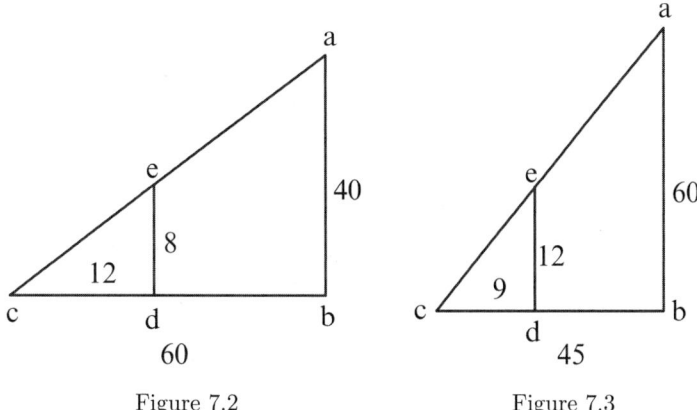

Figure 7.2   Figure 7.3

[3] Suppose you wish to measure trees in a forest to serve as masts for ships; then proceed this way. Get a rod whose length equals your eyeball height. Lay it on the ground perpendicular to the base of the tree. Now step back and forth moving the rod until the line of sight from the far end of the rod goes to the top of the tree.[14] So, whatever the distance is from the far end of the rod to the base of the tree, that is the height of the tree

---

[13] In the figure, 60 measures cb. Similarly in fig. 7.3.
[14] This really requires two rods of equal length and perpendicular to each other; see Figure 7.5. Remember also: the line of sight is from your eye on the ground to the top of the tree.

[see Figure 7.4]. For example: let the height of the tree be *ab*, the rod *cd*, and the height of the measurer be *de* with his eye at *e* so that the line of sight *ae* goes through *c*. And then *ab* will be [to *bc*][15] as *ed* is to *dc*, as shown above. If you want to use some geometric finesse on the problem, then stand a little away from the tree and shoot two arrows at it, a piece of string being attached to each arrow. One arrow goes to the top of the tree, the other to its base. Then bring the two ends of the strings together at some point on the ground, so that a right triangle is formed with the tree. The tree is the cathete, the string along the ground is its base, and the other string is the hypotenuse, against which is the right angle [see Figure 7.5]. For example, let the tree be line *ab*, the string {**p. 204**} of the lower arrow *bc*, and the other string *ac*. Since you know the lengths of the two strings, subtract the square on string *bc* from the square on string *ac*. What remains is the square on the tree *ab*. For instance: let the string *ac* be 50 cubits and string *bc* 30. Then subtract 900 (the square of 30) from 2500 (the square of 50). What remains for the square on *ab* is 1600, the root of which is 40 for the height of the tree *ab*.

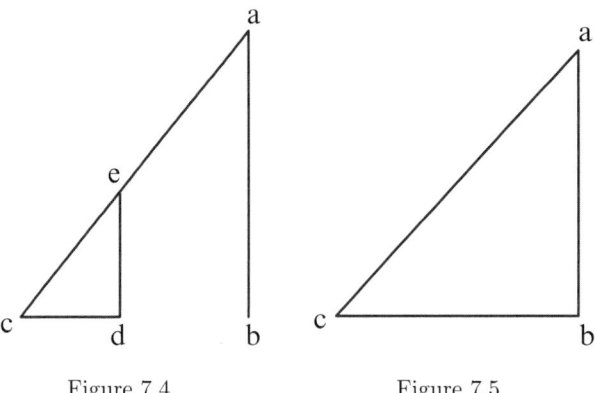

Figure 7.4   Figure 7.5

[4] We can find the measure of any height by a wooden triangle, provided a cathete has been produced from one of the angles in the triangle, and the base on which the cathete falls is on the plane. For example: we want to measure the height *ab* with wooden triangle *ecf* that has cathete *ed* [see Figure 7.6]. Set triangle *ecf* on the plane to which the altitude has been drawn, so that line *ed* stands orthogonal to the plane. Then put your eye at some point *h* on side *ec* of the triangle, and look through point *e*. If your line of sight goes through *e* to *a*, then as *cd* is to *de*, so *cb* is to *ba*. If your line of sight goes through *e* to some point between *a* and *b*, move the triangle back from

---

[15] Text and manuscript read "erit *eb* sicut *ed* ad *dc*" which is incorrect; there is no line *eb* (203.36, *f.* 131v.18–19).

the height *ab*. And if your line of sight is above *ab*, then move the triangle closer. Always have cathete *ed* perpendicular to the plane, propping up the triangle on the plane with rocks. Do this until your eye can see *a* through *e*. Then, when this is done, as I said, *cd* is to *de* as *cb* is to *ba*. If *cd* is 3 units of whatever measure, *de* 4, and *cb* 30 paces, then it follows that the height *ab* is 40 paces. Because *ed* is a third more than *cd*, it follows that *ab* is similarly a third longer than *cb*. Or, if *cb* were multiplied by *ed* and divided by *cd*, the same 40 arises for the height *ab*.

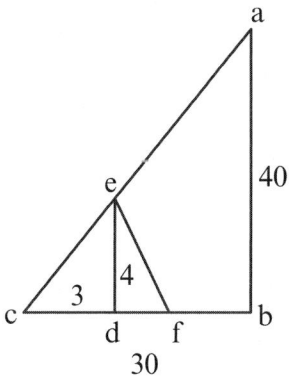

Figure 7.6

[5] Because heights can be measured easily, carefully, and fairly with a quadrant, also called an oroscope, I have taken care to include precise instructions for its construction and use [see Figure 7.7]. [For its construction:] from point *a* as the center, draw at right angles two equal lines, *ab* and *ac*. Let one of the lines *ab* or *ac* create arc *bdc*, and let it extend a bit beyond to point *e*. Then extend line *ab* to point *g*, and make line *eg* equidistant from line *ac*. Divide angle *bac* in two equal parts by line *ad*. From point *d* draw cathetes *dh* and *di* to the two lines, *ab* and *ac*. Consequently it will be shown that quadrilateral *dhai* is both equilateral and rectangular, because the three angles at points *a*, *h*, and *i* are right angles. Hence the remaining angle *hdi* is also a right angle. Since the four angles of every quadrilateral equals four right angles, and because angle *bac* was divided in two equal parts by line *ad*, each of the angles *dab* and *dac* is half a right angle. Because angles *ahd* and *aid* are right angles, therefore each of the angles *adh* and *adi* is half a right angle. Consequently triangles *hda* and *ida* are isosceles. Because they are orthogonal, whence line *ad* subtends the right angles of triangles *ahd* and *aid*, its [square] is twice [the square of] any of the sides *ah*, *dh*, *id*, or *ai*. Therefore the four sides equal one another, and quadrilateral *ahdi* is a square. At point *a* attach a string weighted with a piece of lead that can hang across the arc *bdc*. Divide each of the sides *dh* and *di* in 12 or 60 equal parts each of which I mark with a symbol as seen in similar instruments.

And thus you have a perfect quadrant. Note also line *eg*, both points *e* and *g* being marks or apertures of some kind, so that when you want to measure a height away from where you are standing, you can look through *e* {**p. 205**}, raise or lower the quadrant from point *b* until your line of sight through *e* and *g* meets the summit of the height you want to measure. Then the line of sight becomes line *eg* of which we spoke before. Now consider the point where the aforementioned weighted string falls on lines *hd* and *di*. If the line falls on point *d*, then the altitude equals the distance between you and the foot of the object, increased by your height. If the line falls on some point *k* between points *d* and *i* on line *di*, then the ratio of the distance between you and the height to the excess of the height over your height is as *ki* is to *ia*; that is, as *ki* is to *di*, a ratio of a part or parts to the whole. And if the string falls along line *dh*, say at point *l*, then the ratio of the distance to the residue of the altitude is as line *ah* to line *hl*, that is as line *hd* is to line *hl*. And this is the ratio of the whole to the part or parts.

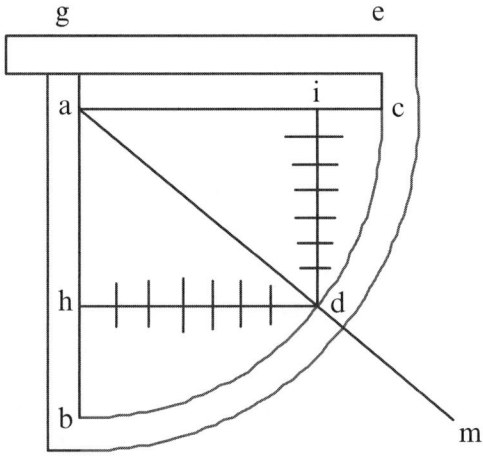

Figure 7.7

[6] To show this geometrically: let the string represented by line *am* fall first on point *d*, *m* being the lead weight [see Figure 7.8]. Because line *am* hangs from the high point *a*, the string is equidistant from the height to be measured. The string is perpendicular to the ground upon which the altitude stands. Know that the string goes up until it meets your line of sight from *e* through *g* to the summit of the object. Know further that from point *e* (where your eye is) another line is dropped equidistant from the altitude of the object and equal to your height. This line fills out the altitude you are going to measure. And then, your line of sight makes a triangle similar to the triangle made by the string and the two sides of the square described for the quadrant.

7 Measuring Heights, Depths, and Longitude of Planets 351

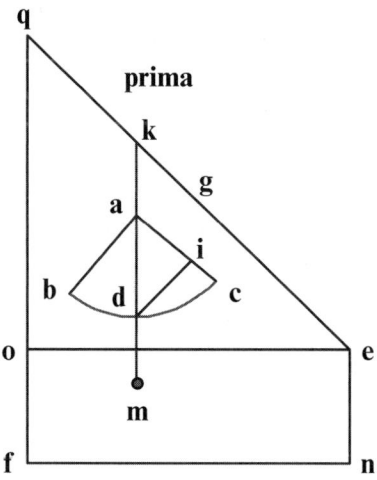

Figure 7.8

[7][16] For example: let *fq* be the altitude to be measured with *q* the summit to where your line of sight goes from point *e* through *g*. Understand that line *en* equals your height, that *eo* is equidistant from line *nf* on the ground, and that *of* also equals *en* your height. Wherefore triangle *eoq* is similar to triangle *dia* in the quadrant. For the demonstration: given string *ma* meeting line *eo* at point *k*, and that line *eo* is equidistant from line *fn*. Wherefore angle *dai* equals angle *dke*.[17] Because string *adm* is equidistant from altitude *qf* and line *eq* intersects them, angle *dke* equals angle *fqe*.[18] But angle *dke* has been shown equal to angle *dai*. Whence angle *dai* equals angle *oqe*, because angle *eoq* equals angle *aid*. Both of these are right angles because line *eo* is equidistant from line *nf*, and angle *qfn* is a right angle. Whence angle *eoq* is a right angle and angle *aid* in the quadrant is a right angle. The remaining angle *oeq* equals angle *adi*. Therefore triangle *qoe* is similar to triangle *aid*. Therefore as *di* is to *ia*, so *eo* is to *oq*. Now *di* equals *ia*. Therefore *qo* equals *oe* or *fn* the distance between you and the foot of the altitude *fq*. If we add the height *fo* (or *en* your height) to this, then the altitude *fn* is known. And this is what we wanted to show.

[8] If the string falls at point *k* on line *di*, triangle *qoe* is similar to triangle *uik*. And so angle *kai* equals angle *oqe* [see Figure 7.9]. And the angles at *o* {p. 206} and *i* are right angles. Whence the other angle *aki* equals angle *oeq*.

---

[16] The letters in [7] stand corrected without further notice unless the manuscript makes a correction; see 205.24–41.

[17] Figure 7.9 confuses the proof because it shows points *d* and *m* separate from each other. In reality they are atop each other. Furthermore, point *k* and *r* are copunctual; hence, the use of *r* is dropped.

[18] Text has *.f.e*; manuscript has *.f.q.e.* (205.34–35, f. 123r.12).

352   Fibonacci's *De Practica Geometrie*

Therefore as *ki* is to *ia* (or *di*), so *eo* is to *oq*. Therefore whatever part *ki* is of *ia*, *eo* is the same part of *oq*. If therefore *ik* is half of *ai* (this is half of *di*), then *eo* is half of *oq*; and if one is a third, the other is a third; one two thirds, the other two thirds, and so on. Perhaps this can be understood better in numbers. Let the ground distance *nf* equal to *eo* be 50 ells, and *ik* 37 of the 60 equal parts in *di* that equal *ai*. Whence as 37 is to 60, so *eo* or 50 is to *oq*. Whence multiply 50 by 60 and divide by 37 to obtain $\frac{3}{37}$ 81 ells for the altitude *oq*. Adding *of* your height to this, the true height *fq* is known, as required.

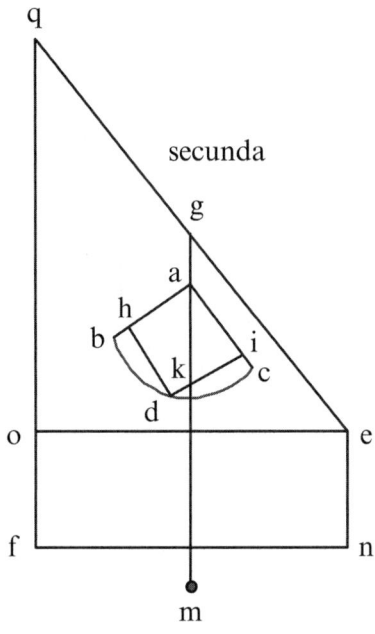

Figure 7.9

[9] Again let the string (line *am*) fall on point *l* on side *hd*, and extend string *ma* to point *r*. Then angle *mre* will equal angle *oqe*. And angle *mai* will equal *oqe*. And because *ahdi* is a tetragon, the opposite sides *ai* and *dh* are equidistant. Because line *al* intersects the equidistant lines *ai* and *hd*, then angle *lai* equals angle *alh* and angle *oqe*. Likewise and conversely as shown in the first book of Euclid,[19] angle *alh* equals angle *oqe*. Now angles *qoe* and *ahl* are right angles [see Figure 7.10]. The remaining angle *oeq* is equal to the other remaining angle *hal*. Therefore triangle *qoe* is similar to triangle *ahl* in the quadrant. But similar triangles have proportional sides.[20]

---

[19] *Elements*, I.29.
[20] Ibid., VI.4.

7 Measuring Heights, Depths, and Longitude of Planets   353

Whence as *ah* is to *hl*, so *eo* is to *oq*. Therefore whatever multiple *ah* is of *hl*, the same multiple is *eo* (that is, distance *nf*) of *oq*. In order to see this better: let *hl* be 39 parts of the 60 parts of *hd*, and let *nf* be 105 cubits. Proportionally therefore, as 60 is to 39 (which can be reduced to 20 to 13), so 105 is to the height *oq* [see Figure 7.11]. Whence multiply 13 by 21 (a fifth of 105) and then divide by 4 (a fifth of 20), and you will get $\frac{1}{4}$ 68 for the height *oq* {**p. 207**}.

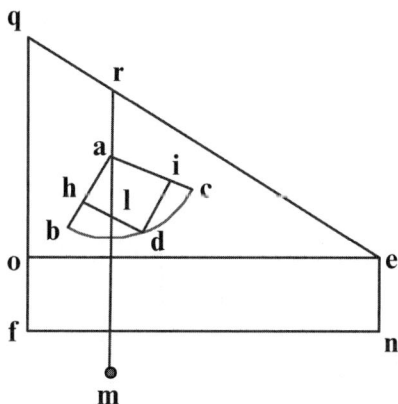

Figure 7.10

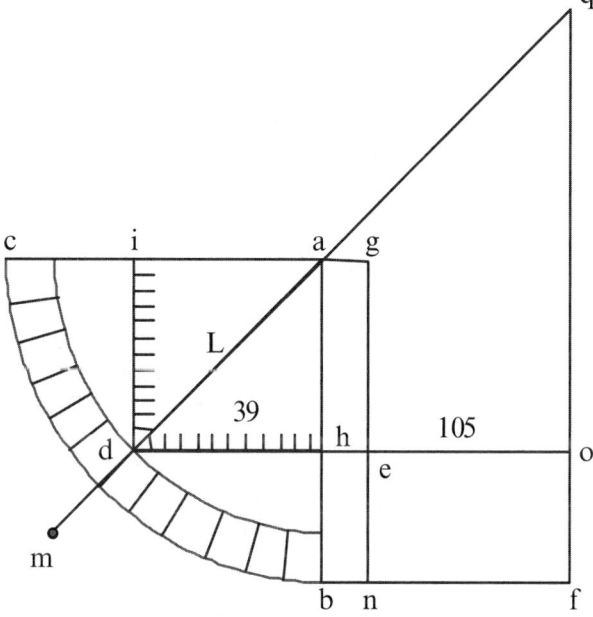

Figure 7.11

354   Fibonacci's *De Practica Geometrie*

[211][21] Now a knowledge of geometry offers a method easier than the foregoing. For those who wish to find the measure of arcs by known chords, I have composed the following table. I constructed the ordinates of 66 known arcs in it. Further I have paired each with its chord in terms of rods, feet, inches, and points. Recall that one rod is 6 feet, a foot is 18 inches, and an inch is 20 points; or, a rod is 108 inches or 2160 points. Be it understood that those 66 arcs are within a semicircle whose diameter is 42 rods, and each chord drawn in the semicircle is the chord of two equal arcs if the chord is the diameter, or of unequal arcs if the chord is not the diameter. Therefore I have set aside two arcs for each chord, as is shown in the following table [Table 7.1, Table of Arcs and Chords].

[212] {p. 97} The table offers a method of finding arcs of circles that, I propose, is better than what is known. If two unequal arcs are in the same circle, then the ratio of the larger arc to its chord is greater than the ratio of the smaller arc to its chord. This is obvious to the eye of anyone who consults the table: the ratio of the arc of a semicircle to its chord the diameter is 66 to 42 or 11 to 7, while the ratio of the seventh part of the circle to its cord is like 22 to 21. Indeed the ratio 11 to 7 is greater than the ratio of 22 to 21. So it is with all the arcs in the table, the ratio of the larger arc to its chord exceeds the ratio of the smaller arc to its chord.

[213] To show this geometrically, consider circle *abgd* [see Figure 7.12], in which lie two unequal arcs, *ab* and *bg*, with *bg* the larger arc.[22] Draw chord *ag* for arc *abg*, and bisect angle *abg* in equal parts by line *bd* that also cuts the chord at point *e*. Finally draw straight lines *ad* and *gd*. Since angle *abg* was divided in two equal parts by line *bd*, angle *abd* equals angle *dbg*. Consequently, their arcs *ad* and *dg* are equal. Now by Euclid III, equal angles standing on equal circumferences are equal to one another whether they stand at the centers or at the circumferences.[23] Therefore straight line *ad* equals straight line *dg*. Add straight line *db* to both. Then the two lines *ad* and *db* equals the two lines *bd* and *dg*. But *ba* is a smaller base than *bg*. Whence angle *gdb* is larger than angle *bda*. Or, because arc length *gb* is larger than arc length *ba*, the angle determined by lines *gd* and *db* is larger than the angle determined by lines *bd* and *da*. Whence as arc length *gb* is to arc length *ba*, so angle *gdb* is to angle *bda*. And because straight lines *ad* and *dg* are equal, if a cathete is dropped from point *d* onto line *ag*, it falls on the middle of the line. Therefore, let it fall between points *e* and *g* onto line *eg*. Because *ge* is longer than *ae*, then as line *gb* is to line *ba*, so *ge* is to *ea*. Now let a cathete fall

---

[21] Here begins the excerpt from Chapter 3, with its own numbering; see (95.32–100.10).

[22] After *Almagest*, I.10.6.

[23] Fibonacci omitted the first three words of the proposition, even though the omission does not affect the veracity of his statement (97.16–18). *Elements*, III.27: "*In equal circles* angles standing on equal circumferences are equal to one another whether they stand at the centers or at the circumferences."

Table 7.1 Table of Arcs and Chords

Arcus pertice	Arcus pertice	Corde pertice	Ar pedes	Cun vncie	In puncta	Arcus pertice	Arcus pertice	Corde pertice	Ar pedes	CV vncie	VM puncta
1	131	0	5	17	17	34	98	30	2	6	17
2	130	1	5	17	13	35	97	31	0	8	5
3	129	2	5	17	4	36	96	31	4	8	7
4	128	3	5	17	2	37	95	32	2	5	15
5	127	4	4	12	10	38	94	33	0	1	9
6	126	5	5	16	7	39	93	34	3	13	0
7	125	6	5	14	5	40	92	35	1	4	15
8	124	7	5	12	9	41	91	35	4	12	10
9	123	8	5	8	16	42	90	36	2	0	0
10	122	9	5	7	8	43	89	36	5	3	5
11	121	10	5	4	2	44	88	37	2	4	6
12	120	11	4	17	18	45	87	37	5	3	2
13	119	12	4	13	6	46	86	38	1	17	15
14	118	13	4	7	16	47	85	38	4	12	13
15	117	14	4	1	0	48	84	38	1	4	0
16	116	15	3	11	18	49	83	39	3	11	15
17	115	16	3	3	12	50	82	39	5	17	2
18	114	17	2	12	8	51	81	40	2	2	1
19	113	18	2	0	15	52	80	40	4	2	10
20	112	19	1	8	12	53	79	40	0	0	11
21	111	20	0	13	18	54	78	40	1	14	5
22	110	21	0	0	0	55	77	41	3	7	8
23	109	21	5	2	16	56	76	41	4	16	2
24	108	22	4	4	5	57	75	41	0	4	12
25	107	23	3	4	8	58	74	41	1	8	1
26	106	24	2	3	2	59	73	41	2	9	0
27	105	25	1	0	6	60	72	41	3	7	14
28	104	25	5	16	2	61	71	41	4	9	2
29	103	26	4	8	0	62	70	41	4	15	10
30	102	27	3	0	3	63	69	41	5	6	9
31	101	28	1	9	7	64	68	41	5	12	17
32	100	28	5	16	4	65	67	41	5	6	14
33	99	29	4	3	9	66	66	42	0	0	0

356   Fibonacci's *De Practica Geometrie*

from point d to point h. Because angle dhe is a right angle, straight line de is longer than straight line dh, and line da is longer than de. Both lines di[24] and df equal line de. Now, with point d as the center and de or di as the radius, draw arc ief. Now sector die is greater than right triangle dhe, and sector dif is smaller than triangle dea. Whence the ratio of sector die to sector def is greater than the ratio of triangle dhe to triangle dea. But the ratio of sector die to sector def is as the ratio of angle ide to angle edf.[25] Therefore the ratio of angle ide to angle idf is larger than the ratio of right triangle dhe to triangle dae.[26] But the ratio of triangle hde to triangle ade is as the ratio of straight line he to ea, since both triangles are under the same altitude from d to h. And the perpendicular cathete dh belongs to both triangles, hde and ade. Therefore the ratio of angle ide to angle eda is a greater ratio than that of straight line he to straight line ea. And if we compose the ratios, the ratio of angle ida to angle eda is greater than the ratio of straight line ah to straight line ae. But angle gdi equals angle adh, and straight line gh equals straight line ah. Whence the ratio of angle gdi to angle eda is greater than the ratio of straight line gh to straight line ae. But the ratio of angle ide to angle ade was found to be larger than the ratio of straight line he to straight line ea. Whence the ratio of the whole angle gde to angle eda is greater than the ratio of straight line {p. 98} ge to straight line ea. But the ratio of angle gde to angle ade is as arc gb to arc ba. And the ratio of ge to ea is as the ratio of chord gb to chord ba. Therefore the ratio of arc gb to arc ba is greater than the ratio of chord gb to chord ba. Alternately therefore the ratio of arc gb to chord gb is greater than the ratio of arc ba to chord ba. And this we had to show.

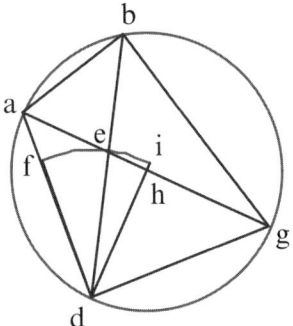

Figure 7.12

[214] With this understood, if you wish to find the arc by its chord when you know both the chord and the diameter of a circle, multiply the chord by

---

[24] Line di contains point h.
[25] Text has *rectilineum angulum edf*; context requires an acute angle (97.33).
[26] Triangle dhe is the right triangle not triangle dae, as stated in the text (97.34).

7 Measuring Heights, Depths, and Longitude of Planets   357

42 the diameter in the table, and divide the product by the diameter of the known circle. What results will be a chord in the table similar to the given chord. Take its arc from the table and multiply by the diameter of the given circle. Divide the product by the diameter of the table, and you will have the arc of the chord you want. For instance, given circle abg with diameter bg equal to 10 rods. Let the given chord ab be 5 rods. And you want to know [the measure of] arc abe. Multiply chord ab by 42, then divide by the diameter bg to obtain 21 for [the measure of a] chord in the table similar to chord ab. Look for it in the table and take the arc [that belongs to it] indirectly. Note that the arc is less than a semi circle. Now aeb the arc you seek is also less than a semi circle, the minor arc of a chord measuring 22 rods. So multiply it by 10 the diameter bg and divide the product by 42. The outcome is $\frac{5}{21}$ 5 for arc aeb.

[215][27] Likewise: let the chord ab be 8 rods, 3 feet, and $\frac{2}{7}$ 16 inches, the diameter bg being 10. So as we said: multiply 8 rods, 3 feet, and $\frac{2}{7}$ 16 inches by 42, and divide the product by 10 to get 36 rods and 2 feet for a chord in the table corresponding to the given chord ba. [This is adjacent to] 42. Multiply 42 by 10 the diameter of the circle and divide the product by 42 to obtain 10 for the measure of minor arc aeb. If you wish to know the measure of the major arc $bdga$: multiply the 36 rods and 2 feet (adjacent to the 90) by 10 and divide by 42 the diameter for the table to obtain $\frac{3}{7}$ 21 rods for the measure of the arc in the major semicircle bdg.

[216] Again, consider circle $abgd$ whose diameter ag is 12 with chord ad measuring 6 rods and 1 foot [see Figure 7.13]. You wish to know the measure of arc afd which is less than a semicircle. Multiply the 6 and 1 foot by 42 to get 259 rods. Divide this by 12 (that is ag) to get 21 rods, 3 feet, and 9 inches for the chord in the table that should be similar to chord ad. This chord, however, is not in the table. Rather it falls between the chord for 22 rods and the chord for 23 rods, namely between 21 and 21 rods, 5 feet, 2 inches, and 16 points. So in order to have a chord in the column of chords that I computed, we use a geometric figure [Figure 7.14]. Consider semicircle $ezitk$ whose diameter ek is 42, the diameter for the table. Take from it arc ez equal to 22 and arc et equal to 23. Draw chords ez and et so that chord ez will be 21 rods and chord et 21 rods, 5 feet, 2 inches, and 16 points as appear in the table. Between these chords will fall a computable chord

---

[27] The paragraph in [215] is garbled in the printed text that is faithful to the manuscript. I suspect miscopying by one or more scribes. What I offer above is, to my thinking, the most acceptable sense for: "Item sit corda ab perticarum 8 et pedum 2 et unciarum $\frac{2}{7}$ 16, et dyameter bg sit 10. Ut diximus: multiplica ergo perticas 8 et pedes 3 et uncias $\frac{2}{7}$ 16 per 42, et summam diuides per 10. Venient 36 et pedes 2, que sunt corda tabularum similes corde date ba. Quare arcum eius minorem, si minorem uis scire, uel maiorem; si de maiori, scilicet ex arcu $bdga$ notitiam uis habet. Et est minor arcus ipsius corde 42, que multiplica per 10, scilicet per dyametrum bg, et quod prouenerit, diuide per dyametrum tabularum, qui est 42, et prouenient 10 pro arcu aeb. Et si maiorum arcum corde perticarum 36 et pedum 2 que est 90, multiplica per 10 et summam diuiserimus per 42. Uenient $\frac{3}{7}$ 21 pro arcu majoris semicirculi bdg" (90. 21–29).

whose arc we look for in the table. Now we know that the arc falls between arc *ez* and arc *et*. Further it falls on point *i*. So draw chord *ei* whose measure is 21 rods, 3 feet, and {**p. 99**} and 9 inches, as found above. Using all of this we want to find the measure of arc *ei*. We know from the foregoing that the ratio of arc *ez* to chord *ez* is less than the ratio of arc *ei* to chord *ei*. So if we let the ratio of arc *ei* to chord *ei* be the same as the ratio of chord *ez* to arc *ez*, then arc *ei* will be 22 rods, 3 feet, and 12 inches. We found this by dividing the product of chord *ez* and chord *ei* by arc *ez*. But the ratio of arc *ei* to chord *ei* is greater than the ratio of arc *ez* to chord *ez*. Therefore arc *ei* is more than what we found, namely, 22 rods, 3 feet, and 12 inches. So, if we make the ratio of arc *ei* to chord *ei* as the ratio of arc *et* to chord *et*, then arc *ei* becomes 22 rods, 4 feet, 4 inches, and 13 points. But the ratio of arc *ei* to chord *ei* is less than the ratio of arc *et* to chord *et*. Hence arc *ei* is less than 22 rods, 4 feet, 4 inches, and 13 points. And as found above arc *ei* is more than 22 rods, 3 feet, and 12 inches. So first, we take half the difference between 22 rods, 3 feet, and 12 inches and 22 rods, 4 feet, 4 inches, 13 points. Then we add the half to 22 rods, 3 feet, and 12 inches. The sum gives us a close approximation to the measure of arc *ei*.

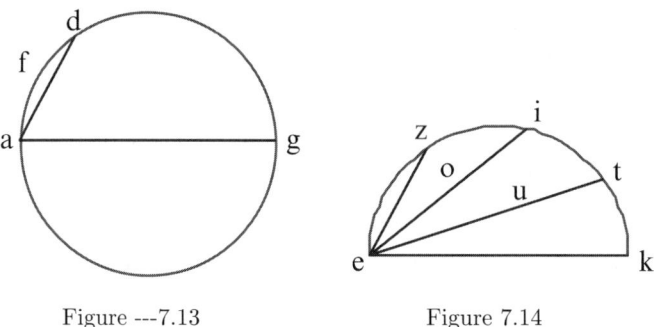

Figure ---7.13   Figure 7.14

[217] Another way: add arcs *ez* and *et* to get 45, and add chords *ez* and *et* to get 42 rods, 5 feet, 2 inches, and 16 points. Now divide this into the product of 45 and chord *ei* (21 rods, 3 feet, 9 inches) and we will have 22 rods, 4 feet, 1 inch, and 6 points for arc *ei*. Yet another way: subtract from chords *ei* and *et* the measure of chord *ez*. What remains from chord *ei* is the measure of *oi* which is $\frac{1}{2}$ 3 feet. And what remains from chord *et* is the measure of *nt* which is 5 feet, 2 inches, and 16 points. Then we set the ratio of arc *zi* (1 rod) to the ratio of *oi* to *nt*. That is, we multiply arc *tz* (1 rods equal to 2160 points) by *oi* (1260 points), divide the product by *nt* (1856 points), and end with 4 feet, 1 inch, and 6 points for the measure of arc *zi*. If we add to this arc *ez* (22 rods), we will have 22 rods, 4 feet, 1 inch, and 6 points for arc *ezi* which is similar to the desired arc *ad* in the figure above. So, if we multiply it by a sixth of diameter *ag* (2) and divide by a sixth of the diameter of the table (7) we obtain 6 rods, 2 feet, 15 inches, and 15 points for arc *ad*.

7 Measuring Heights, Depths, and Longitude of Planets   359

[218] If you wish to use chord *ad* to find arc *abd* which is larger than a semicircle, then subtract chord *ei* from chord *et* to leave 596 points. Multiply this by arc *tz* (2160 points), divide the product by *nt* (1856 points), and subtract 596 points, to leave 1 foot, 16 inches, 14 points for arc *ti*. If you would add 109 rods the major arc from the table whose chord is line *et*, you would have 109 rods, 1 foot, 16 inches, and 14 points for the arc in the table. If we multiply this by 12 and divide by 42, or if we were to multiply it by a sixth of 12 and divide the product by a sixth of 42, we would have 31 rods, 1 foot, 7 inches, and $\frac{6}{7}$ 6 points for arc *abd*.

[219][28] If you wish to use the given arc *afd* to find the unknown chord *ad*, multiply arc *afd* by the diameter of the table, divide the product by the diameter *ag*, and you will have 22 rods, 4 feet, 1 inch, and 6 points for the afore mentioned arc *ezi* which is the arc falling between arc *ez* and arc *ext* in the table. So in order for us to find the chord of {p. 100} *ei* that is similar to chord *ad* in the given circle *abgd* from arc *ezi* already computed, subtract arc *ez* and what remains is 4 feet, 1 inch, and 6 points. In points this is 1466 points. Multiply this by the difference between chord *ez* and chord *et* as found in the table. And that difference is line *nt* which is 1856 points. Divide the sum by the points in one rods (by arc *tiz*), leaving $\frac{1}{2}$ 2 feet for line *oi*. Add this to line *eo* the equal of chord *ez*, and you will have 22 rods and $\frac{1}{2}$ 3 feet for chord *ei* what is similar to your chord *ad*. Whence multiply by a sixth of 12 and divide by a sixth of 42, and what results is 6 rods and 1 foot for chord *ad*. And thus, as we have said, when the diameters of circles are known, we can find their unknown arcs by given chords, et cetera.

---

[28] This is the last paragraph of the excerpt from Chapter 3.

8
Geometric Subtleties

COMMENTARY

Fibonacci enlarged upon a remarkable collection of problems from *On the Pentagon and Decagon*, the second part of the *Algebra* of Abū Kāmil.[1] The problems seek the measures of the sides of pentagons and decagons, and of the radii and diameters of their circumscribing and inscribing circles. Lengths of various line-segments and areas of crucial triangles are among the given information. Fibonacci showed his expertise in problem solving by using two methods, algebra and proportionality.

For the second time he used the word *equation* that he had created in *Liber abaci*.[2] Nearly all the problems lead to quadratic equations of the form $ax^2 = c$ or $ax^2 + bx = c$. For equations in the latter form, I have footnoted the crucial equation in modern notation as this analysis of problem [16] illustrates. The problem seeks the measure of the diameter D of a circle about a pentagon, given the measure of a pentagonal side. Figure 8.a shows the completed construction of the side of the pentagon (gk) and of the decagon (hk) within circle agd with center at h, and the midpoint of hf at e. The measure of pentagonal side gk is 10, and the radius gh is set equal to a thing, here x. Note the appearance of the Pythagorean theorem, which occurs several times without further comment in Chapter 8.

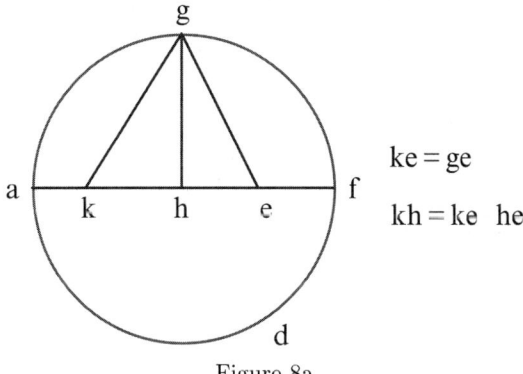

Figure 8a

---

[1] Richard Lorch (1993) published a careful and detailed analysis of this work that was invaluable for my study. I appended references to his analysis wherever I could find passages that parallel those in Boncompagni's transcription of Fibonacci's text. A comparison of the two texts shows that Fibonacci made Abū Kāmil's thought his own.

[2] *De practica geometrie* 210.43; *Liber abaci*, 407.4.

(1) $(gk)^2 = 100 = (gh)^2 + (kh)^2$

(2) $ = x^2 + [\sqrt{\left(\frac{5}{4}x^2\right)} - \frac{x}{2}]^2$

(3) $ = \frac{5}{2}x^2 - \sqrt{(\frac{5}{4}x^4)}$

(4) $(\frac{1}{2} + \sqrt{\frac{1}{20}})\,[100 = \frac{5}{2}x^2 - \sqrt{(\frac{5}{4}x^2)}]$

(5) $50 + \sqrt{500} = x^2 = (\frac{D}{2})^2$

(6) $200 + \sqrt{8000} = D^2$

In step (4) Fibonacci[3] used the multiplier

$$\left(\frac{1}{2} + \sqrt{\frac{1}{20}}\right)$$

to restore the coefficient of x^2 to unity. It speaks to his mastery of fractions that he could find the appropriate multiplicative inverse to accomplish the necessary restoration. You will see this many times in the text.

After solving the first proposition algebraically, to measure the side of an inscribed pentagon, Fibonacci offered a proportional method in [2] to measure pentagonal side ab as shown in Figure 8.b.

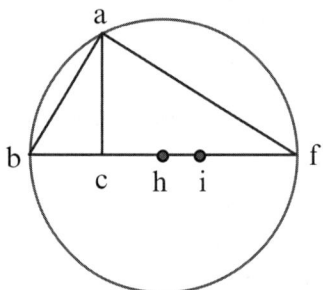

Figure 8b

In circle baf, the measure of diameter bf is 10, the center of the circle is at h, the measure of bi is 5 times hi, and line ac is perpendicular to bf, Therefore

(1) $\dfrac{bi}{ci} = \dfrac{ci}{ih}$ \qquad by construction

(2) $\dfrac{bi}{ih} = \dfrac{bi^2}{ci^2}$

(3) $bi^2 = 5ci^2$ \qquad Elements, XIII.2

---

[3] The problem in Lorch's resource is defective; most of the solution including the multiplicative inverse is missing. See his (1993), 239–240. Of course, Fibonacci's resource might have been complete; so laurels must be shared with Abū Kāmil.

8 Geometric Subtleties   363

(4) $(6\tfrac{1}{4})(1\tfrac{1}{4}) = 7\tfrac{13}{16}$

(5) Hence, $bc = bi - \sqrt{ci^2}\mathbb{Z}$

(6) $bc = 6\tfrac{1}{4} - \sqrt{7\tfrac{13}{16}}$

(7) $\dfrac{fb}{ba} = \dfrac{ba}{bc}$ Elements, VI.8.

(8) $ba^2 = (fb)(bc)$

(9) $ba^2 = (10)(6\tfrac{1}{4} - \sqrt{7\tfrac{13}{16}})$

(10) $ba^2 \sim 34\tfrac{1}{4}$.

In [1] Fibonacci continued the ability of scholars of centuries before and contemporaneous with himself to work with numbers without being able to identify them by nonnumeric names; in this case, using the modern word coefficient. Consider, "... we reach *census* of *census* and 3125 drachmas equal to 125 *census*. Consequently, we multiply $\tfrac{1}{2}$ 62 or half of the *census* and get $\tfrac{1}{4}$ 3906. From this subtract 3125 to leave $\tfrac{1}{4}$ 781." In the expression "125 *census*" the number 125 is an arithmetic concept, the number of *census*, not a scalar multiplier. The procedure for solving the equation is to complete the square. Hence, half the "coefficient" of the third term is required. Both Fibonacci and his source knew what to do, regardless of appropriate vocabulary. Farther on as in [20] the "coefficient" is moved within the radical because roots were not counted. Hence, in order to take one half, Fibonacci squares $\tfrac{1}{2}$ to become $\tfrac{1}{4}$ within the radical, and so on.

Finally at the end of the treatise, following an indeterminate problem in the *Appendix* is a group of geometric problems, [21*] to [33*]. These are prefaced with instructions on where to insert them, namely in this chapter after [20]. I have done this. An overview of the set of fourteen problems reveals that all (this includes [20]) is an organic whole that just does not fit with the penta/decagonal problems, regardless of the solitary reference to a pentagon in the penultimate sentence of [33*]. I question whether the insertion is Fibonacci's work or that of another. A curiosity of this section is the use of sexagesmal numbers. They begin in [28*] and continue through [33*].

SOURCES

As remarked above Fibonacci drew upon the tract *On the Pentagon and Decagon* within the *Algebra* by Abū Kāmil. The following table identifies the propositions in Fibonacci's work that are the same as those in Abū Kāmil's treatise. The numbers represent paragraph numbers, the first from my work, the second from Lorch's work.

1 ↔ 1, 3 ↔ 2, 6 ↔ 3, 8 ↔ 4, 10 ↔ 5, 11 ↔ 7, 13 ↔ 9, 17 ↔ 6,
19 ↔ second part of 10i, 20 ↔ 12, 21 ↔ 15, 22 ↔ 16, 23 ↔ 17,
25 ↔ 18, 26 ↔ 19, 27 ↔ 20.

The following paragraphs in Fibonacci's tract are not in Abū Kāmil's work: 2, 4, 5, 7, 9, 12, 14, 15, 18, and 25. Most of these are alternate methods for solving problems, two others being a numerical example and an additional demonstration. Not appearing in Fibonacci's work are these from Abū Kāmil: 8, 10i (first part), 11i, 11ii, and 13 and 14. Notably missing in Fibonacci's account is Abū Kāmil's 11i that gives directions for finding the side of a quintadecagon inscribed in a circle of stated diameter. Given the minor change in the order of the propositions, the omission of the challenging problem about the quintadecagon, and the addition of a number of additional ways to solve problems, all of this in Fibonacci's treatise, we have an unanswered question. Is Chapter 8 a translation of an Arabic edition of Abū Kāmil's treatise that no longer exists, or is it Fibonacci's own edition of his work?

The simplest hypothesis is that Fibonacci had his own Arabic copy of Abū Kāmil's *Algebra*, translating the section on pentagons and decagons to become his Chapter 8.[4] By comparing a section from Chapter 8 with a conceptually identical section of the Latin *Algebra*, we can appreciate that both translators were studying Arabic copies of the same treatise. The differences between the two selections are shown in *italics*.

From Abū Kāmil: Reduc totum quod habes ad censum, quod est ut ducas totum quod habes in $\frac{e}{5}\frac{2}{5}$ unius dim rad' $\frac{e}{625}\frac{4}{5}$ unius, et erit census equalis 75 drag' et rad' 625 dim radice 3125 drag' et dim rad' 1123 drag'. Et illud est 100 drag' dim rad' 8000, et illud est quadratum linee AB, quod est unum de lateribus decagoni continentis in circulo scito cuius diameter est 10 ex numero. Et hoc est quod voluimus exponere.[5]

From Fibonacci: reduc igitur hec omnia ad census, quod est ut multiplces ea per $\frac{2}{25}$ minus radice $\frac{4}{3125}$: *ex ductu quidem $\frac{2}{25}$ minus radice $\frac{4}{3125}$ in census $\frac{5}{8}$ 15 et in radicem censum census $\frac{5}{8}\frac{6}{8}$ 48 eveniet census*. Et duc $\frac{1}{2}$ 937 minus radice $\frac{1}{4}$ 48829 in $\frac{2}{25}$ minus radice $\frac{1}{4}$. Et erit census equalis 75 et radici 625 drachmarum minus radice 3125 dragme et minus radice 1125 dragme, quod totum est 100 minus radice 8000 quod radix est latus ab, quod est unum ex lateribus decagoni circundanti circulum, cuius dyameter est 10.[6]

Additionally, I compared the two texts for essential verbal fidelity. Major differences are recorded in the footnotes with appropriate changes in Boncompagni's text as needed. Finally, at the end of the treatise, following

---

[4] Lorch (1993), 215–252.
[5] Lorch 238.108–114.
[6] Boncompagni 210.11–17.

8 Geometric Subtleties   365

an indeterminate problem in the *Appendix*, is a group of problems. These last are prefaced with instructions on where to insert them, namely in this chapter after [20]. This has been done.

TEXT[7]

[1] **{p. 207}** *To find the side of a pentagon inscribed within a circle.*[8] Consider the equilateral and equiangular pentagon *abgde* inscribed in circle *abgde* with diameter 10. Draw the pentagonal chord *ag* and the diameter *bf* cutting the chord in two equal parts at point *c*, whereby angle *bca* is a right angle. I make the diameter *fb* to be 10 units and call the pentagonal side thing. By drawing line *af*, I create the right triangle *baf* because it is inscribed in semicircle *baf*. The cathete *ac* is drawn from the right angle to the base. Whence the product of *fb* and *cb* equals the product of *ba* and itself. Multiplying *ab* or thing by itself produces a *census*. Dividing this by *fb* or 10 we obtain $\frac{1}{10}$ *census* for *cb*. Subtracting therefore the square of $\frac{1}{10}$ *census* from the square on the side *ba* (namely the square on *cb* from the *census*), what remains for the square on line *ac* is a *census* less $\frac{1}{100}$ *census* of a *census*. Because *ag* is twice *ac*, the square on *ag* is four times the square on *ac* [see Figure 8.1]. Whence we quadruple the square on line *ac* to obtain 4 *census* less $\frac{1}{25}$ *census* of *census* for the square on chord *ag*. Because quadrilateral *agde* is inscribed in circle *abgd*, the product of *de* and *ga* with the product of *gd* and *ea* is as the product of the diameters of the quadrilateral, the one being *ge* and the other *da*. One of these equals line *ag* since any one of them is a chord of a pentagonal angle. Therefore the product of *de* and *ag* with the product of *gd* and *ae* is as the product of *ag* by itself. But *ag* by itself gives rise to 4 *census* less $\frac{1}{25}$ *census* of *census*. Therefore by multiplying *de* by *ga* and *gd* by *ea*, we obtain 4 *census* less $\frac{1}{25}$ *census* of *census*. If you take from this the *census* that is the product of *gd* and *ae* or of a thing in a thing, what remains is 3 *census* less $\frac{1}{25}$ *census* of *census* for the product of *de* and *ag*. Whence if we divide 3 *census* less $\frac{1}{25}$ *census* of *census* by a thing or *de*, we get 3 things less $\frac{1}{25}$ cube for the measure of chord *ga*. Therefore we multiply 3 things less $\frac{1}{25}$ cube by itself to obtain 9 *census* and $\frac{1}{625}$[9] cube of cube less $\frac{6}{25}$ *census* of *census* that equal 4 *census* less $\frac{1}{25}$ *census* of *census*. Adding to both parts $\frac{6}{25}$ *census* of *census* and subtracting 4 *census* from both, what we have left is 5 *census* $\frac{1}{625}$ cube of cube equals 5 *census* of *census*. Now we divide all of this by a *census* to obtain 5 drachma and $\frac{1}{625}$ *census* of *census* equal to 5 *census*. Reducing all of this to one *census* of *census* and multiplying all by 625, we reach a *census* of *census* and 3125 drachmas equal to 125 *census*.[10] Consequently we multiply $\frac{1}{2}$ 62 or half of a

---

[7] "Incipit distinctio octaua de quibusdam subtilitatibus geometricis" (207.2, f. 134r.1). Lorch 234.19–235.45; Boncompagni 207.8–35.

[8] "Circuli, cuius diameter sit 10, uolo latus penthagonicum inuenire" (207.2, f. 134r.2).

[9] Text has $\frac{1}{25}$; manuscript has $\frac{1}{625}$ (207.26, f. 134r.32).

[10] $x^4 + 3125 = 125x^2$. (207.31–32).

census by itself to get $\frac{1}{4}$ 3906. From this subtract 3125 to leave $\frac{1}{4}$ 781. From this subtract the root of $\frac{1}{2}$ 62 to leave $\frac{1}{2}$ 62 less than the root of $\frac{1}{4}$ 781 for the measure of the *census*, the root of which is a thing, the pentagonal side.

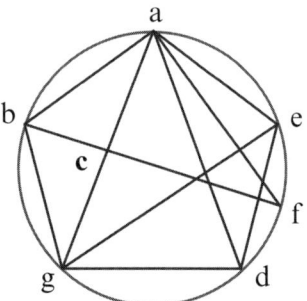

Figure 8.1

[2] Another way: Considering diameter *bf* and its center *h*, add to the radius *bh* a fourth of itself or *hi* so that the total length *bi* is $\frac{1}{4}$ 6 [see Figure 8.2]. As shown in the thirteenth book of Euclid,[11] as *bi* is to *ic*, so *ic* is to *ih*. Whence as the first *bi* is to the third *ih*, so the square on line *bi* is to the square on line *ci*. Now *bi* is five times *ih*. Whence the square on *bi*, that is $\frac{1}{16}$ 39, is five times the square on *ci*. Therefore *ci* is the root of $\frac{13}{16}$ 7. Removing this from *bi* leaves $\frac{1}{4}$ 6 less the root of $\frac{13}{16}$ 7 for the measure of line *bc*. Now because *fb* is to *ba* as *ba* is to *bc*, the product of *ba* and itself equals the product of *fb* and *bc*. Now *fb* is 10. If we multiply this by *bc* or $\frac{1}{4}$ 6 less the root of $\frac{13}{16}$ 7 {**p. 208**}, we get $\frac{1}{2}$ 62 less the root of $\frac{1}{4}$ 781 for the square on line *ba*, the root of which is the pentagonal side *ab*. After subtracting the root of $\frac{1}{4}$ 781 from $\frac{1}{2}$ 62, what remains is a little more than $\frac{1}{2}$ 34. Its approximate root is 6 less $\frac{1}{8}$, a close approximation of the measure of *ab*.

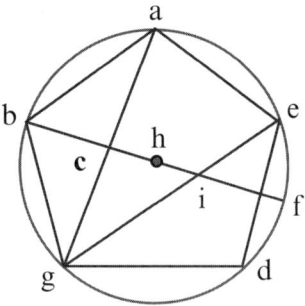

Figure 8.2

---

[11] *Elements*, XIII.2.

[3] *To find the length of the side of a decagon inscribed in a circle.*[12] Consider a circle *abg* of diameter 10: divide the arc of semicircle *abcf* in five equal parts: *ab*, *bc*, *cd*, *de*, and *ef*. Draw these lines: *ab*, *bc*, *cd*, *de*, and *ef*. Each of these is the side of a decagon. Now draw lines *ac* and *df*. Each of these is the side of a pentagon. Draw lines *ad* and *cf*. Because arcs *ac* and *fd* are equal, add arc *cd* to both. Thus arc *acd* equals arc *cdf*. Because line *ad* equals line *cf*, there exists in circle *abg* quadrilateral *cafd* with diameters *cf* and *ad*. Whence the product of *af* and *cd* with the product of *ac* and *fd* equals the product of *cf* and *ad*.[13] But the product of *cf* and *ad* equals the product of *cf* and itself [see Figure 8.3]. With all this understood, call the decagonal side *ad* a thing and multiply it by 10 or *af* to get 10 a things. Add this to the product of *ca* and *df* which is the square on *ca* (which we found above to be $\frac{1}{2}$ 62 less the root of $\frac{1}{4}$ 781), and we have 10 things and $\frac{1}{2}$ 62 drachmas less the root of $\frac{1}{4}$ 781. This equals the square on line *cf*. Add to this $\frac{1}{2}$ 62 drachmas less the root of $\frac{1}{4}$ 781 the square on side *ca* to get 10 things and 125 drachmas less two roots of $\frac{1}{4}$ 781. This is the root of 3125 that equals 100 drachmas or the square on diameter *af*. Now angle *fca* is a right angle, because it is inscribed in semicircle *abf*. Add therefore to both parts the root of 3125 and subtract from them 125, so that 10 things remains equal to the root of 3125 less 25 drachmas. Now divide the root of 3125 less 25 by 10 to obtain the root of $\frac{1}{4}$ 31 less $\frac{1}{2}$ 2 drachmas for the measure of a thing, namely the decagonal side *cd*.

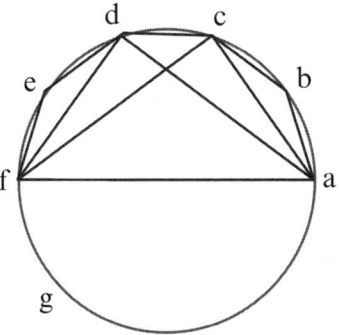

Figure 8.3

[4] *To find the side of a decagon another way, given the diameter of a pentagon.*[14] Again, let the diameter of circle *abg* be 10 and its center at *i* with right angles formed by line *ig*. Divide *if* in two equal parts at *h* and draw line *gh*. Locate *hl* [on *ha*] equal to *gh*. Draw line *gl* [Figure 8.4]. I say that *il* is the

---

[12] "Et si uis habere notitiam lateris decagoni cadentis in circulo" (208.5). Lorch 235.46–236.64; Boncompagni 208.5–27.
[13] *Almagest* I.10(3).
[14] "Aliter inuenies latus decagoni et penthagoni per dyametrum notum" (208.26, f. 135r.10).

side of a decagon and *gl* the side of a pentagon, both inscribed in circle *abfg*. The proof follows. Because line *fi* has been divided in two equal segments at point *h* and line *il* added to it,[15] the product of *il* and *fl* with the square on line *ih* is as the product of *hl* and itself. But line *hl* equals line *gh*. Therefore the product of *li* and *fl* with the product of *hi* and itself is as the product of *hg* and itself. But the product of *hg* and itself is as the product of *gi* and itself and the product of *ih* and itself. Therefore the product of *il* and *fl* with the square on line *hi* is as the squares on lines *gi* and *ih*. Subtracting the square on line *ih* from both leaves the product of *il* and *fl* as the square on line *gi*. But line *gi* equals line *fi*. Therefore plane *il* by *fl* is as the product of *fi* and itself. Whence proportionally, as *lf* is to *fi* so *fi* is to *il*. Therefore line *fl* has been divided in mean and extreme ratio. The longer segment is *fi* the side of a hexagon. The shorter segment is *il* the side of a decagon, as Euclid showed.[16] Because the side of a pentagon can be constructed from the side of a hexagon and a decagon, *gi* being the side of a hexagon and *il* the side of a decagon, line *gl* is the side of a pentagon.

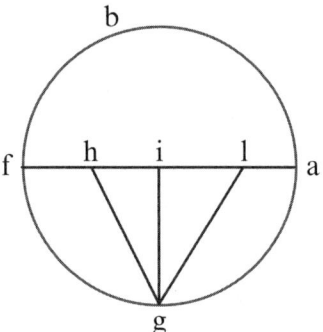

Figure 8.4

[5] To show this numerically: let the diameter *af* be 10. Whence *gi* is 5 and *ih* is $\frac{1}{2}$ 2. Multiply *gi* in itself and *ih* in itself {**p. 209**}, you get 25 and $\frac{1}{4}$ 6 which added together yield $\frac{1}{4}$ 31 for the square on line *gh* equal to the square on line *hl*. Whence the root of $\frac{1}{4}$ 31 measures line *hl*. If you subtract $\frac{1}{2}$ 2 (line *hi*) from this, what remains is the root of $\frac{1}{4}$ 31 less $\frac{1}{2}$ 2 for line *il* the side of the decagon, as we found above. If you multiply *il* by itself and *gi* by itself, you will get $\frac{1}{2}$ 62 less the root of $\frac{1}{4}$ 781 for line *gl*, the side of the pentagon, as we said above.

[6] *To find the measure of the side of a pentagon circumscribed about a given circle.*[17] Given circle *thm* with center *e* and diameter 10. Circumscribe

---

[15] The text has "et .*ei*. addita est recta .*il*"; *ei* is the pronoun object of "addita est" because it lacked the bracketing periods, and the context requires the pronoun (208.31, f. 135r.16).

[16] *Elements*, XIII.9.

[17] "Et quando uolueris scire latus penthagoni circundantis circulum scitum." (209.7, f. 135v.1). Lorch 236.65–237.91; Boncompagni 209.7–31.

pentagon *abcdf* about this circle. Draw lines *ea*, *eb*, *eh*, and chord *th* of the circumscribed pentagon [see Figure 8.5]. Cut the chord in two equal parts by line *eb*; the angles at *l* are right angles. The square of the whole chord *th* is $\frac{1}{2}$ 62 less the root of $\frac{1}{4}$ 781. Hence the square on *tl* (half of chord *th*) is a fourth of that, namely $\frac{5}{8}$ 15 less the root of $\frac{5}{8}\frac{6}{8}$ 48. If the square is subtracted from 25 the square on the radius, $\frac{3}{8}$ 9 and the root of $\frac{5}{8}\frac{6}{8}$ 48 remain for the square on line *el*. Because triangle *eta* is similar to triangle *etl*, since both have right angles and angles *aet* and *tel* are equal, the ratio of *at* to *te* is as the ratio of *tl* to *le*. Because *ab* is twice *at* and *th* twice *tl*, as *ab* (the first term) is to *te* (the second term), so *tl* (the third term) is to *le* (the fourth term). Whence the product of *ab* and *el* equals the product of *te* and *th*. Let side *ab* be a thing, square it to become a *census*, multiply it by $\frac{3}{8}$ 9 and the root of $\frac{5}{8}\frac{6}{8}$ 48 (the square on *el*) to get $\frac{3}{8}$ 9 *census* and the root of $\frac{5}{8}\frac{6}{8}$ 48 *census* of *census* which equals what came from multiplying 25 (the square on radius *te*) by $\frac{1}{2}$ 62 less the root of $\frac{1}{4}$ 781 (the square of chord *th*). From this multiplication you got $\frac{1}{2}$ 1562 less the root of $\frac{1}{4}$ 488281. Reduce all you have to one *census*. That is, multiply what you have by $\frac{6}{25}$ of a drachma less the root of $\frac{4}{125}$ of a drachma. Now the product of $\frac{6}{25}$ of a drachma less the root of $\frac{4}{125}$ of a drachma and $\frac{3}{8}$ 9 *census* and the root of $\frac{5}{8}\frac{6}{8}$ 48 *census* of a *census* results in one *census*.[18] Then multiply $\frac{1}{2}$ 1562 less the root of $\frac{1}{4}$ 488281 by the root of $\frac{4}{125}$ of a drachma to reach 375 drachmas and the root of 15625 drachmas less the root of 78125 drachmas less the root of 28125 drachmas. I make all of this to be 500 less the root of 200000. Therefore one *census* equals 500 less the root of 20000, the root of which is line *ab*, one of the sides of the pentagon circumscribed about the circle of diameter 10.

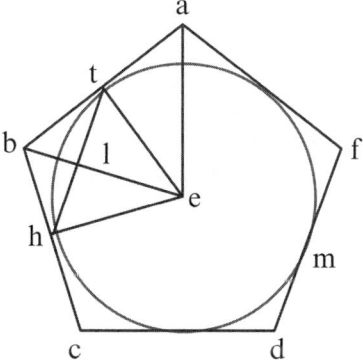

Figure 8.5

---

[18] $(\frac{6}{25} - \sqrt{\frac{4}{125}})(\frac{3}{8}9x^2 + \sqrt{\frac{53}{64}x^4}) = x^2(209.25 - 26)$.

[7] There is a faster way of doing this. Triangle *etb* is a right triangle because of cathete *tl*. So, the ratio of the square on *be* to the square on *et* is as the square on *et* to the square on *el*. Hence, the product of the square on *be* and the square on *rl* equals the product of the square on *et* and itself which is 625. So, let *be* be a thing, multiply it by itself to get one *census* which you multiply by $\frac{3}{8}$ 9 and the root of $\frac{5}{8}\frac{6}{8}$ 48 (the square on *el*) to get $\frac{3}{8}$ 9 and the root of $\frac{5}{8}\frac{6}{8}$ 48 *census* of *census* which equals 625 drachmas. Reduce this to one *census*, that is multiply 625 by $\frac{6}{25}$ of a drachma less the root of $\frac{1}{125}$ to get 150 less the root of 12500 which equals one *census*, the square on *be*. If you subtract 25 (the square on *et*) from this, what remains is 125 less the root of 12500 for the square on *tb*. Quadruple this to 500 less the root of 200000 (the square on *ab*), as we found above.

[8] *To measure the side of the decagon circumscribed about a circle.*[19] Given a decagon circumscribed about circle *def* of diameter 10, identify *ab* and *bg* as two sides of the decagon tangent to circle *def* {**p. 210**} at points *d* and *e*. With *h* the center of the circle, draw lines *de*, *ha*, *hd*, and *hb*. According to what has been said above for the previous figure, *ab* is to *dh* as *de* is to *hl* [see Figure 8.6]. Whence the product of the square on *ab* by the square on *hl* is as the product of the square on *de* by the square on *dh*.[20] Chord *de* is the side of the decagon inscribed in circle *def*. We found its measure to be the root of $\frac{1}{4}$ 31 less $\frac{1}{2}$ 2. Line *dl* is half of this and equals the root of $\frac{13}{16}$ 7 less $\frac{1}{4}$ 1 which squared becomes $\frac{3}{8}$ 9 less the root of $\frac{5}{8}\frac{6}{8}$ 48. Subtracting this from 25 (the square on *dh*) leaves $\frac{5}{8}$ 15 and the root of $\frac{5}{8}\frac{6}{8}$ 48 (for the square on *hl*). If we multiply this square by *census* the square on *ab*, we obtain $\frac{5}{8}$ 15 *census* and the root of $\frac{5}{8}\frac{6}{8}$ 48 *census* of *census*. It equals the product of 25 (the square on *dh*) and $\frac{3}{8}$ 9 less the root of $\frac{5}{8}\frac{6}{8}$ 48 (the square on *dl*)[21] or $\frac{1}{2}$ 937 less the root of $\frac{1}{4}$ 48828. Reduce all of this to one *census* by multiplying it by $\frac{2}{25}$ less the root of $\frac{4}{8215}$. Multiplying $\frac{2}{25}$ less the root of $\frac{4}{8215}$ first by $\frac{5}{8}$ 15 *census* and then by the root of $\frac{5}{8}\frac{6}{8}$ 48 *census* of *census* produces one *census*. Then multiply $\frac{1}{2}$ 937 less the root of $\frac{1}{4}$ 48828 by $\frac{2}{25}$ less the root of $\frac{4}{8215}$[22] to obtain one *census* equal to 75 and the root of 625 drachmas less the root of 3125 drachmas less the root of 1125 drachmas, all totaled equaling 100 less the root of 8000. The root of this is side *ab*, one of the decagonal sides about the circle of diameter 10.

---

[19] "Et si uis inuenire mensuram lateris decagoni circundantis circulum...." (209.41–43, f. 136r.4–5). Lorch 237.92–238.114; Boncompagni 209.41–210.17.

[20] This is my interpretation of "Vnde multiplicatio quadrati, et quadratum *ab* in *hl* est sicut multiplicatio quadrati; et quadratum *de* in *dh*" (210.3–5).

[21] The clause "It equals ... on *dl*)" is not in Lorch nor in the Arabic text. Furthermore, "25" should be "100."

[22] "Multiplying $\frac{2}{25}$ ... root of $\frac{4}{8215}$ " is not in Lorch 238.110.

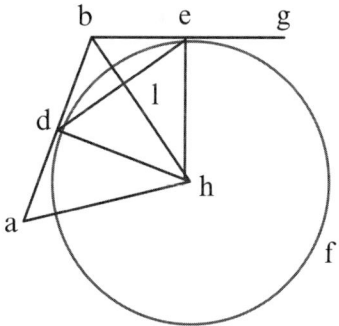

Figure 8.6

[9] Another way: let hb be a thing, square it to get one *census* which you multiply by $\frac{5}{8}$ 15 and the root of $\frac{5\ 6}{8\ 8}$ 48 (the square on hl) to get $\frac{5}{8}$ 15 *census* and the root of $\frac{5\ 6}{8\ 8}$ 48 *census* of *census* that equals the square on dh or 625 drachmas. Multiply all of this by $\frac{2}{25}$ less the root of $\frac{4}{8215}$, and you get one *census* equal to 50 less the root of 500 drachmas for the square on line hb. If you subtract 25 (the square on hd) from this, what is left for the square on line db is 25 less the root of 500. Quadruple this to 100 less the root of 8000 for the square on line ab, as we found above. Now that we have shown how to find the sides of pentagons and decagons inscribed and circumscribed about a circle of given diameter, it remains to find the diameter given the same sides.

[10] [*To measure the diameter of a circle, given the measure of a pentagonal chord in a circle.*][23] Inscribe a pentagon within a circle and let one of the sides of pentagon $abgde$ be 10. The diameter of the circle is dc.[24] Draw lines ge, eb, and bd to create quadrilateral $ebgd$ within circle $abgd$ [see Figure 8.7]. Whence the product of eb and dg with the product of gb and de equals the product of eg and db, that is, the square on eg. Hence, let line eb be a thing and multiply it by dg to get 10 a things. Add 100 (the product of bg and de) to this to get 100 and 10 things equal to the square on eg. Now eg is a thing equal to eb. Therefore 100 and 10 things equals one *census*. Whence a thing is the root of 125 and 5 drachmas, the length of line ge. Its half is gl the root of $\frac{1}{4}$ 31 and $\frac{1}{2}$ 2 drachmas. Squaring this yields $\frac{1}{2}$ 37 and the root of $\frac{1}{4}$ 781. Subtracting this from 100 (the square of yd) leaves $\frac{1}{2}$ 62 less the root of $\frac{1}{4}$ 781 for the square on line dl. So if you multiply cd by dl, it is the same as squaring de because cathete el makes right angles with the base in triangle ced. Whence if you make diameter cd a thing and multiply its square or a *census* by the square on line dl, you get $\frac{1}{2}$ 62 *census* less the root of $\frac{1}{4}$ 781 *census* of *census* which equals the square of the square on de or 10000 drachmas.

---

[23] "ut si dicamus: corda quinte circulj est 10, quanta est longitudo dyametri" (210.27, f. 136v.2–3). Lorch 238.115–239.135; Boncompagni 210.26–43.

[24] At this line in the margin is the remark "as proved above in Chapter 3 by Euclid VI.3" (210.29, f. 136v).

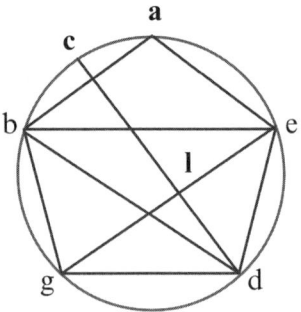

Figure 8.7

[11] Now there is another way to reach this equation[25] {**p. 211**}. For example: in a circle of diameter 10 let the length of the side of a pentagon be the root of $\frac{1}{2}$ 62 less the root of $\frac{1}{4}$ 781. I want a pentagonal side of length 10. Hence, the diameter of another circle is required. Consequently let the ratio of $\frac{1}{2}$ 62 less the root of $\frac{1}{4}$ 781 to 100 (the square of the diameter) be equal to the ratio of 100 (the square of the given pentagonal side) to the square of the unknown diameter. Let the unknown diameter be a thing. Multiply its square or a *census* by $\frac{1}{2}$ 62 less the root of $\frac{1}{4}$ 781 to obtain $\frac{1}{2}$ 62 *census* less the root of $\frac{1}{4}$ 781 *census* of *census* equal to 10000 drachmas (the product of 100 and 100), as we found above. You can reduce all of this to one *census* by multiplying it by $\frac{1}{50}$ and the root of $\frac{1}{2125}$ drachma to reach 200 and the root of 8000 for the square of diameter *cd*. The root of this is a bit more than 17 for *dc* the diameter that we sought.

[12] In another way: the chord of a pentagonal angle and the desired chord of the circle are five times the length of the side of a hexagon or of the radius, as noted in the fourteenth book of Euclid.[26] So, square the root of 125 and 5 drachmas (the length of *eg*) to get 150 and the root of 12500. Add this to 100 (the square of side *gd*) to obtained 250 and the root of 12500, all of which equals five times the square on the radius. A fifth of that, then, is 50 and the root of 500 for the square on the radius. Four times this is 200 and the root of 8000 for the square on the diameter, as we found above.

[13] [*To measure the diameter of a circle, given the decagonal chord within the circle.*][27] Likewise, if you want to find the diameter of a circle for which $\frac{1}{10}$ of the chord is 10, the ratio of the chord to the diameter of its circle

---

[25] Lorch 240.148–165; Boncompagni 210.3–211.11. Fibonacci adapted the Latin word *equatio* for use in algebra.

[26] *Elements*, XIV.3; see Ch. 6 [9].

[27] "Item si uis inuenire dyametrum circuli cuius $\frac{e}{10}$ corda sit 10 ex numeris" (211.18–19, f. 137r.5–6). Lorch 241.178–190; Boncompagni 211.18–31.

equals the ratio of some similar chord [of another circle] to its diameter. Now the measure of the decagon is the root of $\frac{1}{4}$ 31 less $\frac{1}{2}$ 2 for a circle of diameter 10.[28] If you make the unknown diameter a thing and multiply it by the root of $\frac{1}{4}$ 31 less $\frac{1}{2}$ 2, you will get the root of $\frac{1}{4}$ 31 *census* less $\frac{1}{2}$ 2 things equal to 100. This comes about by multiplying 10 the known chord by the given diameter and then adding to both parts $\frac{1}{2}$ 2 things so that you have 100 and $\frac{1}{2}$ 2 things equal to the root of $\frac{1}{4}$ 31 *census*.[29] Square all of this to get $\frac{1}{4}$ 31 *census* equal to $\frac{1}{4}$ 6 *census* and 10000 drachmas less 500 things.[30] Then subtract $\frac{1}{4}$ 6 *census* from both sides and add 500 things to both sides so that you have 25 *census* and 500 things equal to 10000 drachmas. Finally reduce all of this to one *census* by dividing it by 25. The result is one *census* and 20 things equal to 400 drachmas. Solving this by algebra[31] produces a thing equal to the root of 500 drachmas and 10.[32] This is the length you sought for the diameter.

[14] In another way: consider the attached circle *abcdefg* with diameter *af*. Within half of it are five equal chords: *ab*, *bc*, *cd*, *de*, and *ef*, each a decagonal chord in circle *acg*. From the endpoints of the diameter draw lines *ac* and *fd*, each of which is the side of a pentagon inscribed in the same circle because each is subtends a decagonal angle. Draw lines *ad* and *cf*. Let the measure of each decagonal chord be 10 [see Figure 8.8]. Now set the diameter equal to two things. Multiply them by 10 (chord *cd*) to obtain 20 things. Multiply line *ac* by line *fd* (the same as squaring a pentagonal side) to obtain one *census* and 100 drachmas. Now, I say that since the pentagonal side equals the sum of the hexagonal and decagonal sides, and this side is the hexagonal thing, and its power is one *census*, and the decagonal side is 10, and its power is 100, and adding one *census* and 100 to 20 things, there is one *census* and 100 and 20 things, all of which came from multiplying *af* by *cd* and *ac* by *fd*. This is equal to the product of *ad* and {p. 212} *cf* for the square on line *cf*. If we add one *census* and 100 (the square on line *ac*) to this, we will have 2 *census*, 20 things, and 200 drachmas equal to 4 *census*,[33] the square on the diameter. So, subtract 2 *census* from both sides and halve what remains for one *census* equal to 10 things and 100 drachmas. Solve this in the algebraic manner and you will get half the diameter *af*, that is a thing, the root of 125 and 5 drachmas. Doubling this yields the root of 500 and 10 drachmas for the length of the diameter we sought, just as we found above.

---

[28] See [3] above.
[29] $100 + 2.5x = 31.25\, x^2$ (211.25).
[30] Text has *radicibus*; context requires *rebus* 211.27.
[31] "age in his secundum algebram" (211.30).
[32] As above, this is an abstract number, "ex numeris", which I leave unmodified (211.31).
[33] $2x^2 + 20x + 200 = 4x^2$ (212.2–3).

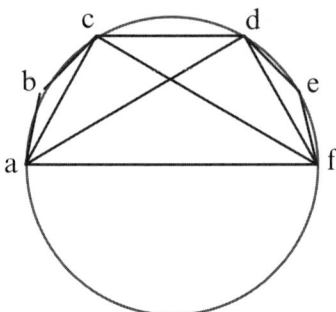

Figure 8.8

[15] In another way: in the same circle with center h draw line gh at right angles to the diameter af. Make hi half of fh, and draw line gi to make ik its equal [see Figure 8.9]. Drawing line gk makes it a pentagonal side in circle acf, according to what was established above. hk will be the side of a decagon and gh the side of a hexagon. Let gh be a thing. Whence ih is half of a thing. Squaring gh produces one *census*. Squaring ih produces a fourth of a *census*. And thus for the square on line gi (or for the square on like ik), we have $\frac{1}{4}$ 1 *census*. Therefore ik is the root of $\frac{1}{4}$ 1 *census*. Subtracting ih from this leaves hk the root of $\frac{1}{4}$ 1 *census* less the root of $\frac{1}{2}$ or hi equal to 10 drachmas, because the decagonal side was made equal to 10. And so, if we add half a thing to both sides, then the root of $\frac{1}{4}$ 1 *census* equals 10 and half a thing. Squaring all of this produces $\frac{1}{4}$ 1 *census* equal to 10 things, 100 drachmas, and $\frac{1}{4}$ *census*. So if you subtract $\frac{1}{4}$ *census* from both sides, what remains is one *census* equal to 100 and 10 things. Thus the thing that is half of the diameter of circle abg is 5 and the root of 125 drachmas. Doubling this produces 10 and the root of 500 drachmas for the length of diameter af. This is the way to find the diameter of any circle, namely the diameter adds the root of five times the square of the decagonal side to the side of decagon that is drawn to it. For instance, if a decagonal side measures 6, the diameter of the circle will be 6 and the root of 180 (the root of five times the square of six). Understand that this is the way to proceed in all similar cases.

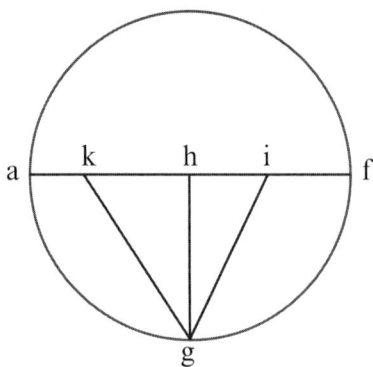

Figure 8.9

[16] We can use the same figure to find the diameter, given the side of a pentagon. For example: let gk be the side of a pentagon inscribed in circle agd and of length 10.[34] Its square is 100 and equal to two squares, on lines gh and hk. Now gh is the radius and side hk is a decagonal chord in the circle. Let line gh be a thing; square it to make one *census*. Square hk (the root of $\frac{1}{4}$ 1 *census* less $\frac{1}{2}$ thing) to reach $\frac{1}{2}$ 1 *census* less the root of $\frac{1}{4}$ 1 *census* of *census*. Add all of this to one *census* (that is the square on line gh) to obtain $\frac{1}{2}$ 2 *census* less the root of $\frac{1}{4}$ 1 *census* of *census* that equals 100 drachmas. Restore all of this to one *census* by multiplying it by $\frac{1}{2}$ and the root of $\frac{1}{20}$ a drachma to obtain 50 and the root of 500 drachmas equal to a *census* (the square on half the diameter). Four times this or 200 and the root of 8000 is [the square of][35] the length of the diameter. Note that if you know the measure of the side of the pentagon in any circle, all you have to do is multiply the square of its side by $\frac{1}{2}$ and the root of $\frac{1}{20}$ of a drachma to obtain the square of half the diameter.

[17] *To find the length of the diameter of a circle circumscribed by an equilateral and equiangular pentagon.*[36] Given a regular pentagon $abgde$ with side of 10 and m the center of a circle. Half the diameter of the circle containing[37] the pentagon is line am. The square of $\frac{1}{2}$ the diameter is 50 and the root of 500 [see Figure 8.10]. Subtract 25 (the square on ac) from this to leave {p. 213} 25 and the root of 500 for the square on cm. Its quadruple, 100 and the root of 8000, is the square on the diameter of the circle contained by pentagon $abcde$. Given the square of the diameter [ab] of the circumscribing circle $(200 + \sqrt{8000})$: if you want the square of the diameter [cd] of the inscribed circle, subtract the square on the side to leave $100 + \sqrt{8000}$.[38]

---

[34] Refer to Figure 8.11 which shows how to construct hk one side of the pentagon, because the pentagon is not actually inscribed in the circle.

[35] Added to correct the text (212.36, f. 138r.6).

[36] "Et si vis habere notitiam dyametri circulj circundati a penthagono equilatero et equiangulo" (212.40, f. 138r.10–11). Lorch 239.136–240.144; Boncompagni 212.40–213.23. In this passage, Fibonacci's Arabic text was apparently fuller than Lorch's Latin text.

[37] In solving the problem, Fibonacci following Abū Kāmil used a second circle circumscribing the pentagon. Further, Lorch's text shows an omission <....> before "half the diameter ..." Hence, the only way to make mathematical sense of the clause, "et medietas dyametri circulj continentis penthagonum linea am, quod est $\frac{1}{2}$ dyametri eius 50 et radix 500," is as I show the translation.

[38] (213.3, f. 138r.10). This sentence is not in Lorch's text. My translation is based on the preceding sentence and the text: "Et si volueris ex quadrato dyametri continentis penthagonum, quod est 20 (*sic*) et radix 8000 penthagoni, remanebunt similiter 100 et radix 8000 pro quadrato quesiti dyametri." Keeping the context in mind, there may be another way to translate this sentence.

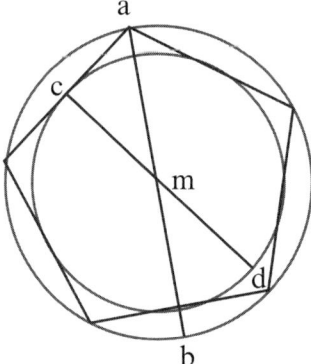

Figure 8.10

[18] *The square on every diameter of a circle containing an equilateral and equiangular figure equals the square on the side of that figure and the square on the diameter of the circle contained by that figure.* The proof follows: consider the adjacent circle *abc* in which is inscribed the equilateral and equiangular triangle *abc* [see Figure 8.11]. From its center *d* draw lines *da*, *db*, and *dc*. Divide all the sides in two equal parts at points *e*, *f*, and *g*, and draw from the center these lines, *de*, *df*, and *dg*. Each one of these lines is a cathete to a side of triangle *abc* and equal to one another. Also lines *da*, *db*, and *dc* are equal to one another as radii of circle *abc*. And because lines *de*, *df*, and *dg* are equal to one another, each of them is a radius of the circle circumscribed by triangle *abc*. Because the squares on lines *de* and *eb* equal the square on line *db*, four times the squares *de* and *eb* equal four times the square on *db*. But four times the square on *de* is the square of twice *de*, the diameter of the circle circumscribed by triangle *abc*. Also, four times the square on line *eb* is the square on line *ab*. And four times the square on radius *db* is the square on the diameter of circle *abc* containing triangle *abc*. Therefore the square on the diameter of the circle contained by the equilateral triangle with the square on the side of the triangle containing the circle equals the square on the diameter of the circle containing that triangle itself. This is shown for any multilateral, equilateral, and equiangular figure circumscribed about a circle or inscribed within a circle.[39] And this is what I wanted to show.

---

[39] See Lorch 240.147; Boncompagni 215.21–25.

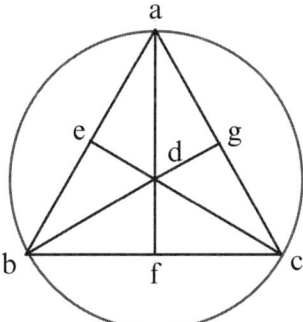

Figure 8.11

[19] *To find the diameter of a circle circumscribed by a regular decagon, given the diameter of a circle circumscribing the decagon.*[40] Given a circle of diameter 10 and the root of 500 about a regular decagon that in turn is about another circle: by squaring the diameter from which 100 (the square on the decagonal side) can be subtracted, what remains is 500 and the root of 20000 for the square on the required diameter.

[20] [*To measure the perpendicular of an equilateral triangle, given the sum of its area and the perpendicular.*][41] If we say to you that the equilateral triangle with its perpendicular is 10 and you want to know the measure of its perpendicular, we choose triangle *abg* with perpendicular *ad*[42] [see Figure 8.12]. Let it be a thing and *db* is the root of $\frac{1}{3}$ *census*. Any side *ab*, *bg*, or *ga* will be the root of $\frac{1}{3}$ 1 *census*. Whence, if we multiply *ad* by *bd* (that is, $\frac{1}{3}$ *census*), we get for the area of triangle *abg* the root of $\frac{1}{3}$ *census* of *census*. If we add to this the perpendicular *ad*, we have a thing and the root of $\frac{1}{3}$ *census* of *census* equal to 10 drachmas. Restore the root of $\frac{1}{3}$ *census* of *census* to the root of a *census* of *census* by multiplying it by the root of 3 to produce the root of a *census* of *census*. Thus one *census* and the root of 3 *census* equals[43] the root of 300.[44] Take half the root of 3 *census* to make the root of $\frac{3}{4}$. Square this to get $\frac{3}{4}$ that you add to the root of 300 to make $\frac{3}{4}$ and the root of 300. Take its root, subtract the $\frac{3}{4}$, and what is left is line *ad*, the perpendicular in triangle *abg*. And this is what we wanted to find.

---

[40] "Rursus si habere uis notitiam dyametri circulj circundati a decagono equilatero et equiangulo, cum dyameter circulj circundantis ipsum decagonum sit radix 500 et 10" (213.23–25, f. 138v.5–7). Lorch 241.201–205; Boncompagni 213.23–27.

[41] "Et si dicamus tibi: trianguli equilateri cum sua perpendiculari est 10 ex numeris, et queritur quanta est perpendicularis" (213.27–28, f. 138v.11). Lorch 244.257–245.270; Boncompagni 213.27–40.

[42] The sense of this sentence is repeated in a second sentence in the manuscripts and omitted here (213.28–30).

[43] Text has *quales*; context requires *equales* (213.37).

[44] $x^2 + \sqrt{3x^2} = \sqrt{300}$ (213.38).

378   Fibonacci's *De Practica Geometrie*

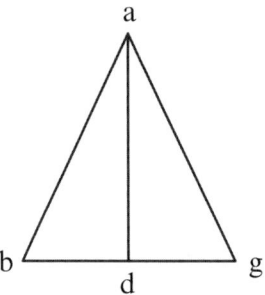

Figure 8.12

[21*][45] [*To find the measure of a line parallel to the known base of an [equilateral] triangle and given the area of the enclosed trapezoid.*] Given triangle *abc* with any side equal to 10. Draw quadrilateral *dbce* with side *de* equidistant from side *bc*. Let the measure of the area of the quadrilateral be 10. You want to know the measure of line *de* [see Figure 8.13]. First, in triangle *abc* draw the perpendicular *af*, the square on which will be 75, that is $\frac{3}{4}$ of the square on one side. The square on line *bf* that is half of side *bc* is a third of the square on the perpendicular *af*. Whence, if we multiply *bf* by the root of 75 (5 times the root of 75), we will get the root of 1875 for the measure of triangle *abc*. Because line *de* is equidistant from line *bc*, triangle *ade* is similar to triangle *abc*. Therefore triangle *ade* is equilateral. Whence, if we make side *de* a thing (that is, one of the equal sides of triangle *ade*), then its square is a *census*. Consequently the square on perpendicular *ag* is $\frac{3}{4}$ *census*. A third of this is $\frac{1}{4}$ *census* equal to the square on line *dg*. If we multiply this by the square on perpendicular *ag*, we will get $\frac{3}{16}$ *census* of *census* for the square of triangle *ade*. Therefore the measure of triangle *ade* is the root of $\frac{3}{16}$ *census* of *census*. If we add this to the area of quadrilateral *dbce* that is 10, we will have 10 and the root of $\frac{3}{16}$ *census* of *census* for the whole triangle *abc*. But the area of triangle *abc* equals the root of 1875. Therefore 10 and the root of $\frac{3}{16}$ *census* of *census* equals the root of 1875. If we subtract 10 from both sides, what remains is the root of $\frac{3}{16}$ *census* of *census* equal to the root of 1875 less 10. Return all of this to one *census* by multiplying it by the root of $\frac{1}{3}$ 5. Because, if you multiply $\frac{3}{16}$ *census* of *census* by the root of $\frac{1}{3}$ 5, you will obtain $\frac{16}{16}$ *census* of *census* (that is, the root of one *census* of *census*), and the root of a *census* of *census* is one *census*. And if you multiply the root of 1875 by the root of $\frac{1}{3}$ 5, you will get the root of 10000 that is 100. And if you multiply the root of $\frac{1}{3}$ 5 by 10, you

---

[45] Paragraphs [21*] through [33*] conclude Boncompagni's transcription of *De practica geometrie* as it appears in his edition. They are introduced by "Subscripte triangulorum questiones ponende sunt in antecedenti quinterno post triangulum equilaterum, cuius mensura cum sua perpendiculari est 10" (218.16–17, f. 141v.17). The problem in the previous quire is at [20], namely here. The transfer has been made to reflect the text that Fibonacci wrote.

will obtain the root of $\frac{1}{3}$ 533. Therefore one *census* (that is, the square on line *de*) is 100 less the root of $\frac{1}{3}$ 533, the root of which is line *de*. And this is what I wanted to find. And because triangle *ade* has equal sides, the square on side *ad* is also 100 less the root of $\frac{1}{3}$ 533. Whence, if the root is removed from 10, what remains for line *db* is 10 less the root of the difference **{p. 219}** between the root of $\frac{1}{3}$ 533 and 100 drachmas.

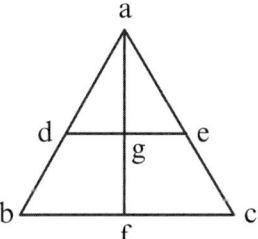

Figure 8.13

[22*] If cathete *dh* is drawn from point *d* and you wish to find its measure, make it a thing. Take half the sides *de* and *be*, and multiply the half by a thing. The result equals 10, the area of quadrilateral *dbce* [see Figure 8.14]. For example: let the square on side *de* be 100 less the root of $\frac{1}{3}$ 533. Hence the square on half of line *de* is 25 less the root of $\frac{1}{3}$ 33. Therefore half of line *be* is the root of its square. Half of side *bc* is 5. Thus for the half of the sides *de* and *bc* there is 5 drachmas and the root of all 25 less the root of $\frac{1}{3}$ 33.[46] Multiply this by a thing, the cathete *dh*. The product of 5 and a thing is 5 things. The product of a thing and the root of all 25 less the root of $\frac{1}{3}$ 33 is the root of all 25 *census* less the root of $\frac{1}{3}$ 33 *census* of *census*. All of this equals 10. Subtract 5 things from each side to leave the root of 25 *census* less the root of $\frac{1}{3}$ 33 *census* of *census* equal to 10 less 5 things. Square all of this to get all 25 *census* less the root of $\frac{1}{3}$ 33 *census* of *census* equal to 25 *census* and 100 drachmas less 100 things. Next, add 100 things and the root of $\frac{1}{3}$ 33 *census* of *census* to both sides. Subtract from both sides 25 *census*. What remains is 100 drachmas and the root of $\frac{1}{3}$ 33 *census* of *census* equal to 100 things. Again square 100 things and the root of $\frac{1}{3}$ 33 *census* of *census* to obtain $\frac{1}{3}$ 33 *census* of *census* and 1000 drachmas and 200 roots of $\frac{1}{3}$ 33 *census* of *census* equal to 10000 *census* (from squaring 100 *census*). Reduce this to *census* of *census* by dividing all that you have by $\frac{1}{3}$ 33 to yield *census* and 300 drachmas and 6 roots of $\frac{1}{3}$ 33 *census* of *census*. These roots are one root of 1200 *census* of *census* that equals 300 *census*. Taking the root of 1200 *census* of *census* from both sides leaves *census* of *census* and 300 drachmas equal to 300 *census* less the root of 1200 *census* of *census*.

---

[46] Awkward in English and Latin, the expression symbolically is $5 + \sqrt{25 + \sqrt{\frac{1}{3}33}}$.

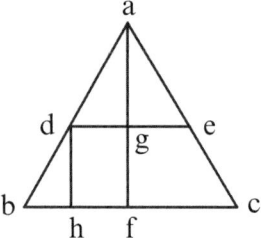

Figure 8.14

[23*] Thus the problem has been reduced to one of the six algebra problems, namely, *census* and number equal to roots. I shall demonstrate this by means of a figure [Figure 8.15]. Let the square *ik* represent a *census* of *census*, any side of which is a *census*. Let the plane *lm* equal 300 drachmas. Add line *mo* to line *tm*, so that the whole line *to* represents 300. Fill out the plane *no*. Therefore, the whole plane *io* is 300 *census* with *to* equal to 300 and *it* equal to a *census*. Plane *im* is a *census* of *census* and 300 drachmas equal to 300 *census* less the root of 1200 *census* of *census*. And so plane *no* is the root of 1200 *census* of *census*. Whence it is necessary that line *mo* represent the root of 1200 drachmas, because, when a *census* is multiplied by the root of 1200, the result is the root of 1200 *census* of *census*. Consequently, if *mo* is taken from all of *to* or 300, what remains is line *tm*, 300 less the root of 1200 drachmas. So divide this in two equal parts at point *p* to get *tp* or 150 less the root of 300 drachmas. Line *tm* therefore has been divided in two equal parts at point *p* and in two unequal parts at point *k*. Whence, if the plane of *tk* by *km* (that is, *lk* by *km*) is removed from the square on line *tp*, what remains is the square on line *kp*. Hence, we square 150 less the root of 300 (or *tp*) to get 22800 less the root of 6750000. Taking 300 or the product of *lk* by *km* from this, what remains is 22500 less the root of 6750000 for the square on line *pk*. If the root of this is taken from *tp*, what remains for the measure of *tk* is 150 less the root of 300 less the root of the difference between the root of 6750000 and 22500 drachmas. The overall remainder or *census tk* {**p. 220**} is very close to $\frac{1}{7}$ 1. The root of this is about $\frac{1}{15}$ 1, the measure of the desired cathete *dh*. I want to prove this.

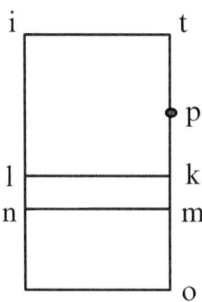

Figure 8.15

[24*] Let the following be given: triangle *abg*, any one of whose sides is 10, and within which is constructed an oblong rectangle *cdef* and another oblong rectangle *hikl* similar to the first. Sides *cf* and *hl* are equidistant from one another and the base *bg* [see Figure 8.16]. The measure of each rectangle is 10. Query: how long are sides *cf* and *hl*? First draw perpendicular *amno*. Let root of a thing represent a side, *cf* or *hl*,[47] which are equidistant from each other and from line *bg*. Further one of the perpendiculars, *am* or *an*, is the root of $\frac{3}{4}$ thing. Subtracting one of these lines from the root of 75 (perpendicular *ao*) leaves the root of 75 less the root of $\frac{3}{4}$ thing for one of the lines *mo* or *no*. Because the area of rectangle *cdef* is the product of *dc* and *cf* (that is, *om* and *cf*), multiplying *om* by *cf* (that is, the root of 75 less the root of $\frac{3}{4}$ thing by the root of a thing) equals 10.[48] Multiplying the root of a thing by the root of 75 drachmas produces the root of 75 things. The multiplication of the root of a thing by the lesser root of $\frac{3}{4}$ thing produces the lesser[49] root of $\frac{3}{4}$ *census*. The one number equals 10. And if we multiply *on* by *hl* (that is, the root of 75 less the root of $\frac{3}{4}$ thing by the root of a thing), we will obtain in like way the root of 75 things less the root of $\frac{3}{4}$ *census*. This also equals 10. Therefore add to both parts the root of $\frac{3}{4}$ *census* to leave the root of 75 things equal to 10 drachmas and the root of $\frac{3}{4}$ *census*. Squaring 10 and the root of $\frac{3}{4}$ *census* produces 100 and $\frac{3}{4}$ *census* and the root of 300 *census* equal to the square of the root of 75 things that is 75 things.

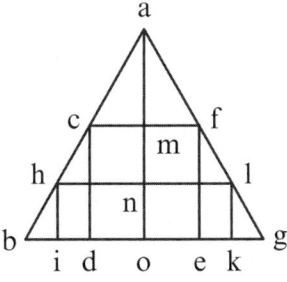

Figure 8.16

[25*] Reduce all of this to one *census* by multiplying by $\frac{1}{3}$ 1. There remains one *census* and $\frac{1}{3}$ 133 equal to 100 things less the root of $\frac{1}{3}$ 533 *census*. Take half the roots (that is, 50 less $\frac{1}{3}$ 133) and square it to get 2633 less 100

---

[47] Text has *hf*; manuscript has *hl* (220.7, f. 142v).
[48] Text has 1017; context requires 10 (220.15).
[49] You may think "negative" as long as you recall that Fibonacci probably did not have the concept of negative numbers. The Latin here is "$\frac{3}{4}$ *census* diminuta" (220.15) that means the positive number to which it is attached is smaller by that amount. It all comes together in the next sentence.

roots of $\frac{1}{3}$ 133. Subtract from this $\frac{1}{3}$ 122 drachmas that are with the *census* to leave 2500 less the root of $\frac{1}{3}$ 1333333. From the root of all of this take half the roots (50 less the root of $\frac{1}{3}$ 133[50]). The remainder is the smaller thing, namely the square on line *cf*. Add the root of what is left between the root of $\frac{1}{3}$ 1333333 and 2500 drachmas to 50 less the root of $\frac{1}{3}$ 133. The result is the square on the greater thing, namely line *hl*. The square on line *cf* is approximately $\frac{7}{9}$ 1, the root of which is $\frac{1}{3}$ 1. And the square on line *hl* is approximately $\frac{1}{9}$ 75, the root of which is $\frac{2}{3}$ 8.

[26*] And if we wish to know the measure of sides *cd* and *hi*,[51] we set either equal to the root of a thing. And because triangles *hbi* and *cbd* are similar to triangle *abo*, so the square on line *bi* is to the square on line *ih* as the square on line *bo* is to the square on line *oa*. But the square on line *bo* is $\frac{1}{3}$ the square on line *oa*. Whence the square on line *bi* is $\frac{1}{3}$ thing, since *ih* is the root of a thing. For the same reason, therefore, the square on line *kg* is $\frac{1}{3}$ thing. Therefore each of the lines *bi* and *kg* equals the root of $\frac{1}{3}$ thing. Subtracting these from all of *bg* (that is from 10), leaves line *ik* equal to 10 less 2 roots of $\frac{1}{3}$ drachma that is the root of $\frac{1}{3}$ 1. Multiplying therefore line *hi* by line *ik* (the root of a thing by 10 less the root of $\frac{1}{3}$ 1) produces the root of 100 things less the root of $\frac{1}{3}$ *census* equal to 10. Similarly, because *bo* is to *oa*, so *hd* to *dc* and *el* to *ef*. Whence each of the lines *bd* and *eg* equals the root of $\frac{1}{3}$. Subtracting these from *bg* leaves *de* equal to 10 less the root of $\frac{1}{3}$ 1. Whence, if we multiply *cd* (the root of a thing) by *de*, what results is {p. 221} the root of 100 things less the root of $\frac{1}{3}$ 1 *census* that is likewise 10. Hence, if the root of $\frac{1}{3}$ 1 *census* is added to both parts, you have 10 and the root of $\frac{1}{3}$ 1 *census* equal to the root of 100 things. And so square 10 and the root of $\frac{1}{3}$ 1 *census* to get 100 and $\frac{1}{3}$ 1 *census* and the root of $\frac{1}{3}$ 533 *census* that equals 100 things (the square of the root of 100 things). Reduce all that you have to one *census* by multiplying it by $\frac{3}{4}$ drachma. This produces one *census* and 75 and the root of 300 *census* equal to 75 things. Subtract from both parts the root of 300 *census* to leave one *census*, 75 drachmas equal to 75 things less the root of 300 *census*. Square half of these roots to obtain $\frac{1}{2}$ 37 less the root of 75 equal to $\frac{1}{4}$ 1481 less 75 roots of 75. From this take the 75 that is with the *census* to leave $\frac{1}{4}$ 1406 less the root of 421875.[52] Take the root of this from $\frac{1}{2}$ 37 less the root of 75 (half the roots). The remainder is the smaller thing, very nearly $\frac{1}{3}$ 1 (the square on line *hi*). Its root is about $\frac{2}{13}$ 1. And this is the length of line *hi*. If we multiply this by *hl* (about $\frac{2}{3}$ 8), we will get very close to 10 for the quadrilateral *hikl*. And this is what we wanted. Similarly, if to the root of $\frac{1}{2}$ 37 less the root of 75 we add the difference between 421875 and $\frac{1}{4}$ 1406 drachmas, we will have the square of the larger thing (line *cd*) about $\frac{1}{4}$ 56. Its

---

[50] Text and manuscript have *3133*; context requires *133* (220.26, f. 143r.13).
[51] Text has *hl*; context requires *hi* (220.32).
[52] Text and manuscript have 481875; 421875 is correct (221.11, f. 143v.8).

8 Geometric Subtleties   383

root (line *cd*) is about $\frac{1}{2}$ 7. Multiply this by $\frac{1}{3}$ 1 (line *cf*) to obtain 10 for the square of *cdef*, as required.

[27*] In another way: We found above that the square on side *cf* together with a thing was 50 less the root of $\frac{1}{3}$ 133, and the root of the difference between the root of $\frac{1}{3}$ 1333333 and 2500 drachmas and the square on *am* was $\frac{3}{4}$ thing. If we take $\frac{3}{4}$ of all the foregoing that is a thing, we will have the square on line *am*. For $\frac{3}{4}$ of 50 is $\frac{1}{2}$ 37. And $\frac{3}{4}$ of the diminished root of $\frac{1}{3}$ 133 is known. If we multiply the square of $\frac{3}{4}$ that is $\frac{9}{16}$ by $\frac{1}{3}$ 133, the product becomes the root. From this multiplication we get the diminished root of 75.[53] Similarly, we take $\frac{9}{16}$ of 2500 and [the root of][54] $\frac{81}{256}$ of $\frac{1}{3}$ 1333333 to get $\frac{1}{4}$ 1406 less the root of 421875. Then take the root of this. Thus we have for the square on line *am* $\frac{1}{2}$ 37 less the root of 75[55] less the root of the difference between the root of 421875 and $\frac{1}{4}$ 1406. Subtract the root of all this from *ao* (the root of 75) to leave $\frac{1}{2}$ 7 and 25 seconds[56] for *mo* equal to *cd*. And *cf* equals about $\frac{1}{3}$ 1 less $\frac{2}{5}$ 4 seconds. And if the difference between the root of 421875 and $\frac{1}{4}$ 1406 is added to $\frac{1}{2}$ 37 less the root of 75, the result is the square on cathete *an*. If the root of this is taken from *ac*, what remains for line *no* equal to line *hi* is about $\frac{1}{6}$ 1.

[28*] Again, in triangle *abg*, one of whose sides is 10, draw quadrilateral *cbef* with right angles at vertices *c*, *f*, and *e* [see Figure 8.17]. Let the measure of the quadrilateral be 10. I want to know the measure of side *fe*. First, draw cathetes *ah* and *ci* in triangle *abg*. Triangles *bic* and *gef* are similar to triangle *ahb*. Whence as the square on side *bh* is to the square on side *ha*, so the square on side *bi* is to the square on side *ic*, and the square on side *ge* to the square on side *ef*. Now the square on side *bh* is a third of the square on side *ha*. Therefore the square on side *bi* is a third of the square on *ic*, and the square on side *ge* is likewise a third of the square on side *ef*. And because the measure of trapezoid *cbef* is found by multiplying *ef* by half the sum of the sides *cf* and *bc*, let side *ef* be the root of a thing. Whence *eg* is the root of $\frac{1}{3}$ thing. Subtracting this from the whole line *bg* leaves {p. 222} 10 less the root of $\frac{1}{3}$ thing for line *be*. And because angles *cfe* and *feb* are right angles, line *cf* is equidistant from line *be*. Since lines *ci* and *fe* are equidistant from one another, *cf* and *ie* are mutually equal as are lines *ei* and *fc*.[57] Whence if either *cf* or *ie* is removed from *bg* or 10, what remains for line *cf* is 10 less two roots of $\frac{1}{3}$ thing. Adding this to 10 less root of $\frac{3}{4}$ thing (the measure of side *be*) produces 20 less three roots of $\frac{1}{3}$ thing for the measure of sides *cf* and *be*. Now if we multiply their half (10 less the root of $\frac{3}{4}$ thing) by line *fe* (the root of a thing), we get the root of 100 things less the root of $\frac{3}{4}$ *census* equal to 10. After adding the root

---

[53] "... prouenit radix 75 diminuta." (221 25)
[54] Context requires this insertion (221.26).
[55] Text and manuscript have 752; context requires 75 (221.27, f. 143v.26).
[56] Base-sixty notation.
[57] Text has *fe*; context requires *fc* (222.3).

of $\frac{3}{4}$ *census* to both parts, we have 10 drachmas and the root of $\frac{3}{4}$ *census* equal to the root of 100 things. Therefore we square 10 and the root of $\frac{3}{4}$ *census* to obtain 100 and $\frac{3}{4}$ *census* and the root of 300 *census* equal to 100 things (the square of the root of 100). Reduce all of this to one *census* by multiplying it by $\frac{1}{3}$ 1.[58] This gives *census* and $\frac{1}{3}$ 133 drachmas and the root of $\frac{1}{3}$ 533 *census* equal to $\frac{1}{3}$ 133 things. Taking the root of $\frac{1}{3}$ 533 *census* from both sides leaves a *census* and $\frac{1}{3}$ 133 drachmas equal to $\frac{1}{3}$ 133 things less the root of $\frac{1}{3}$ 533 *census*. Taking half the roots leaves $\frac{2}{3}$ 66 less the root of $\frac{1}{3}$ 133 drachmas. Square this to produce $\frac{4}{9}$ 4444 and the root of $\frac{1}{2}\frac{2}{9}$ 2270370[59]. Taking the root of the product from the root of half of the roots ($\frac{2}{3}$ 66 less the root of $\frac{1}{3}$ 133) leaves a thing, namely the square on side *ef*. This is how it is computed: subtract the root of $\frac{1}{2}\frac{2}{9}$ 2270370 (about 1539 and 26 minutes, 2 seconds, and 3 thirds) from $\frac{4}{9}$ 4444. The remainder is 2904 and 50 minutes, 37 seconds, and 25 thirds. The root of this is 53 and 53 minutes, 47 seconds, and 46 thirds. Add this to the root of $\frac{1}{3}$ 133 (11 and 32 minutes, 49 seconds, and 23 thirds) to reach 65 and 26 minutes, 36 seconds, and 50 thirds. Subtracting this from $\frac{2}{3}$ 66 leaves 1 and 13 minutes, 23 seconds, and 10[60] thirds for the measure of a thing. The root of this is about 1 and 6 minutes, 21 seconds, and 20 thirds for line *ef*. And this is what I wanted to show.

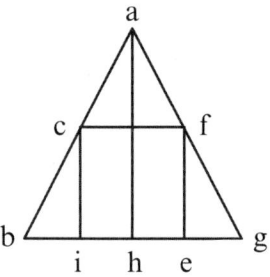

Figure 8.17

[29*] If you want to know the measures of sides *cf* and *be*, there are two ways to find out. First, by approximation: take $\frac{1}{3}$ thing (1 and 13 minutes, 23 seconds, and a third) and you get 22 minutes, 7 seconds, 6 thirds, and 40 fourths for the square on line *eg*. Take the root of this (38 minutes, 18 seconds, 42 thirds) from *bg*. What remains is *be* that is known. If you take twice 38 minutes, 18 seconds, and 42 thirds from 10, what remains is side *cf*. If you want to know the measure of side *cb*, square *ci* and *ib* to obtain $\frac{1}{3}$ 1 things for the square on line *cb*. Then multiply 1 and 13 minutes, 23 seconds, and a third

---

[58] Text and manuscript have 31; context requires $\frac{1}{3}$ 1 as seen in margin (222.13, f. 143r).
[59] Text has $\frac{13}{39}$ 2370370; context requires $\frac{1}{2}\frac{2}{9}$ 2270370 (222.17).
[60] Missing in the text (222.24, f. 144r.35).

by $\frac{1}{3}$ 1 [see Figure 8.18]. If you take the root of this, you will have side cb. For it is half of sides cf and de (9 and 2 minutes, 31 seconds, and 57 thirds). Multiply this by cf (1 and 6 minutes, 21 seconds, and 20 thirds) to obtain 10 and a bit more (2 thirds, 43 fourths, and 36 fifths). In another way: because the square on line eg is $\frac{1}{3}$ of the square on line ef, and the square on line ef is $\frac{1}{3}$ 66 less the root of $\frac{1}{3}$ 133 and the root of the difference between the root of $\frac{1\,2}{2\,9}$ 2270370 and $\frac{4}{9}$ 4444, take $\frac{1}{3}$ of $\frac{2}{3}$ 66 and the roots above, and you will have for the square on line eg, $\frac{2}{9}$ 22 less the root of $\frac{1}{3}\frac{7}{9}$ 14 less the root of the difference between $\frac{1}{8}$ of the root of $\frac{1\,2}{2\,9}$ 2270370[61] and $\frac{1}{9}$ of $\frac{4}{9}$ 4444, that is between the root of $\frac{13472}{8}$ 9263 and $\frac{4\,7}{9\,9}$ 493. After you have taken the root of all this, you will have line eg to be 3999 and about 38 minutes, 27 second, and 42 thirds for the measure of line eg {p. 223}. Because ab is to bh as is cb to bi, then cb is twice the 38 minutes, 27 seconds, and 42 thirds that comes out to 1 and 16 minutes, 55 seconds, and 24 thirds. Subtracting all of this from all of ab, what remains for the measure of line ac (this is cf) is 8 and 43 minutes, 4 seconds, and 36 thirds. Because triangle acf is equilateral and similar to triangle abg, if 38 minutes and 27 seconds are subtracted from 10, what remains for line be is 9 and 21 minutes, 32 seconds, and 18 thirds.

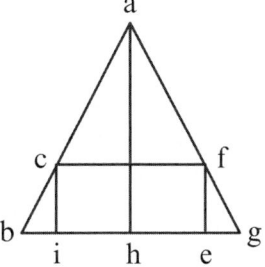

Figure 8.18

[30*] Again, in equilateral triangle abg with side 10 and perpendicular ad, construct quadrilateral $adcf$ with area 10 [see Figure 8.19]. Query: what is the length of line cf? First note that the area of triangle abg is the root of $\frac{3}{4}$ 478, the product of ad by half of dg. Now if 10 (namely quadrilateral $adcf$) is subtracted from the root of $\frac{3}{4}$ 478, what is left is triangle gcf similar to triangle gda. So the square on line cf is thrice the square on line gc. Therefore let a thing be the square on side cf. Consequently, the square on side cg is $\frac{1}{3}$ thing. Next, I multiply cf by half of cg (that is the root of a thing by the root of $\frac{1}{12}$ a thing) to obtain the root of $\frac{1}{12}$ census. Adding this root to the square of $adcf$ gives 10 and the root of $\frac{1}{12}$ census equal to the root of $\frac{3}{4}$ 478. Therefore square 10 and the root of $\frac{1}{12}$ census to reach 100 and $\frac{1}{12}$ census and the root of $\frac{1}{3}$ 33

---

[61] See the second note behind (222.41).

census equal to $\frac{3}{4}$ 478 drachmas. Remove 100 from both sides and multiply what remains by 12 in order to reintegrate to one *census*. Thus we have a *census* and root of 4800 *census* equal to 4425 drachmas. Square half the roots that accompany the *census* to get 1200. Add 4425 to this to reach 5625. After removing the root of 1200 from the root of 5625 (which is 75), the measure of thing remains for line *qf*: 75 less the root of 1200. A third of this is 25 less the root of $\frac{1}{3}$ 133 for the square on line *cg*. If the root of this is subtracted from line *dg* or 5, there remains for line *dc* about $\frac{1}{3}$ 1, for *cf* nearly $\frac{1}{3}$ 6, and for line *ad* $\frac{2}{3}$ 8. Take half the sides *cf* and *da* (which is $\frac{1}{2}$ 7) and multiply it by *dc*[62] or $\frac{1}{3}$ 1 to get 10 for the measure of the square of quadrilateral *adcf*. And this is what I wanted to show.[63]

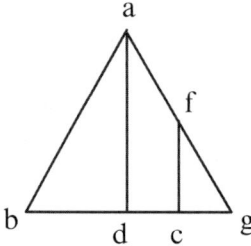

Figure 8.19

[31*][64] Likewise, draw square *defg* in equilateral triangle *abc* with side 10. I want to know the measure of the side of the square [see Figure 8.20]. Because the sides of the square equal one another, side *dg* is equidistant from side *bc*. Whence triangle *adg* is equilateral. With these things understood, set side *dg* equal to side *ad* as the root of a thing. Remove *ad* from *ab* to leave 10 less the root of a thing for line *db*. The square on line *db* is now 100 and a thing less the root of 400 things. Then square *de* to obtain a thing. Square *eb* to get $\frac{1}{3}$ thing.[65] Thus [a thing and][66] $\frac{1}{3}$ thing equal 100 and a thing less the root of 400 things. Take a thing and $\frac{1}{3}$ thing from both sides, add to both sides the root of 400 things, and you have 100 less $\frac{1}{3}$ thing equal to the root of 400 things. Square each side and obtain 10000 and $\frac{1}{9}$ *census* less $\frac{2}{3}$ 66 things equal to 400 things. Add $\frac{2}{3}$ 66 things to each side. Then multiply everything you have by 9 to get one *census* and 90000 drachmas equal to 4200 things. Take half the things (2100) and square it (4410000). Subtract 90000 from this to reach 4320000. The root of this is 2078 and 27 minutes, 48 seconds, and a little

---

[62] Text and manuscript have *de*; context requires *dc* (223.27, f. 145r.14).
[63] But this answer does not fit the question posed at the beginning of [30*].
[64] See Lorch (1993), 219–20 §§ 13 and 14, 245 <13 and 14>.
[65] Text has $\frac{1}{2}$ thing; manuscript has $\frac{1}{3}$ thing (223.35, f. 145r.23).
[66] Context requires insertion (223.25).

less. Subtract this from 2100, leaving a bit more than 21 and 32 minutes, 30 seconds for the measure of a thing, the square on line *de* {**p. 224**}. Its root is about 4 and 38 minutes, 1 second, and 1 third for each side of the square. And so it is for the square in the equilateral triangle *abc*, one of whose sides by what we said above is known for the side of equilateral pentagon *adefg*.

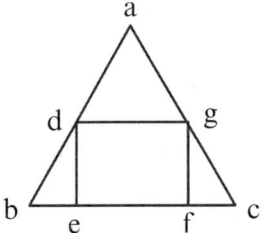

Figure 8.20

[32*] In another way: draw cathete *ai*. Multiply *ai* by *bi* (that is, the root of 75 by 5) to get the root of 1875 drachmas, the area of triangle *abc*. Then, multiply *de* the root of a thing by *be* the root of $\frac{1}{3}$ thing to get the root of $\frac{1}{3}$ census for the areas of triangles *deb* and *gfc* [see Figure 8.21]. Then square *de* to obtain a thing or the area of square *defg*. Again, square *ad* to get a thing from which you subtract the square on line *dh* ($\frac{1}{4}$ thing) to leave $\frac{3}{4}$ thing for the square on cathete *ah*. Then multiply *ah* by *dh* to reach the root of $\frac{3}{16}$ census for the area of triangle *adg*. Then add the root of $\frac{3}{16}$ census to the root of $\frac{1}{3}$ census. This is how you do the addition: square the root of $\frac{1}{3}$ census and the root of $\frac{3}{16}$ census to get $\frac{1}{3}$ census and $\frac{3}{16}$ census that you add together to get $\frac{25}{48}$ census. Multiply the root of $\frac{1}{3}$ census by the root of $\frac{3}{16}$ census to reach the root of $\frac{1}{16}$ census of census which is $\frac{1}{4}$ census. Double it for $\frac{1}{2}$ census. Adding this to $\frac{25}{48}$ census produces $\frac{1}{48}$ 1 census. Its root is the area of triangles *deb*, *feg*, and *adg*.

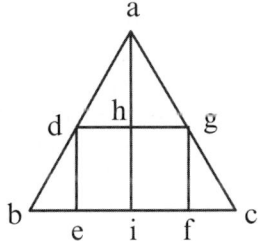

Figure 8.21

[33*] There is another way to find the area. Because triangles *deb* and *ahd* are similar triangles, they have equal angles and proportional sides. Consequently, as *bd* is to *de* (or *da*), so *da* is to *ah*. Further, lines *bd*, *da*, and *ah* are in continued

proportion. Whence, as *bd* (the first term) is to *ah* (the third term) so triangle *dbe* is to triangle *adh*. But, as *bd* is to *ah*, so the square on line *bd* is to the square on line *da*. Now the square on line *bd* is $\frac{1}{3}$ 1 things, and the square on line *da* is a thing. Therefore, as $\frac{1}{3}$ 1 is to 1 (or 4 to 3), so triangle *dbe* is to triangle *adh*. But as triangle *dbe* is to triangle *adh*, so twice triangle *dbe* (which equals triangles *dbe* and *gcf*) is to twice triangle *adh* (which equals triangle *adg*). Therefore as 4 is to 3, so are triangles *dbe* and *gcf* to triangle *adg*. Consequently, as 4 is to 7 (the sum of 4 and 3) so are triangles *dbe* and *gcf* that equal the root of $\frac{1}{3}$ *census* to these three triangles [*deb*, *fcg*, and *adg*]. Hence, if we multiply root of $\frac{1}{3}$ *census* by 7 and divide the product by 4, we get the root of $\frac{1}{48}$ 1 *census* for the area of triangles *deb*, *fcg*, and *adg*. If we add to this the square on *df* that is a thing, we have the root of $\frac{1}{48}$ 1 *census* and a thing for the area of triangle *abc*. We found above that the measure of this area is the root of 1875. Therefore a thing and the root of $\frac{1}{48}$ 1 *census* equal the root of 1875. Subtracting a thing from both parts leaves the root of 1875 less a thing equal to the root of $\frac{1}{48}$ 1 *census*. Squaring both sides produces a *census* and 1875 drachmas less the root of 7500 *census* equal to $\frac{1}{48}$ 1 *census*. Adding the root of 7500 *census* to both parts and subtracting from both one *census* leaves $\frac{1}{48}$ *census* and the root of 7500 *census* equal to 1875 drachmas. Multiplying by 48 restores all this to one *census*. So, the result is one *census* and the root of 17280000 equal to 90000 drachmas. Then, halve the root with the *census* to get the root of 4320000. Square this for 4320000. Add 90000 to get 4410000, the root of which is 2100. From this subtract the root of 4320000 (which is approximately 2078 and 27 minutes, 40 seconds, and a third) to leave 21 and 32 minutes, 20 seconds, and a third to equal a thing. Its root, as we said above, is one of the sides of pentagon *adefg* or of square *defg*. And this is what I wanted to demonstrate.

[34] *To find the side of an equilateral pentagon constructed within a square of given side.*[67] Given 10 the measure of a side of the square, let one of the sides [*ah* or *ac*] of the pentagon be a thing and one of the sides *hd* or *cb* be 10 less a thing {p. 214}. Because the sides *ce* and *hf* are equal, the squares on lines *cb* and *be* equal the squares on lines *hd* and *df*. But *cb* equals *hd*. So *be* remains equal to *fd*. The other lines *eg* and *gf* equal one another because *bg* and *gd* are equal. And because *ef* is a thing, its square is one *census* [see Figure 8.22]. Whence any of the lines *eg* and *gf* is the root of half *census*. Therefore *df* is 10 less the root of half *census*. If we square this we get 100 and $\frac{1}{2}$ *census* less the root of 200 *census*. If we square 10 less a thing (that is, *dh*), we get 100 and *one census* less 20 things. Adding all of this, we have 200 and $\frac{1}{2}$ 1 *census*

---

[67] "In quadrato quidem equilatero et equiangulo *abgd*, cuius unumquodque latus est 10, protractum est penthagonum *acefh* equilaterum. Volo scire longitudinem uniuscuiusque lateris penthagoni" (213.41–43, *f.* 138v.224–25). Lorch 245.279–246.289; Boncompagni 213.41–214.16. The paragraph numbering continues from the insert because the insert belongs here.

less the root of 200 *census* that equals one *census* (this is, the square on line *hf*). By adding 20 things and the root of 200 *census* to both sides and subtracting from both parts the *census*, what remains is 200 and $\frac{1}{2}$ *census* equal to 20 things and the root of 200 *census*. Reduce all of this to one *census* by doubling it, and you produce one *census* and 400 equal to 40 things and the root of 800 *census*.[68] Halve the roots that are 20 and the root of 200, square this to obtain 600 and the root of 320000. Subtract 400 from this to leave 200 and the root of 320000. Subtract the root of this from 20 and the root of 200, and you are left with a thing, the side of the pentagon.

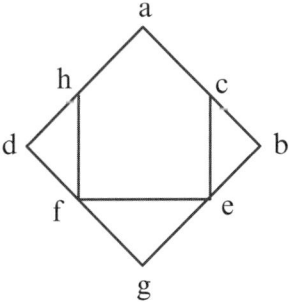

Figure 8.22

[35] *To find the measure of the side of a regular pentagon, given its area.*[69] Given a circle circumscribing pentagon *abgde* with center *c* and area measuring 50 drachmas: draw lines *ca*, *cb*, *cg*, *cd*, and *ce* from its center [see Figure 8.23]. Thus the entire pentagon has been divided into five equal triangles. Select triangle *ace* from these, and the measure of its area is 10. Now, you know that when the side of a pentagon is 10, the square on the diameter of the circumscribing circle is 200 and the root of 8000. This can be done by doubling the square on the side of a pentagon and by multiplying the root of $\frac{4}{5}$ by the square of the same side.[70] Whence if we make the side of the pentagon a thing, we get 2 *census* and the root of $\frac{4}{5}$ *census* of *census* for the square on the diameter of the circle circumscribing the pentagon. Now one of the lines *ca* and *ce* is its radius. The square of any one of them is a fourth part of the square of the same diameter. Therefore the square on line *ca* is $\frac{1}{2}$ *census* and the root of $\frac{1}{20}$ *census* of *census*. If we subtract $\frac{1}{4}$ *census* (the square on line *af*) from this, what remains is $\frac{1}{4}$ *census* and the root of $\frac{1}{20}$ *census* of *census*,

---

[68] $x^2 + 400 = 40x + \sqrt{800x^2}$ (214.13).
[69] "Penthagoni equilateri et equianguli mensura est 50 dragme, quantum est quodlibet latus eius?" (214.17–18, f. 139r.10–11). Lorch 246.290–247.320; Boncompagni 214.17–39.
[70] This sentence and its predecessor exemplify a process that Fibonacci will use immediately. Hence, the 10 in the example should not be confused with the 10 that measures an area.

for the square on the perpendicular *cf*, for it falls at the middle of the base *ae*. Since sides *ac* and *ce* are equal, multiply therefore the square on *cf* by the square on *fa*. That is, multiply $\frac{1}{4}$ *census* and the root of $\frac{1}{20}$ *census* of *census* by $\frac{1}{4}$ *census*. The product is $\frac{1}{16}$ *census* of *census* and the root of $\frac{1}{320}$ *census* of *census* of *census* of *census*.[71] This equals 100, the square of triangle *ace*.[72] Restore, therefore, $\frac{1}{16}$ *census* of *census* to *census* of *census* by multiplying it by 16. Hence you multiply everything by 16 to obtain a *census* of *census* and the root of $\frac{4}{5}$ *census* of *census* of *census* of *census*[73] equal to 1600. Return all of this to a *census* of *census* by multiplying all of it by 5 less the root of 20. The result is a *census* of *census* equal to 8000 and the root of 5120000.[74] The root of this is the measure of a side of the pentagon.[75]

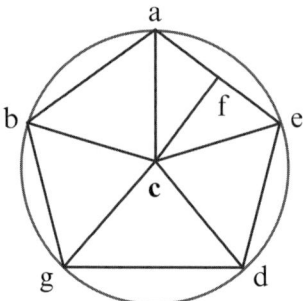

Figure 8.23

[36] *To measure the side of a regular decagon, given its area*.[76] Given the regular decagon of area 100 with side *bg* and center *a* of its circumscribing circle: any one of the lines *ab* and *ag* is the radius of the circle. The area of the triangle *abg* is 10 because it is $\frac{1}{10}$ of the whole decagon {p. 215}. You know that when the side of a decagon is 10 that the diameter of the circle circumscribing the decagon is 10 and the root of 500. Half of this is 5 and the root of 125. Similarly, if we make the side *bg* a thing, then the diameter is a thing and the root of 5 *census*; that is, five times the square of a thing. Whence, the radius *ab* will be $\frac{1}{2}$ thing and the root of $\frac{1}{4}$ 1 *census* [see Figure 8.24]. Then draw the cathete *ad* so that point *d* meets the middle of line *bg*. Whence *bd* is $\frac{1}{2}$ thing. If you square *ab* (that is, $\frac{1}{2}$ thing and the root of $\frac{1}{4}$ 1 *census*), you will get $\frac{1}{2}$ 1

---

[71] The last *census* is missing in (214.32).
[72] $\frac{1}{16}x^4 + \sqrt{\frac{1}{320}x^8} = 100$ (214.33–34).
[73] Context requires x^8 (214.33).
[74] Rather, $8000 - \sqrt{5120000}$.
[75] And, of course, the reader is expected to revert to Chapter 2 to review finding square roots.
[76] "Et si dicamus tibi: mensura decagoni equilateri et equianguli est 100, quanta est longitudo cuiuslibet lateris" (214.33–34, f. 139r.35–139v.1). Lorch 247.321–249.369; Boncompagni 214.41–215.16.

census and the root of $\frac{1}{4}$ 1 *census* of *census*. Subtract from this the square of line *bd* which is $\frac{1}{4}$ *census*. What remains for the cathete is $\frac{1}{4}$ 1 *census* and the root of $\frac{1}{4}$ 1 *census* of *census*. Multiplying this by $\frac{1}{4}$ *census* (the square on line *bd*) leaves $\frac{5}{16}$ *census* of *census* and the root of $\frac{5}{64}$ *census* of *census* of *census* of *census*. This equals 100, the square of the area of triangle *abg*. Multiply the $\frac{5}{16}$ *census* of *census* by $\frac{1}{5}$ 3 to restore it to one *census* of *census*. The whole result is one *census* of *census* less the root of $\frac{4}{5}$ *census* of *census* of *census* equal to 320.[77] Now return this to one *census* of *census* by multiplying everything by 5 less the root of 20. This yields one *census* of *census* equal to 1600 less the root of 2048000.[78] The root of this is a thing, the required measure of the side of the decagon.

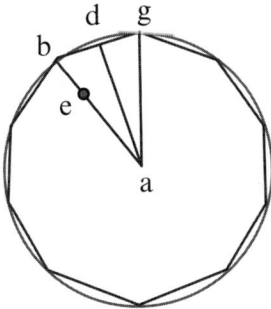

Figure 8.24

[37] There is another method you can use to measure line *ab* the side of a hexagon.[79] Divide the hexagonal side in mean and extreme ratio. Then the longer part will be line *bg* the decagonal side, as Euclid stated.[80] Whence if you subtract from *ab* line *eb* the equal of line *bd*, then the square on line *ae* is five times the square on line *eb*, as stated by Euclid in the same book.[81] But the square on line *eb* is $\frac{1}{4}$ *census* since it is $\frac{1}{2}$ a thing. Because we make line *bg* a thing, the square on line *ae* therefore will be $\frac{1}{4}$ 1 *census*. Thus the measure of the whole line is $\frac{1}{2}$ 40 and the root of $\frac{1}{4}$ 1 *census*, as we found before.

[38] *To measure the side of a pentagon, given the area of an inscribed triangle.*[82] Given 10 as the area of triangle *agd* that is inscribed within pentagon *abgde*:

---

[77] $x^4 - \sqrt{\frac{4}{5}x^8} = 320$. Text has *et*; context requires *diminuta* (Boncompagni 215.13; Lorch 251.421).

[78] Text has 2600; context requires 1600 (Boncompagni 215.15; Lorch 251.424).

[79] That is, the radius of the circumscribing circle.

[80] *Elements*, XIII.9.

[81] *Elements*, XIII.3.

[82] "Et si dicamus tibi: penthagoni *abgde* mensura trianguli *agd* est 10, et uis scire quanta est longitudo linee *gd* que est unum ex lateribus penthagoni" (215.24–25, f. 139v.30–31). Lorch 249.370–250.402; Boncompagni 215.24–37.

let the pentagonal side *gd* be a thing [see Figure 8.25]. Because line *ag* is a chord of a pentagonal angle, line *gd* is a chord of a decagonal angle. Whence, if we cut line *gz* from line *ag* to equal line *gd*, then the ratio of *ag* to *gz* is as the ratio of *gz* to *za*. And if we divide *gz* in two equal parts at point *i*, then the square on line *ai* is five times the square on line *gi*. Now *gi* is half of a thing and its square is $\frac{1}{4}$ *census*. Whence line *ai* is the root of $\frac{1}{4}$ 1 *census*. The whole line *ai* will be the root of $\frac{1}{4}$ 1 *census* and $\frac{1}{2}$ thing. Then draw cathete *at* to fall at the midpoint of line *gd*. Because sides *ag* and *ad* are the upper sides of triangle *abg*, which was the tenth part of the aforementioned decagon, both triangles have equal bases, one of which is a thing. If we proceed in this triangle as we did in that one, we will find in a similar way that one *census* of *census* equals 1600 drachmas less the root of 2048000. Whence the root of the root of this *census* is side *gd* which we set equal to a thing.

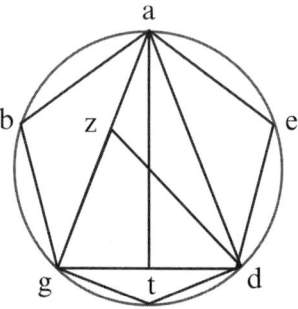

Figure 8.25

[39] *To measure the base-line subtended by a pentagonal angle, given the area of its triangle.*[83] Given 10 the area of triangle *abe* within pentagon *abgde*: set *be* its base-line subtended by an angle of the pentagon equal to a thing. Draw cathete *az* to divide line *be* in two equal parts [see Figure 8.26]. Now by dividing line *be* again in mean and extreme ratio, the longer part is equal to the pentagonal side, as stated in Euclid.[84] Further it is stated that when some line is divided in mean and extreme ratio and half the line is added to the longer part {p. 216}, the square on the added line will be five times the square on half the divided line.[85] And because when line *be* has been divided in mean and extreme ratio, the longer part is line *ab* the pentagonal side. If the half line *bz* (half of *be*) is added to it to make a single line out of both, then the

---

[83] "Et si dicamus: mensuram trianguli *abe*, qui est ex penthagono *abgde*, est 10; quanta est longitudo linee *be*" (215.37–39, f. 140r.10–11). Lorch 250.403–251.425; Boncompagni 215.37–216.16.
[84] *Elements*, XIII.8.
[85] *Elements*, XIII.2.

square on this line composed of lines *ab* and *bz* is five times the square on line *ze*. Now *ze* is $\frac{1}{2}$ thing. Its square is $\frac{1}{4}$ *census* of the line composed of *ab* and *bz*. Whence if line *bz* is removed from its root, what remains for line *ab* is the root of $\frac{1}{4}$ 1 *census* less $\frac{1}{2}$ thing. Squaring this produces $\frac{1}{2}$ 1 *census* less the root of $\frac{1}{4}$ 1 *census* of *census*. Then remove $\frac{1}{4}$ *census* (the square on *bz*) and what remains for the square on line *az* is $\frac{1}{4}$ 1 *census* less the root of $\frac{1}{4}$ 1 *census* of *census*. Multiply this by $\frac{1}{4}$ *census* (the square on line *bz*), to reach $\frac{5}{16}$ *census* of *census* less the root of $\frac{5}{64}$ *census* of *census* of *census* of *census*. This equals 100, the square of the area of triangle *abc*. Multiply all of this by $\frac{1}{5}$ 3 to produce a *census* of *census* less the root of $\frac{4}{5}$ *census* of *census* of *census* of *census*. This equals 320 drachmas.[86] Return all of this to one *census* of *census* by multiplying it by 5 and the root of 20. The result is a *census* of *census* equal to 1600 drachmas and the root of 2048000. The root of this is the square on line *be*.

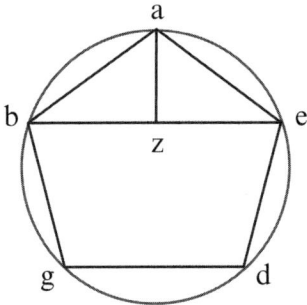

Figure 8.26

[40] *To measure the base-line of a decagonal triangle, given the area of the decagonal triangle*[87] [see Figure 8.27]. Given a regular decagon *abgdeihtc* circumscribed about triangle *cth* the base-line of which is a chord that is $\frac{1}{5}$ of the circle of center *m* circumscribed about a decagon: draw line *tm* to divide line *ch* in two equal parts at point *l*. Make *ch* a thing. Its square is a *census*. The square on the diameter of the circle is 2 *census* and the root of $\frac{4}{5}$ *census* of *census*. Now the square of line *mt* (half the diameter) is $\frac{1}{2}$ *census* less the root of $\frac{1}{20}$ *census* of *census*. Because *ch* the side of the pentagon can be on sides *ta* and *te*, that is on the sides of the hexagon and decagon. We made *ch* a thing; then its square is a *census*. Therefore the squares on lines *mt* and *tc* equal one *census*. Whence if we take $\frac{1}{2}$ *census* and the root of $\frac{1}{20}$ *census* of *census* (the square on line *tm*) from the one *census*, what remains for the

---

[86] $x^4 - \sqrt{\frac{4}{5}x^8} = 320$ (216.13).

[87] "Et si dicamus tibi decagoni *abgdeihtc*, equilateri et equianguli mensuram trianguli *cth* est 10 ex numero, quanta est linea *ch*, que est corda $\frac{e}{5}$ [*circuli*, from Lorch 251.426] circundantis ipsum decagonum" (216.17–19, *f*. 140r.34–140v.1). Lorch 251.426–252.448; Boncompagni 216.17–36.

square on line tc is $\frac{1}{2}$ census less the root of $\frac{1}{20}$ census of census. And if from this we subtract $\frac{1}{4}$ census (the square on line cl), what remains for the square on line tl is $\frac{1}{4}$ census less the root of $\frac{1}{20}$ census of census. Multiplying this by $\frac{1}{4}$ census (the square on cl) produces $\frac{1}{16}$ census of census and the root of $\frac{1}{320}$ census of census of census of census. This equals 100 drachmas. Restore $\frac{1}{16}$ census of census where it is census of census by multiplying it by 16. Consequently multiply all that you have by 16 to get a census of census less the root of $\frac{4}{5}$ census of census of census of census. This equals 1600 drachmas. Return all of this to one census of census by multiplying it by 5 and the root of 20. This produces a census of census equal to 8000 and the root of 51200000. The root of its root is line ch. And this is what I wanted to find.

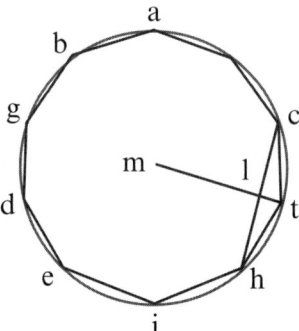

Figure 8.27

Thus end the geometric problems.[88]

---

[88] "Expliciunt questiones geometricales" (216.37, f., 140v.21)

Appendix
Indeterminate Problems with Several Answers

COMMENTARY

A familiar problem introduces the Appendix; find a square number that increased by 5 is still a square number.[1] The triplet (4, 5, 9) comes easily to mind before Fibonacci presented the equation, $x^2 + 5 = (x + 1)^2$. The problem becomes more interesting after varying the equation to $x^2 + 5 = (x + 2)^2$ that produces a second solution. Fibonacci left to the reader the pleasure of forming the obvious generalization. In the text, a geometric approach to the solution limps because the letters for the required diagram (which is missing and I have supplied) seem out of order.

His second and clear method for finding multiple solutions is based on the set of odd numbers: any subset beginning with unity sums to a square number. What follows preys on one's imagination and calculator. Fibonacci ended but did not finish the exercise with basic information for investigating consequences of adding 10 to a square number to make another square number. The activity is enjoyable.

SOURCE

In the preface of *Liber quadratorum* (Book of Squares 1225) Fibonacci describes a challenge posed by Master John of Palermo, "Find a square number from which, when five is added or subtracted, always arises a square number."[2] The Sicilian scholar might have read it in Abū Kāmil's *Algebra*.[3] Clearly Fibonacci had solved part of the problem. By the time he was ready to compose his book on indeterminate analysis he had developed a theory of congruous numbers.[4] This he employed to solve the challenge, rather than

---

[1] This problem and the next are found in Fibonacci's' *Liber quadratorum*; see Sigler (1987), Proposition 17.
[2] Ibid, 3.
[3] Sesiano (1993), 449–450. An earlier claim that Fibonacci had borrowed extensively from *al-Fachri* by al-Karchi was destroyed by Bortolotti (1929–1930), 40.
[4] Sigler (1987), 53–74.

396   Fibonacci's *De Practica Geometrie*

create a proof that would envelope it. The remaining problems and solutions, [3], [4], and [5], may be original.

Regarding the set of propositions in the Appendix of the Boncompagni transcription: I cannot imagine Fibonacci offering Dominic anything other than a complete integral treatise. Consequently, the only reason for the second appendix in manuscript Urbin 292 and other manuscripts of the same group is that an scribe erred in copying from an older manuscript. The second appendix, paragraphs [21*] through [33*], carries the prefatorial remark that these sections belong earlier in the manuscript, with directions on where to place them: "Subscripte triangulorum questiones ponende sunt in antecedenti quinterno post triangulum equilaterum, cuius mensura cum sua perpendiculari est 10" (218.16–17, f. 141v.17). The instruction refers to the proposition at 213.28, Chapter 8, that would require the attachment to be made after [20] and before [21] at 213.41. Consequently, these paragraphs have been transferred to that position. Clearly, the family of manuscripts to which Urbin 292 belongs is lower caste; it does not adequately reflect the treatise finished by Fibonacci.

TEXT[5]

[1] *To find a square number that increased by 5 remains a square number.*[6] There are several ways to do this. You can set the root of the larger number equal to thing and so many drachmas, say 1. After squaring this you have the smaller number [increased] by 5. Hence, square thing and drachma. {p. 217} You get *census*, 2 things, and one drachma equal to *census* and 5 drachmas. Subtract *census* from both parts to leave 2 things equal to 4 drachmas. Therefore thing is 2 drachmas. Its square is 4 the desired number. If we add 5 to it, we get 9 a square number with root 3. If we take thing and 2 drachmas and square it, what results is *census*, 4 things, and 4 drachmas equal to *census* and 5 drachmas. Subtract *census* and 4 things from both parts to leave 4 things equal to 1 drachma. Whence thing is $\frac{1}{4}$ drachma. Its square is $\frac{1}{16}$ drachma for the desired number. If we add 5 to it, we get $\frac{1}{16}$ 5, another square number. Its root is $\frac{1}{4}$ 2.

[2] Another way: let the square number be the square *ag* to which is added the plane *de* or 5. Line *ge* lies as an extension of line *bg*. Let the plane number *ae* be the number that is the product of *ab* and *be*. Transform the number *ae* into

---

[5] "Et incipiunt questiones, quorum solutiones non sunt terminate, hoc est quod non cadunt ad unum terminum tantum, sed ad plures" (216.37–39, f. 140v.21–23).

[6] "Vt est ista in quo proponitur inuenire aliquis quadratus numerus, cui si addatur 5, proueniat inde quadratus numerus" (216.40–41, f. 140v.24). Fibonacci's work here might have led him to Proposition 17 in *Book of Squares*, "Find a number added to a square number and subtracted to a square number yield always a square number." See Sigler (1987), 77–81. Compare with Chapter 3, [141].

a square[7] [see Figure Ap.1]. The numbers ab and be are similar plane numbers. The ratio of [the square from] number ae to the product of ab and be[8] is as square number to square number. Consequently we can find an indefinite number of pairs of numbers having among themselves the ratio of square number to square number from which we can derive what we had proposed. Hence we let the ratio of number eg to number ba (i.e., to number bg) be as 4. The ratio of eg to gb is as 5 is to 4. And the ratio of plane de to square db is as line eg to line gb. Therefore plane de is to square db as 5 is to 4. Whence, if the plane de is 5 and the square db is 4, then the desired square number is 4 to which 5 is added to produce 9, a square number.

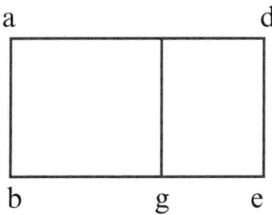

Figure Ap.1

[3] Note that the ratio of plane de or 5 to the square db must be as some square number whose fifth part is square to some other square number. For example: let the ratio of eb to bg be as the ratio of 5 to 1. Therefore the ratio of de or 5 to db or 80 is as the ratio of 1 to 16 the fifth part of 80.[9] Whence the square of db is $\frac{1}{16}$ of 1 with root $\frac{1}{4}$ for the side gb. And thus the total plane ae is $\frac{1}{16}$ 5 with root $\frac{1}{4}$ 2, as we found above.

[4] In other respects: all the square numbers proceed orderly from the sum of [odd] numbers beginning with unity.[10] If we take the odd numbers less than 5, namely 1 and 3, they add to 4, a square number. Now by adding 5 to this, we get 9 a square number. If we multiply some odd square number such as 9 by 5, we get 45. Then, adding all the odd numbers less than 45, that is from 1 to 43, we obtain the square number 484 with root 22. Adding 45 to 484 gives the square number 529. Further, if we divide each square number by the square number [9] by which we multiplied 5, the division will provide for two square numbers, the difference between them being 5, as required. One

---

[7] *Elements* VI, 13.

[8] Text has *eb, ab, ae*; context requires *ae, ab, be*, respectively (217.13).

[9] This is the meaning I derived from "Erit ergo proportio *de* que est 5 ad *db* sicut 80 ad 1 de quibus 80 sextadecima pars" (217.24–25).

[10] This statement by itself would reflect Proposition 4 in *Book of Squares*: "A sequence of squares is produced from the ordered sums of odd numbers which run from 1 to infinity" (ibid, 15–18). What follows is redolent of the introductory problem.

of the squares will be $\frac{7}{9}$ 53 and the other $\frac{7}{9}$ 58. And if you want their roots, divide the square[11] by 9 to get $\frac{1}{3}$ 7. Similarly, if we multiply 5 by some even square (and divide the product)[12] in two or four or more parts so that each of them is an odd number, then we can find more solutions to the original problem. For example: multiply 5 {**p. 218**} by 16 to get 80. Now divide 80 by 2 to get 39 and 41, two odd successive numbers. Next list the odd numbers from 1 that are less than 39. Square the number midway between 1 and 37 to get 361. Add 80 to this and reach 441. If we divide both numbers by 16, we obtain $\frac{9}{16}$ 22 and $\frac{9}{16}$ 27. We will have their roots if we divide 19 and 21 (the roots of 361 and 441) by the root of 16 or 4.

[5] If we take four odd parts of 80, say 17, 19, 21, and 23, each of which is about a fourth of 80 or 20, we find that the sum of the odd numbers less than 17 is 64, a square number with root 8. Adding the square to 80 gives 144. Whence if we divide 64 and 144 and 16 and their roots by 4, we will have what we sought as above.[13] And if we were to say, I took 10 from a square number and what remained was a square number, this problem is similar to the preceding. For if we add 10 to a smaller square number, we obtain the larger square number. So proceed in this problem as we did above and you will find what you want.

---

[11] Text and manuscript have *radicem*; context requires *quadratum* (217.40, f. 141r.31).
[12] Not present in the text (217.41) but required by the following example.
[13] The source of the divisor 4 is not evident. Nor is the reference "as above" obvious.

Bibliography

Primary Resources

A complete list[1] of the Latin copies includes

1. *Belluno, ms. 36, 15 c., *ff.* 2r – 161v[2]

2. *Florence, BN, II.III.22, 16 c., *ff.* 2r – 241v[3]

3. Florence, BN, II.III.23, 16 c., *ff.* 1r – 191v[4]

4. *Florence, BN, II.III.24 (*olim* I.92), 14 c., *ff.* 1r 147v[5]

5. *Paris, BN, lat. 7223, 15 c., *ff.* 1r – 188r[6]

6. *Paris, BN, lat. 10258, 17 c., *pp.* 1–349

7. Paris, BN, Nouv. acqu. lat. 1207, 19 c., *ff.* 1v – 509v

8. *Princeton, UL, Scheide Coll. 32, 16 c., *ff.* 8r – 204[7]

9. *Vatican, lat. 4962, 15 c., *ff.* 4r – 167r.

10. *Vatican, lat. 11589, 16 c., *ff.* 1r – 185v.

11.a [8]*Vatican, Ottob. lat. 1545, 17 c., *ff.* 2r – 341v.

---

[1] Manuscripts noted with an asterisk are the ones I studied. My thanks to Prof. Dr. Menso Folkerts who graciously supplied me with this list and microfilm of two of the manuscripts.
[2] Chapter VIII and appendices are missing.
[3] The manuscript also lacks Chapter VIII and appendices.
[4] This is a copy of Vat. Urb. Latin 259, according to Victor (1979), 22 n. 62.
[5] *f.* 1rv is missing, stolen (?) because of a spectacular *f.* 1 as seen in other manuscripts. The folio was missing when it was acquired by the (Benedictine) Abbey in (15)74, as noted in the manuscript.
[6] The manuscript also lacks Chapter VIII and appendices.
[7] The manuscript stops at the end of paragraph [3] in Chapter VII.
[8] The text was spread over two volumes, 11.a. and 11.b.

11.b. *Vatican, Ottob. lat. 1546, 17c., ff. 2r – 375v.

12. Vatican, Urb. lat. 259, 17c., 175ff.

13. *Vatican, Urb. lat. 292, 14c., ff. 1r – 146r.

Despite the good intentions of Federico Commandino,[9] the transcription by Boncompagni is the first printed transcription of *De practica geometrie*.

The Italian translations of *De practica geometrie* were all edited into a more practical treatise. That is, there is less reliance upon theoretical argument.

1. *Florence, BNC Fondo Principale II. III. iii 198 (*ca.* 1400), ff. 111r – 116v.[10]
2. *Florence, BNC Fondo Palatino 577 (*ca.* 1400), ff. 1r – 149r(!)[11]
3. *Florence, Biblioteca Riccardiana 2186 [*olim* R. III. 25] (1443), ff. 92r – 125v.[12]
4. *Vatican, Ottoboniano 3307 (*ca.* 1465), ff. 355r – 433v.[13]
5. *Vatican, Chigi E.VII. 234 (xiv c.). ff. 1r – 31v.[14]

I used the Riccardiana and Chigi manuscripts as reference points for any difficulties that I had with the Latin copies.

Secondary Resources

Abdeljaouad, M. (2004$_1$), "The eight hundredth anniversary of the death of Ibn al-Yasamin: Bilaterality as part of his thinking and practice." Paper delivered at *Colloque 2004 de Tunis sur l'histoire des mathématiques arabes*.

---

[9] "Commandin, il es vrai, avait publier la *Pratique de la Géometrié; mais la mort l'empêcha d'effectuer ce projet*," in Libri, *Histoire*, pp. 26–27, his resource: Baldi, *Cronica de Matematici*, Urbino 1707, in –4, p. 89.

[10] This is an abridged translation that ends with "...viene $16\frac{1}{4}$ e tanto sia ??? ??? ed fatta (?). Deo Gratias." The text on the microfilm is quite difficult to read because of what appears to be water damage. My foliation corrects Van Egmond (1980), 197.

[11] This translation is not as complete as, say, the Riccardiana; the ending is an incomplete sentence. Nor does its final figure match any figure in ms 2186. Again, the microfilm is not very helpful.

[12] Edited by Arrighi (1966).

[13] Edited by Simi (1995?), 5–8. Although her study provides much information about the codex, her transcription is of "Distinction (Chapter) 5 *Geometric Subtleties*" that contains excerpts from Chapters 4 and 8 of *De practica geometrie*. There is considerable material from other practical geometries, the authors of which she names on p. 3.

[14] This translation is quite different from the Riccardiana, in spelling and content. The first point of difference is their incipits. Riccardiana has "Qui incominica la practica della geometria de M° Lunardo Pisano," Chigi shows "Incipit practica geometrie magistri leonardi pisani."

Bibliography 401

Abdeljaouad, M. (2004_2), "Issues about the status of mathematics teaching in Arab countries—elements of its history and some case studies." Paper delivered at *ICME–10*; available at http://facstaff.uindy.edu/~oaks/Biblio/issues_ Abdeljaouad.PDF.

Abulafia, D. (1994), "The Role of Trade in Muslim–Christian Contact During the Middle Ages," in Agius, D. and Hitchock, R., *The Arab Influence in Medieval Europe.* Reading: Ithaca Press.

Aïssani, D. et al. (1994), "Les mathématiques à Bougie Médiéval et Fibonacci," in Morelli and Tangheroni, *op. cit.*, 67–82.

Allard, A. (1996), "Arabic Mathematics in the Medieval West," in Rashed R., *Encyclopedia of the History of Arabic Science* (London: Routledge), 539–580.

Allard, A. (2001), "The influence of Abū Kāmil's Algebra on the Latin authors of the 12th and 13th centuries," *Journal for the History of Arabic Sciences*, (2001) $12_{1,2}$:83–90.

Archibald, R. (ed. 1915), *Euclid's Book on Division of Figures with a Restoration based on Woepcke's Text and on the* Practica Geometriae *of Leonardo Pisano*. Cambridge: University Press.

Arrighi, G. (ed. 1966), *Leonardo Fibonacci La Practica di Geometria Volgarizzata da Cristofano di Gherardo di Dino cittadino pisano dal codice 2186 della Biblioteca Riccardiana de Firenze*. Pisa: Domus Galilaeana (hereafter: Arrigi, *Cristofano*).

Arrighi, G. (1970), "La fortuna di Leonardo Pisano alla corte di Federico II," in *Dante e la cultura Svava. Atti del convegno di studi tenuta a Melfi. 2–5 novembre 1969*, (Florence: Olschki), 17–31.

Banti, O. (1988), "I trattati tra Pisa e Tunisi dal XII al XIV secolo," in *L'Italia ed i Paesi Mediterranei Vie di comuniczione e scambi commerciali e culturali al tempo delle Repubbliche Marinare* (Pisa: Nisri–Luschi e Pacini), 43–74.

Bartolozzi, M. and R. Franci (1990), "La teorea delle proporzioni nella matematica dell'abaco da Leonardo Pisano a Luca Pacioli," *Bolletinno di Storia delle Scienze Matematiche* X_1, 3–28.

Berggren, J.L. (1995,6,7), "Numbers at Work in Medieval Islam," *Journal for the History of Arabic Science*, 11(no. 1&2):45 51.

Boncompagni, B. (ed. 1857), *Il Liber Abbaci di Leonardo Pisano la lezione del Codice Magliabechiano C.I. 2616*, in *Scritti di Leonardo Pisano matematico del secolo decimoterzo*, Vol. I. Rome: Typografia delle Scienze Matematiche e Fisiche.

Boncompagni, B. (ed. 1862), *La Practica Geometrie di Leonardo Pisano secondo le lezione del Codice Urbinate n. 292 della Biblioteca Vaticana*, in *Scritti di Leonardo Pisano dmatematico del secolo decimoterzo*, Vol. II. Rome: Typografia delle Scienze Matematiche e Fisiche.

Bortolotti, E. (1929–1930), "Le Fonti Arabe di Leonardo Pisano," *Memorie R. Accademia delle scienze dell'Istituto di Bologna*, fis–mat. cl., 8:39–49.

Bubnov, N. (ed. 1899), *Geometria incerti auctoris* in *Gerberti Opera Mathematica*, (Hildesheim: Olms reprint 1963), 310–365.

Busard, H.L.L. (ed. 1968), "L'Algebre au Moyen Age: Le 'Libre Mensurationum' d'Abū Bekr," *Journal des Savants*, Avril–Juin 1968, 64–125.

Busard, H.L.L. (ed. 1987), *The Mediaeval Latin Translation of Euclid's Elements Made Directly From the Greek*. Stutgartt: Franz Steiner Verlag.

Carruthers, M.J. (1990), *The Book of Memory A Study of Memory in Medieval Culture*. Cambridge: Cambridge University Press.

Clagett, M. (1964), *Archimedes in the Middle Ages* I *The Arabo–Latin Tradition*. Madison: University of Wisconsin Press.

Constable, O. (2003), *Housing the Stranger in the Mediterranean World Lodging, Trade, and Travel in Latin Antiquity and the Middle Ages*. Cambridge: University Press.

Curtze, M. (ed. 1902), *Urkunden zur Geschichte der Mathematik im Mittelalter und der Renaissance*. New York: Johnson Reprint, 1968, 1–188.

Datta, B. and A.N. Singh (1935–1938), *History of Hindu Mathematics A Source Book* 2 vols. Bombay: Asia.

Dilke, O.A.W. (1987), *Mathematics and Measurement*. London: University of California Press/British Museum.

Djebbar, A. (2003), "A Panorama of Research on the History of Mathematics in al-Andulus and the Maghreb between the Ninth and Sixteenth Centuries," in Hogendijk and Sabra, *op. cit.*, 309–350.

Drachman, A.G. (1974), "Philo of Byzantium," in Gillispie, C.C. (ed.), *Dictionary of Scientific Biography* X, (New York: 1974), 586–589.

Dunton, M. and R.E. Grimm (trans. 1966), "Fibonacci on Egyptian fractions," *The Fibonacci Quarterly*, 4(4):339–354.

Favaro, A. (1883), "Notizie storico—critiche sulla divisione delle aree," *Memorie del Reale Itsituto veneto di scienze, lettere ed arti*, 22:129–154.

Florio, J. (1611), *Queen Anna's New World of Words or Dictionarie of the Italian and English tongues.* London: Melch, Bradwood.

Folkerts, M. (2003), *Essays on Early Medieval Mathematics.* Burlington, VT: Ashgate.

Folkerts, M. (2004), "Leonardo Fibonacci's knowledge of Euclid's *Elements* and of other mathematical texts," *Bollettino di Storia delle Scienze Matematiche* 24_1:93–113.

Folkerts, M. (2006), *The Development of Mathematics in Medieval Europe The Arabs, Euclid, Regiomontanus.* Burlington, VT: Ashgate.

Folkerts, M. and J. Hogendijk (eds. 1993), *Vestigia Mathematic–Studies in Medieval and Early Modern Mathematics in Honour of H.L.L. Busard.* Amsterdam: Rodopi.

Franci, R. (2002), "Il *Liber Abaci* di Leonardo Fibonacci 1202–2002," in *Bollettino U.M.I. La Matematica nella Società e nella Cultura*, Series VIII, Vol. V–A, Agosto, 293–328.

Grant. E. (ed. 1974), *A Source Book in Medieval Science.* Cambridge. MA: Harvard University Press.

Hahn, N. (ed. 1982), *Medieval Mensuration: Quadrans Vetus and Geometrie Due Sunt Partes Principales.* Philadelphia: Transactions of the American Philosophical Society, Vol., 72, Part 8.

Hay, C. (ed. 1988), *Mathematics from Manuscript to Print 1300–1600.* Oxford: Clarendon Press.

Heath, T. (1921), *A History of Greek Mathematics*, 2 vols. Oxford: Clarendon Press.

Hogendijk, J. (1991), "The geometrical parts of the *Istikmāl* of al-Mu'taman ibn Hūd. An analytical table of contents," *Archives Internatinales d'Histoire des Sciences* 1:207–281.

Hogendijk, J. (1993), "The Arabic version of Euclid's *On Divisions*," in Folkerts and Hogendijk, *op. cit.*, 143–162.

Hogendijk, J. and A. Sabra (ed. 2003), *The Enterprise of Science in Islam New Perspectives*. Cambridge, MA: MIT Press.

Høyrup, J. (1984), "Formative conditions for the development of mathematics in medieval Islam." Paper delivered at the George Sarton Centennial, University of Ghent, 14–17 November 1984.

Høyrup, J. (1986), "Al-Khwarizmi, ibn Turk, and the Liber Mensurationum: On the origins of Islamic algebra," *Erdem* 5 (Ankara), 456–468.

Høyrup, J. (1990), "On parts of parts and ascending continued fractions an investigation of the origins and spread of a peculiar system," *Centaurus* 33:293–324.

Høyrup, J. (1996_1), "The FOUR SIDES AND THE AREA Oblique Light on the Prehistory of Algebra," in R. Calinger (ed.), *Vita Mathematica Historical Research and Integration with Teaching*. Washington, DC: Mathematical Association of America.

Høyrup, J. (1996_2), "Hero, Ps-Hero, and Near Eastern practical geometry an investigation of *metrica, geometrica*, and other treatises." Preprint from Roskilde University Center.

Hughes, B. (2004), "The Geometric algebra of Leonardo da Pisa, a.k.a. Fibonacci," *CUBO* March 2004, 6(1):1–13.

Hughes, B. (2004), "Fibonacci, teacher of algebra: An analysis of Chapter 15.3 of *Liber Abbaci*," *Mediaeval Studies* 2004, 66:313–362.

Hughes, B. (2008), "Toward the Original Treatise, *De practica geometrie*, of Fibonacci." (in preparation).

Knorr, W. (1989), *Textual Studies in Ancient and Medieval Geometry*. Boston: Birkhäuser.

Latham, R. (ed. 1963), *Revised Medieval Latin Word-List from British and Irish Sources*. London: The British Academy.

Levey, M. (1970), "Abraham bar Hiyya ha-Nasi," in Gillispie, C.C. (ed.), *Dictionary of Scientific Biography* I (New York: 1970), 22.

Levey, M. and M. Petruck (eds. 1982), *Kûshyâr ibn Labbân—Principles of Hindu Reckoning*. Madison: University of Wisconsin Press.

Libri, G. (1837), *Histoire des sciences mathématiques en Italie, depuis la renaissance des lettres, jusqu'à la fin du 17e siècle*. Paris.

Lorch, R. (1993), "Abū Kāmil on the pentagon and decagon," in Folkerts and Hogendijk, *op. cit.*, 215–252.

Lorch, R. (1995), "The Arabic transmission of Archimedes' *Sphere and Cylinder* and Eutocius' commentary," in R. Rorch, *Arabic Mathematical Sciences: Instruments, Texts, Transmission* (Brookfield, VT: Variorum), I.

Lüneburg, H. (1993). *Leonardi Pisani Liber Abbaci oder Lesevergnügen eines Mathematikers*. Mannheim: Wissenschaftsverlag, (2nd edition).

Lüneburg, H. (1994), "On the notion of numbers in Leonardo Pisano's Liber Abbaci," in Morelli and Tangheroni, *op. cit.*, 97–108.

Minio-Paluello, L. (1975), "Plato of Tivoli," in *Dictionary of Scientific Biography* IX, 31–33.

Morelli, M. and M. Tangheroni (ed. 1994). *Leonardo Fibonacci Il tempo, le opere, l'eredità scientifica*. Pisa: Pacini Editore.

Oaks, J. and H. Alkhateeb (2005), "*Māl*, enunciations, and the prehistory of Arabic algebra," *Historia Mathematica* 2005 32:426–454.

Pistarino, G. (ed. 1986). *Notai Genovesi in Oltremare Atti Rogati a Tunisi da Pietro Battifoglio (1288–1289)*. Genova: Libreria Bozzi.

Rivest, F. and Zafirov, S. (1998), *Duplication of the Cube* http://www.cs.mcgill.ca/~cs507/àà.

Rosenfeld, B. and E. Ihsanoğlu (ed. 2003), *Mathematicians, Astronomers, and Other Scholars of Islamic Civilization and Their Works (7th–19th c.)*. Istanbul: Research Center for Islamic History, Art and Culture.

Sesiano, J. (1993), "La version latine médiévale de l'Algèbre d'Abū Kāmil," in Folkerts and Hogendijk, *op.cit.*, 315–452.

Sigler, L. (trans. 1987), *Leonardo Pisano Fibonacci the Book of Squares An Annotated Translation into Modern English*. Boston: Academic Press.

Sigler, L. (trans. 2002), *Fibonacci's Liber Abaci—A Translation into Modern English of Leonardo Pisano's Book of Calculation*. New York: Springer-Verlag.

Simi, A. (1993), *Trattato di Geometria Practica* dal Codice IV.18 (sec. V) della Biblioteca Comunale di Siena (Quaderni del Centro Studi della Matematica Medioevale, # 22). Siena: Università degli Studi di Siena.

Simi, A. (ed. 1995?), *Alchuno Chaso Sottile La quinta distinzione della Practicha di Geometria* dal Codice Ottoboniano Latino 3307 della Biblioteca Apostolica Vaticana (Quaderni del Centro Studi della Matematica Medioevale, # 23). Siena: Università degli Studi di Siena.

Simi, A. (1996), "Celerimensura e Strumenti in Manoscritti dei Secoli XIII–XV," in R. Franci, P. Pagli, and L. Toti Rigatelli, *Itinera Mathematica Studi in onore di Gino Arrighi per il suo 90° complanno* (Siena: Centro Studi sulla Matematica Medioevale, Università di Siena), 71–122.

Smith, D. (1925/1953). *The History of Mathematics* II. New York: Dover.

Tangheroni, M. (1994), "Fibonacci, Pisa e il Mediterraneo," in Morelli and Tangheroni, *op.cit.*, 15–34.

Thomas, I (1957), *Selections Illustrating the History of Greek Mathematics with an English Translation*, I. Cambridge, MA: Harvard University Press.

Toubert, P. and A. Paravicini Bagliani (eds. 1994), *Federico II e le scienze*. Palermo: Sellerio.

Tropfke, J. (1980), *Geschichte der Elementar—Mathematik* I, revised eidtion by K. Vogel, K. Reich, and H. Gericke. Berlin: Walter de Gruyter.

Van Egmond, W. (1980), *Practical Mathematics in the Renaissance: A Catalog of Italian Abbacus Manuscripts and Printed Books to 1600*. Florence: Istituto e museo di storia della scienza.

Victor, S. (trans. 1979), *Practical Geometry in the High Middle Ages* Artis Cuiuslibet Consummatio *and the Pratike de Geomertie*. Philadelphia: American Philosophical Society.

Vogel, K. (1971), "Fibonacci," in Gillispie, C.C. (ed.), *Dictionary of Scientific Biography* IV (New York: 1971), 609 and 611.

Zupko, R. (1981), *Italian Weights and Measures from the Middle Ages to the Nineteenth Century*. Philadelphia: American Philosophical Society.

Index of Proper Names and Terms

Abraham, 277
Abraham bar Hiyya, xxiii, 4, 185
Abū Bekr, xxiv, 63
Abū Kāmil, xxii, xxiv, xxix, xxii, 59, 361 ff, 395
al-Andalus, xxi
al-Hassās, xxi
al-'Ibādī, xxii, 258
al-Karāji, xxi, 63
al-Khwārizmī, xxi, xxix, 59
Allard, André, xviii
Almohad Empire, xx
Almohade, xx
al-Qurashi, xxi
al-Sijzī, 184
Ancients, 265
Archibald, Raymond Clare, 183 ff
Archimedes, 63, 65, 182, 258
Archytas of Tarento, 256 ff
Arrighi, Gino, xxxii, xxxiii
al-Tūsī, 258

Banū Mūsā, xxiii, 65, 258, 277
Boncompagni, Baldassarre, xi, *passim*
Bortollotti, Ettore, 258
Bougie, xix, xx, 12
Busard, Hubert, xix

Ceuta, xxi
Chinese *Rule of Five*, 58n
Clagett, Marshall, 258

Commandino, Federico, xxvi
Cristofano di Gherardo di Dino, xxxi ff
Curtze, Maximilian, xxii, 185

Dee, John, 183
destans, 335
dodrans, 335
Domenico, 1

embadum xxiii, 60
Euclid, xxii, *passim, Elements* Books I–X *see footnotes*, Book XI 278, Book XII 280, Book XIII 281, Book XIV 281, Book XV 282n, *On Divisions* xxiv, 184, Special Constructions 282
Eutocius, xviii, xxiii, 258

Favaro, Antonio, 183 ff
Fibonacci, 1, *passim, Liber Abaci* xxiv, 37, 258, *Liber Quadratorum* 395
Folkerts, Menso, xix
fondaco, xx
Franci, Raffaella, xvii n
Frederick II, 1

Gerard of Cremona, 63, 64, 258
Gerbert, xxv, 345
Grant, Edward, 185

Hays, Cynthia, xxviii n
Heath, Thomas L., 184
Hero's formula, 57, 65
Hippocrates of Chios, 256
Hogendijk, Jan, xxiii n, 184n
Høyrup, Jens, 63
Hugh of Saint Victor, xxvii, xxv

Ibn 'Aqnūn, xxi
Ibn al-Mun'im, xxi
Ibn al-Yāsamin, xxi
Ibn Hūd, xxiii, 258
Ibn Hunayn, 258
Ibn Matar, xxii
Ibn Qurra, xxii, 184
Ibn Yusof, xx
Ibn Yussof, xxiv, 65

Leonardo da Pisa, *see* Fibonacci
Lorch, Richard, 361n
Lüneburg, Heinz, xi, 258n

Maghreb, xx, *passim*
Marrakech, xxi
Menelaus, xxxiii, xxv, 65, 258, 277, 345
Merio, Enrico, xx

Miles, *see* Menelaus
modulo 7, 36
Morocco, xx

oroscope, 393

Philo of Byzantium, 256 *ff*
Pisa, xx
Plato, 257 *ff*
Plato of Tivoli, xxiii, 185
Ptolemy, xxiii, 61, 65, 344
Pythagorean Theorem, 57, 343, 361

Savasorda, xxiv
Seville, xxi
Sigler, Laurence, xi, xviii

Table of Arcs and Chords, 61
Theodosius, xxiii, xxv, 277
Tlemcen, xx
Trapezoids, Latin names, 211
Tunis, xxi

Veronese (anon.), 99
Vogel, Kurt, xvii n

Sources and Studies in the History of Mathematics and Physical Sciences

K. Andersen
Brook Taylor's Work on Linear Perspective

K. Andersen
The Geometry of an Art: The History of the Mathematical Theory of Perspective from Alberti to Monge

H.J.M. Bos
Redefining Geometrical Exactness: Descartes' Transformation of the Early Modern Concept of Construction

J. Cannon/S. Dostrovsky
The Evolution of Dynamics: Vibration Theory from 1687 to 1742

B. Chandler/W. Magnus
The History of Combinatorial Group Theory

A.I. Dale
A History of Inverse Probability: From Thomas Bayes to Karl Pearson, Second Edition

A.I. Dale
Most Honourable Remembrance: The Life and Work of Thomas Bayes

A.I. Dale
Pierre-Simon Laplace, Philosophical Essay on Probabilities, Translated from the fifth French edition of 1825, with Notes by the Translator

P. Damerow/G. Freudenthal/P. McLaughlin/J. Renn
Exploring the Limits of Preclassical Mechanics

P. Damerow/G. Freudenthal/P. McLaughlin/J. Renn
Exploring the Limits of Preclassical Mechanics: A Study of Conceptual Development in Early Modern Science: Free Fall and Compounded Motion in the Work of Descartes, Galileo, and Beeckman, Second Edition

P.J. Federico
Descartes on Polyhedra: A Study of the *De Solidorum Elementis*

J. Friberg
A Remarkable Collection of Babylonian Mathematical Texts

B.R. Goldstein
The Astronomy of Levi ben Gerson (1288–1344)

H.H. Goldstine
A History of Numerical Analysis from the 16th Through the 19th Century

H.H. Goldstine
A History of the Calculus of Variations from the 17th Through the 19th Century

G. Graßhoff
The History of Ptolemy's Star Catalogue

Sources and Studies in the
History of Mathematics and Physical Sciences

Continued from previous page

A.W. Grootendorst
Jan de Witt's *Elementa Curvarum Linearum, Liber Primus*

A. Hald
A History of Parametric Statistical Inference from Bernoulli to Fischer 1713–1935

T. Hawkins
Emergence of the Theory of Lie Groups: An Essay in the History of Mathematics 1869–1926

A. Hermann/K. von Meyenn/V.F. Weisskopf (Eds.)
Wolfgang Pauli: Scientific Correspondence I: 1919–1929

C.C. Heyde/E. Seneta
I.J. Bienaymé: Statistical Theory Anticipated

J.P. Hogendijk
Ibn Al-Haytham's *Completion of the Conics*

J. Høyrup
Length, Widths, Surfaces: A Portrait of Old Babylonian Alegbra and Its Kin

B. Hughes
Fibonacci's *De Practica Geometrie*

A. Jones (Ed.)
Pappus of Alexandria, Book 7 of the *Collection*

E. Kheirandish
The Arabic Version of Euclid's *Optics*, Volumes I and II

J. Lützen
Joseph Liouville 1809–1882: Master of Pure and Applied Mathematics

J. Lützen
The Prehistory of the Theory of Distributions

G.H. Moore
Zermelo's Axiom of Choice

O. Neugebauer
A History of Ancient Mathematical Astronomy

O. Neugebauer
Astronomical Cuneiform Texts

F.J. Ragep
Naṣīr al-Dīn al-Ṭūsī's *Memoir on Astronomy*
(al-Tadhkira fī ᶜilm al-hay'a)

B.A. Rosenfeld
A History of Non-Euclidean Geometry

G. Schubring
Conflicts Between Generalization, Rigor and Intuition: Number Concepts Underlying the Development of Analysis in 17th-19th Century France and Germany

Sources and Studies in the History of Mathematics and Physical Sciences

Continued from previous page

J. Sesiano
Books IV to VII of Diophantus' *Arithmetica*: In the Arabic Translation Attributed to Qusṭā ibn Lūqā

L.E. Sigler
Fibonacci's *Liber Abaci*: A Translation into Modern English of Leonardo Pisano's Book of Calculation

J.A. Stedall
The Arithmetic of Infinitesimals: John Wallis 1656

B. Stephenson
Kepler's Physical Astronomy

N.M. Swerdlow/O. Neugebauer
Mathematical Astronomy in Copernicus's De Revolutionibus

G.J. Toomer (Ed.)
Appolonius *Conics* **Books V to VII: The Arabic Translation of the Lost Greek Original in the Version of the Banū Mūsā,** Edited, with English Translation and Commentary by G.J. Toomer

G.J. Toomer (Ed.)
Diocles on Burning Mirrors: The Arabic Translation of the Lost Greek Original, Edited, with English Translation and Commentary by G.J. Toomer

C. Truesdell
The Tragicomical History of Thermodynamics, 1822–1854

K. von Meyenn/A. Hermann/V.F. Weisskopf (Eds.)
Wolfgang Pauli: Scientific Correspondence II: 1930–1939

K. von Meyenn (Ed.)
Wolfgang Pauli: Scientific Correspondence III: 1940–1949

K. von Meyenn (Ed.)
Wolfgang Pauli: Scientific Correspondence IV, Part I: 1950–1952

K. von Meyenn (Ed.)
Wolfgang Pauli: Scientific Correspondence IV, Part II: 1953–1954

Printed in the United States of America